AF323831

ADVANCED ALGEBRA

ADVANCED ALGEBRA

Nanqing Ding
Gongxiang Liu
Qingzhong Ji
Xuejun Guo

Nanjing University, China

NEW JERSEY · LONDON · SINGAPORE · BEIJING · SHANGHAI · HONG KONG · TAIPEI · CHENNAI

Published by

World Scientific Publishing Co. Pte. Ltd.

5 Toh Tuck Link, Singapore 596224

USA office: 27 Warren Street, Suite 401-402, Hackensack, NJ 07601

UK office: 57 Shelton Street, Covent Garden, London WC2H 9HE

Library of Congress Control Number: 2024029680

British Library Cataloguing-in-Publication Data
A catalogue record for this book is available from the British Library.

ADVANCED ALGEBRA

ISBN 978-981-12-9552-2 (hardcover)
ISBN 978-981-12-9553-9 (ebook for institutions)
ISBN 978-981-12-9554-6 (ebook for individuals)

For any available supplementary material, please visit
https://www.worldscientific.com/worldscibooks/10.1142/13910#t=suppl

Desk Editors: Sanjay Varadharajan/Tan Rok Ting

Typeset by Stallion Press
Email: enquiries@stallionpress.com

Preface

This book is an advanced algebra textbook for undergraduate students in comprehensive universities. It is based on the original handouts of advanced algebra, fully draws on the advantages of the commonly used textbooks of advanced algebra and linear algebra in colleges and universities at home and abroad, and conforms to the requirements of the "Three-Three System" talent training system of undergraduate education in Nanjing University.

In 2009, Nanjing University creatively launched the "**Three-Three** System" talent training mode of undergraduate education, namely, **three** training stages of "general category training, professional training and multiple training" and **three** development paths of "professional academic, cross-compound and employment and entrepreneurship" designed for different types of talent training needs.

In 2014, Nanjing University's undergraduate teaching reform of the "Three-Three System" won a special prize for national teaching achievements.

In 2016, Nanjing University continued to construct "Ten-Hundred-Thousand" high-quality courses under the guidance of "running the best undergraduate education in China" and improving teaching quality. The advanced algebra course was successfully selected for the first batch of "one hundred" high-quality courses at Nanjing University.

In 2017, Nanjing University began recruiting and classifying students by big academic subjects, further improving the "Three-Three System" talent training model characterized by "personalized training, independent choice, and diversified development".

In 2020, the advanced algebra course successfully passed the high-quality course expert panel assessment. The advanced algebra textbook is an important part of the high-quality course.

To give students a preliminary and clear understanding of the research objects, fundamental ideas, and basic methods in algebra in the two semesters of advanced algebra learning, we abide by the following three "principles" in writing this book:

(1) **Stimulating interest:** We start the study of each chapter by introducing the research background, development context, and some notable achievements in mathematics. For example, at the beginning of the first chapter, we briefly introduce the long history of integers and polynomials, from Pythagoras' research on the divisibility of integers in the 6th century BCE to the formal introduction of negative numbers and addition and subtraction algorithms in "The Nine Chapters on the Mathematical Art"; from solving quadratic equations in one variable in the time of the ancient Babylonians to the solutions in radicals of the cubic and quartic equations in one variable given by Girolamo Cardano, and then to the proof by Niels Henrik Abel and Évariste Galois that there is no solution in radicals for a general polynomial equation in one variable of degree greater than four. This arrangement can fully stimulate students' curiosity and thirst for learning and let them naturally walk into the palace of knowledge.

(2) **Inspiring thinking:** The generation of mathematical concepts depends on insight after long-term thinking! So, we must teach students more than just what we already know. We should also inspire students to discover new knowledge. For example, in Chapter 2, when introducing Jacobson's lemma, we particularly introduce the relevant history and research ideas of this lemma. Through this specific example, students can understand the formation and development of knowledge, thus enlightening their thinking and understanding and cultivating their exploration spirit.

(3) **Developing ability:** The contents of this book follow the principle of going from simple to deep and from concrete to abstract.

We emphasize the **three "link up"** — the contents **link up** with the high school programmes, **link up** with the follow-up curriculum, and **link up** with scientific research and pay attention to the **three "their own"** — students put forward **their own** questions, find out **their own** examples, and give **their own** proofs, aiming to help students develop problem awareness, questioning spirit and critical thinking, to build their knowledge structure, to cultivate their logical thinking ability and problem-finding and problem-solving ability, and to lay seeds for their professional and lifelong learning. For example, in Chapter 1, we first introduce the arithmetic properties of integers and then present the general polynomial theory. In particular, we illustrate the difference between a polynomial in one indeterminate and a polynomial function through the field of p elements. This approach will help students complete the transition from high school to university, improve their understanding and reasoning ability, and lay a foundation for their subsequent courses. As another example, in the last section of Chapter 6, we briefly introduce integer matrices with similar properties to λ-matrices. This arrangement is not only to echo back and forth but also to guide students to learn independently, promote students to complete the organic integration of relevant knowledge, and cultivate their ability for sustainable development.

This book pays special attention to the promotion of traditional Chinese mathematical culture and highlights the contribution of Chinese people to mathematics. For example, the book has repeatedly introduced the brilliant achievements of our ancient mathematical masterpiece, "The Nine Chapters on the Mathematical Art", in different fields of mathematical research. It introduces the Chinese remainder theorem, the only one named after a country in mathematics with critical applications in algebra and analysis. It also presents Hua's identity, named after the famous contemporary mathematician Luogeng Hua. These mathematical achievements and the introduction of mathematicians will strengthen students' confidence in Chinese culture and enhance their national pride.

Under the background of the "strong foundation plan" of basic disciplines and following the requirements of the first-class undergraduate courses of the Ministry of Education of China, we have prepared this book according to the following standards.

High level: In selecting the book's content, we pay more attention to the organic integration of students' knowledge, ability and quality and focus on cultivating students' comprehensive ability and advanced thinking to solve complex problems. For example, a special orthogonal matrix of order 3 corresponds to a rotation in a 3-dimensional space, a difficult topic in advanced algebra. We introduce the concept of a rotation from two different angles: Euler angles and quaternions. This treatment combines algebra and geometry, abstract and concrete. It is convenient for students to understand the related content and conducive to cultivating their ability to solve complex problems with the knowledge they have learned.

Innovating: This book strives to reflect the cutting-edge and contemporary nature of the discipline and meet the needs of talent cultivation in the new era. For example, this book first introduces the famous theorem on the eigenvector-eigenvalue identity, which shows that the relative phases between the components of any eigenvectors of an $n \times n$ Hermitian matrix A can be computed from the eigenvalues of A and its n submatrices. In 2019, Terence Tao worked with three physicists to discover and prove the theorem. Many media have reported this achievement. Tao was surprised to find a theorem important enough to be written in an advanced algebra textbook but needed to be noticed by algebraists. Later, it was discovered that this theorem has been repeatedly found in the past two hundred years but has yet to be paid attention to by algebraists.

Challenging: Some thinking questions and complex problems in this book require the students to work hard to solve them. For example, Exercise 39 of the first chapter is a problem that combines trigonometric functions and polynomial functions. Students are familiar with trigonometric functions, but to solve this problem thoroughly, they must understand polynomial functions better and be clever enough. The teaching practice shows that the appropriate degree of challenge can stimulate students' interest in learning, cultivate their ability to explore, and discover new knowledge and enable them to enjoy the joy of success through the extensive use of existing knowledge or the creation of new knowledge.

The book arranges some thinking questions, typical examples, and many remarks and footnotes to cultivate students' inquiry learning ability in and out of class, give full play to students' potential, and broaden their thinking and knowledge. Some remarks are the interpretation, supplementation, and expansion of corresponding definitions, theorems, or conclusions, and some are relevant knowledge of historical background or development trends. The footnotes mainly introduce mathematicians in different historical periods involved in the appropriate content. This arrangement will help students understand the relevant mathematical history and research results, guide them to pay attention to and read the relevant literature (including the literature in the book's bibliography), and let students feel that they are not so far away from the mathematical front and the great mathematicians as they think!

The book consists of nine chapters. Nanqing Ding prepares the first and second chapters. Gongxiang Liu organizes the third and fourth chapters. Qingzhong Ji organizes the fifth and sixth chapters. Xuejun Guo organizes the seventh, eighth, and ninth chapters. Nanqing Ding is responsible for compiling the whole book.

At the end of each chapter, there are exercises with different difficulties. Some of these exercises are selected from the references in the book's bibliography, and some are compiled by ourselves. The exercises with "*" indicate that the problem has specific difficulties.

This book can be used as a textbook for the undergraduate advanced algebra course. The content without "*" is completed in one academic year under 5–6 class hours (including tutorial class) every week. The content with "*" is for teachers to choose or students to learn by themselves. This book can also be used as a reference for teaching advanced or linear algebra courses.

In compiling this book, we have received strong support from the School of Mathematics and the Undergraduate School of Nanjing University, as well as funding from the outstanding teaching team of the "Qinglan Project" in Jiangsu universities and the high-quality curriculum construction funds of Nanjing University.

We are deeply grateful to Professors Hourong Qin, Xiaosheng Zhu, and Weixue Shi from Nanjing University and Professor Jianlong Chen from Southeast University for their kind support and help.

We are sincerely pleased to express our thanks to the following people who have read the book: Professor Yuanlin Li from Brock University in Canada, Professors Fuhai Zhu, Ting Wu, and Lizhen Qin from Nanjing University, Professor Huixiang Chen from Yangzhou University, Professors Jianhua Zhou, Xiaoxiang Zhang, and Liang Shen from Southeast University, Professors Chungang Ji and Haiyan Zhou from Nanjing Normal University, Professor Haiyan Zhu from Zhejiang University of Technology, Professor Yuxian Geng from Jiangsu University of Technology, Professor Jiangsheng Hu from Hangzhou Normal University, Professor Yongduo Wang from Lanzhou University of Technology, Professor Li Liang from Lanzhou Jiaotong University, Professor Yunling Kang from Nanjing Audit University, Professor Yanyan Gao from Nanjing Institute of Technology, and the students in our advanced algebra classes at Nanjing University (in particular, Shengzhuo Cong and Yifu Wang). The numerous suggestions, comments, and criticisms of these people greatly improved the manuscript.

We also thank Editors Zhongxing Zhang and Qing Liang from Science Press for their excellent work on the Chinese version of this book.

Finally, we thank Professor Fuzhen Zhang from Nova Southeastern University in the USA and Professor Yiqaing Zhou from the Memorial University of Newfoundland in Canada for recommending this book to World Scientific, and the staff of World Scientific for their interest and unfailing cooperation during the publication of the English version of this book.

About the Authors

Nanqing Ding is a full Professor of Mathematics at Nanjing University. His research concerns homological algebra, ring and module theory.

Gongxiang Liu is a full Professor of Mathematics at Nanjing University. His research concerns Hopf algebras, quantum groups and tensor categories, representation theory of finite dimensional algebras.

Qingzhong Ji is a full Professor of Mathematics at Nanjing University. His research concerns algebraic number theory and algebraic K-theory.

Xuejun Guo is a full Professor of Mathematics at Nanjing University. His research concerns number theory and algebraic K-theory.

Contents

Chapter 1

Integers and Polynomials

1.1 Introduction

There is a long history of integers, which have been one of the leading mathematical research objects. For example, Pythagoras[1] already studied the divisibility of integers in the 6th century BCE. Euclid[2] first proved that there are infinitely many prime numbers, gave a method of finding the greatest common divisor of two positive integers, and established the basic theory for the divisibility of integers in the 3rd–4th century BCE. There are many results on integers in famous ancient Chinese mathematics books. For example, the concept of negative numbers was first introduced in the chapter "Equations" in "The Nine Chapters on the Mathematical Art",[3] and the rules of addition and subtraction of positive and negative numbers are given in the book.

On the other hand, polynomials originated from the study of algebraic equations. In the time of the ancient Babylonians (about 1894–1595 BCE), people already knew how to solve quadratic equations with one variable. In the 12th century, Al-Samawal,[4] born to a Jewish family in Bagdad, already established an algorithm for polynomials in his most famous treatise "al-Bahir fi'l-jabr", meaning "The brilliant

---

[1]Pythagoras, 580–500 BCE, Greek philosopher and mathematician.

[2]Euclid, 330–275 BCE, Greek mathematician.

[3]A Chinese mathematics book, composed by several generations of scholars from the 10th–2nd century BCE, its latest stage being from the 2nd century CE.

[4]Ibn Yahya al-Maghribi Al-Samawal, 1130–1180, Arabian mathematician.

in algebra". The solution of the quartic was published together with that of the cubic by Cardano[5] in the book "Ars Magna" in 1545. In the first half of the 19th century, Abel[6] and Galois[7] showed that polynomials of degree greater than four have no solutions in radicals in the general case. Gauss[8] proved that every nonconstant polynomial in one indeterminate with complex coefficients has at least one complex root in 1799. Polynomials are fundamental concepts in advanced algebra and play an important role in studying advanced algebra and other branches of mathematics.

This chapter introduces arithmetic properties and congruences of integers, and then polynomials with coefficients in a number field are investigated.

1.2 Division Algorithm for Integers

The set of integers consists of zero (0), the positive integers $(1, 2, 3, \ldots)$, and the negative integers $(-1, -2, -3, \ldots)$. A natural number is a nonnegative integer.

As usual, the set of natural numbers is often denoted by the symbol $\mathbb{N}$, while $\mathbb{N}^*$ is used for the positive integers.

The set of integers is often denoted by $\mathbb{Z}$. It is said that Noether[9] first used the letter "Z" to denote the ring of integers (the set of integers is itself a number ring) in 1921 because "Z" stands initially for the German word Zahlen ("numbers"). Since then, the blackboard bold ($\mathbb{Z}$) letter "Z" is often used to denote the set of integers.

There are many proof styles in mathematics, and mathematical induction is one of them. Mathematical induction is mainly used to prove that a statement involving natural numbers is true.

There are many statements involving natural numbers, e.g.,

(1) $1 + 2 + \cdots + n = \frac{n(n+1)}{2}$ for all integers $n \geqslant 1$,

(2) $1^3 + 2^3 + \cdots + n^3 = \left(\frac{n(n+1)}{2}\right)^2$ for all integers $n \geqslant 1$,

(3) $1 + 3 + 5 + \cdots + (2n - 1) = n^2$ for all integers $n \geqslant 1$,

---

[5]Girolamo Cardano, 1501–1576, Italian mathematician.
[6]Niels Henrik Abel, 1802–1829, Norwegian mathematician.
[7]Évariste Galois, 1811–1832, French mathematician.
[8]Carl Friedrich Gauss, 1777–1855, German mathematician.
[9]Emmy Noether, 1882–1935, German mathematician.

(4) the sum of the measures of the interior angles of a convex n-sided polygon in a Euclidean space is $(n-2) \times 180°$ for all integers $n \geqslant 3$,

(5) $2^n > n^2$ for all integers $n \geqslant 5$,

(6) $2^n > n^3$ for all integers $n \geqslant 10$.

To check whether the statement is true for all numbers n greater than or equal to a certain number n_0, it is impossible to check every n; it is also not appropriate to check the first 100, 1000, or even 10,000 natural numbers. We need strict mathematical reasoning — **mathematical induction**.

The first known proof by mathematical induction appeared in the 16th century. Maurolico[10] used the induction hypothesis to prove that the sum of the first n odd positive integers is n^2 in 1575, unraveling the mystery of mathematical induction. It was not until the 19th century that the term "mathematical induction" was introduced by De Morgan[11] in 1836.

We collect two forms of mathematical induction and the so-called well-ordering principle here for future reference. They are equivalent in the context of the Peano[12] axioms.

Theorem 1.2.1 (First form of induction). *Fix an integer n_0 and let $S(n)$ be a family of statements, one for each integer $n \geqslant n_0$. Suppose that*

(1) $S(n_0)$ is true and

(2) for every $k \geqslant n_0$, if $S(k)$ is true, then $S(k+1)$ is true.

Then $S(n)$ is true for all integers $n \geqslant n_0$.

Theorem 1.2.2 (Second form of induction). *Fix an integer n_0 and let $S(n)$ be a family of statements, one for each integer $n \geqslant n_0$. Suppose that*

(1) $S(n_0)$ is true and

(2) for every $m \geqslant n_0$, if $S(k)$ is true for all k with $n_0 \leqslant k \leqslant m$, then $S(m+1)$ is true.

Then $S(n)$ is true for all integers $n \geqslant n_0$.

---

[10] Francesco Maurolico, 1494–1575, Italian mathematician.

[11] Augustus De Morgan, 1806–1871, English mathematician.

[12] Giuseppe Peano, 1858–1932, Italian mathematician.

Theorem 1.2.3 (Well-ordering principle). *Fix an integer n_0 and let*

$$S = \{n \mid n \in \mathbb{Z} \text{ and } n \geqslant n_0\}.$$

Then every nonempty subset of S contains a smallest member.

In particular, every nonempty set of natural numbers contains the smallest member.

Applying the well-ordering principle, we establish the following fundamental property of integers that we use often.

Theorem 1.2.4 (Division algorithm for integers). *Let $m, n \in \mathbb{Z}$ with $n \neq 0$. Then there exist unique $q, r \in \mathbb{Z}$ such that $m = qn + r$, where $0 \leqslant r < |n|$.*

Proof. Since $n \neq 0$, we may assume that $n > 0$ (otherwise, we consider $-n$). We begin with the existence portion of the theorem.

First of all, we give an intuitive proof. We divide the number axis by the multiples of n into intervals of length n, m must be on an endpoint or lie between two endpoints of an interval, as shown in the following:

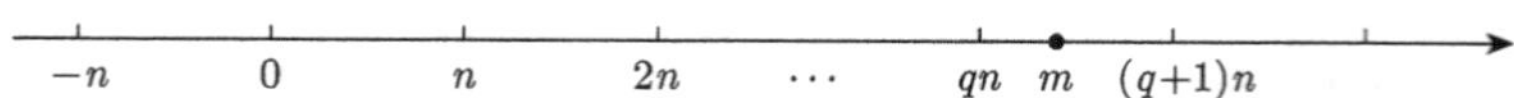

Thus, there is $q \in \mathbb{Z}$ with $m \in [qn, (q+1)n)$. Let $r = m - qn$, then $m = qn + r$, where $0 \leqslant r < n$.

The proof above suggests the following proof by the well-ordering principle. Consider the set $S = \{m - kn \mid k \in \mathbb{Z} \text{ and } m - kn \geqslant 0\}$.

If $0 \in S$, then there is $k \in \mathbb{Z}$ such that $m = kn$. Let $q = k, r = 0$, then $m = qn + r$.

Now assume $0 \notin S$. In this case, $m \neq 0$ (if $m = 0$, then $0 = 0 - 0n = m - 0n \in S$, a contradiction). If $m > 0$, then $m = m - 0n \in S$; if $m < 0$, then $m - (2m)n = m(1 - 2n) \in S$. Thus S is a nonempty set of natural numbers. By the well-ordering principle, S contains a smallest integer, say $r = m - qn$ for some $q \in \mathbb{Z}$. Thus $m = qn + r$ and $r \geqslant 0$, so all that remains to be proved is $r < n$.

If $r \geqslant n$, then

$$m - (q+1)n = m - qn - n = r - n$$

so that $0 \leqslant m - (q+1)n < r$. Hence, $m - (q+1)n$ is an element of S that is smaller than r, contradicting r being the smallest integer in S. Therefore, $0 \leqslant r < n$.

To establish the uniqueness of q and r, let us suppose that

$$m = q_1 n + r_1 = q_2 n + r_2,$$

where $q_i, r_i \in \mathbb{Z}$ and $0 \leqslant r_i < n, i = 1, 2$. Then $(q_1 - q_2)n = r_2 - r_1$. If $q_1 \neq q_2$, then $q_1 - q_2 \neq 0$. We may assume that $q_1 - q_2 > 0$ so that $q_1 - q_2 \geqslant 1$ (for $q_1 - q_2$ is an integer). Thus, since $n > 0$, $r_2 - r_1 = (q_1 - q_2)n \geqslant n$. But $r_2 - r_1 < n$, a contradiction. We conclude that $q_1 = q_2$ and hence $r_1 = r_2$. $\qquad\square$

If m and n are integers with $n \neq 0$, then the integers q and r occurring in the division algorithm are respectively called the **quotient** and the **remainder** upon dividing m by n.

Definition 1.2.5. Let $m, n \in \mathbb{Z}$.

(1) We say that n **divides** m if there is $k \in \mathbb{Z}$ such that $m = kn$. In this case, we say that n is a **divisor** (or **factor**) of m, and m is a **multiple** of n. If n divides m, we write $n|m$; if n does not divide m, we write $n \nmid m$.

(2) A **common divisor** of m and n is an integer d such that d is a divisor of m and d is a divisor of n. An integer d is the **greatest common divisor** of m and n if d is a common divisor of m and n, and any common divisor of m and n is a divisor of d.

(3) A **common multiple** of m and n is an integer l such that l is a multiple of m and l is a multiple of n. An integer l is the **least common multiple** of m and n if l is a common multiple of m and n, and any common multiple of m and n is a multiple of l.

Remark 1.2.6. By the definition above, we have the following:

(1) In the definition of "$n|m$", n may be equal to 0. In this case, $m = 0$. So we can say "$0|0$". If $n \neq 0$, then $n|m$ if and only if the remainder upon dividing m by n is 0.

(2) Every integer divides 0; every integer divides itself; ± 1 divides every integer.

(3) If $m|n$ and $n|m$, then $m = \pm n$; if $m|n, n|k$, then $m|k$; if $m|n, m|k$, then $m|(sn + tk)$ for all integers s and t.

(4) If d_1 and d_2 both are the greatest common divisors of m and n, then $d_1 = \pm d_2$. If $m|n$, then m is the greatest common divisor of m and n. In particular, m is the greatest common divisor of m and 0. Of course, 0 is the greatest common divisor of 0 and 0. If m, n are not both 0, then every greatest common divisor of m and n is not equal to 0. We denote the positive greatest common divisor of m and n by (m, n).

(5) If l_1 and l_2 both are least common multiples of m and n, then $l_1 = \pm l_2$. If $m = 0$ or $n = 0$, then 0 is the unique common multiple of m and n, and hence 0 is the least common multiple of m and n by definition. If neither m nor n is 0, we often use $[m, n]$ to denote the positive least common multiple of m and n.

(6) Let $m_1, m_2, \ldots, m_k$ be integers with $k \geqslant 2$. We may define the greatest common divisors and least common multiples of $m_1, m_2, \ldots, m_k$ similar to the case for two integers. As usual, $(m_1, m_2, \ldots, m_k)$ and $[m_1, m_2, \ldots, m_k]$ will stand for the positive greatest common divisor and the positive least common multiple of $m_1, m_2, \ldots, m_k$, respectively.

Example 1.2.7. Suppose $m, n \in \mathbb{Z}$ are not both 0. If there are $q, r \in \mathbb{Z}$ such that $m = qn + r$, prove that $(m, n) = (n, r)$.

Proof. Since $(m, n)|m$, $(m, n)|n$ and $r = m - qn$, we have $(m, n)|r$. Thus, by definition, $(m, n)|(n, r)$. Similarly, $(n, r)|(m, n)$. So $(m, n) = (n, r)$. $\qquad\square$

Theorem 1.2.8. *Any two integers m and n have the greatest common divisor d, which is unique up to a sign; moreover, d is a **combination** of m and n, i.e., there exist $u, v \in \mathbb{Z}$ such that*

$$d = um + vn \ (\textbf{\textit{Bézout's}}^{13} \textbf{\textit{identity}}).$$

Proof. The uniqueness (up to a sign) part of the theorem is clear by Remark 1.2.6 (4). It remains to prove the existence.

If $n|m$, then n is the greatest common divisor of m and n, and $n = 0m + n$.

Suppose $n \nmid m$. In this case, we may assume $n \neq 0$ (if $n = 0$, then $m|n$, we are done). For convenience, we may also suppose that $n > 0$.

---

[13]Étienne Bézout, 1730–1783, French mathematician.

Apply Theorem 1.2.4 successively as follows:

$$
\begin{aligned}
m &= q_1 n + r_1, & 0 &< r_1 < n, \\
n &= q_2 r_1 + r_2, & 0 &< r_2 < r_1, \\
r_1 &= q_3 r_2 + r_3, & 0 &< r_3 < r_2, \\
&\ \ \vdots & &\ \ \vdots \\
r_{s-3} &= q_{s-1} r_{s-2} + r_{s-1}, & 0 &< r_{s-1} < r_{s-2}, \\
r_{s-2} &= q_s r_{s-1} + r_s, & 0 &< r_s < r_{s-1}, \\
r_{s-1} &= q_{s+1} r_s + r_{s+1}, & r_{s+1} &= 0.
\end{aligned}
\tag{1.1}
$$

Note that there is a positive integer s with $r_s \neq 0$ but $r_{s+1} = 0$ because the remainders $r_1, r_2, \ldots$ form a strictly decreasing sequence of nonnegative integers. Thus, by Example 1.2.7, we have

$$(m, n) = (n, r_1) = (r_1, r_2) = \cdots = (r_{s-1}, r_s) = r_s.$$

We find coefficients u and v with $d = um + vn$ by working upward from the bottom of the list.

From the penultimate equation in (1.1), $d = r_s = r_{s-2} - q_s r_{s-1}$ is a combination of r_{s-2} and r_{s-1}. Combining this with the equation immediately above it, $r_{s-1} = r_{s-3} - q_{s-1} r_{s-2}$, gives

$$d = r_s = r_{s-2} - q_s(r_{s-3} - q_{s-1} r_{s-2}) = (-q_s) r_{s-3} + (1 + q_s q_{s-1}) r_{s-2},$$

a combination of r_{s-3} and r_{s-2}. Continuing this way, we get integers u, v such that $d = un + vn$. $\qquad\square$

The method used for finding the greatest common divisor of two integers in the proof of the theorem above is called the **Euclidean algorithm** or **Euclid's algorithm**.

Definition 1.2.9. We say that two integers m and n are **relatively prime** (or **coprime**) if $(m, n) = 1$. Let $m_i \in \mathbb{Z}$, $i = 1, 2, \ldots, t$, where $t \geqslant 2$. We say that $m_1, m_2, \ldots, m_t$ are relatively prime (or coprime) if $(m_1, m_2, \ldots, m_t) = 1$; and $m_1, m_2, \ldots, m_t$ are said to be **pairwise coprime** if m_i and m_j are coprime for every pair i, j with $1 \leqslant i \neq j \leqslant t$.

The following proposition gives some properties of relative primeness.

Proposition 1.2.10. *Let m, n, k be integers.*

(1) *m and n are relatively prime if and only if ± 1 are the only common divisors of m and n.*

(2) *m and n are relatively prime if and only if there are integers u and v so that $um + vn = 1$.*

(3) *If m and n are not both 0, then $\left(\frac{m}{(m,n)}, \frac{n}{(m,n)} \right) = 1$.*

(4) *If $m|nk$ and $(m,n) = 1$, then $m|k$.*

(5) *If m and n are positive integers, then $[m, n] = \frac{mn}{(m,n)}$.*

(6) *If $m|k, n|k$, then $[m, n]|k$, in particular, if $(m, n) = 1$, then $mn|k$.*

(7) *If $(m, k) = 1$ and $(n, k) = 1$, then $(mn, k) = 1$.*

Proof. (1) follows from the definition of relative primeness.

(2) Necessity. If m and n are relatively prime, then, from Theorem 1.2.8, there are integers u and v such that $um + vn = 1$.

Sufficiency. If $1 = um + vn$ for some integers u and v and d is a common divisor of m and n, then d divides $um + vn = 1$, so $d = 1$ or -1. Hence, m and n are relatively prime.

(3) By Theorem 1.2.8, there are integers u and v with $um + vn = (m, n)$. Note that $(m, n) \neq 0$. Thus $u\frac{m}{(m,n)} + v\frac{n}{(m,n)} = 1$, and so, by (2), $\left(\frac{m}{(m,n)}, \frac{n}{(m,n)} \right) = 1$.

(4) Since $(m, n) = 1$, there are integers u and v with $um + vn = 1$, and so $umk + vnk = k$. Since $m|nk$, it follows that $m|k$.

(5) Let $d = (m, n)$, $m = dm_1$, $n = dn_1$. Then, by (3), $(m_1, n_1) = 1$. Put $l = \frac{mn}{(m,n)}$, then $l = dm_1n_1 = n_1m = m_1n$. So l is a common multiple of m and n. If k is a common multiple of m and n, then there are $k_1, k_2 \in \mathbb{Z}$ such that $k = k_1m = k_2n$, and so $k_1m_1d = k_2n_1d$, whence $k_1m_1 = k_2n_1$. Since $(m_1, n_1) = 1$, $m_1|k_2$ by (4). Thus there is $k_3 \in \mathbb{Z}$ with $k_2 = k_3m_1$, and hence $k = k_3m_1n = k_3l$, i.e., k is a multiple of l. So $[m, n] = \frac{mn}{(m,n)}$.

(6) follows from the definition of lowest common multiples, and (5).

(7) Since $(m, k) = 1$ and $(n, k) = 1$, there exist $u, v, s, t \in \mathbb{Z}$ such that $um + vk = 1$ and $sn + tk = 1$. Therefore, $(um + vk)(sn + tk) = 1$, i.e., $(us)mn + (vsn + (um + vk)t)k = 1$. Thus, by (2), $(mn, k) = 1$.

$\square$

Corollary 1.2.11. *Let k, n, m_i be integers, $i = 1, 2, \ldots, t$.*

(1) *Suppose $(m_i, k) = 1, i = 1, 2, \ldots, t$, then $(m_1 m_2 \cdots m_t, k) = 1$.*
(2) *Suppose $m_i | n$, $i = 1, 2, \ldots, t$, then $[m_1, m_2, \ldots, m_t] | n$, in particular, $m_1 m_2 \cdots m_t | n$ if $m_1, m_2, \ldots, m_t$ are pairwise coprime.*

Definition 1.2.12. Let p be an integer with $|p| > 1$. If ± 1, $\pm p$ are the only divisors of p, then p is called a **prime** (or **prime number**). Usually, primes are assumed to be positive.

Let $p > 1$ be an integer. Then, it is easily seen that p is a prime if and only if p does not factor into the product of two positive integers, each smaller than p.

Theorem 1.2.13. *Let $p > 1$ be an integer. The following statements are equivalent:*

(1) *p is a prime.*
(2) *For any integer m, we always have $p | m$ or $(p, m) = 1$.*
(3) *For any two integers m and n, if $p | mn$, then $p | m$ or $p | n$.*

Proof. (1) $\Longrightarrow$ (2). Let m be an integer and $(p, m) = d$, then $d | p$. Since p is a prime, $d = 1$ or $d = p$. Therefore, $p | m$ or $(p, m) = 1$.

(2) $\Longrightarrow$ (3). Let m, n be two integers and $p | mn$. If $p \nmid m$, then $(p, m) = 1$ by (2). Thus $p | n$ by Proposition 1.2.10 (4).

(3) $\Longrightarrow$ (1). If p is not a prime, then there are integers p_1, p_2 such that $p = p_1 p_2$ and $1 < p_1, p_2 < p$. Obviously, $p | p_1 p_2$, but $p \nmid p_1$ and $p \nmid p_2$, contradicting (3). So (1) holds. $\square$

Corollary 1.2.14. *Let $p, n_i \in \mathbb{Z}$, $i = 1, 2, \ldots, t$. If p is a prime and $p | n_1 n_2 \cdots n_t$, then $p | n_i$ for some i with $1 \leqslant i \leqslant t$.*

The following theorem shows that primes are building blocks of integers.

Theorem 1.2.15 (Fundamental theorem of arithmetic). *Every integer $n > 1$ is either a prime or a product of primes. This product is unique, except for the order in which the factors appear,* *that is, if n has factorizations*

$$n = p_1 p_2 \cdots p_t = q_1 q_2 \cdots q_s,$$

where the p's and q's are primes, then $s = t$ and the q's may be reindexed so that $q_i = p_i$ for all i.

Proof. First, we prove the existence.

For $n = 2$, the existence is obvious.

Let $n \geqslant 2$ be an integer. Suppose the existence is true for all k with $2 \leqslant k \leqslant n$. Now consider $n + 1$. If $n + 1$ is prime, then the existence is trivial. If $n + 1$ is not prime, then $n + 1$ factors as $n + 1 = n_1 n_2$, where $2 \leqslant n_1, n_2 \leqslant n$. By the inductive hypothesis, $n_1 = p_1 \cdots p_r$, a product of primes, and also $n_2 = q_1 \cdots q_l$, a product of primes. So

$$n + 1 = n_1 n_2 = p_1 \cdots p_r q_1 \cdots q_l,$$

a product of primes. By the second form of induction, the existence of the theorem is true.

Next, we prove the uniqueness. Suppose that

$$n = p_1 p_2 \cdots p_t = q_1 q_2 \cdots q_s,$$

where the p's and q's are primes.

We do an induction on the number t of prime factors in one factorization.

If $t = 1$, then $n = p_1 = q_1 q_2 \cdots q_s$. Since p_1 is a prime, $s = 1$ and $p_1 = q_1$. So the uniqueness is true for $t = 1$.

Let $t > 1$ be an integer. Assume that the uniqueness is proved when the number of prime factors in one factorization is $t - 1$.

Next, we consider the case when the number of prime factors in one factorization is t.

Since $p_1 p_2 \cdots p_t = q_1 q_2 \cdots q_s$, $p_1 | q_1 q_2 \cdots q_s$, and so p_1 divides one of $q_1, q_2, \ldots, q_s$. We may assume that $p_1 | q_1$, and hence $p_1 = q_1$ because q_1 is itself a prime. Canceling p_1, we have $p_2 \cdots p_t = q_2 \cdots q_s$. By the inductive hypothesis, $s - 1 = t - 1$, i.e., $s = t$, and $q_2, \ldots, q_t$ may be reindexed so that $p_2 = q_2, \ldots, p_t = q_t$. By the first form of induction, the uniqueness of the theorem is true. $\qquad\square$

1.3   Congruence for Integers

This section is devoted to defining and studying the notion of congruence for integers, which is related to divisibility.

Definition 1.3.1. Let $a, b, m \in \mathbb{Z}$ and $m \neq 0$. We say that a and b are **congruent modulo** m, denoted by $a \equiv b \pmod{m}$, if a and b have the same remainders when divided by m, where m is called the **modulus**, "$\equiv$" is the **congruence symbol**.

The notion of congruence is due to the famous mathematician Euler,[14] who introduced congruence in the study of the integers in the 18th century, and the congruence symbol was developed by Carl Friedrich Gauss in his book "Disquisitiones Arithmeticae", published in 1801.

Remark 1.3.2. By the above definition, we have the following properties (the proofs are left to the reader):

(1) Reflexive: $a \equiv a \pmod{m}$ for all integers a.
(2) Symmetric: for all integers a, b, if $a \equiv b \pmod{m}$, then $b \equiv a \pmod{m}$.
(3) Transitive: for all integers a, b, c, if $a \equiv b \pmod{m}$ and $b \equiv c \pmod{m}$, then $a \equiv c \pmod{m}$.
(4) $a \equiv b \pmod{m}$ if and only if $m \mid (a - b)$ if and only if $a = b + km$, where k is an integer.
(5) For all integers a, b, c, d, if $a \equiv b \pmod{m}$ and $c \equiv d \pmod{m}$, then

$$a + c \equiv b + d \pmod{m}, \quad ac \equiv bd \pmod{m}.$$

(6) If $ac \equiv bd \pmod{m}$ and $(d, m) = 1$, then $a \equiv b \pmod{m}$.

Remark 1.3.3. A relation $\sim$ on a set X is called an **equivalence relation** if it is **reflexive** ($a \sim a$ for all $a \in X$), **symmetric** (if $a \sim b$, then $b \sim a$ for all $a, b \in X$), and **transitive** (if $a \sim b$ and $b \sim c$, then $a \sim c$ for all $a, b, c \in X$). For example, congruence "$\equiv$" is an equivalence relation on $\mathbb{Z}$; the similarity of triangles is an

---

[14]Leonhard Euler, 1707–1783, Swiss mathematician.

equivalence relation on the set of triangles. But the less than relation "$<$" is not an equivalence relation on $\mathbb{Z}$.

Each equivalence relation provides a partition of the underlying set into disjoint equivalence classes. Two elements of the given set are equivalent to each other if and only if they belong to the same equivalence class. For example, congruence modulo m partitions the set $\mathbb{Z}$ of integers into equivalence classes (for congruence is an equivalence relation), integers in the same equivalence class are congruent, and integers belonging to different equivalence classes are not congruent.

Usually, one assumes the modulus $m > 1$ because $a \equiv b \pmod 1$ is true for every pair of integers a and b.

Let a be an integer. Given an integer $m > 1$, the set consisting of all the integers congruent to a modulo m is called the **congruence class** of a modulo m, denoted by $\bar{a}$, i.e.,

$$\bar{a} = \{b \in \mathbb{Z} \mid b \equiv a \pmod m\} = \{a + km \mid k \in \mathbb{Z}\}.$$

By the division algorithm, every integer is congruent modulo m to exactly one of $0, 1, \ldots, m-1$, and hence $\mathbb{Z}$ is partitioned into m congruence classes: $\bar{0}, \bar{1}, \ldots, \overline{m-1}$. The set of congruence classes modulo m is usually denoted by $\mathbb{Z}/m\mathbb{Z}$, i.e., $\mathbb{Z}/m\mathbb{Z} = \{\bar{0}, \bar{1}, \ldots, \overline{m-1}\}$.

Let $a, b \in \mathbb{Z}$. Then $\bar{a} = \bar{b}$ if and only if $a \equiv b \pmod m$. If the integer x_1 is a solution of the congruence $ax \equiv b \pmod m$, i.e., $ax_1 \equiv b \pmod m$, then it is easily seen that all the integers congruent to x_1 modulo m are solutions of the congruence.

Theorem 1.3.4. *Let $a, b, m \in \mathbb{Z}$ with $m > 1$, and let $a \not\equiv 0$ $\pmod m$. Then the congruence equation $ax \equiv b \pmod m$ has a solution in $\mathbb{Z}$ if and only if $(a, m) \mid b$.*

Proof. Necessity. By hypothesis, there is $x \in \mathbb{Z}$ with $ax \equiv b \pmod m$, and hence there is an integer k such that $b = ax + km$. It follows that $(a, m) \mid b$.

Sufficiency. Since $(a, m) \mid b$, there is an integer t such that $b = t(a, m)$. By Theorem 1.2.8, there are integers u, v with $(a, m) = ua + vm$. Therefore, $b = tua + tvm$, and so tu satisfies the congruence $ax \equiv b \pmod m$. $\qquad\qquad\square$

Corollary 1.3.5. *Let $a, m \in \mathbb{Z}$ with $m > 1$, and let $(a, m) = 1$. Then the congruence equation $ax \equiv 1 \pmod{m}$ has a unique solution modulo m, that is, if x_1 and x_2 are integers satisfying $ax \equiv 1 \pmod{m}$, then $x_1 \equiv x_2 \pmod{m}$.*

Proof. By Theorem 1.3.4, the equation $ax \equiv 1 \pmod{m}$ has a solution in $\mathbb{Z}$. If the integers x_1 and x_2 are solutions of the equation $ax \equiv 1 \pmod{m}$, then $ax_1 \equiv ax_2 \pmod{m}$. Thus, there is an integer k with $a(x_1 - x_2) = km$. Since $(a, m) = 1$, $m \mid (x_1 - x_2)$. So $x_1 \equiv x_2 \pmod{m}$. $\qquad\qquad\square$

Next, we consider a system of congruences

$$x \equiv a_1 \pmod{m_1}, x \equiv a_2 \pmod{m_2}, \ldots, x \equiv a_k \pmod{m_k}.$$

The earliest known statement of a system of congruences, as a problem with specific numbers, appeared in "Sunzi Suanjing"[15]:

There are certain things whose number is unknown. If we count them by threes, we have two left over; by fives, we have three left over; and by sevens, two are left over. How many things are there?

The problem above is to find all solutions in $\mathbb{Z}$ of the system of congruences $x \equiv 2 \pmod{3}, x \equiv 3 \pmod{5}, x \equiv 2 \pmod{7}$.

Generalizing the solution to the problem provided in this book, we have the following theorem.

Theorem 1.3.6 (Chinese remainder theorem). *Let k be an integer greater than 1. If $m_1, m_2, \ldots, m_k$ are pairwise coprime positive integers, and $a_1, a_2, \ldots, a_k$ are any integers, then there exist integers a satisfying the system of congruences*

$$\begin{cases} x \equiv a_1 \pmod{m_1}, \\ x \equiv a_2 \pmod{m_2}, \\ \quad\vdots \qquad \vdots \\ x \equiv a_k \pmod{m_k}. \end{cases}$$

Moreover, if both integers a and b satisfy the system of congruences, then

$$a \equiv b \pmod{m_1 m_2 \cdots m_k}.$$

---

[15] An ancient Chinese mathematical treatise written during 3rd–5th centuries CE.

Proof. Since $m_1, m_2, \ldots, m_k$ are pairwise coprime,

$$(m_i, m_1 m_2 \cdots m_{i-1} m_{i+1} \cdots m_k) = 1,$$

and hence there are $s_i, t_i \in \mathbb{Z}$ such that

$$1 = s_i m_i + t_i m_1 m_2 \cdots m_{i-1} m_{i+1} \cdots m_k, 1 \leqslant i \leqslant k.$$

Put $r_i = 1 - s_i m_i = t_i m_1 m_2 \cdots m_{i-1} m_{i+1} \cdots m_k$, then

$$r_i \equiv 1 \ (\mathrm{mod} \ m_i), r_i \equiv 0 \ (\mathrm{mod} \ m_j), j \neq i, 1 \leqslant i, j \leqslant k.$$

Let $a = a_1 r_1 + a_2 r_2 + \cdots + a_k r_k$, then

$$a \equiv a_i \ (\mathrm{mod} \ m_i), \ 1 \leqslant i \leqslant k.$$

If both integers a and b satisfy the system of congruences, then

$$m_1 | (a - b), m_2 | (a - b), \ldots, m_k | (a - b).$$

Note that $m_1, m_2, \ldots, m_k$ are pairwise coprime. So $m_1 m_2 \cdots m_k | (a - b)$, i.e., $a \equiv b \ (\mathrm{mod} \ m_1 m_2 \cdots m_k)$. $\qquad\square$

Example 1.3.7. Find all solutions of the system of congruences

$$x \equiv 2 \ (\mathrm{mod} \ 3), \ x \equiv 3 \ (\mathrm{mod} \ 5), \ x \equiv 2 \ (\mathrm{mod} \ 7).$$

Solution. First, $(3, 35) = 1, 1 = 12 \times 3 + (-1) \times 35$. Put $r_1 = -35$.
    Second, $(5, 21) = 1, 1 = (-4) \times 5 + 21$. Put $r_2 = 21$.
    Third, $(7, 15) = 1, 1 = (-2) \times 7 + 15$. Put $r_3 = 15$.
    Thus, $x = 2 \times (-35) + 3 \times 21 + 2 \times 15 = 23$ is a solution of the system.
    It is easy to see that all solutions are $23 + 105 \times k$, where $k \in \mathbb{Z}$, and 23 is the smallest positive solution. $\qquad\square$

1.4   Complex Numbers and Number Fields

Definition 1.4.1. A **complex number** is a number that can be expressed in the form $a + bi$, where a and b are real numbers, and $i = \sqrt{-1}$ is the **imaginary unit**.

Remark 1.4.2. Carl Friedrich Gauss was the first mathematician to use the term "complex numbers" in 1831. Descartes[16] coined the terms "imaginary numbers" and "real numbers" in "La Géométrie" in 1637. Leonhard Euler began to use the letter i to denote the imaginary unit in 1777, and Gauss used the symbol i systematically and made it universal.

Let $z = a + bi$ be a complex number. Then a is called the **real part** and b is called the **imaginary part** of z. If $b \neq 0$, then z is called an **imaginary number**. In particular, z is said to be a **purely imaginary number** if $a = 0$ and $b \neq 0$.

A real number a can be regarded as a complex number $a + 0i$, whose imaginary part is 0. The **complex conjugate** of the complex number $z = a + bi$ is given by $\overline{z} = a - bi$.

It is common to write a for $a + 0i$, bi for $0 + bi$, and 0 for $0 + 0i$.

Two complex numbers $z_1 = a + bi$ and $z_2 = c + di$ are equal, denoted by $z_1 = z_2$, if and only if $a = c$ and $b = d$. So $a + bi = 0$ if and only if $a = b = 0$.

Definition 1.4.3. We define the addition, subtraction, multiplication, and division of two complex numbers as follows:

$$(a + bi) + (c + di) = (a + c) + (b + d)i,$$

$$(a + bi) - (c + di) = (a - c) + (b - d)i,$$

$$(a + bi)(c + di) = (ac - bd) + (bc + ad)i,$$

$$\frac{a + bi}{c + di} = \frac{ac + bd}{c^2 + d^2} + \frac{bc - ad}{c^2 + d^2}i, \quad \text{if } c + di \neq 0,$$

where a, b, c, d are real numbers.

It is easily seen that the sum, difference, product, and quotient (with a nonzero denominator) of two complex numbers are still complex numbers.

As is known to all, every real number corresponds to a single point on the real number line, and vice versa. Similarly, there is a one-to-one correspondence between complex numbers and points in the **complex plane** established using the horizontal axis for the

---

[16] René Descartes, 1596–1650, French mathematician.

real part and the vertical axis for the imaginary part. The complex number $a + bi$ can be identified with the point (a, b) in the complex plane.

Complex numbers can also be represented in polar form, which associates each complex number with its distance from the origin (its magnitude or modulus) and a particular angle known as the argument of the complex number.

Definition 1.4.4. Let $z = a + bi$ be a complex number. Then z corresponds to a point $Z(a, b)$ on the complex plane. The length r of the vector $\overrightarrow{OZ}$ pointing from the origin O to the point Z is called the **modulus** (or **absolute value**) of z, the angle θ measured counterclockwise from the positive real axis to the vector $\overrightarrow{OZ}$ is called the **argument** of z. Obviously, $a = r \cos\theta, b = r \sin\theta, r = \sqrt{a^2 + b^2}$.

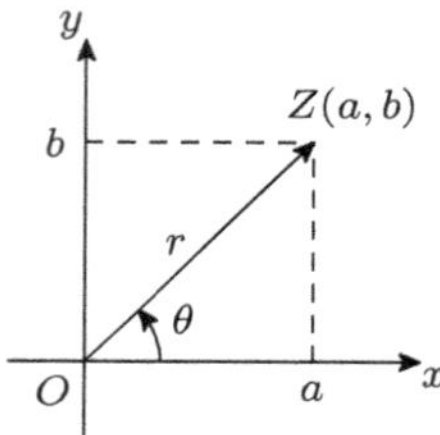

Remark 1.4.5.

(1) The argument for the complex number 0 may be an arbitrary value.
(2) The argument of a nonzero complex number has infinitely many values, and the difference between any two of these values is an integer multiple of 2π.
(3) The modulus of $z = a + bi$ is often denoted by $|z|$. It is easy to see that

$$|z| = \sqrt{a^2 + b^2}, |z|^2 = a^2 + b^2 = z\bar{z}.$$

Definition 1.4.6. Let $z = a + bi$ be a complex number with r the modulus of z and θ an argument of z. Then $z = r(\cos\theta + i\sin\theta)$ is called the **trigonometric form** of z.

Let r and θ be as above. By **Euler's formula** $e^{i\theta} = \cos\theta + i\sin\theta$, every complex number z can be written as $z = re^{i\theta}$, which is called the **exponential form** of z.

Note that nonzero complex numbers written in the trigonometric form are equal if and only if they have the same modulus and their arguments differ by an integer multiple of 2π.

Theorem 1.4.7.

(1) *Let* $z_1 = r_1(\cos\theta_1 + i\sin\theta_1)$
    and $z_2 = r_2(\cos\theta_2 + i\sin\theta_2)$

   be the trigonometric forms of z_1 *and* z_2, *respectively.*

   (1.1) $z_1 z_2 = r_1 r_2 \left(\cos(\theta_1 + \theta_2) + i\sin(\theta_1 + \theta_2)\right),$

   (1.2) $\frac{z_1}{z_2} = \frac{r_1}{r_2}(\cos(\theta_1 - \theta_2) + i\sin(\theta_1 - \theta_2))$, *where* $z_2 \neq 0$.

(2) *Let* n *be a positive integer and* $z = |z|(\cos\theta + i\sin\theta)$ *be the trigonometric form of* z.

   (2.1) $z^n = |z|^n(\cos(n\theta) + i\sin(n\theta))$, *in particular, if* $z = \cos\theta + i\sin\theta$, *then* $(\cos\theta + i\sin\theta)^n = \cos(n\theta) + i\sin(n\theta)$, *which is called* **De Moivre's**[17] **formula.**

   (2.2) *The equation* $x^n = z$ *has exactly* n *roots:*

   $$\sqrt[n]{|z|}\left(\cos\frac{\theta + 2k\pi}{n} + i\sin\frac{\theta + 2k\pi}{n}\right), \quad k = 0, 1, 2, \ldots, n-1.$$

   In particular, the n *roots of the equation* $x^n = 1$ *are given by*

   $$\cos\frac{2k\pi}{n} + i\sin\frac{2k\pi}{n}, \quad k = 0, 1, 2, \ldots, n-1.$$

Proof.   (1) By hypothesis,

$$z_1 z_2 = (r_1(\cos\theta_1 + i\sin\theta_1))(r_2(\cos\theta_2 + i\sin\theta_2))$$
$$= r_1 r_2((\cos\theta_1 \cos\theta_2 - \sin\theta_1 \sin\theta_2)$$
$$+ i(\cos\theta_1 \sin\theta_2 + \sin\theta_1 \cos\theta_2))$$
$$= r_1 r_2(\cos(\theta_1 + \theta_2) + i\sin(\theta_1 + \theta_2)).$$

---

[17]Abraham De Moivre, 1667–1754, French mathematician.

Suppose $z_2 \neq 0$, then

$$\frac{z_1}{z_2} = \frac{r_1}{r_2}(\cos\theta_1 + i\sin\theta_1)(\cos\theta_2 - i\sin\theta_2)$$

$$= \frac{r_1}{r_2}(\cos\theta_1 + i\sin\theta_1)(\cos(-\theta_2) + i\sin(-\theta_2))$$

$$= \frac{r_1}{r_2}(\cos(\theta_1 - \theta_2) + i\sin(\theta_1 - \theta_2)).$$

(2) By (1.1), we have $z^n = |z|^n(\cos(n\theta) + i\sin(n\theta))$, i.e., (2.1) holds.

It remains to show (2.2). Let $u = r(\cos\alpha + i\sin\alpha)$ be a root of the equation $x^n = z$, where r is the modulus of u and α an argument of u. Then $r^n(\cos(n\alpha) + i\sin(n\alpha)) = |z|(\cos\theta + i\sin\theta)$. It follows that

$$\begin{cases} r^n = |z| \\ n\alpha = \theta + 2k\pi \end{cases}, \text{ and so } \begin{cases} r = \sqrt[n]{|z|} \\ \alpha = \dfrac{\theta + 2k\pi}{n} \end{cases}, \quad k \in \mathbb{Z}.$$

Thus

$$u = \sqrt[n]{|z|}\left(\cos\frac{\theta + 2k\pi}{n} + i\sin\frac{\theta + 2k\pi}{n}\right), \quad k \in \mathbb{Z}.$$

If $k = 0, 1, 2, \ldots, n-1$, then the arguments of u belong to the interval $\left[\frac{\theta}{n}, \frac{\theta}{n} + 2\pi\right)$ and they are different so that the corresponding values of u are different. Since the sine and cosine functions are periodic with a period of 2π, the equation $x^n = z$ has exactly n roots:

$$\sqrt[n]{|z|}\left(\cos\frac{\theta + 2k\pi}{n} + i\sin\frac{\theta + 2k\pi}{n}\right), \quad k = 0, 1, 2, \ldots, n-1.$$

In particular, taking $z = 1 = \cos 0 + i\sin 0$ gives the n roots of the equation $x^n = 1$:

$$\cos\frac{2k\pi}{n} + i\sin\frac{2k\pi}{n}, \quad k = 0, 1, 2, \ldots, n-1. \qquad \square$$

Definition 1.4.8. Let n be a positive integer. An n**th root of unity** is a complex number α satisfying the equation $\alpha^n = 1$. An nth root α of unity is said to be **primitive** if it is not a kth root of unity for some smaller k, that is, if $\alpha^n = 1$ and $\alpha^k \neq 1$ for $k = 1, 2, \ldots, n-1$.

Remark 1.4.9. By Theorem 1.4.7, every complex number z has exactly n nth roots, where n is a positive integer, and on the complex plane all the nth roots of z are at the vertices of a regular n-sided polygon inscribed in the circle with radius $\sqrt[n]{|z|}$ and center at the origin. In particular, the n nth roots of unity are $\zeta_0 = 1$, $\zeta_1 = \xi$, $\zeta_2 = \xi^2, \ldots, \zeta_{n-1} = \xi^{n-1}$, where $\xi = \cos\frac{2\pi}{n} + i\sin\frac{2\pi}{n}$, and all the nth roots of unity are at the vertices of a regular n-sided polygon inscribed in the unit circle. ζ_1 is a primitive nth root of unity.

Question. Let $n > 1$ be an integer and $1 \leqslant t \leqslant n - 1$. When is ζ_t a primitive nth root of unity?

Definition 1.4.10. Let F be a set of complex numbers containing 0 and 1. If the sum, difference, product, and quotient (with a nonzero denominator) of any two complex numbers from F are still in F, then F is called a **number field**.

By definition, a set F of complex numbers is a number field if and only if F contains a nonzero element and F is closed under addition, subtraction, multiplication, and division (with a nonzero denominator).

Example 1.4.11.

(1) The set $\mathbb{Q}$ of rational numbers forms the **field of rational numbers**. The set $\mathbb{R}$ of real numbers forms the **field of real numbers**. The set $\mathbb{C}$ of complex numbers forms the **field of complex numbers**. The set $\mathbb{Z}$ of integers cannot be a number field.
(2) $F = \mathbb{Q}[\sqrt{2}] = \{a + b\sqrt{2} \mid a, b \in \mathbb{Q}\}$ is a number field, however, $F_1 = \mathbb{Z}[\sqrt{2}] = \{a + b\sqrt{2} \mid a, b \in \mathbb{Z}\}$ cannot be a number field.
(3) $F = \mathbb{Q}[i] = \{a + bi \mid a, b \in \mathbb{Q}\}$ is a number field.

Proposition 1.4.12.

(1) *Let F be a number field. Then $\mathbb{Q} \subseteq F \subseteq \mathbb{C}$.*
(2) *If F is a number field and $\mathbb{R} \subsetneq F$, then $F = \mathbb{C}$.*

Proof. (1) follows from the definition.

(2) Since $\mathbb{R} \subsetneq F$, there exists $z \in F$ but $z \notin \mathbb{R}$. Let $z = a + bi$, where $a, b \in \mathbb{R}$. Since $z \notin \mathbb{R}$, $b \neq 0$. Thus $i = \frac{1}{b}(z - a) \in F$. It follows that any complex number $c + di \in F$. So $F = \mathbb{C}$. $\qquad\square$

By Proposition 1.4.12, we know that $\mathbb{Q}$ is the smallest number field, i.e., $\mathbb{Q}$ is contained in every number field, and $\mathbb{C}$ is the largest number field, i.e., $\mathbb{C}$ contains every number field.

Definition 1.4.13. A nonempty subset R of $\mathbb{C}$ is called a **number ring** if R is closed under addition, subtraction, and multiplication.

Example 1.4.14.

(1) $R = \{0\}$ is a number ring.
(2) Every number field is a number ring.
(3) The set $\mathbb{Z}$ of integers form a number ring, which is called the **ring of integers**.
(4) The set $\mathbb{N}$ of natural numbers cannot be a number ring.

Proposition 1.4.15. *If $R \neq \{0\}$ is number ring, then R contains infinitely many elements.*

Proof. Since $R \neq \{0\}$, there exists a nonzero $r \in R$, and so $2r$, $3r$, ..., mr, ... are all in R, where $m \in \mathbb{N}^*$. Note that $mr \neq kr$ when $m \neq k$. So R is an infinite set. $\qquad\square$

We have just introduced the notions of number fields and number rings. Propositions 1.4.12 and 1.4.15 show that every number field or nonzero number ring contains infinitely many elements. Next, let's take an example to illustrate that there is an algebraic system containing only a finite number of elements in which addition, subtraction, and multiplication can be defined. Moreover, the associative, commutative, and distributive laws hold for addition and multiplication.

Let $m > 1$ be a fixed but arbitrary integer. Then the set of congruence classes modulo m is $\mathbb{Z}/m\mathbb{Z} = \{\overline{0}, \overline{1}, \ldots, \overline{m-1}\}$, where

$$\overline{r} = \{r + km \mid k \in \mathbb{Z}\}, \ r = 0, 1, \ldots, m-1.$$

We define addition, subtraction, and multiplication in $\mathbb{Z}/m\mathbb{Z}$ as follows:

$$\begin{cases} \overline{a} + \overline{b} = \overline{a + b}, \\ \overline{a} - \overline{b} = \overline{a - b}, \\ \overline{a}\,\overline{b} = \overline{ab}, \end{cases}$$

where $\overline{a}, \overline{b} \in \mathbb{Z}/m\mathbb{Z}$.

It is easy to check that the definition above is independent of the choice of representatives, that is, if $\bar{a} = \bar{a'}, \bar{b} = \bar{b'}$, then $\overline{a+b} = \overline{a'+b'}, \overline{a-b} = \overline{a'-b'}, \overline{ab} = \overline{a'b'}$. Moreover, $\mathbb{Z}/m\mathbb{Z}$ is closed under addition, subtraction, and multiplication, and the associative, commutative, and distributive laws hold for addition and multiplication; $\bar{0}$ is the **identity for addition** (i.e., $\bar{0} + \bar{a} = \bar{a}$ for any $\bar{a} \in \mathbb{Z}/m\mathbb{Z}$), and $\bar{1}$ is the **identity for multiplication** (i.e., $\bar{1}\bar{a} = \bar{a}$ for any $\bar{a} \in \mathbb{Z}/m\mathbb{Z}$).

With the operations of addition and multiplication as above, $\mathbb{Z}/m\mathbb{Z}$ is often called the **ring of residue classes modulo** m.

Let p be a prime. We'll prove that there is a unique $\bar{c} \in \mathbb{Z}/p\mathbb{Z}$ such that $\bar{b}\bar{c} = \bar{a}$ for any $\bar{a}, \bar{b} \in \mathbb{Z}/p\mathbb{Z}$ with $\bar{b} \neq \bar{0}$. In fact, since $1 \leqslant b \leqslant p - 1$, $(b, p) = 1$. By Corollary 1.3.5, there exists $u \in \mathbb{Z}$ with $bu \equiv 1 \pmod{p}$, and hence $bua \equiv a \pmod{p}$. Let $c = ua$. Then $\bar{b}\bar{c} = \overline{bc} = \overline{bua} = \bar{a}$. Suppose that $\overline{c_1} \in \mathbb{Z}/p\mathbb{Z}$ also satisfies the equation $\bar{b}\overline{c_1} = \bar{a}$, then $\bar{b}(\bar{c} - \overline{c_1}) = \bar{0}$. Note that $\overline{ub} = \overline{ub} = \bar{1}$, so $\bar{c} - \overline{c_1} = \bar{1}(\bar{c} - \overline{c_1}) = \overline{ub}(\bar{c} - \overline{c_1}) = \overline{u}\bar{0} = \bar{0}$, i.e., $\bar{c} = \overline{c_1}$. Thus, we define $\bar{c}$ to be the quotient upon dividing $\bar{a}$ by $\bar{b}$, denoted by $\bar{c} = \frac{\bar{a}}{\bar{b}}$. Therefore, $\mathbb{Z}/p\mathbb{Z}$ is closed under addition, subtraction, multiplication, and division (with a nonzero denominator), and it is usually called the **field of** p **elements**, denoted by $\mathbb{F}_p$. In particular, taking $p = 2$ gives the field of 2 elements, $\mathbb{F}_2 = \{\bar{0}, \bar{1}\}$, consisting of the sets of even and odd integers.

Question. If m is not a prime, is it possible that $\mathbb{Z}/m\mathbb{Z}$ is a field?

1.5 Polynomials in One Indeterminate and Their Operations

The polynomial theory is an important part of advanced algebra. It is relatively independent and self-contained. It is not based on other contents of advanced algebra but provides the theoretical basis for the other parts. The polynomial theory also has critical applications in other branches: Polynomial functions are used to approximate continuous functions in mathematical analysis; the solution set of polynomial equations is the primary research object in algebraic geometry; there are polynomial algorithms in computer science, polynomial codes in coding theory, and so on.

Definition 1.5.1. Let F be a number field and x an **indeterminate**. A **polynomial** over F in the indeterminate x is a formal expression

$$a_0 + a_1 x + a_2 x^2 + \cdots + a_n x^n,$$

where $n \in \mathbb{N}$, $a_0, a_1, \ldots, a_n \in F$.

In the following, we often use $f(x), g(x), \ldots$ or $f, g, \ldots$ to denote polynomials (over F in the indeterminate x).

By convention, $x^0 = 1, x^1 = x, 1x^k = x^k$. So $a_0 = a_0 x^0, a_1 x = a_1 x^1$. Thus a polynomial $f(x) = a_0 + a_1 x + a_2 x^2 + \cdots + a_n x^n$ can be rewritten as

$$f(x) = a_0 x^0 + a_1 x^1 + a_2 x^2 + \cdots + a_n x^n = \sum_{i=0}^{n} a_i x^i, \qquad (1.2)$$

where $a_i x^i$ is called the **term** of ith degree (or, simply, a term) of $f(x)$, and a_i is called the **coefficient** of the term $a_i x^i$, $i = 0, 1, \ldots, n$. In particular, a_0 is called the **constant term** of $f(x)$, and $a_i x^i$ is called a **monomial** over F, $i = 0, 1, \ldots, n$.

In (1.2), if $a_n \neq 0$, then one calls n the **degree** of $f(x)$, and one denotes the degree n by $\deg(f(x))$, in this case, $f(x)$ is called a polynomial of degree n, $a_n x^n$ is called the **leading term** of $f(x)$, and a_n is said to be the **leading coefficient** of $f(x)$. If $a_n = 1$, we say that $f(x)$ is a **monic polynomial**. If $a_0 = a_1 = \cdots = a_n = 0$, i.e., all coefficients of $f(x)$ are 0, then $f(x)$ is called the **zero polynomial**.

Remark 1.5.2.

(1) The zero polynomial is the only polynomial with no degree. When the symbol $\deg(f(x))$ is used, we always assume that $f(x)$ is not the zero polynomial. In some references, the degree of the zero polynomial is defined as $-\infty$. In this case, by convention, it is assumed that $(-\infty) + (-\infty) = -\infty$, $(-\infty) + n = -\infty$, and $-\infty < n$ for all natural numbers n.

(2) In a polynomial, by convention, the terms with 0 coefficients may be removed or added arbitrarily. Therefore, the zero polynomial is unique and is denoted by 0. Moreover, any polynomial $f(x)$

may be written in the form of adding infinite terms:

$$f(x) = \sum_{i=0}^{n} a_i x^i = \sum_{i=0}^{\infty} a_i x^i,$$

where $a_i = 0$ when $i > n$.

(3) Let $a \in F$, then $a = ax^0$. Therefore, if $a \neq 0$, then a is a **polynomial of degree zero**; if $a = 0$, then a is the zero polynomial.

(4) The zero polynomial and a polynomial of degree zero are different! The zero polynomial is the one whose coefficients are all zero, i.e., the number 0 in F; a polynomial of degree zero is a nonzero number in F.

(5) The polynomial we define here is a formal expression in one indeterminate. When this indeterminate is a variable or an unknown, it is the polynomial (function) we learned in high school. Generally speaking, polynomials differ from the functions determined by polynomials (see Section 1.9).

Definition 1.5.3. Two polynomials $f(x)$ and $g(x)$ over a number field F are **equal**, denoted by $f(x) = g(x)$, if the corresponding coefficients of the terms of the same degree of $f(x)$ and $g(x)$ are equal.

Let $f(x), g(x)$ be polynomials over a number field F. By Remark 1.5.2 (2), we may assume

$$f(x) = \sum_{i=0}^{\infty} a_i x^i, \text{ where } a_i \in F, \text{ only finitely many } a_i \neq 0,$$

$$g(x) = \sum_{i=0}^{\infty} b_i x^i, \text{ where } a_i \in F, \text{ only finitely many } b_i \neq 0.$$

Then $f(x) = g(x)$ if and only if $a_i = b_i, i = 0, 1, 2, \ldots$.

Let $F[x]$ be the set of all polynomials over a number field F, that is,

$$F[x]$$

$$= \left\{ f(x) \,\middle|\, f(x) = \sum_{i=0}^{\infty} a_i x^i, \text{ where } a_i \in F, \text{only finitely many } a_i \neq 0 \right\}.$$

Then $F \subsetneq F[x]$. Therefore, polynomials are a wider class of mathematical research objects than ordinary numbers.

Next, we consider the addition, subtraction, and multiplication operations of polynomials.

Definition 1.5.4. Suppose that

$$f(x) = \sum_{i=0}^{\infty} a_i x^i, \text{ where } a_i \in F, \text{ only finitely many } a_i \neq 0,$$

$$g(x) = \sum_{i=0}^{\infty} b_i x^i, \text{ where } b_i \in F, \text{ only finitely many } b_i \neq 0.$$

The **sum** of $f(x)$ and $g(x)$ is $f(x) + g(x) = \sum_{i=0}^{\infty} (a_i + b_i) x^i$.

The **difference** between $f(x)$ and $g(x)$ is

$$f(x) - g(x) = \sum_{i=0}^{\infty} (a_i - b_i) x^i.$$

The **product** of $f(x)$ and $g(x)$ is $f(x)g(x) = \sum_{i=0}^{\infty} c_i x^i$, where $c_i = \sum_{k+l=i} a_k b_l$, $i = 0, 1, 2, \ldots$.

The proof of the following proposition is straightforward, so it is omitted.

Proposition 1.5.5. *Let F be a number field and $f(x), g(x), h(x) \in F[x]$.*

(1) *Addition of polynomials satisfies the following:*

    (1.1) ***Commutative law:*** *$f(x) + g(x) = g(x) + f(x)$.*

    (1.2) ***Associative law:***

$$(f(x) + g(x)) + h(x) = f(x) + (g(x) + h(x)).$$

    (1.3) *The zero polynomial 0 is the **identity for addition:** $0 + f(x) = f(x)$.*

    (1.4) *Negation produces the **additive inverse:***
        *For any polynomial $f(x) = a_0 + a_1 x + a_2 x^2 + \cdots + a_n x^n$, there is a polynomial $-f(x) = -a_0 - a_1 x - a_2 x^2 - \cdots - a_n x^n$, called the **negative** of $f(x)$, such that $f(x) + (-f(x)) = 0$.*

(2) *Multiplication of polynomials satisfies the following:*

    (2.1) **Commutative law:** $f(x)g(x) = g(x)f(x)$.

    (2.2) **Associative law:** $(f(x)g(x))h(x) = f(x)(g(x)h(x))$.

    (2.3) *The constant polynomial* 1 *is the* **identity for multiplication:** $1f(x) = f(x)$.

    (2.4) **Cancellation law:**
        If $f(x)g(x) = f(x)h(x)$ *and* $f(x) \neq 0$, *then* $g(x) = h(x)$.

(3) **Distributive law** *of polynomial multiplication:*

$$f(x)(g(x) + h(x)) = f(x)g(x) + f(x)h(x).$$

Remark 1.5.6. Let $k \in F$ and $f(x) = a_0 + a_1 x + a_2 x^2 + \cdots + a_n x^n \in F[x]$. Then $kf(x) = ka_0 + ka_1 x + ka_2 x^2 + \cdots + ka_n x^n$ is the product of the polynomials $g(x) = k$ and $f(x)$. The polynomial $kf(x)$ is also called the **scalar multiplication** of the number k with the polynomial $f(x)$.

It is easily seen that

$$k(f(x)g(x)) = (kf(x))g(x) = f(x)(kg(x)), \quad (-1)f(x) = -f(x)$$

for all $k \in F, f(x), g(x) \in F[x]$.

Remark 1.5.7.

(1) The discussion above shows that $F[x]$ is closed under addition, subtraction, and multiplication, and the associative, commutative, and distributive laws hold for addition and multiplication. So $F[x]$ is often called the **ring of polynomials over F in one indeterminate.**

(2) Let F be a number field and x an indeterminate. A **formal power series over F in one indeterminate** is a formal expression

$$a_0 + a_1 x + a_2 x^2 + \cdots + a_n x^n + a_{n+1} x^{n+1} + \cdots,$$

where $a_i \in F, i = 0, 1, \ldots, n, \ldots$.

Let $F[[x]]$ be the set of all formal power series over F in one indeterminate over a number field F, that is,

$$F[[x]] = \left\{ f(x) \;\middle|\; f(x) = \sum_{i=0}^{\infty} a_i x^i, \text{ where each } a_i \in F \right\}.$$

The sum, difference, and product of two formal power series in $F[[x]]$ are defined in just the same way as the sum, difference, and product of two polynomials in $F[x]$. Then $F[[x]$ is closed under addition, subtraction, and multiplication, and the addition and multiplication operations satisfy the commutative, associative, and distributive laws. So $F[[x]]$ is usually called the **ring of formal power series over F in one indeterminate**. The discussion on power series is beyond the scope of this book. Readers can refer to the references $[19, 20, 22]$.

By definition, the operations and degrees of polynomials have the following relationship.

Proposition 1.5.8. *Let F be a number field and $f(x), g(x) \in F[x]$. Then*

(1) $\deg(f(x) \pm g(x)) \leqslant \max\{\deg(f(x)), \deg(g(x))\};$
(2) $\deg(f(x)g(x)) = \deg(f(x)) + \deg(g(x)).$

1.6   Division Algorithm for Polynomials

It can be seen from the discussion in Section 1.5 that, in the ring $F[x]$ of polynomials over a number field F in one indeterminate, we can perform the operations of addition, subtraction, and multiplication. Still, the result obtained by dividing two polynomials is not necessarily a polynomial; that is to say, the inverse operation of multiplication — division is not generally available. When is the result of dividing one polynomial by another still a polynomial? This question is our concern. For this reason, we first discuss the division algorithm for polynomials.

Theorem 1.6.1 (Division algorithm for polynomials). *Let F be a number field and $f(x), g(x) \in F[x]$ with $g(x) \neq 0$. Then there exist unique $q(x), r(x) \in F[x]$ such that $f(x) = q(x)g(x) + r(x)$, where $r(x) = 0$, or $\deg(r(x)) < \deg(g(x))$.*

Proof.   We first prove the existence of such $q(x)$ and $r(x)$. If $f(x) = 0$, define $q(x) = r(x) = 0$, and we are done. If $\deg(g(x)) = 0$, i.e., $g(x) = c$ is a nonzero number, we may choose $q(x) = \frac{1}{c}f(x), r(x) = 0$.

Next, we assume that

$$f(x) = \sum_{i=0}^{n} a_i x^i, \ n \geqslant 0, a_n \neq 0, \ g(x) = \sum_{j=0}^{m} b_j x^j, \ m > 0, b_m \neq 0.$$

We'll do an induction on the degree n of $f(x)$.

If $n = 0$, then $q(x) = 0, r(x) = f(x)$ will do.

Let $n \geqslant 1$. Suppose that the existence holds for polynomials with degree $< n$. Now consider the case when the degree of a polynomial equals n. If $n < m$, define $q(x) = 0, r(x) = f(x)$, and we are done. Suppose $n \geqslant m$. Let $f_1(x) = f(x) - \frac{a_n}{b_m} x^{n-m} g(x)$. If $f_1(x) = 0$, then $q(x) = \frac{a_n}{b_m} x^{n-m}$, $r(x) = 0$ will do. Suppose $f_1(x) \neq 0$, then $\deg(f_1(x)) < n$. By the inductive hypothesis, for the polynomials $f_1(x)$, $g(x)$, there are $q_1(x), r_1(x) \in F[x]$ with $f_1(x) = q_1(x)g(x) + r_1(x)$, where $r_1(x) = 0$, or $\deg(r_1(x)) < \deg(g(x))$. Thus, $f(x) = \left(\frac{a_n}{b_m} x^{n-m} + q_1(x)\right) g(x) + r_1(x) = q(x)g(x) + r(x)$, where $q(x) = \frac{a_n}{b_m} x^{n-m} + q_1(x), r(x) = r_1(x)$. By the second form of induction, the existence of the theorem holds.

To prove uniqueness of $q(x)$ and $r(x)$, assume that there are $q_2(x), r_2(x) \in F[x]$ such that $f(x) = q_2(x)g(x) + r_2(x)$, where $r_2(x) = 0$, or $\deg(r_2(x)) < \deg(g(x))$, then $q(x)g(x) + r(x) = q_2(x)g(x) + r_2(x)$. It follows that $(q(x) - q_2(x))g(x) = r_2(x) - r(x)$. If $q(x) \neq q_2(x)$, then $q(x) - q_2(x) \neq 0$, and so $r_2(x) - r(x) \neq 0$. Thus $\deg(r_2(x) - r(x)) = \deg((q(x) - q_2(x))g(x)) \geqslant \deg(g(x))$, contradicting $\deg(r_2(x) - r(x)) < \deg(g(x))$. Therefore, $q(x) = q_2(x)$ and hence $r(x) = r_2(x)$, as required. $\qquad\square$

Remark 1.6.2. If $f(x)$ and $g(x)$ are polynomials with $g(x) \neq 0$, then the polynomials $q(x)$ and $r(x)$ occurring in the division algorithm are respectively called the **quotient** and the **remainder** upon dividing $f(x)$ by $g(x)$.

From the uniqueness in the division algorithm, we know that the quotient and remainder upon dividing $f(x)$ by $g(x)$ will not change due to the extension of the coefficient field, that is to say, if $f(x), g(x)$ are polynomials over a number field F, and F is contained in a larger number field $\overline{F}$, then $f(x), g(x)$ are also polynomials over $\overline{F}$. It follows from the division algorithm that whether $f(x), g(x)$ are regarded as polynomials in $F[x]$ or $\overline{F}[x]$, the quotient and remainder upon dividing $f(x)$ by $g(x)$ are the same.

Let $f(x) = a_n x^n + a_{n-1} x^{n-1} + \cdots + a_2 x^2 + a_1 x + a_0 \in F[x]$ and $a \in F$. By the division algorithm, we may assume that the quotient and remainder upon dividing $f(x)$ by $x - a$ are $b_{n-1} x^{n-1} + b_{n-2} x^{n-2} + \cdots + b_1 x + b_0$ and r (a constant), respectively. Thus

$$a_n x^n + a_{n-1} x^{n-1} + \cdots + a_2 x^2 + a_1 x + a_0$$
$$= (b_{n-1} x^{n-1} + b_{n-2} x^{n-2} + \cdots + b_1 x + b_0)(x - a) + r$$
$$= b_{n-1} x^n + (b_{n-2} - ab_{n-1}) x^{n-1} + \cdots + (b_1 - ab_2) x^2$$
$$+ (b_0 - ab_1) x + r - ab_0,$$

and so $b_{n-1} = a_n, b_{n-2} = a_{n-1} + ab_{n-1}, \ldots, b_0 = a_1 + ab_1, r = a_0 + ab_0$.

The above relationship can be arranged in the following format, and then the quotient and remainder upon dividing $f(x)$ by $x - a$ can be obtained.

a	a_n	a_{n-1}	$\cdots$	a_1	a_0
		ab_{n-1}	$\cdots$	ab_1	ab_0
	a_n	$a_{n-1} + ab_{n-1}$	$\cdots$	$a_1 + ab_1$	$a_0 + ab_0$
	$\shortparallel$	$\shortparallel$	$\cdots$	$\shortparallel$	$\shortparallel$
	b_{n-1}	b_{n-2}	$\cdots$	b_0	r

This calculation format is usually called the **synthetic division**.

Example 1.6.3.

(1) Find the quotient and remainder upon dividing x^4 by $x - 1$.
(2) Express x^4 as the sum of powers of $x - 1$.

Solution. (1) By the synthetic division, we have the table as follows:

$$
\begin{array}{c|ccccc}
1 & 1 & 0 & 0 & 0 & 0 \\
\hline
 & 1 & 1 & 1 & 1 & 1
\end{array}
$$

Therefore, upon dividing x^4 by $x - 1$, the quotient is $x^3 + x^2 + x + 1$, and the remainder is 1.

(2) By the synthetic division, we have the table as follows:

$$
\begin{array}{c|ccccc}
1 & 1 & 0 & 0 & 0 & 0 \\
\hline
1 & 1 & 1 & 1 & 1 & 1 \\
1 & 1 & 2 & 3 & 4 \\
1 & 1 & 3 & 6 \\
1 & 4 \\
\end{array}
$$

So $x^4 = (x-1)^4 + 4(x-1)^3 + 6(x-1)^2 + 4(x-1) + 1$. $\qquad\square$

Definition 1.6.4. Let $f(x), g(x) \in F[x]$. We say that $g(x)$ **divides** $f(x)$, denoted by $g(x)|f(x)$, if there exists $q(x) \in F[x]$ such that $f(x) = q(x)g(x)$. In this case, $g(x)$ is called a **divisor** (or **factor**) of $f(x)$, and $f(x)$ is called a **multiple** of $g(x)$.

We write $g(x) \nmid f(x)$ if $g(x)$ does not divide $f(x)$.

Remark 1.6.5.

(1) In the definition of "$g(x)|f(x)$", $g(x)$ may be equal to 0. In this case, $f(x) = 0$. So we can say "$0|0$".
(2) If $f(x), g(x) \in F[x]$ and $g(x) \neq 0$, then $g(x)|f(x)$ if and only if the remainder upon dividing $f(x)$ by $g(x)$ is 0.
(3) Every polynomial divides itself.
(4) Every polynomial divides 0.
(5) Polynomials of degree 0 (or nonzero constants) divide every polynomial.
(6) The divisibility relation between two polynomials does not change due to the extension of the coefficient field.
(7) If $g(x)|f(x)$ and $g(x) \neq 0$, then the quotient upon dividing $f(x)$ by $g(x)$ may be denoted by $\frac{f(x)}{g(x)}$.

Proposition 1.6.6.

(1) *Let $f(x), g(x) \in F[x]$. Then $f(x)|g(x)$ and $g(x)|f(x)$ if and only if $f(x) = cg(x)$, where c is a nonzero constant.*
(2) *If $f(x)|g(x)$ and $g(x)|h(x)$, then $f(x)|h(x)$.*
(3) *Let $f(x)|g_i(x), i = 1, 2, \ldots, n$. Then $f(x)\left|\sum_{i=1}^{n} u_i(x)g_i(x)\right.$ for all $u_i(x) \in F[x], i = 1, 2, \ldots, n$, where $\sum_{i=1}^{n} u_i(x)g_i(x)$ is often called a **combination** of $g_1(x), g_2(x), \ldots, g_n(x)$.*

Proof. This result follows from the definition of divisibility.   □

Example 1.6.7. Let d and n be positive integers. Prove that

$$(x^d - 1)|(x^n - 1) \text{ if and only if } d|n.$$

Proof. Sufficiency. Since $d|n$, $n = qd$ for some positive integer q. Thus

$$x^n - 1 = (x^d)^q - 1 = (x^d - 1)((x^d)^{q-1} + (x^d)^{q-2} + \cdots + x^d + 1).$$

So $(x^d - 1)|(x^n - 1)$.

Necessity. **Method 1** (Using proof by contradiction and the result of sufficiency).

From $(x^d - 1)|(x^n - 1)$ we know that $d \leqslant n$. Suppose $d \nmid n$, then $n = qd + r$, where q is a positive integer and $0 < r < d$. Thus

$$x^n - 1 = x^{qd+r} - 1 = x^{qd+r} - x^r + x^r - 1 = x^r(x^{qd} - 1) + x^r - 1.$$

By hypothesis, $(x^d - 1)|(x^n - 1)$. By the sufficiency, $(x^d - 1)|(x^{qd} - 1)$. Therefore, $(x^d - 1)|(x^r - 1)$, and so $d \leqslant r$, a contradiction.

Method 2 (Using the division algorithm).

Since $(x^d - 1)|(x^n - 1)$, $d \leqslant n$. Thus $n = qd + r$, where q is a positive integer and $0 \leqslant r < d$, and hence

$$x^n - 1 = (x^{n-d} + x^{n-2d} + \cdots + x^{n-qd})(x^d - 1) + x^{n-qd} - 1$$

$$= (x^{n-d} + x^{n-2d} + \cdots + x^{n-qd})(x^d - 1) + x^r - 1,$$

as shown in the following figure.

$$
\require{enclose}
\begin{array}{r}
x^{n-d} + x^{n-2d} + \quad \cdots \quad + x^{n-qd} \phantom{{}-1} \\[2pt]
x^d - 1 \enclose{longdiv}{\; x^n \qquad\quad - 1 } \\[2pt]
\underline{x^n \quad - x^{n-d} } \\[2pt]
x^{n-d} \quad - 1 \\[2pt]
\underline{x^{n-d} \quad - x^{n-2d} } \\[2pt]
x^{n-2d} - 1 \\[2pt]
\ddots \\[2pt]
\underline{} \\[2pt]
x^{n-qd} - 1
\end{array}
$$

Therefore, upon dividing $x^n - 1$ by $x^d - 1$, the quotient is

$$q(x) = x^{n-d} + x^{n-2d} + \cdots + x^{n-qd},$$

and the remainder is $r(x) = x^r - 1$. Since $(x^d - 1)|(x^n - 1)$, $r(x) = x^r - 1 = 0$. We conclude that $r = 0$, and hence $d|n$.

Method 3 (Using the second form of induction).
From $(x^d - 1)|(x^n - 1)$ we know $d \leqslant n$.
If $n = d$, the necessity is true.
Let $m > d$ and assume that the necessity is true for all k with $d \leqslant k < m$, i.e., $(x^d - 1)|(x^k - 1) \Longrightarrow d|k$.
Next we show that $(x^d - 1)|(x^m - 1) \Longrightarrow d|m$.
In fact, $x^m - 1 = x^m - x^{m-d} + x^{m-d} - 1 = x^{m-d}(x^d - 1) + x^{m-d} - 1$.
Since $(x^d - 1)|(x^m - 1)$, $(x^d - 1)|(x^{m-d} - 1)$. Clearly, $d \leqslant m - d < m$.
By the inductive hypothesis, $d|(m-d)$. Note that $d|d$, so $d|(m-d+d)$, that is, $d|m$. By the second form of induction, the necessity of the theorem is true. $\qquad\square$

Definition 1.6.8. Let $f(x), g(x), d(x) \in F[x]$.

(1) If $d(x)|f(x)$ and $d(x)|g(x)$, then $d(x)$ is called a **common divisor** of $f(x)$ and $g(x)$.
(2) If $d(x)$ is a common divisor of $f(x)$ and $g(x)$, and every common divisor of $f(x)$ and $g(x)$ is a divisor of $d(x)$, then $d(x)$ is said to be the **greatest common divisor** of $f(x)$ and $g(x)$.

Remark 1.6.9.

(1) If $f(x)|g(x)$, then $f(x)$ is the greatest common divisor of $f(x)$ and $g(x)$. In particular, $f(x)$ is the greatest common divisor of $f(x)$ and 0. Of course, 0 is the greatest common divisor of 0 and 0.
(2) If $d_1(x)$ is the greatest common divisor of $f(x)$ and $g(x)$, then $d_2(x)$ is also the greatest common divisor of $f(x)$ and $g(x)$ if and only if $d_2(x) = cd_1(x)$, where c is a nonzero constant. So, the greatest common divisors of two polynomials are unique up to a nonzero constant. Given two polynomials $f(x)$ and $g(x)$, not both 0, we use $(f(x), g(x))$ to denote the monic greatest common divisor of $f(x)$ and $g(x)$.

Lemma 1.6.10. *Suppose $f(x), g(x) \in F[x]$ are not both 0. If there are $q(x), r(x) \in F[x]$ with $f(x) = q(x)g(x) + r(x)$, prove that $(f(x), g(x)) = (g(x), r(x))$.*

Proof. By hypothesis, $r(x) = f(x) - q(x)g(x)$. Since $(f(x), g(x))|f(x)$ and $(f(x), g(x))|g(x)$, $(f(x), g(x))|r(x)$. It follows that $(f(x), g(x))|(g(x), r(x))$. Similarly, $(g(x), r(x))|(f(x), g(x))$. So $(f(x), g(x)) = (g(x), r(x))$. $\square$

Theorem 1.6.11. *Any two polynomials $f(x), g(x) \in F[x]$ have the greatest common divisor $d(x)$, which is unique up to the multiplication by a nonzero constant in F; moreover, $d(x)$ is a **combination** of $f(x)$ and $g(x)$, i.e., there exist $u(x), v(x) \in F[x]$ such that*

$$d(x) = u(x)f(x) + v(x)g(x) \ (\textbf{Bézout's identity}).$$

Proof. The uniqueness (up to a nonzero constant) part of the theorem is clear by Remark 1.6.9 (2). It remains to prove the existence.

If $g(x)|f(x)$, then $g(x)$ is the greatest common divisor of $f(x)$ and $g(x)$, and $g(x) = 0f(x) + g(x)$.

Suppose $g(x) \nmid f(x)$. In this case, we may assume $g(x) \neq 0$ (if $g(x) = 0$, then $f(x)|g(x)$, we are done).

Apply Theorem 1.6.1 successively as follows:

$$
\begin{aligned}
f(x) &= q_1(x)g(x) + r_1(x), & \deg(r_1(x)) &< \deg(g(x)), \\
g(x) &= q_2(x)r_1(x) + r_2(x), & \deg(r_2(x)) &< \deg(r_1(x)), \\
r_1(x) &= q_3(x)r_2(x) + r_3(x), & \deg(r_3(x)) &< \deg(r_2(x)), \\
&\ \ \vdots & &\ \ \vdots \\
r_{s-3}(x) &= q_{s-1}(x)r_{s-2}(x) + r_{s-1}(x), & \deg(r_{s-1}(x)) &< \deg(r_{s-2}(x)), \\
r_{s-2}(x) &= q_s(x)r_{s-1}(x) + r_s(x), & \deg(r_s(x)) &< \deg(r_{s-1}(x)), \\
r_{s-1}(x) &= q_{s+1}(x)r_s(x) + r_{s+1}(x), & r_{s+1}(x) &= 0.
\end{aligned}
$$

$$(1.3)$$

Note that there is a positive integer s with $r_s(x) \neq 0$ but $r_{s+1}(x) = 0$ because the degrees of the remainders $r_1(x), r_2(x), \ldots$ form a strictly decreasing sequence of nonnegative integers. Thus, by Lemma 1.6.10, we have

$$(f(x), g(x))$$
$$= (g(x), r_1(x)) = (r_1(x), r_2(x)) = \cdots = (r_{s-1}(x), r_s(x)) = c_s r_s(x),$$

where c_s is a nonzero constant such that $c_s r_s(x)$ is monic. Then $r_s(x)$, the last nonzero remainder, is a greatest common divisor of $f(x)$ and $g(x)$.

Next, we show that $r_s(x)$ is a combination of $f(x)$ and $g(x)$.

We find coefficients $u(x)$ and $v(x)$ with $r_s(x) = u(x)f(x) + v(x)g(x)$ by working upward from the bottom of the list.

From the penultimate equation in (1.3),

$$r_s(x) = r_{s-2}(x) - q_s(x)r_{s-1}(x)$$

is a combination of $r_{s-2}(x)$ and $r_{s-1}(x)$. Combining this with the equation immediately above it, $r_{s-1}(x) = r_{s-3}(x) - q_{s-1}(x)r_{s-2}(x)$, gives

$$\begin{aligned} r_s(x) &= r_{s-2}(x) - q_s(x)(r_{s-3}(x) - q_{s-1}(x)r_{s-2}(x)) \\ &= (-q_s(x))r_{s-3}(x) + (1 + q_s(x)q_{s-1}(x))r_{s-2}(x), \end{aligned}$$

a combination of $r_{s-3}(x)$ and $r_{s-2}(x)$. Continuing this way, we get $u(x), v(x) \in F[x]$ such that $r_s(x) = u(x)f(x) + v(x)g(x)$. $\square$

The method used for finding the greatest common divisor of two polynomials in the proof of the theorem above is called the **Euclidean algorithm** or **Euclid's algorithm**. From the proof of this theorem, we know that the monic greatest common divisor of two polynomials does not change due to the extension of the coefficient field.

Example 1.6.12. Let

$$f(x) = x^4 + 3x^3 - x^2 - 4x - 3, \ \ g(x) = 3x^3 + 10x^2 + 2x - 3.$$

Find $(f(x), g(x))$ and $u(x), v(x)$ such that

$$u(x)f(x) + v(x)g(x) = (f(x), g(x)).$$

Solution. We apply the Euclidean algorithm and obtain

$$f(x) = q_1(x)g(x) + r_1(x),$$

where $q_1(x) = \dfrac{1}{3}x - \dfrac{1}{9}$, $r_1(x) = -\dfrac{5}{9}x^2 - \dfrac{25}{9}x - \dfrac{10}{3}$,

$$g(x) = q_2(x)r_1(x) + r_2(x),$$

where $q_2(x) = -\dfrac{27}{5}x + 9$, $r_2(x) = 9x + 27$,

$$r_1(x) = q_3(x)r_2(x), \quad \text{where } q_3(x) = -\dfrac{5}{81}x - \dfrac{10}{81}.$$

Thus $(f(x), g(x)) = \frac{1}{9}r_2(x) = x + 3$, and

$$r_2(x) = g(x) - q_2(x)r_1(x) = g(x) - q_2(x)(f(x) - q_1(x)g(x))$$

$$= -q_2(x)f(x) + (1 + q_2(x)q_1(x))g(x)$$

$$= \left(\frac{27}{5}x - 9\right) f(x) + \left(-\frac{9}{5}x^2 + \frac{18}{5}x\right) g(x),$$

Hence $(f(x), g(x)) = \left(\frac{3}{5}x - 1\right) f(x) + \left(-\frac{1}{5}x^2 + \frac{2}{5}x\right) g(x)$.
Let $u(x) = \frac{3}{5}x - 1$, $v(x) = -\frac{1}{5}x^2 + \frac{2}{5}x$. Then

$$u(x)f(x) + v(x)g(x) = (f(x), g(x)). \qquad \square$$

Remark 1.6.13. It can be seen from the above example that when using the Euclidean algorithm to find the greatest common divisor of two polynomials, there appear some fractional coefficients, which bring some troubles to the calculation because we need to find $u(x), v(x)$. On the other hand, if we only have to find the greatest common divisor, we may use appropriate nonzero constants to multiply the dividends and divisors to avoid the appearance of fractional coefficients when applying the Euclidean algorithm. Of course, doing so will cause changes in quotients and remainders. However, since different greatest common divisors can differ by a nonzero constant, doing so does not affect finding the monic greatest common divisor.

As far as Example 1.6.12 is concerned, the following calculation can be made:

$$3f(x) = xg(x) + (-1)(x^3 + 5x^2 + 9x + 9),$$

$$g(x) = 3(x^3 + 5x^2 + 9x + 9) + (-5)(x^2 + 5x + 6),$$

$$x^3 + 5x^2 + 9x + 9 = x(x^2 + 5x + 6) + 3(x + 3),$$

$$x^2 + 5x + 6 = (x + 2)(x + 3).$$

So $(f(x), g(x)) = x + 3$.

Corollary 1.6.14. *Let* $f(x), g(x), d(x) \in F[x]$. *If* $f(x), g(x)$ *are not both* 0, *then* $d(x)$ *is the greatest common divisor of* $f(x)$ *and* $g(x)$ *if and only if* $d(x)$ *is a common divisor of* $f(x)$ *and* $g(x)$ *of the largest degree.*

Proof. Since $f(x), g(x)$ are not both 0, all common divisors of $f(x)$ and $g(x)$ are not 0.

Necessity. Suppose that $d(x)$ is a greatest common divisor of $f(x)$ and $g(x)$. If $h(x)$ is a common divisor of $f(x)$ and $g(x)$, then $h(x)|d(x)$, and so $\deg(h(x)) \leqslant \deg(d(x))$. It follows that $d(x)$ is a common divisor of $f(x)$ and $g(x)$ of the largest degree.

Sufficiency. Let $d_0(x)$ be the greatest common divisor of $f(x)$ and $g(x)$. Since $d(x)$ is a common divisor of $f(x)$ and $g(x)$ by hypothesis, $d(x)|d_0(x)$, and hence $\deg(d(x)) \leqslant \deg(d_0(x))$. Note that $\deg(d_0(x)) \leqslant \deg(d(x))$, and so $\deg(d_0(x)) = \deg(d(x))$. It follows that $d(x) = cd_0(x)$ for some nonzero constant c since $d(x)|d_0(x)$. Thus, $d(x)$ is the greatest common divisor of $f(x)$ and $g(x)$. $\square$

Corollary 1.6.15. *Let* $f(x), g(x), d(x) \in F[x]$. *If* $f(x), g(x)$ *are not both* 0, *then* $d(x)$ *is the greatest common divisor of* $f(x)$ *and* $g(x)$ *if and only if* $d(x)$ *is of the lowest degree among all nonzero polynomials of the form* $u(x)f(x) + v(x)g(x)$ *with* $u(x), v(x) \in F[x]$.

Proof. Necessity. Let $d(x)$ be a greatest common divisor of $f(x)$ and $g(x)$. According to Theorem 1.6.11, there are $k(x), l(x) \in F[x]$ such that $d(x) = k(x)f(x) + l(x)g(x)$. Note that and $d(x)|(u(x)f(x) + v(x)g(x))$ for any $u(x), v(x) \in F[x]$, and so $\deg(d(x)) \leqslant \deg(u(x)f(x) + v(x)g(x))$ if $u(x)f(x) + v(x)g(x) \neq 0$. Thus, $d(x)$ is of the lowest degree among all nonzero polynomials of the form $u(x)f(x) + v(x)g(x)$ with $u(x), v(x) \in F[x]$.

Sufficiency. Let $d_0(x)$ be the greatest common divisor of $f(x)$ and $g(x)$. By hypothesis, $d_0(x)|d(x)$, and hence $\deg(d_0(x)) \leqslant \deg(d(x))$. By Theorem 1.6.11, there are $k_0(x), l_0(x) \in F[x]$ such that $d_0(x) = k_0(x)f(x) + l_0(x)g(x)$, and so $\deg(d(x)) \leqslant \deg(d_0(x))$. Therefore, $\deg(d_0(x)) = \deg(d(x))$, and hence $d(x) = cd_0(x)$ for some nonzero constant c. So $d(x)$ is the greatest common divisor of $f(x)$ and $g(x)$.

$\square$

Definition 1.6.16. Two polynomials $f(x), g(x) \in F[x]$ are said to be **relatively prime** (or **coprime**) if $(f(x), g(x)) = 1$. Let $f_i(x) \in F[x]$, $i = 1, 2, \ldots, t$, where $t \geqslant 2$. We say that $f_1(x), f_2(x), \ldots, f_t(x)$ are **pairwise coprime** if $f_i(x)$ and $f_j(x)$ are coprime for every pair i, j with $1 \leqslant i \neq j \leqslant t$.

By definition, $f(x)$ and $g(x)$ are relatively prime if and only if $f(x)$ and $g(x)$ only have nonzero constants as their common divisors if and only if greatest common divisors of $f(x)$ and $g(x)$ are nonzero constants. It is easy to see that the relative prime relation of two polynomials will not change due to the extension of the coefficient field.

Proposition 1.6.17. *Let* $f(x), g(x), h(x) \in F[x]$.

(1) $(f(x), g(x)) = 1$ *if and only if there are* $u(x), v(x) \in F[x]$ *such that* $u(x)f(x) + v(x)g(x) = 1$.

(2) *If* $f(x), g(x)$ *are not both* 0, *then* $\left(\frac{f(x)}{(f(x),g(x))}, \frac{g(x)}{(f(x),g(x))} \right) = 1$.

(3) *If* $f(x)|g(x)h(x)$ *and* $(f(x), g(x)) = 1$, *then* $f(x)|h(x)$.

(4) *If* $f(x)|h(x), g(x)|h(x)$ *and* $(f(x), g(x)) = 1$, *then* $f(x)g(x)|h(x)$.

(5) *If* $(f(x), g(x)) = 1$ *and* $(f(x), h(x)) = 1$, *then* $(f(x), g(x)h(x)) = 1$.

Proof. (1) The necessity follows from Theorem 1.6.11.

For the sufficiency, suppose $\varphi(x)|f(x), \varphi(x)|g(x)$, then $\varphi(x)|1$. It follows that $f(x)$ and $g(x)$ only have nonzero constants as their common divisors, i.e., $(f(x), g(x)) = 1$.

(2) By Theorem 1.6.11, there are $u(x), v(x) \in F[x]$ such that $u(x)f(x) + v(x)g(x) = (f(x), g(x))$. Note that $(f(x), g(x)) \neq 0$, and so

$$u(x)\frac{f(x)}{(f(x), g(x))} + v(x)\frac{g(x)}{(f(x), g(x))} = 1.$$

Thus $\left(\frac{f(x)}{(f(x),g(x))}, \frac{g(x)}{(f(x),g(x))} \right) = 1$ by (1).

(3) Since $(f(x), g(x)) = 1$, there are $u(x), v(x) \in F[x]$ such that $u(x)f(x) + v(x)g(x) = 1$. Thus $u(x)f(x)h(x) + v(x)g(x)h(x) = h(x)$. Since $f(x)|g(x)h(x)$ and $f(x)|f(x)$, $f(x)|h(x)$.

(4) Since $f(x)|h(x)$, there is $f_1(x) \in F[x]$ with $h(x) = f_1(x)f(x)$. Note that $g(x)|h(x)$ and $(f(x), g(x)) = 1$, so $g(x)|f_1(x)$ by (3). Thus there is $k(x) \in F[x]$ with $f_1(x) = k(x)g(x)$. Consequently, $h(x) = k(x)f(x)g(x)$, and so $f(x)g(x)|h(x)$.

(5) Since $(f(x), g(x)) = 1, (f(x), h(x)) = 1$, there are $u(x)$, $v(x)$, $k(x)$, $l(x) \in F[x]$ such that

$$u(x)f(x) + v(x)g(x) = 1, \text{ and } k(x)f(x) + l(x)h(x) = 1.$$

Thus $(u(x)f(x) + v(x)g(x))(k(x)f(x) + l(x)h(x)) = 1$, i.e.,

$$(u(x)(k(x)f(x) + l(x)h(x)) + v(x)g(x)k(x))f(x)$$
$$+ v(x)l(x)g(x)h(x) = 1.$$

So $(f(x), g(x)h(x)) = 1$ by (1). $\qquad\qquad\square$

Corollary 1.6.18. *Let $g(x), f_i(x) \in F[x], i = 1, 2, \ldots, n$. If $f_i(x)|g(x)$, $i = 1, 2, \ldots, n$, and $f_1(x), f_2(x), \ldots, f_n(x)$ are pairwise coprime, then $f_1(x)f_2(x) \cdots f_n(x)|g(x)$.*

Example 1.6.19. Let n be a positive integer. Prove that

$$x(x + 1)(2x + 1)|((x + 1)^{2n} - x^{2n} - 2x - 1).$$

Proof. First, since

$$(x + 1)^{2n} - x^{2n} - 2x - 1$$
$$= (x + 1)^{2n} - 1 - x(x^{2n-1} + 2)$$
$$= x\left((x + 1)^{2n-1} + (x + 1)^{2n-2} + \cdots + (x + 1) + 1\right)$$
$$- x\left(x^{2n-1} + 2\right),$$

$x|((x + 1)^{2n} - x^{2n} - 2x - 1)$.

Second, note that

$$(x + 1)^{2n} - x^{2n} - 2x - 1$$
$$= \left((x + 1)^2\right)^n - \left(x^2\right)^n - (2x + 1)$$
$$= (2x + 1)\left((x + 1)^{2(n-1)} + (x + 1)^{2(n-2)}x^2 + \cdots + x^{2(n-1)}\right)$$
$$- (2x + 1),$$

so $(2x + 1)|((x + 1)^{2n} - x^{2n} - 2x - 1)$.

Finally, by the equality

$$(x + 1)^{2n} - x^{2n} - 2x - 1$$
$$= (x + 1)^{2n} - x^{2n} + 1 - 2x - 2$$
$$= (x + 1)^{2n} - \left(x^{2n} - 1\right) - 2(x + 1)$$
$$= (x + 1)^{2n} - \left((x^2)^n - 1\right) - 2(x + 1)$$
$$= (x + 1)^{2n} - \left(x^2 - 1\right)\left(x^{2(n-1)} + x^{2(n-2)} + \cdots + x^2 + 1\right)$$
$$- 2(x + 1),$$

we have $(x + 1)|((x + 1)^{2n} - x^{2n} - 2x - 1)$.

Therefore, $x(x + 1)(2x + 1)|((x + 1)^{2n} - x^{2n} - 2x - 1)$ since x, $x + 1$, $2x + 1$ are pairwise coprime. $\qquad\square$

Definition 1.6.20. Let $f(x), g(x), m(x) \in F[x]$.

(1) If $f(x)|m(x)$ and $g(x)|m(x)$, then $m(x)$ is called a **common multiple** of $f(x)$ and $g(x)$.
(2) If $m(x)$ is a common multiple of $f(x)$ and $g(x)$, and every common multiple of $f(x)$ and $g(x)$ is a multiple of $m(x)$, then $m(x)$ is said to be the **least common multiple** of $f(x)$ and $g(x)$.

Remark 1.6.21.

(1) If one of $f(x)$ and $g(x)$ is 0, then 0 is the only common multiple of $f(x)$ and $g(x)$. Therefore, 0 is their least common multiple by definition.
(2) The least common multiples of two polynomials are unique up to nonzero constants. For two polynomials $f(x)$ and $g(x)$, which

are both nonzero, $[f(x), g(x)]$ is often used to denote the monic least common multiple of $f(x)$ and $g(x)$.

(3) The monic least common multiples of two polynomials will not change due to the extension of the coefficient field.

Theorem 1.6.22. *Let $f(x), g(x) \in F[x]$ be monic polynomials. Then*

$$[f(x), g(x)] = \frac{f(x)g(x)}{(f(x), g(x))}.$$

Proof. Suppose that $d(x) = (f(x), g(x))$, $f(x) = f_1(x)d(x)$, and $g(x) = g_1(x)d(x)$, then $(f_1(x), g_1(x)) = 1$. Let $m(x) = \frac{f(x)g(x)}{(f(x), g(x))}$. Then $m(x) = f_1(x)g_1(x)d(x) = f_1(x)g(x) = g_1(x)f(x)$. So $m(x)$ is a common multiple of $f(x)$ and $g(x)$.

Let $h(x)$ be a common multiple of $f(x)$ and $g(x)$. Then there exist $f_2(x), g_2(x) \in F[x]$ such that $h(x) = f_2(x)f(x) = g_2(x)g(x)$, and hence $f_2(x)f_1(x)d(x) = g_2(x)g_1(x)d(x)$. Note that $d(x) \neq 0$, and so $f_2(x)f_1(x) = g_2(x)g_1(x)$. Since $(f_1(x), g_1(x)) = 1$, $f_1(x)|g_2(x)$. Thus there is $k(x) \in F[x]$ with $g_2(x) = k(x)f_1(x)$ so that $h(x) = k(x)f_1(x)g(x) = k(x)m(x)$. Therefore, every common multiple of $f(x)$ and $g(x)$ is a multiple of $m(x)$. Note that $m(x)$ is a monic. So $[f(x), g(x)] = m(x) = \frac{f(x)g(x)}{(f(x), g(x))}$. $\square$

Proposition 1.6.23. *Let $f(x), g(x), m(x) \in F[x]$. If $f(x), g(x)$ are both nonzero, then $m(x)$ is the least common multiple of $f(x)$ and $g(x)$ if and only if $m(x)$ is of the lowest degree among all nonzero common multiples of $f(x)$ and $g(x)$.*

Proof. The necessity follows from the definition.

For the sufficiency, assume that $m(x)$ is of the lowest degree among all nonzero common multiples of $f(x)$ and $g(x)$, and suppose $m_0(x)$ is the least common multiple of $f(x)$ and $g(x)$, then $m_0(x)|m(x)$, and hence $\deg(m_0(x)) \leqslant \deg(m(x))$. It is easy to see that $m_0(x)$ is a nonzero common multiple of $f(x)$ and $g(x)$. Thus $\deg(m(x)) \leqslant \deg(m_0(x))$ by hypothesis. It follows that $\deg(m(x)) = \deg(m_0(x))$, and so $m(x) = cm_0(x)$ is the least common multiple of $f(x)$ and $g(x)$, where c is a nonzero constant. $\square$

For n polynomials $f_1(x), f_2(x), \ldots, f_n(x) \in F[x]$ $(n \geqslant 2)$, one can define their greatest common divisors, relative primeness, and least common multiples in the way similar to those for two polynomials.

In the following, we use the symbols $(f_1(x), f_2(x), \ldots, f_n(x))$ and $[f_1(x), f_2(x), \ldots, f_n(x)]$ to denote the monic greatest common divisor and the monic least common multiple of $f_1(x), f_2(x), \ldots, f_n(x)$, respectively. For example, the polynomials $f_1(x), f_2(x), \ldots, f_n(x)$ are called relatively prime (or coprime) if $(f_1(x), f_2(x), \ldots, f_n(x)) = 1$. Note that if $f_1(x), f_2(x), \ldots, f_n(x)$ are pairwise coprime, then they are relatively prime, but the converse is not true.

Similar to congruence for integers, there is a congruence for polynomials.

Definition 1.6.24. Let $a(x), b(x), m(x) \in F[x]$ with $m(x) \neq 0$. We say that $a(x)$ and $b(x)$ are **congruent modulo** $m(x)$, denoted by

$$a(x) \equiv b(x) \pmod{m(x)},$$

if $a(x)$ and $b(x)$ have the same remainders when divided by $m(x)$.

The properties of congruence for polynomials are very similar to those for congruence for integers. For example, by the argument in the proof of Theorem 1.3.6, we get

Theorem 1.6.25 (Chinese remainder theorem). *Let k be an integer greater than 1. If $m_1(x)$, $m_2(x), \ldots, m_k(x)$ are pairwise coprime polynomials, and $a_1(x), a_2(x), \ldots, a_k(x)$ are any polynomials, then there exist polynomials $a(x)$ satisfying the system of congruences*

$$\begin{cases} f(x) \equiv a_1(x) \pmod{m_1(x)}, \\ f(x) \equiv a_2(x) \pmod{m_2(x)}, \\ \qquad \vdots \qquad\qquad \vdots \\ f(x) \equiv a_k(x) \pmod{m_k(x)}. \end{cases}$$

Moreover, if both polynomials $a(x), b(x)$ satisfy the system of congruences, then $a(x) \equiv b(x) \pmod{m_1(x)m_2(x) \cdots m_k(x)}$.

Example 1.6.26. Let m and n be positive integers. Prove that

$$(x^m - 1, x^n - 1) = x^d - 1,$$

where $d = (m, n)$.

Proof. Since $d = (m, n)$, there are $m_1, n_1, s, t \in \mathbb{Z}$ such that

$$m = m_1 d, \quad n = n_1 d, \quad d = sm + tn.$$

Note that $d \leqslant m, d \leqslant n$, and so s and t are not both positive integers. If $t = 0$ or $s = 0$, then $d = m$ or $d = n$. Thus $m \mid n$ or $n \mid m$, and the result holds (see Example 1.6.7).

Let's assume that one of s and t is positive and the other is negative, say, $s > 0, t < 0$. Since

$$x^m - 1 = (x^d - 1)(x^{d(m_1-1)} + x^{d(m_1-2)} + \cdots + x^d + 1),$$

$$x^n - 1 = (x^d - 1)(x^{d(n_1-1)} + x^{d(n_1-2)} + \cdots + x^d + 1),$$

$x^d - 1$ is a common divisor of $x^m - 1$ and $x^n - 1$.

Suppose that $h(x)$ is a common divisor of $x^m - 1$ and $x^n - 1$, then $h(x) \mid ((x^{sm} - 1) - (x^{-tn} - 1))$. Since

$$(x^{sm} - 1) - (x^{-tn} - 1) = x^{sm} - x^{-tn} = x^{-tn}(x^{sm+tn} - 1) = x^{-tn}(x^d - 1),$$

$h(x) \mid x^{-tn}(x^d - 1)$. Note that $(x, x^m - 1) = 1$ and $h(x) \mid (x^m - 1)$, and so $(h(x), x) = 1$. Thus $(h(x), x^{-nt}) = 1$, and hence $h(x) \mid (x^d - 1)$.

In summary, $(x^m - 1, x^n - 1) = x^d - 1$. $\qquad\qquad\square$

Example 1.6.27. Find a polynomial $f(x)$ of the lowest degree such that the remainders upon dividing $f(x)$ by $(x - 1)^2$ and $(x - 2)^3$ are $2x$ and $3x$, respectively.

Solution. **Method 1** (By definition).

Let $f(x) = q_1(x)(x - 1)^2 + 2x = q_2(x)(x - 2)^3 + 3x$, where $q_1(x), q_2(x)$ are polynomials, then

$$q_1(x)(x - 1)^2 + 2x$$

$$= q_2(x)\left((x - 1) - 1\right)^3 + 3x$$

$$= q_2(x)\left((x - 1)^3 - 3(x - 1)^2 + 3(x - 1) - 1\right) + 3x.$$

Thus $\left(q_1(x) - q_2(x)(x - 1) + 3q_2(x)\right)(x - 1)^2 = q_2(x)(3x - 4) + x$ so that

$$(x - 1)^2 \mid (q_2(x)(3x - 4) + x).$$

If $q_2(x)(3x - 4) + x = 0$, then $q_2(x)(3x - 4) = -x$, and so $q_2(x) = c$ is a nonzero constant. However, from the equality $c(3x - 4) = -x$ we get $c = 0$, a contradiction! So $q_2(x)(3x - 4) + x \neq 0$. It follows that $\deg(q_2(x)(3x - 4) + x) \geqslant \deg(x - 1)^2 = 2$, and hence $\deg(q_2(x)) \geqslant 1$, i.e., the lowest degree of $q_2(x)$ is 1.

Suppose that $q_2(x) = ax + b$, then $(ax + b)(3x - 4) + x = 3a(x - 1)^2$, i.e.,

$$3ax^2 + (3b - 4a + 1)x - 4b = 3ax^2 - 6ax + 3a.$$

Thus $\begin{cases} 3b - 4a + 1 = -6a, \\ -4b = 3a, \end{cases}$ and so $\begin{cases} a = 4, \\ b = -3. \end{cases}$ Consequently,

$$f(x) = (4x - 3)(x - 2)^3 + 3x = 4x^4 - 27x^3 + 66x^2 - 65x + 24.$$

Method 2 (By Chinese remainder theorem).
Since

$$(x - 2)^3 = ((x - 1) - 1)^3$$

$$= (x - 1)^3 - 3(x - 1)^2 + 3(x - 1) - 1$$

$$= (x - 4)(x - 1)^2 + 3x - 4,$$

$$(x - 1)^2 = \left(\frac{1}{3}x - \frac{2}{9}\right)(3x - 4) + \frac{1}{9},$$

$$\frac{1}{9} = (x - 1)^2 - \left(\frac{1}{3}x - \frac{2}{9}\right)(3x - 4)$$

$$= (x - 1)^2 - \left(\frac{1}{3}x - \frac{2}{9}\right)\left((x - 2)^3 - (x - 4)(x - 1)^2\right)$$

$$= \left(1 + \left(\frac{1}{3}x - \frac{2}{9}\right)(x - 4)\right)(x - 1)^2 - \left(\frac{1}{3}x - \frac{2}{9}\right)(x - 2)^3$$

$$= \left(\frac{1}{3}x^2 - \frac{14}{9}x + \frac{17}{9}\right)(x - 1)^2 - \left(\frac{1}{3}x - \frac{2}{9}\right)(x - 2)^3.$$

Thus $1 = (3x^2 - 14x + 17)(x - 1)^2 - (3x - 2)(x - 2)^3$.

Let $f_1(x) = -(2x)(3x - 2)(x - 2)^3 + (3x)(3x^2 - 14x + 17)(x - 1)^2$. Then

$$f_1(x) \equiv 4x^4 - 27x^3 + 66x^2 - 65x + 24 \pmod{(x - 1)^2(x - 2)^3}.$$

Therefore, $f(x) = 4x^4 - 27x^3 + 66x^2 - 65x + 24$ is of the lowest degree among the polynomials satisfying the required condition. $\qquad\square$

1.7 Unique Factorization for Polynomials

We know that any integer n larger than 1 can be factored into the product of finite prime factors. In this section, we discuss the similar properties of polynomials. To discuss the factorization of polynomials, we first introduce the concept of "prime polynomials", often called irreducible polynomials.

Definition 1.7.1. Let F be a number field and $p(x) \in F[x]$ with $\deg(p(x)) \geqslant 1$. If $p(x)$ cannot be expressed as the product of two polynomials in $F[x]$ with a lower degree than that of $p(x)$, then $p(x)$ is said to be an **irreducible polynomial** over F. If $p(x)$ is not an irreducible polynomial, then $p(x)$ is called a **reducible polynomial**.

Remark 1.7.2.

(1) Whether a polynomial is reducible depends on the nature of its coefficient field.
(2) The zero polynomial or a polynomial of degree 0 cannot be said to be reducible or irreducible.
(3) A polynomial of degree one is always irreducible.
(4) Let $p(x) \in F[x]$ and $\deg(p(x)) \geqslant 1$. Then $p(x)$ is irreducible if and only if $p(x)$ only has nonzero constants and $cp(x)$ as divisors of $p(x)$, where c is a nonzero constant.

Proposition 1.7.3. *Let* $p(x) \in F[x]$ *and* $\deg(p(x)) \geqslant 1$. *The following statements are equivalent:*

(1) *$p(x)$ is irreducible.*
(2) *For any* $f(x) \in F[x]$, *we always have* $p(x)|f(x)$ *or* $(p(x), f(x)) = 1$.
(3) *For any* $g(x), h(x) \in F[x]$, *if* $p(x)|g(x)h(x)$, *then* $p(x)|g(x)$ *or* $p(x)|h(x)$.

Proof. (1) $\Longrightarrow$ (2). Let $f(x) \in F[x]$. Suppose $d(x) = (p(x), f(x))$, then $d(x)|p(x)$. Since $p(x)$ is irreducible, $d(x) = 1$ or $d(x) = \frac{1}{c}p(x)$, where c is the leading coefficient of $p(x)$. It follows that $(p(x), f(x)) = 1$, or $p(x)|f(x)$.

  (2) $\Longrightarrow$ (3). Let $g(x), h(x) \in F[x]$ and $p(x)|g(x)h(x)$. If $p(x) \nmid g(x)$, then $(p(x), g(x)) = 1$ by (2). So $p(x)|h(x)$.

(3) $\implies$ (1). Suppose that $p(x)$ is not irreducible, i.e., $p(x)$ is reducible, then there are $p_1(x), p_2(x) \in F[x]$ such that $p(x) = p_1(x)p_2(x)$, where

$$0 < \deg(p_i(x)) < \deg(p(x)), i = 1, 2.$$

Clearly, $p(x)|p_1(x)p_2(x)$, but $p(x) \nmid p_1(x)$ and $p(x) \nmid p_2(x)$, contradicting (3). So (1) holds. $\qquad\square$

Corollary 1.7.4. *Let* $p(x), f_i(x) \in F[x], i = 1, 2, \ldots, n$. *If* $p(x)$ *is irreducible, and* $p(x)|f_1(x)f_2(x) \cdots f_n(x)$, *then* $p(x)|f_i(x)$ *for some* i *with* $1 \leqslant i \leqslant n$.

Theorem 1.7.5 (Unique factorization theorem). *Let* F *be a field. Every* $f(x) \in F[x]$ *of degree* $\geqslant 1$ *is an irreducible polynomial or a product of irreducible polynomials. This product is unique, except for the order in which the factors appear, that is, if* $f(x)$ *has factorizations*

$$n = p_1(x)p_2(x) \cdots p_t(x) = q_1(x)q_2(x) \cdots q_s(x),$$

where the $p(x)$*'s and* $q(x)$*'s are irreducible polynomials, then* $s = t$ *and the* $q(x)$*'s may be reindexed so that* $p_i(x) = c_i q_i(x)$ *with* c_i *a nonzero constant for all* i.

Proof.   This theorem can be proved in the same way as in Theorem 1.2.15 by noting that the role of "nonzero constants" in the polynomial ring $F[x]$ is similar to that of "1, -1" in the ring $\mathbb{Z}$ of integers. $\qquad\square$

By Theorem 1.7.5, if $f(x) \in F[x]$ with $\deg(f(x)) \geqslant 1$, then $f(x)$ has a factorization $f(x) = ap_1^{r_1}(x)p_2^{r_2}(x) \cdots p_t^{r_t}(x)$, which is called the **standard factorization** of $f(x)$, where a is the leading coefficient of $f(x)$, $p_1(x), p_2(x), \ldots, p_t(x)$ are different irreducible monic polynomials over F, and $r_1, r_2, \ldots, r_t$ are positive integers.

Although the factorization theorem is fundamental in theory, it does not give a specific method to factorize polynomials. There is no universally feasible method to factorize polynomials in general!

Theoretically, we can use the unique factorization theorem to find the greatest common divisors and the least common multiples of polynomials.

Let $f(x), g(x) \in F[x]$ with $\deg(f(x)) \geqslant 1, \deg(g(x)) \geqslant 1$. We may assume that

$$f(x) = ap_1^{n_1}(x)p_2^{n_2}(x)\cdots p_t^{n_t}(x), \quad g(x) = bp_1^{m_1}(x)p_2^{m_2}(x)\cdots p_t^{m_t}(x),$$

where a and b are the leading coefficients of $f(x)$ and $g(x)$, respectively, $p_1(x), p_2(x), \ldots, p_t(x)$ are different irreducible monic polynomials over F, n_i, m_i are natural numbers, $i = 1, 2, \ldots, t$.

Define $k_i = \min\{n_i, m_i\}$, $l_i = \max\{n_i, m_i\}$, $i = 1, 2, \ldots, t$. Then

$$(f(x), g(x)) = p_1^{k_1}(x)p_2^{k_2}(x)\cdots p_t^{k_t}(x),$$

$$[f(x), g(x)] = p_1^{l_1}(x)p_2^{l_2}(x)\cdots p_t^{l_t}(x).$$

Example 1.7.6. Let $f(x), g(x) \in F[x]$. Prove that $f(x)|g(x)$ if and only if $f^2(x)|g^2(x)$.

Proof. The necessity is clear. We only need to prove the sufficiency.

Method 1 (Using the unique factorization theorem).

If $f(x)$ or $g(x)$ is 0 or a polynomial of zero degree, then it is easily seen that $f(x)|g(x)$.

Next, let $\deg(f(x)) \geqslant 1, \deg(g(x)) \geqslant 1$. So, we may assume that

$$f(x) = ap_1^{n_1}(x)p_2^{n_2}(x)\cdots p_t^{n_t}(x), \quad g(x) = bp_1^{m_1}(x)p_2^{m_2}(x)\cdots p_t^{m_t}(x),$$

where a and b are the leading coefficients of $f(x)$ and $g(x)$ respectively, $p_1(x), p_2(x), \ldots, p_t(x)$ are different irreducible monic polynomials over F, and n_i, m_i are natural numbers, $i = 1, 2, \ldots, t$. It follows that

$$f^2(x) = a^2 p_1^{2n_1}(x)p_2^{2n_2}(x)\cdots p_t^{2n_t}(x),$$

$$g^2(x) = b^2 p_1^{2m_1}(x)p_2^{2m_2}(x)\cdots p_t^{2m_t}(x).$$

Since $f^2(x)|g^2(x)$ and $(p_i(x), p_j(x)) = 1, 1 \leqslant i \neq j \leqslant t$, $p_i^{2n_i}(x)|p_i^{2m_i}(x)$. Thus $2n_i \leqslant 2m_i$, i.e., $n_i \leqslant m_i, i = 1, 2, \ldots, t$. So $f(x)|g(x)$.

Method 2 (Using properties of relative primeness for polynomials).

Suppose that $(f(x), g(x)) = d(x)$, $f(x) = f_1(x)d(x)$, $g(x) = g_1(x)d(x)$. Then $(f_1(x), g_1(x)) = 1$. Since $f^2(x)|g^2(x)$, $f_1^2(x)|g_1^2(x)$.

Note that $(f_1^2(x), g_1^2(x)) = 1$. Therefore, $f_1^2(x)$, hence $f_1(x)$, is a nonzero constant. Consequently, $f(x)|g(x)$. $\square$

1.8 Multiple Factors

The previous section has proved that every polynomial $f(x)$ of degree $\geqslant 1$ over a number field F can be factored into the product of finite irreducible polynomials over F. This section discusses the multiplicity of irreducible factors of $f(x)$.

Definition 1.8.1. Let F be a field, $f(x), p(x) \in F[x]$, and let k be a natural number. If $p(x)$ is an irreducible polynomial such that $p^k(x)|f(x)$ and $p^{k+1}(x) \nmid f(x)$, then $p(x)$ is called a k-**multiple factor** of $f(x)$, and k is called the **multiplicity** of $p(x)$.

Remark 1.8.2.

(1) If $k = 0$, then $p(x)$ is not a factor of $f(x)$ at all.
(2) If $k = 1$, then $p(x)$ is called a **simple factor** of $f(x)$.
(3) If $k \geqslant 2$, then $p(x)$ a **multiple factor** of $f(x)$.

To determine whether a polynomial has multiple factors, we define the derivative of polynomials.

Definition 1.8.3. Let $f(x) = a_n x^n + a_{n-1} x^{n-1} + \cdots + a_2 x^2 + a_1 x + a_0$ belong to $F[x]$. The **derivative** of $f(x)$, denoted by $f'(x)$ or $(f(x))'$, is the polynomial $na_n x^{n-1} + (n-1)a_{n-1} x^{n-2} + \cdots + 2a_2 x + a_1$ in $F[x]$.

From the definition above, we know that the derivatives of polynomials are still polynomials, and the derivative here is consistent with the derivative of a polynomial function in mathematical analysis in form. However, in mathematical analysis, the derivative $f'(x)$ of a polynomial function $f(x)$ is given by the definition of the derivative of a function, while the derivative $f'(x)$ here is directly defined by the formal expression of $f(x)$.

In general, $f'(x)$ is called the **first derivative** of $f(x)$, the derivative of $f'(x)$, denoted by $f''(x)$, is called the **second derivative** of $f(x)$, ..., and the derivative of $f^{(k-1)}(x)$, denoted by $f^{(k)}(x)$, is called the k**th derivative** of $f(x)$, i.e., $f^{(k)}(x) = (f^{(k-1)}(x))'$, where $k \geqslant 2$ is an integer.

By the above definition, $f'(x) = 0$ if $f(x)$ is a constant; suppose $\deg(f(x)) = n$, then $\deg(f'(x)) = n-1$, $f^{(n)}(x)$ is a nonzero constant, and $f^{(n+1)}(x) = 0$.

The following proposition follows directly from the definition.

Proposition 1.8.4. *Let $f(x), g(x) \in F[x]$ and $c \in F$. Then*

(1) $(f(x) + g(x))' = f'(x) + g'(x)$;
(2) $(cf(x))' = cf'(x)$;
(3) $(f(x)g(x))' = f'(x)g(x) + f(x)g'(x)$;
(4) $(f^n(x))' = nf^{n-1}(x)f'(x)$.

Theorem 1.8.5. *Let $f(x), p(x) \in F[x]$ with $p(x)$ irreducible and let k be a positive integer. Then $p(x)$ is a k-multiple factor of $f(x)$ if and only if $p(x)$ is a factor of $f(x)$ and $p(x)$ is a $(k-1)$-multiple factor of $f'(x)$.*

Proof. Necessity. Suppose $p(x)$ is a k-multiple factor of $f(x)$, then there is $g(x) \in F[x]$ with $f(x) = p^k(x)g(x)$, where $p(x) \nmid g(x)$. It is obvious that $p(x)$ is a factor of $f(x)$. Taking the derivative of $f(x)$ gives

$$f'(x) = kp^{k-1}(x)p'(x)g(x) + p^k(x)g'(x)$$
$$= p^{k-1}(x)(kp'(x)g(x) + p(x)g'(x))$$

so that $p^{k-1}(x)|f'(x)$. Let $h(x) = kp'(x)g(x) + p(x)g'(x)$, then $f'(x) = p^{k-1}(x)h(x)$. If $p^k(x)|f'(x)$, then $p(x)|h(x)$. It follows that $p(x)|kp'(x)g(x)$. But $(p(x), kp'(x)) = 1$, and hence $p(x)|g(x)$, a contradiction. Therefore, $p^k(x) \nmid f'(x)$, and so $p(x)$ is a $(k-1)$-multiple factor of $f'(x)$.

Sufficiency. Since $p(x)$ is a factor of $f(x)$, we may assume that $p(x)$ is an s-multiple factor of $f(x)$, where s is a positive integer. From the necessity, $p(x)$ is an $(s-1)$-multiple factor of $f'(x)$. Thus $s-1 = k-1$, i.e., $s = k$. So $p(x)$ is a k-multiple factor of $f(x)$. $\square$

From Theorem 1.8.5, we have the following result.

Corollary 1.8.6. *Let k be a positive integer. If $p(x)$ is a k-multiple factor of $f(x)$, then $p(x)$ is a common factor of $f'(x)$, $f''(x), \ldots, f^{(k-1)}(x)$, but $p(x)$ is not a factor of $f^{(k)}(x)$. In particular, a simple factor of $f(x)$ is not a factor of $f'(x)$.*

Remark 1.8.7. Let $f(x) \in F[x]$. Then a $(k-1)$-multiple factor of $f'(x)$ need not be a k-multiple factor of $f(x)$ for any integer $k \geqslant 2$. For example, if $f(x) = x^k + 1$, then $f'(x) = kx^{k-1}$. Clearly, x is a $(k-1)$-multiple factor of $f'(x)$, however, x is not a k-multiple factor of $f(x)$.

Corollary 1.8.8. *If $p(x)$ is a k-multiple factor of $f(x)$, where k is a positive integer, then $p(x)$ is a $(k-1)$-multiple factor of $(f(x), f'(x))$.*

Theorem 1.8.9. *Let $f(x), p(x) \in F[x]$ with $p(x)$ irreducible and let $k \geqslant 2$ be an integer. Then $p(x)$ is a k-multiple factor of $f(x)$ if and only if $p(x)$ is a $(k-1)$-multiple factor of $(f(x), f'(x))$.*

Proof. The necessity follows from Corollary 1.8.8.

For the sufficiency, note that $k - 1 \geqslant 1$, and so $p(x)$ is a factor of $f(x)$. We may assume that $p(x)$ is an s-multiple factor of $f(x)$, where s is a positive integer, then $p(x)$ is an $(s-1)$-multiple factor of $(f(x), f'(x))$ by Corollary 1.8.8. Thus, $s - 1 = k - 1$, i.e., $s = k$. So $p(x)$ is a k-multiple factor of $f(x)$. $\qquad\square$

Corollary 1.8.10. *A polynomial $f(x) \in F[x]$ has no multiple factors if and only if $(f(x), f'(x)) = 1$.*

By Theorem 1.8.9, if $f(x) = ap_1^{r_1}(x)p_2^{r_2}(x)\cdots p_t^{r_t}(x)$ is the standard factorization of $f(x)$, then $d(x) = (f(x), f'(x))$ has a factorization of the form

$$d(x) = p_1^{r_1-1}(x)p_2^{r_2-1}(x)\cdots p_t^{r_t-1}(x).$$

Hence

$$\frac{f(x)}{(f(x), f'(x))} = ap_1(x)p_2(x)\cdots p_t(x),$$

the **squarefree part** of $f(x)$. It is obvious that $f(x)$ and $\frac{f(x)}{(f(x),f'(x))}$ have the same irreducible factors in $F[x]$, except that in $\frac{f(x)}{(f(x),f'(x))}$ all the irreducible factors have multiplicity 1. Therefore, we get an effective method to remove the multiplicity of an irreducible factor of $f(x)$: First use the Euclidean algorithm to find $(f(x), f'(x))$, then the quotient upon dividing $f(x)$ by $(f(x), f'(x))$ is the polynomial without multiple factors.

Example 1.8.11. Let $f(x) \in F[x]$ and $\deg(f(x)) = n \geqslant 1$. Then $f'(x)|f(x)$ if and only if there are $a, b \in F$ with $f(x) = a(x - b)^n$.

Proof. Sufficiency. Since $f(x) = a(x - b)^n$, $f'(x) = na(x - b)^{n-1}$. So $f'(x)|f(x)$.

Necessity. Since $f'(x)|f(x)$, we may assume that $f(x) = \frac{1}{n}f'(x)(x - b)$, where $b \in F$. It follows that $(f(x), f'(x)) = \frac{1}{na}f'(x)$, where $a \in F$ is the leading coefficient of $f(x)$, and hence $\frac{f(x)}{(f(x),f'(x))} = a(x - b)$. Note that $f(x)$ and $\frac{f(x)}{(f(x),f'(x))}$ have the same irreducible factors, and $\deg(f(x)) = n$. So $f(x) = a(x - b)^n$. $\square$

1.9 Polynomial Functions

Definition 1.9.1. Let F be a number field and $f(x) = \sum_{i=0}^{n} a_i x^i \in F[x]$. For any $\alpha \in F$, define $f(\alpha) = \sum_{i=0}^{n} a_i \alpha^i \in F$, which is called the **value** of $f(x)$ at $x = \alpha$. The function $f : F \to F$ given by $\alpha \mapsto f(\alpha)$ is called the **polynomial function** defined by $f(x)$.

Proposition 1.9.2.

(1) *Let $f(x), g(x) \in F[x]$, $h_1(x) = f(x) + g(x)$, $h_2(x) = f(x)g(x)$. Then $h_1(\alpha) = f(\alpha) + g(\alpha)$, $h_2(\alpha) = f(\alpha)g(\alpha)$ for all $\alpha \in F$.*
(2) *(**Remainder theorem**) Let $f(x) \in F[x]$ and $\alpha \in F$. Then there is $q(x) \in F[x]$ such that $f(x) = q(x)(x - \alpha) + f(\alpha)$.*
(3) *(**Factor theorem**) Let $f(x) \in F[x]$ and $\alpha \in F$. Then*

$$(x - \alpha)|f(x) \ \text{if and only if} \ f(\alpha) = 0.$$

Proof. (1) follows from straightforward verification.

(2) By the Euclidean algorithm, there exist $q(x) \in F[x]$ and $r \in F$ such that $f(x) = q(x)(x - \alpha) + r$. Setting $x = \alpha$ in this formula, we obtain $r = f(\alpha)$. So $f(x) = q(x)(x - \alpha) + f(\alpha)$.

(3) comes from (2). $\square$

Definition 1.9.3. Let F and $\overline{F}$ be number fields with $F \subseteq \overline{F}$ and let $f(x) \in F[x]$. If $\alpha \in \overline{F}$ and $f(\alpha) = 0$, then α is called a **root** of $f(x)$ in $\overline{F}$. Let k be a natural number. If $x - \alpha$ is a k-multiple factor of $f(x)$, then α is called a k-**multiple root** of $f(x)$ in $\overline{F}$, and k is called the **multiplicity** of α.

Remark 1.9.4.

(1) If $k = 0$, α is not a root of $f(x)$ at all.
(2) If $k = 1$, α is called a **simple root** of $f(x)$.
(3) If $k \geqslant 2$, α is called a **multiple root** of $f(x)$.
(4) Let $f(x) \in F[x]$. If $f(x)$ has a k-multiple root in F, then $f(x)$ has a k-multiple factor; the converse is not true. For example, $f(x) = (x^2 + 1)^2$ has a multiple factor over $\mathbb{Q}$, but $f(x)$ has no roots in $\mathbb{Q}$.

Theorem 1.9.5. *Every polynomial of degree n ($n \geqslant 0$) over a number field F has at most n roots in F (multiple roots are counted by multiplicity).*

Proof.   Let $f(x) \in F[x]$ with $\deg(f(x)) = n \geqslant 0$.

If $n = 0$, the conclusion is obvious.

Suppose that $n > 0$, then $f(x)$ can be factored into a product of irreducible factors in $F[x]$ by the unique factorization theorem. Proposition 1.9.2 (3) shows that the number of roots of $f(x)$ in F is equal to the number of factors of degree 1 in the factorization; this number is certainly not more than n. $\qquad\square$

Theorem 1.9.6. *Let F be a number field and $f(x), g(x) \in F[x]$, and let $\deg(f(x)) \leqslant n, \deg(g(x)) \leqslant n$. If there are $n+1$ different numbers $\alpha_1, \alpha_2, \ldots, \alpha_{n+1}$ in F such that $f(\alpha_i) = g(\alpha_i)$, $i = 1, 2, \ldots, n + 1$, then $f(x) = g(x)$.*

Proof.   Let $h(x) = f(x) - g(x)$. Then $h(\alpha_i) = 0, i = 1, 2, \ldots, n+1$. If $f(x) \neq g(x)$, then $h(x) \neq 0$ so that $0 \leqslant \deg(h(x)) \leqslant n$. Therefore, by Theorem 1.9.5, $h(x)$ has at most n roots in F; this is a contradiction. So $f(x) = g(x)$. $\qquad\square$

By definition, every polynomial in $F[x]$ determines a polynomial function. Whether different polynomials define different polynomial functions? The answer is "yes", as shown by the following result.

Corollary 1.9.7. *Let F be a number field and $f(x), g(x) \in F[x]$. Then*

$$f(x) = g(x) \text{ if and only if } f = g, \text{ i.e., } f(\alpha) = g(\alpha) \text{ for all } \alpha \in F.$$

Proof.   The necessity is obvious.

For the sufficiency, we assume that $f(x) \neq g(x)$ and $h(x) = f(x) - g(x)$, then $h(x) \neq 0$, and $h(\alpha) = 0$ for all $\alpha \in F$, i.e., every number in F is a root of $h(x)$. According to Theorem 1.9.5, $h(x)$ has at most finitely many roots in F, but there are infinitely many numbers in F; this is a contradiction. $\qquad\square$

Remark 1.9.8. Let $\mathbb{F} = \{0, 1\}$ be the field of two elements. Similar to the case of the number field, we can define the polynomial ring $\mathbb{F}[x]$ over $\mathbb{F}$. In this case, Corollary 1.9.7 does not hold. For example, let $f(x) = x + 1, g(x) = x^2 + 1 \in \mathbb{F}[x]$. It is clear that $f(x) \neq g(x)$, but $f = g$.

Example 1.9.9. With the help of roots of polynomials, prove Examples 1.6.7, 1.6.26, and 1.6.19, that is, prove that

(1) $(x^d - 1) \mid (x^n - 1)$ if and only if $d \mid n$, where d, n are positive integers.
(2) $(x^m - 1, x^n - 1) = x^d - 1$, where m, n are positive integers, $d = (m, n)$.
(3) $x(x + 1)(2x + 1) \mid ((x + 1)^{2n} - x^{2n} - 2x - 1)$, where n is a positive integer.

Proof.   (1) Sufficiency. Since $d \mid n$, every root of $x^d - 1$ is a root of $x^n - 1$. Note that $x^d - 1$ has no multiple roots, and so $(x^d - 1) \mid (x^n - 1)$.

Necessity. Since $e^{\frac{2\pi i}{d}} = \cos \frac{2\pi}{d} + i \sin \frac{2\pi}{d}$ is a root of $x^d - 1$, $\left(x - e^{\frac{2\pi i}{d}}\right) \mid (x^d - 1)$ by the factor theorem. By hypothesis, $(x^d - 1) \mid (x^n - 1)$, hence $\left(x - e^{\frac{2\pi i}{d}}\right) \mid (x^n - 1)$. Thus, $e^{\frac{2n\pi i}{d}} = \cos \frac{2n\pi}{d} + i \sin \frac{2n\pi}{d} = 1$. It follows that $\frac{n}{d}$ is an integer, i.e., $d \mid n$.

(2) Since $d = (m, n)$, there are $s, t \in \mathbb{Z}$ such that $d = sm + tn$. If α is a root of $(x^m - 1, x^n - 1)$, then α is a common root of $x^m - 1$ and $x^n - 1$. Thus $\alpha^m = \alpha^n = 1$, and hence $\alpha^d = \alpha^{sm+tn} = 1$, i.e., α is a root of $x^d - 1$. Therefore, every root of $(x^m - 1, x^n - 1)$ is a root of $x^d - 1$. Obviously, every root of $x^d - 1$ is a root of $(x^m - 1, x^n - 1)$. Note that both $(x^m - 1, x^n - 1)$ and $x^d - 1$ are monic polynomials having no multiple roots, so $(x^m - 1, x^n - 1) = x^d - 1$.

(3) Let $f(x) = (x + 1)^{2n} - x^{2n} - 2x - 1$. Then

$$f(0) = f(-1) = f\left(-\frac{1}{2}\right) = 0.$$

Thus $x|f(x)$, $(x+1)|f(x)$, $\left(x+\frac{1}{2}\right)\big|\,f(x)$ by the factor theorem. Since x, $x+1$, $x+\frac{1}{2}$ are pairwise coprime, $x(x+1)\left(x+\frac{1}{2}\right)\big|\,f(x)$. So

$$x(x+1)(2x+1)\,\big|\,((x+1)^{2n} - x^{2n} - 2x - 1). \qquad \square$$

Remark 1.9.10. The proofs of Example 1.9.9 (1) and (2) are done in $\mathbb{C}[x]$, based on the fact that both the divisibility relation and the monic greatest common divisor of polynomials do not change due to the extension of the coefficient field.

1.10   Polynomials with Complex, Real, and Rational Coefficients

We have just discussed polynomials with coefficients in a general number field. Now, we consider the case where the coefficient field is the field of complex, real, and rational numbers, respectively. First, consider polynomials over the complex number field. We start with the following:

Theorem 1.10.1 (Fundamental theorem of algebra). *Every polynomial in $\mathbb{C}[x]$ of degree $\geqslant 1$ has at least one root in $\mathbb{C}$.*

Carl Friedrich Gauss first proved this theorem in 1799. It has many proofs; the theory of complex variable functions gives a relatively simple one. Here, we omit its proof without loss of continuity.

By Proposition 1.9.2 (3), the fundamental theorem of algebra can be equivalently stated as follows: Every polynomial in $\mathbb{C}[x]$ of degree $\geqslant 1$ has at least one factor of degree 1 over $\mathbb{C}$. It follows that irreducible polynomials over $\mathbb{C}$ are precisely polynomials of degree 1. By Theorem 1.7.5, we have the following theorem.

Theorem 1.10.2 (Unique factorization of polynomials over $\mathbb{C}$). *Every polynomial in $\mathbb{C}[x]$ of degree $\geqslant 1$ can be factored uniquely as the product of polynomials of degree 1.*

Therefore, if $f(x) \in \mathbb{C}[x]$ and $\deg(f(x)) = n \geqslant 1$, then $f(x)$ has the standard factorization

$$f(x) = a(x - \alpha_1)^{n_1}(x - \alpha_2)^{n_2} \cdots (x - \alpha_t)^{n_t},$$

where a is the leading coefficient of $f(x)$, $\alpha_1, \alpha_2, \ldots, \alpha_t$ are different complex numbers, $n_1, n_2, \ldots, n_t$ are positive integers, and $n_1 + n_2 + \cdots + n_t = n$. It follows that every polynomial in $\mathbb{C}[x]$ of degree n has precisely n roots in $\mathbb{C}$ (multiple roots are counted by multiplicity).

Let $f(x) = x^n + a_1 x^{n-1} + a_2 x^{n-2} + \cdots + a_k x^{n-k} + \cdots + a_{n-1} x + a_n \in \mathbb{C}[x]$, and let $x_1, x_2, \ldots, x_n$ be n complex roots of $f(x)$. Then

$$f(x) = (x - x_1)(x - x_2) \cdots (x - x_n)$$

$$= x^n + \left(-\sum_{i=1}^{n} x_i \right) x^{n-1} + \left(\sum_{1 \leqslant i < j \leqslant n} x_i x_j \right) x^{n-2} + \cdots$$

$$+ \left((-1)^k \sum_{1 \leqslant i_1 < i_2 < \cdots < i_k \leqslant n} x_{i_1} x_{i_2} \cdots x_{i_k} \right) x^{n-k} + \cdots$$

$$+ (-1)^n x_1 x_2 \cdots x_n.$$

Thus, we have the following formulas that relate the coefficients of $f(x)$ to sums and products of its roots:

$$(1.4) \qquad \begin{cases} \displaystyle\sum_{i=1}^{n} x_i = -a_1, \\[2mm] \displaystyle\sum_{1 \leqslant i < j \leqslant n} x_i x_j = a_2, \\[2mm] \qquad \vdots \quad \vdots \\[2mm] \displaystyle\sum_{1 \leqslant i_1 < i_2 < \cdots < i_k \leqslant n} x_{i_1} x_{i_2} \cdots x_{i_k} = (-1)^k a_k, \\[2mm] \qquad \vdots \quad \vdots \\[2mm] x_1 x_2 \cdots x_n = (-1)^n a_n. \end{cases}$$

These formulas are usually called **Viète's**[18] **formulas.**

Next, we discuss polynomials over the real number field.

Theorem 1.10.3. *Complex roots of a polynomial over $\mathbb{R}$ occur in pairs: If α is a k-multiple complex root of a polynomial $f(x) \in \mathbb{R}[x]$, then the complex conjugate $\overline{\alpha}$ is also a k-multiple complex root of $f(x)$.*

---

[18]François Viète, 1540–1603, French mathematician.

Proof. Let $f(x) = a_n x^n + a_{n-1} x^{n-1} + \cdots + a_{n-k} x^{n-k} + \cdots + a_1 x + a_0 \in \mathbb{C}[x]$. Define $\overline{f}(x) = \overline{a_n} x^n + \overline{a_{n-1}} x^{n-1} + \cdots + \overline{a_{n-k}} x^{n-k} + \cdots + \overline{a_1} x + \overline{a_0}$, then $\overline{f}(x) \in \mathbb{C}[x]$.

If α is a k-multiple complex root of $f(x)$, then $f(x) = \overline{f}(x) = (x - \overline{\alpha})^k \overline{g}(x)$, where $g(x) \in \mathbb{C}[x], g(\alpha) \neq 0$. It follows that $\overline{f}(x) = (x - \overline{\alpha})^k \overline{g}(x)$, where $\overline{g}(\overline{\alpha}) \neq 0$. Suppose that $f(x) \in \mathbb{R}[x]$, then

$$f(x) = \overline{f}(x) = (x - \overline{\alpha})^k \overline{g}(x).$$

Therefore, if α is a k-multiple complex root of a polynomial $f(x) \in \mathbb{R}[x]$, then $\overline{\alpha}$ is also a k-multiple complex root of $f(x)$. $\qquad\square$

The following corollary is an immediate consequence of the theorem above.

Corollary 1.10.4. *Every polynomial in $\mathbb{R}[x]$ of odd degree has at least one root in $\mathbb{R}$.*

Theorem 1.10.5. *Monic irreducible polynomials over $\mathbb{R}$ are exactly the following two types:*

(1) *polynomials of the form $x - \alpha$ with $\alpha \in \mathbb{R}$,*
(2) *polynomials of the form $x^2 + bx + c$ with $b, c \in \mathbb{R}$ and $b^2 - 4c < 0$.*

Proof. It is easy to see that the two types of polynomials in the theorem are irreducible over $\mathbb{R}$.

Next, let $p(x)$ be a monic irreducible polynomial over $\mathbb{R}$. By the fundamental theorem of algebra, $p(x)$ has a complex root $\alpha = s + ti$, where $s, t \in \mathbb{R}$, and hence $(x - \alpha)|p(x)$ in $\mathbb{C}[x]$.

(1) If $t = 0$, then $\alpha = s \in \mathbb{R}$. Note that both $p(x)$ and $x - \alpha$ belong to $\mathbb{R}[x]$, and so $(x - \alpha)|p(x)$ in $\mathbb{R}[x]$. Since $p(x)$ is a monic irreducible polynomial over $\mathbb{R}$, $p(x) = x - \alpha$ with $\alpha \in \mathbb{R}$.

(2) If $t \neq 0$, then $\overline{\alpha} = s - ti$ is also a root of $p(x)$, and so $(x - \overline{\alpha})|p(x)$. Since $x - \alpha$ and $x - \overline{\alpha}$ are relatively prime,

$$(x - \alpha)(x - \overline{\alpha})|p(x)$$

in $\mathbb{C}[x]$. Note that $(x - \alpha)(x - \overline{\alpha}) = x^2 - (2s)x + s^2 + t^2 \in \mathbb{R}[x]$. Thus $(x - \alpha)(x - \overline{\alpha})|p(x)$ in $\mathbb{R}[x]$. Since $p(x)$ is a monic irreducible polynomial in $\mathbb{R}[x]$,

$$p(x) = (x - \alpha)(x - \overline{\alpha}) = x^2 - (2s)x + s^2 + t^2.$$

Define $b = -2s, c = s^2 + t^2$, then $p(x) = x^2 + bx + c$, where $b, c \in \mathbb{R}$ and $b^2 - 4c = 4s^2 - 4(s^2 + t^2) = -4t^2 < 0$. $\qquad\square$

We have the following theorem by specializing Theorem 1.7.5 to the real number field.

Theorem 1.10.6 (Unique factorization of polynomials over $\mathbb{R}$). Every polynomial in $\mathbb{R}[x]$ of degree ≥ 1 can be factored uniquely as the product of polynomials of the form $x - \alpha$ with $\alpha \in \mathbb{R}$, and polynomials of the form $x^2 + bx + c$ with $b, c \in \mathbb{R}$ and $b^2 - 4c < 0$.

Example 1.10.7. Find the standard factorizations of $x^4 + 2$ over $\mathbb{C}$ and $\mathbb{R}$.

Solution. Since $-2 = 2(\cos \pi + i \sin \pi)$, the four complex roots of $x^4 + 2$ are as follows:

$$\alpha_1 = \sqrt[4]{2}\left(\cos\frac{\pi}{4} + i\sin\frac{\pi}{4}\right) = \frac{1}{\sqrt[4]{2}}(1 + i),$$

$$\alpha_2 = \sqrt[4]{2}\left(\cos\frac{3\pi}{4} + i\sin\frac{3\pi}{4}\right) = \frac{1}{\sqrt[4]{2}}(-1 + i),$$

$$\alpha_3 = \sqrt[4]{2}\left(\cos\frac{5\pi}{4} + i\sin\frac{5\pi}{4}\right) = \frac{1}{\sqrt[4]{2}}(-1 - i),$$

$$\alpha_4 = \sqrt[4]{2}\left(\cos\frac{7\pi}{4} + i\sin\frac{7\pi}{4}\right) = \frac{1}{\sqrt[4]{2}}(1 - i).$$

Therefore, the standard factorization of the polynomial $x^4 + 2$ over $\mathbb{C}$ is

$$\left(x - \frac{1}{\sqrt[4]{2}}(1 + i)\right)\left(x - \frac{1}{\sqrt[4]{2}}(1 - i)\right)\left(x - \frac{1}{\sqrt[4]{2}}(-1 + i)\right)$$
$$\times \left(x - \frac{1}{\sqrt[4]{2}}(-1 - i)\right).$$

It follows that $x^4 + 2 = (x^2 - \sqrt[4]{8}x + \sqrt{2})(x^2 + \sqrt[4]{8}x + \sqrt{2})$, which is the standard factorization of the polynomial $x^4 + 2$ over $\mathbb{R}$. $\square$

Example 1.10.8. Let F be a number field and $f(x) \in F[x]$. Prove that $f(x)$ is a constant if there is a nonzero $a \in F$ such that $f(x) = f(x + a)$.

Proof. Suppose that $f(x)$ is not a constant, then $\deg(f(x)) \geqslant 1$. By the fundamental theorem of algebra, $f(x)$ has a complex root α, i.e., there is $\alpha \in \mathbb{C}$ with $f(\alpha) = 0$. Since $f(x) = f(x+a)$,

$$\alpha, \ \alpha + a, \ \alpha + 2a, \ \ldots, \ \alpha + ma, \ \ldots, \ \alpha + na, \ \ldots$$

are all roots of $f(x)$. Note that $f(x)$ has only finitely many roots, so there are $m, n \in \mathbb{Z}$ such that $m \neq n$ and $\alpha + ma = \alpha + na$. Therefore, $(m-n)a = 0$, and hence $a = 0$, a contradiction. So $f(x)$ is a constant.
$\square$

Now, let's discuss polynomials over the rational number field. The previous discussion shows that irreducible polynomials over the complex number field are exactly polynomials of degree 1, and irreducible polynomials over the real number field are at most degree 2. However, we prove that irreducible polynomials of any degree exist over the rational number field.

Let $f(x) \in \mathbb{Q}[x]$. Take an appropriate integer c such that $cf(x)$ is a polynomial with integer coefficients, and let d be the greatest common divisor of the coefficients of $cf(x)$. Then, the coefficients of $\frac{c}{d}f(x)$ are coprime. So, we introduce the following definition.

Definition 1.10.9. Let $g(x) = b_n x^n + b_{n-1} x^{n-1} + \cdots + b_1 x + b_0$ be a nonzero polynomial with integer coefficients. If the coefficients of $g(x)$ are coprime, i.e., $(b_n, b_{n-1}, \ldots, b_1, b_0) = 1$, then $g(x)$ is called a **primitive polynomial**.

For example, $x^2 + 2x + 1, 3x^4 + 6x + 2$ are primitive polynomials.

Theorem 1.10.10. *Any nonzero polynomial $f(x)$ with rational coefficients can be expressed as the product of a rational number and a primitive polynomial: $f(x) = r_1 f_1(x)$ with $r_1 \in \mathbb{Q}$ and $f_1(x)$ a primitive polynomial, and this expression is unique up to a sign.*

Proof. Let $f(x) \in \mathbb{Q}[x]$ be a nonzero polynomial. Then, take an appropriate integer c such that $cf(x)$ is a polynomial with integer coefficients, and let d be the greatest common divisor of the coefficients of $cf(x)$. It follows that $cf(x) = df_1(x)$, i.e., $f(x) = r_1 f_1(x)$, where $r_1 = \frac{d}{c} \in \mathbb{Q}$ and $f_1(x)$ is a primitive polynomial.

Suppose that $f(x) = r_2 f_2(x)$ with $r_2 \in \mathbb{Q}$ and $f_2(x)$ a primitive polynomial, then $r_1 f_1(x) = r_2 f_2(x)$. Thus $f_1(x) = \frac{r_2}{r_1} f_2(x)$.

Let $\frac{r_2}{r_1} = \frac{m}{n}$, where $m, n \in \mathbb{Z}$ and $(m, n) = 1$, then

$$n f_1(x) = m f_2(x).$$

Since $(m, n) = 1$, n must divide all the coefficients of $f_2(x)$. But $f_2(x)$ is primitive, so $n = \pm 1$. Since $f_1(x)$ is primitive, by the same argument, $m = \pm 1$. So $r_1 = \pm r_2, f_1(x) = \pm f_2(x)$. $\qquad\square$

Theorem 1.10.11 (Gauss's lemma). *The product of two primitive polynomials is also primitive.*

Proof. Let $f(x) = a_n x^n + \cdots + a_1 x + a_0$ and $g(x) = b_m x^m + \cdots + b_1 x + b_0$ be primitive polynomials with $a_n \neq 0, b_m \neq 0$, and let

$$h(x) = f(x)g(x) = d_{n+m}x^{n+m} + \cdots + d_1 x + d_0.$$

If $h(x)$ is not primitive, then $d_{n+m}, \ldots, d_1, d_0$ has a common divisor different from ± 1. Hence, there must be a prime p, a common divisor of all coefficients of $h(x)$. Since $f(x)$ is primitive, there exists i with $0 \leqslant i \leqslant n$ such that

$$p | a_0, p | a_1, \ldots, p | a_{i-1}, p \nmid a_i.$$

Note that $g(x)$ is also primitive, so there is j with $0 \leqslant j \leqslant m$ such that

$$p | b_0, p | b_1, \ldots, p | b_{j-1}, p \nmid b_j.$$

Considering the coefficient of the term x^{i+j} in $h(x)$, we have

$$d_{i+j} = a_i b_j + a_{i-1}b_{j+1} + a_{i-2}b_{j+2} + \cdots + a_{i+1}b_{j-1} + a_{i+2}b_{j-2} + \cdots.$$

In the equation above, $p | d_{i+j}$, and p divides all terms on the right except the first term $a_i b_j$. It follows that $p | a_i b_j$, and hence $p | a_i$ or $p | b_j$, a contradiction. So $h(x)$ is a primitive polynomial. $\qquad\square$

Corollary 1.10.12. *If a nonzero polynomial with integer coefficients can be factored into a product of two polynomials with rational coefficients of lower degree, then it must be factored into a product of two polynomials with integer coefficients of lower degree.*

Proof. Let $f(x)$ be a nonzero polynomial with integer coefficients and $f(x) = g(x)h(x)$, where $g(x), h(x) \in \mathbb{Q}[x]$, and the degrees of $g(x)$ and $h(x)$ are smaller than the degree of $f(x)$.

Suppose that $f(x) = af_1(x), g(x) = rg_1(x), h(x) = sh_1(x)$ with $a \in \mathbb{Z}$, $r, s \in \mathbb{Q}$ and $f_1(x), g_1(x), h_1(x)$ primitive, then $af_1(x) = (rs)g_1(x)h_1(x)$. By Theorem 1.10.11, $g_1(x)h_1(x)$ is primitive. It follows that $rs = \pm a \in \mathbb{Z}$ by Theorem 1.10.10. So $f(x) = g(x)h(x) = (rs)g_1(x)h_1(x)$ is a product of two polynomials with integer coefficients of lower degree. $\square$

Corollary 1.10.13. *Let $f(x), g(x)$ be polynomials with integer coefficients and let $g(x)$ be primitive. If there is a polynomial $h(x)$ with rational coefficients such that $f(x) = g(x)h(x)$, then $h(x)$ must be a polynomial with integer coefficients.*

Proof. Let $f(x) = af_1(x), h(x) = rh_1(x)$ with $a \in \mathbb{Z}$, $r \in \mathbb{Q}$ and $f_1(x), h_1(x)$ primitive. Then $af_1(x) = rg(x)h_1(x)$, and hence $r = \pm a \in \mathbb{Z}$. So $h(x) = rh_1(x)$ is a polynomial with integer coefficients.

$\square$

The discussion above shows that the factorization problem of polynomials with rational coefficients can be reduced to that of polynomials with integral coefficients. Therefore, finding the rational roots of a polynomial with rational coefficients can be reduced to finding the rational roots of a polynomial with integer coefficients.

The following theorem gives a method for finding all rational roots of a polynomial with integer coefficients.

Theorem 1.10.14 (Descartes' rational root theorem). *Let*

$$f(x) = a_n x^n + \cdots + a_1 x + a_0$$

be a polynomial with integer coefficients and $a_n \neq 0$. Suppose that $\frac{r}{s}$ is a rational root of $f(x)$, where r, s are in $\mathbb{Z}$ with $(r, s) = 1$. Then $s \mid a_n$ and $r \mid a_0$.

In particular, all rational roots of a monic polynomial with integer coefficients, if any, are integers.

Proof. Since $\frac{r}{s}$ is a rational root of $f(x)$, $\left(x - \frac{r}{s}\right) \big| f(x)$ in $\mathbb{Q}[x]$, i.e., there is $g(x) \in \mathbb{Q}[x]$ with $f(x) = \left(x - \frac{r}{s}\right)g(x)$, and hence $f(x) = (sx - r)\frac{1}{s}g(x)$. Note that $f(x)$ is a polynomial with integer coefficients and $sx - r$ is primitive, so $\frac{1}{s}g(x)$ is a polynomial with integer coefficients.

Let $\frac{1}{s}g(x) = b_{n-1}x^{n-1} + \cdots + b_1 x + b_0$, where $b_i \in \mathbb{Z}, i = 0, 1, \ldots, n-1$. Then $a_n x^n + \cdots + a_1 x + a_0 = (sx - r)(b_{n-1}x^{n-1} + \cdots + b_1 x + b_0)$, and clearly $a_n = sb_{n-1}, a_0 = r(-b_0)$. So $s|a_n$ and $r|a_0$. $\square$

Example 1.10.15. Find all rational roots of the following polynomials and determine their multiplicities:

(1) $2x^3 + 5x^2 + 9x - 6$;
(2) $x^4 - x^2 - 2x + 2$.

Solution. (1) The only possible rational roots of the polynomial

$$f(x) = 2x^3 + 5x^2 + 9x - 6$$

are $\pm 1, \pm 2, \pm 3, \pm 6, \pm \frac{1}{2}, \pm \frac{3}{2}$. By the synthetic division, $\frac{1}{2}$ is a root of multiplicity 1 of $f(x)$, and the rest are not roots of $f(x)$.
(2) The only possible rational roots of the polynomial

$$g(x) = x^4 - x^2 - 2x + 2$$

are $\pm 1, \pm 2$. By the synthetic division, 1 is a root of multiplicity 2 of $g(x)$, and the rest are not roots of $g(x)$. $\square$

Example 1.10.16. Prove that the polynomial $f(x) = x^3 - x + 2$ is irreducible over $\mathbb{Q}$.

Proof. Suppose $f(x)$ is reducible over $\mathbb{Q}$, then it has at least a factor of degree 1, i.e., $f(x)$ has a rational root. However, the only possible rational roots of $f(x)$ are $\pm 1, \pm 2$. It is straightforward to check that $\pm 1, \pm 2$ are not roots of $f(x)$. So $f(x)$ is irreducible over $\mathbb{Q}$. $\square$

Remark 1.10.17. Let $f(x) \in \mathbb{Q}[x]$.

(1) If $\deg(f(x)) = 1$, then $f(x)$ is irreducible over $\mathbb{Q}$, but $f(x)$ has a rational root.
(2) If $2 \leqslant \deg(f(x)) \leqslant 3$, then $f(x)$ is irreducible over $\mathbb{Q}$ if and only if $f(x)$ has no rational roots.
(3) Suppose $\deg(f(x)) > 3$. If $f(x)$ is irreducible over $\mathbb{Q}$, then $f(x)$ has no rational roots; the converse is false. The condition that $f(x)$ has no rational roots only implies $f(x)$ has no factors of degree 1, but $f(x)$ may have factors of degree greater than 1, and so $f(x)$ may be reducible. For example, $(x^2 + 1)^2$ has no rational roots but is reducible over $\mathbb{Q}$.

The following theorem tells us irreducible polynomials of any degree exist over the rational number field.

Theorem 1.10.18 (Eisenstein's[19] criterion). *Let*

$$f(x) = a_n x^n + a_{n-1} x^{n-1} + \cdots + a_1 x + a_0$$

be a polynomial with integer coefficients. If there exists a prime number p such that (i) $p \nmid a_n$, (ii) $p \mid a_i, i = 0, 1, \ldots, n-1$, and (iii) $p^2 \nmid a_0$, then $f(x)$ is irreducible over the rational numbers.

Proof. Suppose that $f(x)$ is reducible over the rational numbers. Then $f(x)$ can be factored into a product of two polynomials with rational coefficients of lower degree. So it must be factored into a product of two polynomials with integer coefficients of lower degree:

$$f(x) = (b_m x^m + \cdots + b_1 x + b_0)(c_l x^l + \cdots + c_1 x + c_0),$$

where all b_i, c_j are integers and $0 < m, l < n$, $m + l = n$. Thus $a_n = b_m c_l, a_0 = b_0 c_0$.

By the assumption, $p \nmid a_n$ so that $p \nmid b_m$ and $p \nmid c_l$. Note that $p \mid a_0$ and $p^2 \nmid a_0$, so it is possible that $p \mid b_0$ or $p \mid c_0$, but not both. Without loss of generality, we may assume that $p \mid b_0$ but $p \nmid c_0$. Since $p \mid b_0$ and $p \nmid b_m$, there exists k with $0 < k \leqslant m < n$ such that $p \mid b_0, p \mid b_1, \ldots, p \mid b_{k-1}, p \nmid b_k$.

Considering the coefficient of the term x^k in $f(x)$, we have

$$a_k = b_k c_0 + b_{k-1} c_1 + \cdots + b_0 c_k.$$

By the hypothesis (ii) of the criterion, $p \mid a_k$. Since $p \mid b_i$, $i = 0, 1, \ldots, k - 1$, $p \mid b_k c_0$. Thus $p \mid b_k$ or $p \mid c_0$, a contradiction. So $f(x)$ is irreducible over the rational numbers. $\square$

Remark 1.10.19.

(1) Let p be a rime number. By Theorem 1.10.18, $x^n + p$ is irreducible over the rational numbers for any positive integer n. So irreducible polynomials of any degree exist in $\mathbb{Q}[x]$.

---

[19] Ferdinand Gotthold Max Eisenstein, 1823–1852, German mathematician.

(2) Theorem 1.10.18 gives a sufficient condition for a polynomial with integer coefficients to be irreducible over the rational numbers, but this condition is not necessary. For instance, the polynomial $x^3 - x + 2$ in Example 1.10.16 is irreducible over $\mathbb{Q}$, but there is no prime numbers p satisfying the condition of Theorem 1.10.18.

Remark 1.10.20.

(1) Theorem 1.10.18 is the earliest and probably best known irreducibility criterion. There are two standard proofs of Eisenstein's criterion. Schönemann[20] gave one proof in 1846, and Eisenstein gave another proof in 1950. So Eisenstein's criterion is also known as the Schönemann–Eisenstein criterion; see [7, 11] for details.

(2) Let the decimal representation of a prime number p be

$$p = a_n 10^n + a_{n-1} 10^{n-1} + \cdots + a_1 10 + a_0,$$

where $n \geqslant 1$. Then the polynomial

$$f(x) = a_n x^n + a_{n-1} x^{n-1} + \cdots + a_1 x + a_0$$

is irreducible over the rational numbers. For example, $2017, 19997$ are prime numbers, so the polynomials $2x^3 + x + 7$ and $x^4 + 9x^3 + 9x^2 + 9x + 7$ are irreducible over rational numbers. Interested readers can refer to [33, 35].

Example 1.10.21. Let $\frac{q}{p}$ be a rational root of a polynomial $f(x)$ with integer coefficients, where $p, q \in \mathbb{Z}$ and $(p, q) = 1$. Prove that $(pm - q) \mid f(m)$ for any integer m.

In particular, $(p - q) \mid f(1)$, $(p + q) \mid f(-1)$.

Proof. Since $\frac{q}{p}$ is a rational root of $f(x)$, $\left(x - \frac{r}{s}\right) \mid f(x)$ in $\mathbb{Q}[x]$, i.e., there is $h(x) \in \mathbb{Q}[x]$ with $f(x) = (x - \frac{q}{p})h(x)$, and hence $f(x) = (px - q)h_1(x)$, where $h_1(x) = \frac{1}{p}h(x) \in \mathbb{Q}[x]$. Since $f(x)$ is a polynomial with integer coefficients and $sx - r$ is primitive, $h_1(x)$ is a polynomial with integer coefficients. Thus $f(m) = (pm - q)h_1(m)$, and so $(pm - q) \mid f(m)$ for any integer m.

In particular, letting $m = 1$ gives $(p - q) \mid f(1)$, and taking $m = -1$ yields $(p + q) \mid f(-1)$. $\qquad\square$

---

[20]Theodor Schönemann, 1812–1868, German mathematician.

Example 1.10.22. Let p be a prime number, and let

$$f(x) = 1 + x + \frac{1}{2!}x^2 + \cdots + \frac{1}{(p-1)!}x^{p-1} + \frac{1}{p!}x^p.$$

Prove that $f(x)$ is irreducible over the rational numbers.

Proof. First, $f(x)$ can be transformed into a polynomial with integer coefficients. Let $g(x) = p!f(x)$. Then

$$g(x) = p! + (p!)x + \frac{p!}{2!}x^2 + \cdots + \frac{p!}{(p-1)!}x^{p-1} + x^p.$$

Thus $g(x)$ is irreducible by Eisenstein's criterion, since $p \nmid 1$, $p \mid \frac{p!}{k!}$ for $1 \leqslant k \leqslant p-1$ and $p^2 \nmid p!$. Hence, $f(x)$ is irreducible over the rational numbers. $\square$

Example 1.10.23. Let p be a prime number, and let

$$f(x) = x^{p-1} + x^{p-2} + \cdots + x + 1.$$

Prove that $f(x)$ is irreducible in $\mathbb{Q}[x]$.

Proof. Here, the coefficients 1's prevent Eisenstein's criterion from applying directly. Note that

$$x^p - 1 = (x-1)(x^{p-1} + x^{p-1} + \cdots + x + 1) = (x-1)f(x).$$

Then letting $x - 1 = y$ or $x = y + 1$, we have

$$
\begin{aligned}
yf(y+1) \\
&= (y+1)^p - 1 \\
&= y^p + C_p^1 y^{p-1} + \cdots + C_p^k y^{p-k} + \cdots + C_p^{p-2} y^2 + C_p^{p-1} y + 1 - 1 \\
&= y(y^{p-1} + C_p^1 y^{p-2} + \cdots + C_p^k y^{p-k-1} + \cdots + C_p^{p-2} y + C_p^{p-1}),
\end{aligned}
$$

and so $f(y+1) = y^{p-1} + C_p^1 y^{p-2} + \cdots + C_p^k y^{p-k-1} + \cdots + C_p^{p-2} y + C_p^{p-1}$.

Let

$$g(y) = f(y+1)$$
$$= y^{p-1} + C_p^1 y^{p-2} + \cdots + C_p^k y^{p-k-1} + \cdots + C_p^{p-2} y + C_p^{p-1}.$$

Since $p \nmid 1$, $p | C_p^k, k = 1, 2, \ldots, p-1$ and $p^2 \nmid C_p^{p-1}$, $g(y)$ is irreducible by Eisenstein's criterion. Hence $f(x)$ is irreducible in $\mathbb{Q}[x]$.

In fact, if $f(x)$ is reducible in $\mathbb{Q}[x]$, then there are $k(x), l(x) \in \mathbb{Q}[x]$ such that $f(x) = k(x)l(x)$, where $0 < \deg(k(x)), \deg(l(x)) < \deg(f(x))$. It follows that

$$g(y) = f(y+1) = k(y+1)l(y+1),$$

where $0 < \deg(k(y+1)), \deg(l(y+1)) < \deg(g(y))$, contradicting the fact that $g(y)$ is irreducible in $\mathbb{Q}[x]$. $\qquad\square$

Example 1.10.24. Let n, r be positive integers, and let $p_1, p_2, \ldots, p_r$ be different prime numbers. If $n > 1$, prove that $\sqrt[n]{p_1 p_2 \cdots p_r}$ is an irrational number.

Proof. Method 1 (Applying the fundamental theorem of arithmetic).

Suppose $\sqrt[n]{p_1 p_2 \cdots p_r}$ is a rational number. Let $\sqrt[n]{p_1 p_2 \cdots p_r} = \frac{t}{s}$, where s, t are positive integers and $(s, t) = 1$. Then $s^n p_1 p_2 \cdots p_r = t^n$.

If $s = 1$, then $p_1 p_2 \cdots p_r = t^n$ and so $t > 1$. By the fundamental theorem of arithmetic, $t = q_1 q_2 \cdots q_k$, where $q_1, q_2, \ldots, q_k$ are prime numbers. Therefore,

$$p_1 p_2 \cdots p_r = q_1^n q_2^n \cdots q_k^n.$$

Since $n > 1$, there are at least two identical prime numbers on the right side of the above equation, contradicting the hypothesis that $p_1, p_2, \ldots, p_r$ are different prime numbers.

If $s \neq 1$, then there is a prime number p such that $p|s$ by the fundamental theorem of arithmetic. Note that $s^n p_1 p_2 \cdots p_r = t^n$, and so $p|t^n$. Hence $p|t$, contradicting the hypothesis that $(s, t) = 1$. So $\sqrt[n]{p_1 p_2 \cdots p_r}$ is an irrational number.

Method 2 (Applying Eisenstein's criterion).

Let $f(x) = x^n - p_1 p_2 \cdots p_r$. Then, Eisenstein's criterion makes $f(x)$ irreducible in $\mathbb{Q}[x]$. Clearly, $\sqrt[n]{p_1 p_2 \cdots p_r}$ is a root of $f(x)$, i.e., $f(\sqrt[n]{p_1 p_2 \cdots p_r}) = 0$. If $\sqrt[n]{p_1 p_2 \cdots p_r}$ is a rational number, then $f(x)$ has a rational root. Since $n > 1$, $f(x)$ is reducible in $\mathbb{Q}[x]$, a contradiction. So $\sqrt[n]{p_1 p_2 \cdots p_r}$ is an irrational number. $\qquad\square$

1.11*   Real Roots of Polynomials with Real Coefficients

The discussion in Section 1.10 shows that the roots of polynomials with complex coefficients are complex numbers. On the other hand, we can determine all its rational roots for a polynomial with rational coefficients. This section discusses the real roots of polynomials with real coefficients. Here, we mainly introduce Descartes' rule of signs and Sturm's[21] theorem. To this end, let's first introduce the concept of sign variations.

Definition 1.11.1. Let $q_1, q_2, \ldots, q_m$ be a sequence of real numbers. If $q_j q_{j+1} < 0$, then we say that there is a **sign variation** between q_j and q_{j+1}; if $q_{j+1} = \cdots = q_{j+s-1} = 0$, but $q_j q_{j+s} < 0$, then we say that there is a sign variation between $q_j, q_{j+1}, \ldots, q_{j+s}$. We define **the number of sign variations** of the sequence $q_1, q_2, \ldots, q_m$ as the total number of sign variations in this sequence.

Let $q_1, q_2, \ldots, q_m$ be a sequence of real numbers. By definition, we have the following:

(1) If $q_1 q_m > 0$, then the number of sign variations of $q_1, q_2, \ldots, q_m$ is even.
(2) If $q_1 q_m < 0$, then the number of sign variations of $q_1, q_2, \ldots, q_m$ is odd.

To prove Descartes' rule of signs, we first give a lemma.

Lemma 1.11.2. *Let* $f(x) = a_0 x^n + a_1 x^{n-1} + \cdots + a_{n-1} x + a_n \in \mathbb{R}[x]$, $a_0 > 0$ *and* $a_n \neq 0$. *If* $f(x)$ *has* p *positive roots (multiple roots are counted by multiplicity), then* p *is even if and only if* $a_n > 0$; p *is odd if and only if* $a_n < 0$.

---

[21] Jacques Charles Francois Sturm, 1803–1855, French mathematician.

Proof. Without loss of generality, we may assume that $\alpha_1, \alpha_2, \ldots, \alpha_k$ are all the different positive roots of $f(x)$ with $\alpha_1 < \alpha_2 < \cdots < \alpha_k$, and their multiplicities are respectively $r_1, r_2, \ldots, r_k$, then

$$f(x) = (x - \alpha_1)^{r_1}(x - \alpha_2)^{r_2} \cdots (x - \alpha_k)^{r_k} g(x),$$

where $r_1 + r_2 + \cdots + r_k = p$, $g(x) \in \mathbb{R}[x]$ and $g(x)$ has no positive roots. Note that $g(x) = f(x)$ if $p = 0$. Since $\deg(g(x)) = n - p$, we may write

$$g(x) = b_0 x^{n-p} + b_1 x^{n-p-1} + \cdots + b_{n-p-1} x + b_{n-p},$$

then $b_0 = a_0$ and

$$(-1)^p \alpha_1^{r_1} \alpha_2^{r_2} \cdots \alpha_k^{r_k} b_{n-p} = (-1)^{r_1+r_2+\cdots+r_k} \alpha_1^{r_1} \alpha_2^{r_2} \cdots \alpha_k^{r_k} b_{n-p} = a_n.$$

$$(1.5)$$

Since $a_n \neq 0$, $b_{n-p} \neq 0$.

Suppose that $g(0) = b_{n-p} < 0$. Note that $g(x)$ and b_0 are both positive if x is big enough, so $g(x)$ has a positive root by Bolzano's[22] theorem in mathematical analysis, contradicting the fact that $g(x)$ has no positive roots. Hence $b_{n-p} > 0$. Thus, by (1.5), p is even if and only if $a_n > 0$; p is odd if and only if $a_n < 0$. $\qquad\square$

Theorem 1.11.3 (Descartes' rule of signs). *Let*

$$f(x) = a_0 x^n + a_1 x^{n-1} + \cdots + a_{n-1} x + a_n \in \mathbb{R}[x],$$

where $a_0 > 0$ and $a_n \neq 0$. If $f(x)$ has p positive roots (multiple roots are counted by multiplicity), and the number of sign variations of $a_0, a_1, \ldots, a_{n-1}, a_n$ is μ, then (1) $p \leqslant \mu$; (2) $\mu - p$ is even.

Proof. From $a_0 > 0$, we know that μ is an even number if $a_n > 0$ and μ is an odd number if $a_n < 0$. Therefore, according to Lemma 1.11.2, the parity of p and μ is the same, so (2) is true.

Next, we show (1) $p \leqslant \mu$ by an induction on the degree n of $f(x)$.

If $n = 1$, then $f(x) = a_0 x + a_1 = a_0(x + \frac{a_1}{a_0})$. Thus $f(x)$ has a positive root if and only if $a_0 a_1 < 0$. So $p \leqslant \mu$.

---

[22]Bernard Bolzano, 1781–1848, Czech mathematician.

Let $n > 1$. Choose an integer m with $0 \leqslant m \leqslant n - 1$ such that $a_m \neq 0$, but $a_{m+1} = \cdots = a_{n-1} = 0$, then $f(x) = a_0 x^n + a_1 x^{n-1} + \cdots + a_m x^{n-m} + a_n$, whence

$$f'(x) = na_0 x^{n-1} + (n-1)a_1 x^{n-2} + \cdots + (n-m)a_m x^{n-m-1}.$$

Let μ' be the number of sign variations of the sequence

$$na_0, (n-1)a_1, \ldots, (n-m)a_m.$$

Obviously, $\mu = \mu'$ if $a_n a_m > 0$ and $\mu = \mu' + 1$ if $a_n a_m < 0$. If $p = 0$, i.e., $f(x)$ has no positive roots, then $p \leqslant \mu$.

Let $p > 0$. Suppose that $\alpha_1, \alpha_2, \ldots, \alpha_k$ are all the different positive roots of $f(x)$ with $\alpha_1 < \alpha_2 < \cdots < \alpha_k$, and the multiplicities of $\alpha_1, \alpha_2, \ldots, \alpha_k$ are respectively $r_1, r_2, \ldots, r_k$, then $r_1 + r_2 + \cdots + r_k = p$. Thus, $\alpha_1, \alpha_2, \ldots, \alpha_k$ are respectively the roots of multiplicity $r_1 - 1, r_2 - 1, \ldots, r_k - 1$ of $f'(x)$. On the other hand, according to Rolle's[23] theorem in mathematical analysis, $f'(x)$ has at least one root on the open interval (α_j, α_{j+1}), $j = 1, 2, \ldots, k - 1$. Therefore, $f'(x)$ has at least $r_1 - 1 + r_2 - 1 + \cdots + r_k - 1 + k - 1 = p - 1$ roots on the closed interval $[\alpha_1, \alpha_k]$.

If $a_n a_m < 0$, then $\mu = \mu' + 1$, and so the number of positive roots of $f'(x)$ is $\leqslant \mu'$ by the inductive hypothesis. Since $f'(x)$ has at least $p - 1$ positive roots, $p - 1 \leqslant \mu'$. So $p \leqslant \mu' + 1 = \mu$.

If $a_n a_m > 0$, then $\mu = \mu'$. By the inductive hypothesis, $p - 1 \leqslant \mu' = \mu$. Since $\mu - p$ is even, $p - 1 \neq \mu$. Thus $p - 1 \leqslant \mu - 1$, i.e., $p \leqslant \mu$, as desired. $\qquad\square$

Remark 1.11.4. Let $g(x) = f(-x)$, then the number of positive roots of $g(x)$ is the number of negative roots of $f(x)$.

Example 1.11.5. Let $f(x) = x^4 - x^3 - 1$. Find the number of real roots of $f(x)$.

Proof. The sequence of coefficients of $f(x) = x^4 - x^3 - 1$ is $1, -1, 0, 0, -1$, and its number of sign variations is 1. So the number p of positive roots of $f(x)$ is $\leqslant 1$. Since $1 - p$ is even, $p = 1$. Therefore, $f(x)$ has precisely one positive root. Note that $f(-x) = x^4 + x^3 - 1$ and the number of sign variations of the sequence of coefficients of

---

[23] Michel Rolle, 1652–1719, French mathematician.

$f(-x)$ is 1. Hence, $f(-x)$ has precisely one positive root, i.e., $f(x)$ has one negative root. So, the number of real roots of $f(x)$ is 2 (one positive root and one negative root). $\square$

Definition 1.11.6. Let $f(x) \in \mathbb{R}[x], a, b \in \mathbb{R}$ and $a < b$. A **Sturm sequence** for $f(x)$ on the interval $[a, b]$ is a sequence

$$f_0(x) = f(x), f_1(x), \ldots, f_m(x) \tag{1.6}$$

of nonzero real coefficient polynomials headed by $f(x)$ such that

(1) the last polynomial $f_m(x)$ has no roots on $[a, b]$,
(2) if $f_j(c) = 0$ $(0 < j < m)$ for some $c \in [a, b]$, then $f_{j-1}(c) f_{j+1}(c) < 0$,
(3) if $f(c) = 0$ for some $c \in (a, b)$, then $f(x)f_1(x)$ is monotonically increasing near $x = c$, i.e., there is a sufficiently small $\delta > 0$ such that $f(x)f_1(x) < 0$ for $x \in (c - \delta, c)$ and $f(x)f_1(x) > 0$ for $x \in (c, c + \delta)$.

The condition (2) above implies that two consecutive polynomials in (1.6) do not have any common root on $[a, b]$.

Let $f(x) \in \mathbb{R}[x]$ with $\deg(f(x)) \geqslant 1$. Put $f_0(x) = f(x)$, $f_1(x) = f'(x)$. By the Euclidean algorithm, we have the following:

$$
\begin{aligned}
f_0(x) &= q_1(x)f_1(x) - f_2(x), &\quad& \deg(f_2(x)) < \deg(f_1(x)), \\
f_1(x) &= q_2(x)f_2(x) - f_3(x), &\quad& \deg(f_3(x)) < \deg(f_2(x)), \\
&\ \ \vdots &\quad& \ \ \vdots \\
f_{m-2}(x) &= q_{m-1}(x)f_{m-1}(x) - f_m(x), &\quad& \deg(f_m(x)) < \deg(f_{m-1}(x)), \\
f_{m-1}(x) &= q_m(x)f_m(x).
\end{aligned}
$$

$$\tag{1.7}$$

Theorem 1.11.7. *If $f(x)$ has no multiple roots in the field of complex numbers, then the sequence*

$$f_0(x) = f(x), f_1(x), \ldots, f_m(x) \tag{1.8}$$

*of polynomials constructed in (1.7) is a Sturm sequence for $f(x)$ (on any interval), which is called **the standard Sturm sequence** for $f(x)$.*

Proof. Since $f(x)$ has no multiple roots in the field of complex numbers, $(f(x), f'(x)) = 1$ by Corollary 1.8.10. From the construction in (1.7), we have

$$
\begin{aligned}
1 &= (f(x), f'(x)) \\
&= (f_0(x), f_1(x)) = (f_1(x), f_2(x)) = \cdots \\
&= (f_{m-1}(x), f_m(x)) = c_m f_m(x),
\end{aligned}
$$

where $f_m(x)$ is a nonzero constant, and $c_m = \frac{1}{f_m(x)}$.

Next, we show that the sequence (1.8) satisfies the three conditions in the definition of a Sturm sequence:

(1) Since $f_m(x)$ is a nonzero constant, $f_m(x)$ has no roots on $[a, b]$.

(2) Suppose $f_j(c) = 0$ $(0 < j < m)$ for some $c \in [a, b]$. If $f_{j-1}(c) = 0$, then c is a root of $(f_{j-1}(x), f_j(x)) = (f(x), f'(x))$. It follows that c is a multiple root of $f(x)$; this is a contradiction. So $f_{j-1}(c) \neq 0$. Similarly, we have $f_{j+1}(c) \neq 0$. Since $f_{j-1}(x) = q_j(x)f_j(x) - f_{j+1}(x)$, it follows that $f_{j-1}(c) = -f_{j+1}(c)$. So $f_{j-1}(c)f_{j+1}(c) < 0$.

(3) If $f(c) = 0$ for some $c \in (a, b)$, then $f(x) = (x - c)q(x)$, where $q(c) \neq 0$. Thus

$$
\begin{aligned}
f(x)f_1(x) &= (x - c)q(x)\big(q(x) + (x - c)q'(x)\big) \\
&= (x - c)\big(q^2(x) + (x - c)q(x)q'(x)\big).
\end{aligned}
$$

Let $g(x) = q^2(x) + (x - c)q(x)q'(x)$. Then $f(x)f_1(x) = (x - c)g(x)$.

Note that $g(c) = q^2(c) > 0$, so $f(x)f_1(x)$ is monotonically increasing near $x = c$, as required. $\square$

Let $f_0(x) = f(x), f_1(x), \ldots, f_m(x)$ be a Sturm sequence for a polynomial $f(x)$ with real coefficients. If c is a real number, the number of sign variations of the sequence $f_0(c), f_1(c), \ldots, f_m(c)$ is often denoted by $V_c(f)$.

Theorem 1.11.8 (Sturm's theorem). *Let* $f(x) \in \mathbb{R}[x]$ *with* $\deg(f(x)) \geq 1$, *and let* $a, b \in \mathbb{R}$ *with* $a < b$ *such that* $f(a)f(b) \neq 0$. *If* $f(x)$ *has no multiple roots in the field of complex numbers and*

$$
f_0(x) = f(x), f_1(x), \ldots, f_m(x)
$$

is a Sturm sequence for $f(x)$ *on* $[a, b]$, *then the number of real roots of* $f(x)$ *on* (a, b) *is equal to the difference* $V_a(f) - V_b(f)$.

Here, we omit the proof of Sturm's theorem. Instead, interested readers can refer to [22, 34, 40].

Remark 1.11.9.

(1) If $f_0(x), f_1(x), \ldots, f_m(x)$ is a Sturm sequence for $f(x)$, then

$$c_0 f_0(x), c_1 f_1(x), \ldots, c_m f_m(x)$$

is a Sturm sequence for $c_0 f(x)$, where $c_0, c_1, \ldots, c_m$ are positive constants.

(2) If $f(x)$ has multiple roots, we may replace $f(x)$ with $\frac{f(x)}{(f(x), f'(x))}$. Let $g(x) = \frac{f(x)}{(f(x), f'(x))}$, then $g(x)$ has no multiple roots, and $g(x)$ and $f(x)$ have precisely the same roots (regardless of multiplicity) by the discussion in Section 1.8. According to Sturm's theorem, $V_a(g) - V_b(g)$ is the number of real roots of $g(x)$ on (a, b), that is, the number of different real roots of $f(x)$ on (a, b).

(3) Sturm gave Sturm's theorem in 1829. With this theorem, the number of different real roots of a real coefficient polynomial can be calculated, and the root can also be separated (for a real root, we can locate an interval containing only this real root and no other real roots). It is said Sturm himself often expressed his pride in his achievements in this way. After telling the students about the proof of the theorem, he added, "This is the theorem named after me"!

Example 1.11.10. Find the number of real roots of

$$f(x) = x^4 - 6x^2 - 4x + 2.$$

Solution. $f'(x) = 4x^3 - 12x - 4 = 4(x^3 - 3x - 1)$.
Let $f_0(x) = f(x), f_1(x) = x^3 - 3x - 1$, then

$$f_0(x) = x f_1(x) - (3x^2 + 3x - 2).$$

Let $f_2(x) = 3x^2 + 3x - 2$, then $f_1(x) = \left(\frac{1}{3}x - \frac{1}{3}\right) f_2(x) - \frac{1}{3}(4x + 5)$.

Let $f_3(x) = 4x + 5$, then $f_2(x) = \left(\frac{3}{4}x - \frac{3}{16}\right) f_3(x) - \frac{7}{16}$.

Let $f_4(x) = 1$, then $f_0(x), f_1(x), f_2(x), f_3(x), f_4(x)$ is a Sturm sequence for $f(x)$.

Let x take some values and calculate the signs of the corresponding values taken by each polynomial in the Sturm sequence. We have the following list:

x	$f_0(x)$	$f_1(x)$	$f_2(x)$	$f_3(x)$	$f_4(x)$	the number of sign variations
$-\infty$	$+$	$-$	$+$	$-$	$+$	4
0	$+$	$-$	$-$	$+$	$+$	2
$+\infty$	$+$	$+$	$+$	$+$	$+$	0

So $f(x)$ has four real roots (two positives and two negatives). $\square$

Example 1.11.11. Find the number of real roots of

$$f(x) = 1 + \frac{x}{1} + \frac{x^2}{2!} + \cdots + \frac{x^{n-1}}{(n-1)!} + \frac{x^n}{n!}.$$

Solution. If $f(x)$ has a real root, then the real root must lie within an open interval $(-M, -\epsilon)$, where M is a sufficiently large positive number, and ϵ is a small enough positive number.

Taking the derivative of $f(x)$ gives $f'(x) = 1 + \frac{x}{1} + \frac{x^2}{2!} + \cdots + \frac{x^{n-1}}{(n-1)!}$.

Let $f_0(x) = f(x), f_1(x) = f'(x)$, then

$$f_0(x) = f_1(x) + \frac{x^n}{n!} = f_1(x) - \left(-\frac{x^n}{n!}\right).$$

Let $f_2(x) = -\frac{x^n}{n!}$, then $f_0(x), f_1(x), f_2(x)$ is a Sturm sequence for $f(x)$ on $[-M, -\epsilon]$ (the proof is left to the reader).

Similar to Example 1.11.10, we have the following list:

x	$f_0(x)$	$f_1(x)$	$f_2(x)$	the number of sign variations	
				n is even	n is odd
$-M$	$(-1)^n$	$(-1)^{n-1}$	$(-1)^{n-1}$	1	1
$-\varepsilon$	$+$	$+$	$(-1)^{n-1}$	1	0

It follows that $f(x)$ has no negative roots if n is even, and $f(x)$ has a negative root if n is odd. Therefore, $f(x)$ has no real roots if n is even, and $f(x)$ has a real root if n is odd. $\qquad\qquad\square$

1.12* Polynomials in n Indeterminates

We have discussed polynomials in one indeterminate and their basic properties. Now, we briefly introduce the concept of multivariate polynomials. Suppose that F is a number field, and x is an indeterminate. Then, we have defined the polynomial ring $F[x]$ with coefficients in F. Let y be another indeterminate. Similarly, we can define the polynomial ring $F[x][y]$ with coefficients in $F[x]$, whose elements are called polynomials over F in the two indeterminates x and y. For example, $x^2+y^2-1, x^5+y^4-3x^2y^3+6xy^2-3x+4y+2$ are both polynomials over F in x and y. Polynomials over F in n indeterminates can be defined as follows by induction.

Definition 1.12.1. Let F be a number field and $n \geqslant 2$ an integer, and let $x_1, x_2, \ldots, x_n$ be n indeterminates. **The ring** $F[x_1, x_2, \ldots, x_n]$ **of polynomials over** F **in** n **indeterminates** is defined by

$$F[x_1, x_2, \ldots, x_n] = F[x_1, x_2, \ldots, x_{n-1}][x_n],$$

whose elements are called polynomials over F in the n indeterminates $x_1, x_2, \ldots, x_n$.

Let $a \in F$ and $k_1, k_2, \ldots, k_n$ be natural numbers. The element of the form $ax_1^{k_1} x_2^{k_2} \cdots x_n^{k_n}$ in $F[x_1, x_2, \ldots, x_n]$ is called a **monomial** over F, where a is called the coefficient of the monomial. We often denote a by $a = a_{k_1 k_2 \cdots k_n}$, indicating that a is the coefficient of what kind of monomials. If $a \neq 0$, $k_1 + k_2 + \cdots + k_n$ is called the **degree** of the monomial $ax_1^{k_1} x_2^{k_2} \cdots x_n^{k_n}$. If $a = 0$, the monomial is called the zero monomial (the zero monomial is the only monomial with no degree).

By definition, every polynomial over a number field F in the n indeterminates $x_1, x_2, \ldots, x_n$ is a sum of finite monomials:

$$\sum_{k_1 k_2 \cdots k_n} a_{k_1 k_2 \cdots k_n} x_1^{k_1} x_2^{k_2} \cdots x_n^{k_n}.$$

Two monomials $ax_1^{k_1} x_2^{k_2} \ldots x_n^{k_n}$ and $bx_1^{l_1} x_2^{l_2} \cdots x_n^{l_n}$ are called **like terms** if $ab \neq 0$ and $k_1 = l_1, k_2 = l_2, \ldots, k_n = l_n$. By convention, if a monomial and the zero monomial are like terms, then the monomial must be the zero monomial.

Every polynomial in n indeterminates can be expressed as the sum of finite monomials, in which any two monomials are not like terms, and the highest degree of the nonzero monomials is called the **degree** of this polynomial.

Polynomials in n indeterminates and polynomials in one indeterminate have similar properties. For example, the unique factorization theorem is true for polynomials over a number field F in n indeterminates. The proof of this conclusion is discussed in the course of abstract algebra. Interested readers can refer to [20]. However, there is no division algorithm for polynomials in $F[x_1, x_2, \ldots, x_n]$ if $n \geqslant 2$.

In what follows, we often use $f(x, x_2, \ldots, x_n)$, $g(x_1, x_2, \ldots, x_n), \ldots$ to denote polynomials in n indeterminates, and the degree of $f(x, x_2, \ldots, x_n)$ is denoted by $\deg(f)$.

As we know, the leading term of a polynomial in one indeterminate is the term with the highest degree in the polynomial. Although every polynomial $f(x_1, x_2, \ldots, x_n)$ in n indeterminates can be expressed as the sum of finite monomials, in which any two monomials are not like terms, we cannot call the monomial with the highest degree in the expression the leading term of $f(x_1, x_2, \ldots, x_n)$ (because different types of monomials may have the same degree). For this reason, we need to specify an order of arrangement for the monomials in the expression of $f(x_1, x_2, \ldots, x_n)$ to give the concept of a leading term.

If $a \neq 0$, then each type of monomials $ax_1^{k_1} x_2^{k_2} \cdots x_n^{k_n}$ corresponds bijectively to an n-tuple $(k_1, k_2, \ldots, k_n)$ of natural numbers. To give the order between different types of monomials $ax_1^{k_1} x_2^{k_2} \cdots x_n^{k_n}$, we only need to define an order for n-tuples $(k_1, k_2, \ldots, k_n)$ of natural numbers.

Let $(k_1, k_2, \ldots, k_n)$ and $(l_1, l_2, \ldots, l_n)$ be two n-tuples of natural numbers. We say that $(k_1, k_2, \ldots, k_n)$ precedes $(l_1, l_2, \ldots, l_n)$, denoted by $(k_1, k_2, \ldots, k_n) > (l_1, l_2, \ldots, l_n)$, if there is an integer i with $1 \leqslant i \leqslant n$ such that $k_1 = l_1, k_2 = l_2, \ldots, k_{i-1} = l_{i-1}, k_i > l_i$.

Given any two n-tuples $(k_1, k_2, \ldots, k_n)$ and $(l_1, l_2, \ldots, l_n)$ of natural numbers. From the above definition, we can see that one and only one of the three relations

$$(k_1, k_2, \ldots, k_n) > (l_1, l_2, \ldots, l_n),$$

$$(k_1, k_2, \ldots, k_n) = (l_1, l_2, \ldots, l_n),$$

$$(l_1, l_2, \ldots, l_n) > (k_1, k_2, \ldots, k_n)$$

holds.

Suppose $ab \neq 0$. We say that the monomial $ax_1^{k_1} x_2^{k_2} \cdots x_n^{k_n}$ precedes the monomial $bx_1^{l_1} x_2^{l_2} \cdots x_n^{l_n}$ if $(k_1, k_2, \ldots, k_n) > (l_1, l_2, \ldots, l_n)$.

The above method is obtained by imitating the principle of dictionary arrangement, so it is called the **lexicographical order.**

According to the lexicographical order, after sorting the items of a polynomial in n indeterminates, the first nonzero monomial is called the **leading term** of the polynomial.

For example, by the lexicographical order,

$$f(x_1, x_2, x_3) = x_1 x_3^4 + x_1 x_2 x_3 + x_2^4 x_3^5 + x_3^{10}$$

can be written as

$$f(x_1, x_2, x_3) = x_1 x_2 x_3 + x_1 x_3^4 + x_2^4 x_3^5 + x_3^{10},$$

its leading term is $x_1 x_2 x_3$, but x_3^{10} is the term with the highest degree.

By the lexicographical order, it is straightforward to check that the leading term of $f(x_1, x_2, \ldots, x_n)g(x_1, x_2, \ldots, x_n)$ is equal to the product of the leading term of $f(x_1, x_2, \ldots, x_n)$ and the leading term of $g(x_1, x_2, \ldots, x_n)$ if $f(x_1, x_2, \ldots, x_n) \neq 0, g(x_1, x_2, \ldots, x_n) \neq 0$. So, the product of two nonzero polynomials in n indeterminates is nonzero.

A polynomial $f(x_1, x_2, \ldots, x_n)$ is called a **homogeneous polynomial** of degree i if its nonzero monomials all have the same degree i. It is easy to see that the product of two homogeneous polynomials is still a homogeneous polynomial. Any polynomial $f(x_1, x_2, \ldots, x_n)$ of degree m can be uniquely expressed as the sum of homogeneous polynomials, that is,

$$f(x_1, x_2, \ldots, x_n) = \sum_{i=0}^{m} f_i(x_1, x_2, \ldots, x_n),$$

where $f_i(x_1, x_2, \ldots, x_n)$ is zero or a homogeneous polynomial of degree i. $f_i(x_1, x_2, \ldots, x_n)$ is often called the **homogeneous component** of degree i of $f(x_1, x_2, \ldots, x_n)$.

Definition 1.12.2. A polynomial $f(x_1, x_2, \ldots, x_n)$ in n indeterminates is said to be a **symmetric polynomial** if

$$f(x_1, \ldots, x_i, \ldots, x_j, \ldots, x_n) = f(x_1, \ldots, x_j, \ldots, x_i, \ldots, x_n),$$

for any integers i, j with $1 \leqslant i < j \leqslant n$.

When discussing symmetric polynomials, it is usually necessary to specify the number of indeterminates. For example, $x_1^2 + x_2^2 + x_1 x_2$ is a symmetric polynomial in 2 indeterminates but not a symmetric polynomial in 3 indeterminates.

Note that an important feature of Viète's formulas (1.4) is that the expression on the left is independent of the order of roots, so we naturally introduce the following symmetric polynomials in n indeterminates:

$$\sigma_1 = x_1 + x_2 + \cdots + x_n = \sum_{i=1}^{n} x_i,$$

$$\sigma_2 = x_1 x_2 + x_1 x_3 + \cdots + x_{n-1} x_n = \sum_{1 \leqslant i < j \leqslant n} x_i x_j,$$

$$\vdots$$

$$\sigma_k = \sum_{1 \leqslant i_1 < i_2 < \cdots < i_k \leqslant n} x_{i_1} x_{i_2} \cdots x_{i_k},$$

$$\vdots$$

$$\sigma_n = x_1 x_2 \cdots x_n,$$

which are called **elementary symmetric polynomials**.

Proposition 1.12.3. *If the leading term of the symmetric polynomial $f(x_1, x_2, \ldots, x_n)$ in n indeterminates is $a x_1^{k_1} x_2^{k_2} \cdots x_n^{k_n}$, then*

(1) *$k_1 \geqslant k_2 \geqslant \cdots \geqslant k_n$,*
(2) *both $a \sigma_1^{k_1 - k_2} \sigma_2^{k_2 - k_3} \cdots \sigma_{n-1}^{k_{n-1} - k_n} \sigma_n^{k_n}$ and $f(x_1, x_2, \ldots, x_n)$ have the same the leading term.*

Proof. (1) If there is i such that $k_i < k_{i+1}$, then $f(x_1, x_2, \ldots, x_n)$ has a monomial $a x_1^{k_1} \cdots x_i^{k_{i+1}} x_{i+1}^{k_i} \cdots x_n^{k_n}$, which precedes the leading term, a contradiction.

(2) Since the leading term of the product of polynomials in n indeterminates is equal to the product of the leading terms of polynomials in n indeterminates, the leading term of $a \sigma_1^{k_1 - k_2} \sigma_2^{k_2 - k_3} \cdots \sigma_{n-1}^{k_{n-1} - k_n} \sigma_n^{k_n}$ is

$$a x_1^{k_1 - k_2} (x_1 x_2)^{k_2 - k_3} \cdots (x_1 x_2 \cdots x_{n-1})^{k_{n-1} - k_n} (x_1 x_2 \cdots x_{n-1} x_n)^{k_n}$$

$$= a x_1^{k_1} x_2^{k_2} \cdots x_n^{k_n}. \qquad \square$$

It is easy to see that the sum, difference, and product of symmetric polynomials are symmetric, and the polynomial of symmetric polynomials is still symmetric. In particular, the polynomial of elementary symmetric polynomials is symmetric. One can show that any symmetric polynomial in n indeterminates can be uniquely given by a polynomial expression in terms of elementary symmetric polynomials. That is, there is the following **fundamental theorem of symmetric polynomials.**

Theorem 1.12.4. *Let* $f(x_1, x_2, \ldots, x_n)$ *be a symmetric polynomial in n indeterminates. Then there is a unique polynomial* $\varphi(y_1, y_2, \ldots, y_n)$ *such that* $f(x_1, x_2, \ldots, x_n) = \varphi(\sigma_1, \sigma_2, \ldots, \sigma_n)$.

For the proof of this theorem, readers can refer to [34, 40]. The following example shows how to express a symmetric polynomial by a polynomial expression in terms of elementary symmetric polynomials.

Example 1.12.5. Let $f(x_1, x_2, x_3) = x_1^3 + x_2^3 + x_3^3$. Express $f(x_1, x_2, x_3)$ by a polynomial expression in terms of elementary symmetric polynomials.

Solution. **Method 1** (by elimination of the leading term, the method used to prove the fundamental theorem of symmetric polynomials).

First, the leading term of $f(x_1, x_2, x_3) = x_1^3 + x_2^3 + x_3^3$ is x_1^3, which corresponds to the triple $(3, 0, 0)$.

Let $\varphi_1 = \sigma_1^{3-0}\sigma_2^{0-0}\sigma_3^0 = \sigma_1^3$. Then both $f(x_1, x_2, x_3)$ and φ_1 have the same leading term by Proposition 1.12.3. Put $f_1 = f - \varphi_1$. It follows that

$$f_1 = x_1^3 + x_2^3 + x_3^3 - (x_1 + x_2 + x_3)^3$$

$$= -3(x_1^2 x_2 + x_1^2 x_3 + x_2^2 x_1 + x_2^2 x_3 + x_3^2 x_1 + x_3^2 x_2) - 6x_1 x_2 x_3.$$

Second, the leading term of f_1 is $-3x_1^2 x_2$, which corresponds to the triple $(2, 1, 0)$.

Let $\varphi_2 = -3\sigma_1^{2-1}\sigma_2^{1-0}\sigma_3^0 = -3\sigma_1\sigma_2$, and $f_2 = f_1 - \varphi_2$. Then

$$f_2 = f_1 + 3\sigma_1\sigma_2$$

$$= f_1 + 3(x_1 + x_2 + x_3)(x_1 x_2 + x_1 x_3 + x_2 x_3) = 3x_1 x_2 x_3 = 3\sigma_3.$$

So $f = \varphi_1 + f_1 = \varphi_1 + \varphi_2 + f_2 = \sigma_1^3 - 3\sigma_1\sigma_2 + 3\sigma_3$.

Method 2 (By the method of undetermined coefficients, which is only applicable to homogeneous symmetric polynomials).

According to Method 1, $\varphi_1, \varphi_2, \ldots$ in the required expression completely depend on the leading terms of the symmetric polynomials $f, f_1, \ldots$. These leading terms must meet the following conditions:

(1) The leading term of f precedes that of f_i, and the leading term of f_i precedes that of f_j if $i < j$;
(2) The n-tuple $(k_1, k_2, \ldots, k_n)$ corresponding to each leading term satisfies $k_1 \geqslant k_2 \geqslant \cdots \geqslant k_n$ and $\sum\limits_{i=1}^{n} k_i = \deg(f)$ (for f is homogeneous).

For this example, because $f(x_1, x_2, x_3) = x_1^3 + x_2^3 + x_3^3$, the leading term of $f(x_1, x_2, x_3)$ is x_1^3. It follows that the triples corresponding to the leading terms in f and f_i can only be $(3, 0, 0), (2, 1, 0), (1, 1, 1)$, whose associated powers of σ are respectively

$$\sigma_1^{3-0}\sigma_2^{0-0}\sigma_3^0 = \sigma_1^3, \ \sigma_1^{2-1}\sigma_2^{1-0}\sigma_3^0 = \sigma_1\sigma_2, \ \sigma_1^{1-1}\sigma_2^{1-1}\sigma_3^1 = \sigma_3.$$

Therefore, according to the fundamental theorem of symmetric polynomials, we may assume $f(x_1, x_2, x_3) = \sigma_1^3 + a\sigma_1\sigma_2 + b\sigma_3$.

To find the unknown numbers a and b, we choose some special values of x_1, x_2, x_3 and calculate the corresponding values taken by $\sigma_1, \sigma_2, \sigma_3$ and f. We have the following list:

x_1	x_2	x_3	σ_1	σ_2	σ_3	f
1	1	0	2	1	0	2
1	1	1	3	3	1	3

Therefore, $\begin{cases} 8 + 2a = 2, \\ 27 + 9a + b = 3, \end{cases}$ and hence $\begin{cases} a = -3, \\ b = 3. \end{cases}$

So $f(x_1, x_2, x_3) = \sigma_1^3 - 3\sigma_1\sigma_2 + 3\sigma_3$.

Method 3 (Using the elementary method).

$$f(x_1, x_2, x_3)$$
$$= x_1^3 + x_2^3 + x_3^3$$
$$= x_1^3 + x_2^3 + x_3^3 - 3x_1x_2x_3 + 3x_1x_2x_3$$
$$= (x_1 + x_2 + x_3)(x_1^2 + x_2^2 + x_3^2 - x_1x_2 - x_1x_3 - x_2x_3) + 3x_1x_2x_3$$
$$= (x_1 + x_2 + x_3)\left((x_1 + x_2 + x_3)^2 - 3(x_1x_2 + x_1x_3 + x_2x_3)\right)$$
$$\quad + 3x_1x_2x_3$$
$$= \sigma_1(\sigma_1^2 - 3\sigma_2) + 3\sigma_3$$
$$= \sigma_1^3 - 3\sigma_1\sigma_2 + 3\sigma_3. \qquad \square$$

Let $x_1, x_2, \ldots, x_n$ be n indeterminates. Then

$$D = \prod_{1 \leqslant i < j \leqslant n} (x_i - x_j)^2$$

is an important symmetric polynomial.

According to the fundamental theorem of symmetric polynomials, D can be expressed as a polynomial $D(a_1, a_2, \ldots, a_n)$ in

$$a_1 = -\sigma_1, a_2 = \sigma_2, \ldots, a_k = (-1)^k \sigma_k, \ldots, a_n = (-1)^n \sigma_n.$$

If $x_1, x_2, \ldots, x_n$ take fixed values, then

$$f(x) = (x - x_1)(x - x_2) \cdots (x - x_n)$$

is a polynomial of degree n, whose roots are $x_1, x_2, \ldots, x_n$. Therefore, $f(x) = x^n + a_1 x^{n-1} + \cdots + a_{n-1} x + a_n$ by Viète's formulas (1.4).

It is clear that $f(x)$ has multiple roots if and only if

$$D(a_1, a_2, \ldots, a_n) = \prod_{1 \leqslant i < j \leqslant n} (x_i - x_j)^2 = 0.$$

As usual, $D(a_1, a_2, \ldots, a_n)$ is called the **discriminant** of the polynomial $f(x) = x^n + a_1 x^{n-1} + \cdots + a_{n-1} x + a_n$.

If $n = 2$, then $f(x) = x^2 + a_1 x + a_2$. It follows that

$$D = (x_1 - x_2)^2 = (x_1 + x_2)^2 - 4x_1 x_2 = \sigma_1^2 - 4\sigma_2.$$

So the discriminant of f is $D(a_1, a_2) = (-a_1)^2 - 4a_2 = a_1^2 - 4a_2$.

If $n = 3$, then $f(x) = x^3 + a_1 x^2 + a_2 x + a_3$. Thus

$$D = (x_1 - x_2)^2 (x_1 - x_3)^2 (x_2 - x_3)^2$$

is a homogeneous symmetric polynomial of degree 6, whose leading term is $x_1^4 x_2^2$.

By the method of undetermined coefficients, the triples of natural numbers corresponding to all possible leading items and the associated powers of σ can be listed as follows:

Triples of natural numbers	Associated powers of σ
$(4, 2, 0)$	$\sigma_1^{4-2} \sigma_2^{2-0} \sigma_3^0 = \sigma_1^2 \sigma_2^2$
$(4, 1, 1)$	$\sigma_1^{4-1} \sigma_2^{1-1} \sigma_3^1 = \sigma_1^3 \sigma_3$
$(3, 3, 0)$	$\sigma_1^{3-3} \sigma_2^{3-0} \sigma_3^0 = \sigma_2^3$
$(3, 2, 1)$	$\sigma_1^{3-2} \sigma_2^{2-1} \sigma_3^1 = \sigma_1 \sigma_2 \sigma_3$
$(2, 2, 2)$	$\sigma_1^{2-2} \sigma_2^{2-2} \sigma_3^2 = \sigma_3^2$

Therefore, we may assume

$$D = \sigma_1^2 \sigma_2^2 + a\sigma_1^3 \sigma_3 + b\sigma_2^3 + c\sigma_1 \sigma_2 \sigma_3 + d\sigma_3^2.$$

Similar to Example 1.12.5, we have the following list by taking some special values of x_1, x_2, x_3:

x_1	x_2	x_3	σ_1	σ_2	σ_3	D
1	1	0	2	1	0	0
1	1	1	3	3	1	0
1	1	-1	1	-1	-1	0
2	-1	-1	0	-3	2	0

It follows that
$$\begin{cases} 4 + b = 0, \\ 81 + 27a + 27b + 9c + d = 0, \\ 1 - a - b + c + d = 0, \\ -27b + 4d = 0, \end{cases} \quad \text{and so} \quad \begin{cases} a = -4, \\ b = -4, \\ c = 18, \\ d = -27. \end{cases}$$

Thus $D = \sigma_1^2\sigma_2^2 - 4\sigma_1^3\sigma_3 - 4\sigma_2^3 + 18\sigma_1\sigma_2\sigma_3 - 27\sigma_3^2$.

Since $\sigma_1 = -a_1, \sigma_2 = a_2, \sigma_3 = -a_3$, the discriminant of f is

$$D = a_1^2 a_2^2 - 4a_1^3 a_3 - 4a_2^3 + 18a_1 a_2 a_3 - 27a_3^2.$$

In particular, the discriminant of $x^3 + px + q$ is

$$D = -4p^3 - 27q^2 = -4 \times 27 \left(\frac{q^2}{4} + \frac{p^3}{27} \right).$$

So $x^3 + px + q$ has multiple roots if and only if $\frac{q^2}{4} + \frac{p^3}{27} = 0$.

Exercises

1. Let $n \geqslant 1$ be an integer. Find a formula for $1 + \sum_{k=1}^{n} k! \cdot k$, and use induction to prove your formula is correct.
2. Prove that every positive integer n has a unique factorization $n = 2^s t$, where s is a natural number and t is an odd number.
3. Given integers m and n with $n > 0$. Use induction to prove that there exist integers q and r with $m = qn + r$ and $0 \leqslant r < n$.
4. Prove that $1 + \dfrac{1}{8} + \dfrac{1}{27} + \cdots + \dfrac{1}{n^3} < \dfrac{29}{24}$ for all integers $n \geqslant 1$.

5. Prove that $9|(7^n + 3n - 1)$ for all integers $n \geqslant 1$.

6. Prove that $8|(5^n + 2 \times 3^{n-1} + 1)$ for all integers $n \geqslant 1$.

7. Suppose $m, n \in \mathbb{Z}$ are not both 0. Use the well-ordering principle to prove that there are integers u and v such that $(m, n) = um + vn$.

8. Use the well-ordering principle to prove that every integer $n > 1$ is either a prime or a product of primes.

9. Let $m \geqslant 2$ be an integer. Prove that m is a prime if m cannot be divided by any prime p with $p \leqslant \sqrt{m}$.

10. Prove the six properties in Remark 1.3.2.

11. Let n be a positive integer. Find $(1 + \cos \alpha + i \sin \alpha)^n$.

12. Let n be a positive integer, and let $[r]$ denote the greatest integer less than or equal to a real number r. Find the following sums:

(1) $\displaystyle\sum_{k=0}^{[\frac{n}{2}]} (-1)^k C_n^{2k}$;

(2) $\displaystyle\sum_{k=0}^{[\frac{n-1}{2}]} (-1)^k C_n^{2k+1}$;

(3) $\displaystyle\sum_{k=0}^{n} (-1)^k C_n^k \cos(k+1)x$;

(4) $\displaystyle\sum_{k=0}^{n} (-1)^k C_n^k \sin(k+1)x$.

13. Prove that $P = \{a + b\sqrt[3]{2} + c\sqrt[3]{4} \mid a, b, c \in \mathbb{Q}\}$ is a number field.

14. Find the quotient and remainder upon dividing $f(x)$ by $g(x)$:

(1) $f(x) = 2x^5 - 5x^3 - 8x$, $g(x) = x + 3$;

(2) $f(x) = x^6 - 6x^4 + 12x^2 - 8$, $g(x) = x^2 - x + 2$.

15. Express $f(x)$ as the sum of powers of $x - x_0$, that is the sum of the form $c_0 + c_1(x - x_0) + c_2(x - x_0)^2 + \cdots$:

(1) $f(x) = x^5$, $x_0 = 2$;

(2) $f(x) = x^4 - 2x^2 + 3$, $x_0 = -2$.

16. Find the unknown coefficients in the following polynomials $f(x)$ and $g(x)$ such that $g(x)|f(x)$:

(1) $f(x) = x^4 + 3x^2 + ax + b$, $g(x) = x^2 - 2ax + 2$;

(2) $f(x) = x^3 + px + q$, $g(x) = x^2 + mx - 1$.

17. Find the greatest common divisors of the following groups of polynomials:

    (1) $f(x) = x^4 + x^3 - 3x^2 - 4x - 1, \ g(x) = x^3 + x^2 - x - 1;$
    (2) $f(x) = x^4 - 4x^3 + 1, \ g(x) = x^3 - 3x^2 + 1.$

18. Find polynomials $u(x)$ and $v(x)$ such that $u(x)f(x) + v(x)g(x) = (f(x), g(x))$:

    (1) $f(x) = x^4 - x^3 - 4x^2 + 4x + 1, \ g(x) = x^2 - x - 1;$
    (2) $f(x) = 3x^5 + 5x^4 - 16x^3 - 6x^2 - 5x - 6, \ g(x) = 3x^4 - 4x^3 - x^2 - x - 2;$
    (3) $f(x) = x^3, \ g(x) = (1 - x)^2.$

19. Find the unknown coefficients t and u such that the greatest common divisor of $f(x) = x^3 + (1 + t)x^2 + 2x + 2u$ and $g(x) = x^3 + tx + u$ is of degree 2.

20. Let $f(x), g(x), d(x) \in F[x]$. If $d(x)$ is a common divisor of $f(x)$ and $g(x)$, and $d(x)$ is a combination of $f(x)$ and $g(x)$, prove that $d(x)$ is a greatest common divisor of $f(x)$ and $g(x)$.

21. Let $f(x), g(x), h(x) \in F[x]$. If $f(x)$ and $g(x)$ are not both 0 and $h(x)$ is monic, prove that $(f(x)h(x), g(x)h(x)) = (f(x), g(x))h(x)$.

22. Let $f_1(x), f_2(x), \ldots, f_m(x), g_1(x), g_2(x), \ldots, g_n(x) \in F[x]$. If

$$(f_i(x), g_j(x)) = 1, i = 1, 2, \ldots, m; j = 1, 2, \ldots, n,$$

    prove that $(f_1(x)f_2(x) \cdots f_m(x), g_1(x)g_2(x) \cdots g_n(x)) = 1.$

23. Let $f(x), g(x) \in F[x]$. Prove that

$$(f(x), g(x)) = 1 \text{ if and only if } (f(x)g(x), f(x) + g(x)) = 1.$$

24. Let $f(x), g(x) \in F[x]$ and

$$\deg \left(\frac{f(x)}{(f(x), g(x))} \right) > 0, \deg \left(\frac{g(x)}{(f(x), g(x))} \right) > 0.$$

    Prove that there exist unique $u(x), v(x) \in F[x]$ such that

$$u(x)f(x) + v(x)g(x) = (f(x), g(x))$$

and

$$\deg(u(x)) < \deg\left(\frac{g(x)}{(f(x),g(x))}\right), \ \deg(v(x)) < \deg\left(\frac{f(x)}{(f(x),g(x))}\right).$$

25. Find the least common multiples of the following groups of polynomials:

    (1) $f(x) = x^4 - 4x^3 + 1$, $g(x) = x^3 - 3x^2 + 1$;
    (2) $f(x) = x^4 - x - 1 + i$, $g(x) = x^2 + 1$.

26. Let $f(x), g(x) \in F[x]$ and n a positive integer. If $f(x)$ and $g(x)$ are monic, prove that

$$(f^n(x), g^n(x)) = (f(x), g(x))^n, \ [f^n(x), g^n(x)] = [f(x), g(x)]^n.$$

27. Let $f_i(x) \in F[x], i = 1, 2, \ldots, n$. If $f_1(x), f_2(x), \ldots, f_n(x)$ are not all 0, prove that there are $u_i(x) \in F[x], i = 1, 2, \ldots, n$, such that

$$u_1(x)f_1(x) + u_2(x)f_2(x) + \cdots + u_n(x)f_n(x)$$
$$= (f_1(x), f_2(x), \ldots, f_n(x)).$$

28. Let $f(x) \in F[x]$ be a monic polynomial with $\deg(f(x)) \geqslant 1$. The following statements are equivalent:

    (1) $f(x)$ is a power of an irreducible polynomial.
    (2) for any $g(x) \in F[x]$, we always have $(f(x), g(x)) = 1$, or there is some positive integer m such that $f(x)|g^m(x)$.
    (3) for any $g(x), h(x) \in F[x]$, if $f(x)|g(x)h(x)$, then $f(x)|g(x)$, or there is some positive integer m such that $f(x)|h^m(x)$.

29. Determine whether the following polynomials have multiple factors:

    (1) $f(x) = x^5 - 5x^4 + 7x^3 - 2x^2 + 4x - 8$;
    (2) $f(x) = x^4 + 4x^2 - 4x - 3$.

30. Find multiple factors of the following polynomials:

    (1) $f(x) = x^3 - 7x^2 + 16x - 12$;
    (2) $f(x) = x^4 - 11x^2 + 18x - 8$.

31. Let $f(x) = x^3 + 2x^2 + 2x + 1$ and $g(x) = x^4 + x^3 + 2x^2 + x + 1$. Find the common roots of $f(x)$ and $g(x)$.

32. Find the unknown coefficient t such that $f(x) = x^3 - 3x^2 + tx - 1$ has multiple roots.

33. Find the necessary and sufficient condition for the polynomial $x^3 + px + q$ to have multiple roots.

34. Find A and B such that $(x-1)^2 | (Ax^4 + Bx^2 + 1)$.

35. Prove that $1 + \frac{x}{1} + \frac{x^2}{2!} + \cdots + \frac{x^n}{n!}$ has no multiple roots.

36. Let a be k-multiple root of $f'''(x)$. Prove that a is a $(k+3)$-multiple root of $g(x) = \frac{x-a}{2}(f'(x) + f'(a)) - f(x) + f(a)$.

37. Let $f(x) \in F[x]$ and n be a positive integer. If $(x-1)|f(x^n)$, prove that $(x^n - 1)|f(x^n)$.

38. Let $f_1(x), f_2(x) \in F[x]$ and $(x^2 + x + 1)|(f_1(x^3) + xf_2(x^3))$. Prove that

$$(x-1)|f_1(x), \quad (x-1)|f_2(x).$$

39.* For every positive integer n, prove that there is a polynomial $f_n(x)$ with integer coefficients such that $\cos(nx) = f_n(\cos x)$. If $\cos(nx)$ is replaced by $\sin(nx)$, is there a conclusion similar to the above for every positive integer n? Find all positive integers n such that $\sin(nx) = g_n(\sin x)$, where $g_n(x)$ is a polynomial over the integers.

40. Let n be a positive integer. Find the standard factorization of the polynomial $x^n - 1$ over the fields of complex and real numbers, respectively.

41. Let n be a positive integer. Prove that

(1) $\sin \frac{\pi}{2n} \sin \frac{2\pi}{2n} \cdots \sin \frac{(n-1)\pi}{2n} = \frac{\sqrt{n}}{2^{n-1}}$, where $n > 1$;

(2) $\cos \frac{\pi}{2n+1} \cos \frac{2\pi}{2n+1} \cdots \cos \frac{n\pi}{2n+1} = \frac{1}{2^n}$.

42. Let $f(x)$ be a nonzero polynomial and $n > 1$ an integer. If $f(x)|f(x^n)$, prove that the root of $f(x)$ can only be zero or a root of unity.

43. Let $a_1, a_2, \ldots, a_n$ be n numbers different from each other, and let

$$F(x) = (x - a_1)(x - a_2) \cdots (x - a_n).$$

Prove the following:

(1) $\displaystyle\sum_{i=1}^{n} \frac{F(x)}{(x-a_i)F'(a_i)} = 1$.

(2) For any polynomial $f(x)$, the remainder upon dividing $f(x)$ by $F(x)$ is $\displaystyle\sum_{i=1}^{n} \frac{f(a_i)F(x)}{(x-a_i)F'(a_i)}$.

44. Let $a_1, a_2, \ldots, a_n$ and $F(x)$ be the same as above, and let $b_1, b_2, \ldots, b_n$ be n numbers. Then

$$L(x) = \sum_{i=1}^{n} \frac{b_i F(x)}{(x - a_i) F'(a_i)}$$

satisfies $L(a_i) = b_i, i = 1, 2, \ldots, n$. $L(x)$ is often called the **Lagrange**[24] **interpolation formula**. Use the above formula to find the following:

(1) A polynomial $f(x)$ with $\deg(f(x)) < 4$ such that

$$f(2) = 3, \ f(3) = -1, \ f(4) = 0, \ f(5) = 2;$$

(2) a polynomial $f(x)$ with $\deg(f(x)) = 2$ such that $f(x)$ and $\sin x$ have the same value at $x = 0$, $\frac{\pi}{2}$, π;

(3) a polynomial $f(x)$ with the lowest degree such that

$$f(0) = 31, \ f(1) = 2, \ f(2) = 5, \ f(3) = 10.$$

45. Let $f(x)$ be a polynomial, all of whose coefficients are integers, and let $f(0)$ and $f(1)$ be odd. Prove that $f(x)$ has no integer roots.

46. Find all rational roots of the following polynomials and determine their multiplicities:

(1) $2x^4 - 7x^3 - 9x^2 + 3$;   (2) $x^3 - 6x^2 + 15x - 14$;
(3) $4x^4 - 7x^2 - 5x - 1$;   (4) $x^5 + x^4 - 6x^3 - 14x^2 - 11x - 3$.

47. Prove that the following polynomials are irreducible over the field of rational numbers:

(1) $x^4 - 8x^3 + 12x^2 + 2$;   (2) $x^4 - x^3 + 2x + 1$;
(3) $x^6 + x^3 + 1$;   (4) $x^4 - 10x^2 + 1$;
(5) $x^p + px + 1$, where p is an odd number;
(6) $x^4 + 4kx + 1$, where k is an integer.

48.[*] Let n be a positive integer, and let $a_1, a_2, \ldots, a_n$ be integers different from each other. Prove that the following polynomials are irreducible over the field of rational numbers:

(1) $(x - a_1)(x - a_2) \cdots (x - a_n) - 1$;
(2) $(x - a_1)^2 (x - a_2)^2 \cdots (x - a_n)^2 + 1$.

---

[24] Joseph-Louis Lagrange, 1736–1813, French mathematician.

49.* Let n be a positive integer, and let $a_1, a_2, \ldots, a_n$ be integers different from each other.

   (1) When $n \geqslant 5$, prove that $(x - a_1)(x - a_2) \cdots (x - a_n) + 1$ is irreducible over the field of rational numbers.

   (2) When $2 \leqslant n < 5$, is the conclusion in (1) true?

50. Use Descartes' rule of signs to find the number of real roots of the following equations:

   (1) $x^6 + x^4 - x^3 - 2x - 1 = 0$;

   (2) $x^4 - x^2 + x - 2 = 0$;

   (3) $5x^4 - 4x^3 + 3x^2 - 2x + 1 = 0$.

51. Let n be a positive integer. Find the number of real roots of $f(x) = nx^n - x^{n-1} - x^{n-2} - \cdots - x - 1$.

52. Let a be a nonzero real number. Prove that the polynomial

$$x^n + ax^{n-1} + a^2 x^{n-2} + \cdots + a^{n-1}x + a^n$$

has at most one real root.

53.* Let $f(x) \in \mathbb{R}[x]$ with $\deg(f(x)) \geqslant 1$. If all roots of $f(x)$ are real numbers, prove that all roots of the polynomial $\lambda f(x) + f'(x)$ are real numbers, where λ is a real number.

54. Find the number of real roots of the following polynomials:

   (1) $x^3 + 3x - 1$;

   (2) $x^3 + 3x^2 - 1$;

   (3) $x^3 + px + q$, where p and q are real numbers.

55. Express the following symmetric polynomials by polynomial expressions in terms of elementary symmetric polynomials:

   (1) $x_1^2 x_2 + x_1 x_2^2 + x_1^2 x_3 + x_1 x_3^2 + x_2^2 x_3 + x_2 x_3^2$;

   (2) $(x_1 + x_2)(x_1 + x_3)(x_2 + x_3)$;

   (3) $\displaystyle\sum_{1 \leqslant i < j \leqslant n} x_i^2 x_j^2$.

56. Let $f(x) = x^3 + a_1 x^2 + a_2 x + a_3$. Prove that the three roots of $f(x)$ form an arithmetic sequence if and only if $2a_1^3 - 9a_1 a_2 + 27a_3 = 0$.

57. Let $x_1, x_2, \ldots, x_n$ be n roots of the equation

$$x^n + a_1 x^{n-1} + \cdots + a_{n-1}x + a_n = 0.$$

Prove that every symmetric polynomial in $x_2, \ldots, x_n$ can be expressed as a polynomial in $x_1, a_1, a_2, \ldots, a_{n-1}$.

58. Let $f(x) = (x - x_1)(x - x_2) \cdots (x - x_n) = x^n - \sigma_1 x^{n-1} + \cdots + (-1)^n \sigma_n$, $s_k = x_1^k + x_2^k + \cdots + x_n^k$, $k = 0, 1, 2, \ldots$.

(1) Prove that

$$x^{k+1} f'(x) = (s_0 x^k + s_1 x^{k-1} + \cdots + s_{k-1} x + s_k) f(x) + g(x),$$

where $g(x) = 0$ or $\deg(g(x)) < n$.

(2) Use the equality in (1) to prove **Newton's**[25] **identities**

$$s_k - \sigma_1 s_{k-1} + \sigma_2 s_{k-2} + \cdots + (-1)^{k-1} \sigma_{k-1} s_1 + (-1)^k k \sigma_k = 0,$$

$$\text{where } 1 \leqslant k \leqslant n;$$

$$s_k - \sigma_1 s_{k-1} + \sigma_2 s_{k-2} + \cdots + (-1)^n \sigma_n s_{k-n} = 0,$$

$$\text{where } k > n \geqslant 1.$$

(3) Use Newton's identities to express s_2, s_3, s_4, s_5, s_6 by polynomial expressions in terms of elementary symmetric polynomials.

---

[25]Isaac Newton, 1643–1727, English mathematician and physicist.

Chapter 2

Determinants and Matrices

2.1 Introduction

Determinants and matrices were first developed, along with finding solutions to systems of linear equations. In the ancient Chinese mathematical work "The Nine Chapters on the Mathematical Art", there already appeared examples in which the coefficients of a system of linear equations were expressed in array form to solve the equations. This array can be regarded as the embryonic form of matrices. Matrices formally emerged as a mathematical research object with the development of determinants. Logically speaking, the concept of matrices is before that of determinants, but the actual history is just the opposite. The introduction of determinants can be traced back to the 17th century. Seki[1] and Leibniz[2] almost put forward the concept of determinants at the same time. In the late 17th century, Seki and Leibniz used determinants to determine the number and form of solutions of a system of linear equations. After the 18th century, determinants were studied as an independent mathematical concept. In the 19th century, the study of determinants further developed, and the concept of matrices emerged. In 1812, Cauchy[3] first used the word "determinant" in its modern sense. He was also the first

---

[1]Takakazu Seki, 1642–1708, Japanese mathematician.
[2]Gottfried Wilhelm Leibniz, 1646–1716, German mathematician.
[3]Augustin Louis Cauchy, 1789–1857, French mathematician.

mathematician to arrange a determinant into a square matrix and represent its elements with double subscripts. In 1841, Cayley[4] published the first English contribution to the theory of determinants. In this paper, he used two vertical lines on either side of the array to denote the determinant, a notation which has now become standard.

Modern determinants were first introduced to China at the end of the 19th century. In comparison, the concept of matrices was first seen in Chinese in 1922. The Chinese Mathematical Society reviewed the translation of various terms and formally defined the translation of "determinant" as "Hanglieshi (in Chinese)" and translated "matrix" as "Juzhen (in Chinese)" for the first time in 1935.

This chapter first introduces the determinant theory, then deals with the operations, the rank, and the elementary operations of matrices, and so on.

2.2　Definition of Determinants

The concept of a determinant originates from the problem of solving systems of linear equations.

Let's consider the system of binary linear equations

$$\begin{cases} a_{11}x_1 + a_{12}x_2 = b_1, \\ a_{21}x_1 + a_{22}x_2 = b_2. \end{cases}$$

By the method of elimination, if $a_{11}a_{22} - a_{12}a_{21} \neq 0$, then the system has a unique solution:

$$x_1 = \frac{b_1 a_{22} - a_{12} b_2}{a_{11} a_{22} - a_{12} a_{21}}, \quad x_2 = \frac{a_{11} b_2 - b_1 a_{21}}{a_{11} a_{22} - a_{12} a_{21}}.$$

For the convenience of memory, we introduce the symbol $\begin{vmatrix} a_{11} & a_{12} \\ a_{21} & a_{22} \end{vmatrix}$, which is called a **second-order determinant**, or a **determinant of order 2**, its value is defined as $a_{11}a_{22} - a_{12}a_{21}$, i.e.,

$$\begin{vmatrix} a_{11} & a_{12} \\ a_{21} & a_{22} \end{vmatrix} = a_{11}a_{22} - a_{12}a_{21}.$$

---

[4]Arthur Cayley, 1821–1895, English mathematician.

Thus, the solution above can be described in terms of second-order determinants as follows:

If $\begin{vmatrix} a_{11} & a_{12} \\ a_{21} & a_{22} \end{vmatrix} \neq 0$, then the system has a unique solution:

$$x_1 = \frac{\begin{vmatrix} b_1 & a_{12} \\ b_2 & a_{22} \end{vmatrix}}{\begin{vmatrix} a_{11} & a_{12} \\ a_{21} & a_{22} \end{vmatrix}}, \quad x_2 = \frac{\begin{vmatrix} a_{11} & b_1 \\ a_{21} & b_2 \end{vmatrix}}{\begin{vmatrix} a_{11} & a_{12} \\ a_{21} & a_{22} \end{vmatrix}}.$$

There are similar results for ternary linear equations. Given a system of ternary linear equations

$$\begin{cases} a_{11}x_1 + a_{12}x_2 + a_{13}x_3 = b_1, \\ a_{21}x_1 + a_{22}x_2 + a_{23}x_3 = b_2, \\ a_{31}x_1 + a_{32}x_2 + a_{33}x_3 = b_3. \end{cases}$$

The symbol $\begin{vmatrix} a_{11} & a_{12} & a_{13} \\ a_{21} & a_{22} & a_{23} \\ a_{31} & a_{32} & a_{33} \end{vmatrix}$ is called a **third-order determinant**, or a **determinant of order 3**, its value is defined as

$$a_{11}a_{22}a_{33} + a_{12}a_{23}a_{31} + a_{13}a_{21}a_{32} - a_{11}a_{23}a_{32} - a_{12}a_{21}a_{33} - a_{13}a_{22}a_{31},$$

i.e., $\begin{vmatrix} a_{11} & a_{12} & a_{13} \\ a_{21} & a_{22} & a_{23} \\ a_{31} & a_{32} & a_{33} \end{vmatrix}$

$$= a_{11}a_{22}a_{33} + a_{12}a_{23}a_{31} + a_{13}a_{21}a_{32} - a_{11}a_{23}a_{32} - a_{12}a_{21}a_{33} - a_{13}a_{22}a_{31}.$$

If $d = \begin{vmatrix} a_{11} & a_{12} & a_{13} \\ a_{21} & a_{22} & a_{23} \\ a_{31} & a_{32} & a_{33} \end{vmatrix} \neq 0$, then the system above has a unique solution:

$$x_1 = \frac{d_1}{d}, \quad x_2 = \frac{d_2}{d}, \quad x_3 = \frac{d_3}{d},$$

where $d_1 = \begin{vmatrix} b_1 & a_{12} & a_{13} \\ b_2 & a_{22} & a_{23} \\ b_3 & a_{32} & a_{33} \end{vmatrix}$, $d_2 = \begin{vmatrix} a_{11} & b_1 & a_{13} \\ a_{21} & b_2 & a_{23} \\ a_{31} & b_3 & a_{33} \end{vmatrix}$, $d_3 = \begin{vmatrix} a_{11} & a_{12} & b_1 \\ a_{21} & a_{22} & b_2 \\ a_{31} & a_{32} & b_3 \end{vmatrix}$.

Is there a similar conclusion for a system of linear equations in n unknowns (see Section 3.2 for a general definition)? To answer this question, let's introduce the concept of an nth-order determinant.

We first deal with permutations to define n^{th}-order determinants.

Definition 2.2.1. Let n be a positive integer.

(1) An ordered arrangement $i_1 i_2 \cdots i_n$ of n elements consisting of $1, 2, \ldots, n$ without repetitions is called a **permutation of** $1, 2, \ldots, n$ (or, simply, a **permutation**). We denote the set of all permutations by S_n.

(2) Given a permutation $i_1 i_2 \cdots i_p \cdots i_q \cdots i_n$. If $i_p > i_q$, but $1 \leqslant p < q \leqslant n$, then $i_p i_q$ is called an **inversion**. The total number of inversions of a permutation $i_1 i_2 \cdots i_n$ is called the **inversion number** of $i_1 i_2 \cdots i_n$, denoted by $\tau(i_1 i_2 \cdots i_n)$.

(3) A permutation $i_1 i_2 \cdots i_n$ is said to be **even** if $\tau(i_1 i_2 \cdots i_n)$ is even, or **odd** if $\tau(i_1 i_2 \cdots i_n)$ is odd.

(4) Let $i_1 i_2 \cdots i_k \cdots i_l \cdots i_n$ be a permutation. We get another permutation by interchanging i_k and i_l and keeping all others fixed. Such a transformation is called a **transposition**, denoted by (k, l).

Example 2.2.2. Let $n \geqslant 2$ be an integer. Then
$$\tau(n(n-1)\cdots 21) = (n-1) + (n-2) + \cdots + 2 + 1 = \frac{n(n-1)}{2};$$
$$\tau(135\cdots(2n-1)246\cdots(2n)) = 0+1+2+\cdots+(n-1) = \frac{n(n-1)}{2}.$$

Theorem 2.2.3.

(1) *Every transposition changes the parity of a permutation.*

(2) *Any permutation $i_1 i_2 \cdots i_n$ can be changed to the **natural permutation** $12\cdots n$ by a series of transpositions and vice versa; moreover, the number of transpositions used above has the same parity as the permutation.*

(3) *When $1 \leqslant k \neq l \leqslant n$, the transposition $(k, l) : S_n \to S_n$ is a bijection.*

Proof. (1) First, consider the case where the two interchanged numbers are in adjacent positions. Let $\alpha = i_1 \cdots i_k i_{k+1} \cdots i_n$ be a permutation. Interchanging i_k and i_{k+1} yields the permutation

$\beta = i_1 \cdots i_{k+1} i_k \cdots i_n$. By definition, $\tau(\beta) = \tau(\alpha) \pm 1$. So the parities of α and β are different.

Next, we consider the general situation. Let $\alpha = \cdots j i_1 i_2 \cdots i_s k \cdots$ be a permutation. By interchanging j and k, we get the permutation $\beta = \cdots k i_1 i_2 \cdots i_s j \cdots$.

As shown in the following, β can be achieved by an odd number of successive interchanges of adjacent numbers. Starting from α, we move k to the left by interchanging adjacent numbers until k precedes j. So we get the permutation $\gamma = \cdots k j i_1 i_2 \cdots i_s \cdots$ after $s+1$ steps. Then, starting from γ, we move j to the right by interchanging adjacent numbers until where k was; this requires s steps. Thus, the total number of adjacent interchanges required is $(s+1)+s = 2s+1$, which is always odd. Since a transposition of two adjacent numbers changes the parity of a permutation, it follows that α and β have different parities.

(2) comes from (1).

(3) Let $1_{S_n} : S_n \to S_n$ denote the identity may. Then $(k,l)(k,l) = 1_{S_n}$ by definition. Suppose $\alpha, \beta \in S_n$ and $(k,l)(\alpha) = (k,l)(\beta)$, then

$$\alpha = 1_{S_n}(\alpha) = (k,l)(k,l)(\alpha) = (k,l)(k,l)(\beta) = 1_{S_n}(\beta) = \beta.$$

Thus, (k,l) is an injection. Clearly, $(k,l) : S_n \to S_n$ is a surjection. So $(k,l) : S_n \to S_n$ is a bijection. $\qquad\qquad\square$

Let's introduce the nth-order determinants.

Definition 2.2.4. Let n be a positive integer. For any given n^2 numbers or polynomials a_{ij}, $i, j = 1, 2, \ldots, n$, the square array with vertical lines on either side

$$\begin{vmatrix} a_{11} & a_{12} & \cdots & a_{1n} \\ a_{21} & a_{22} & \cdots & a_{2n} \\ \vdots & \vdots & & \vdots \\ a_{n1} & a_{n2} & \cdots & a_{nn} \end{vmatrix}$$

is called an nth-**order determinant**, or a **determinant of order** n, often denoted by $|a_{ij}|_n$, and its value is defined as the expansion

$$\sum_{j_1 j_2 \cdots j_n} (-1)^{\tau(j_1 j_2 \cdots j_n)} a_{1j_1} a_{2j_2} \cdots a_{nj_n},$$

i.e.,
$$
\begin{vmatrix}
a_{11} & a_{12} & \cdots & a_{1n} \\
a_{21} & a_{22} & \cdots & a_{2n} \\
\vdots & \vdots & & \vdots \\
a_{n1} & a_{n2} & \cdots & a_{nn}
\end{vmatrix}
= \sum_{j_1 j_2 \cdots j_n} (-1)^{\tau(j_1 j_2 \cdots j_n)} a_{1j_1} a_{2j_2} \cdots a_{nj_n}, \text{ where}
$$

the summation $\displaystyle\sum_{j_1 j_2 \cdots j_n}$ is over all permutations $j_1 j_2 \cdots j_n$ of $1, 2, \ldots, n$.

Remark 2.2.5.

(1) The first-order determinant $|a_{11}|_1 = a_{11}$. Let $D = |a_{ij}|_n$. We refer to a_{ij} as the (i, j) **entry** (entry in the ith row and jth column) of D. The index i is called the **row index**, and j is the **column index** of a_{ij}.

(2) By definition, an nth-order determinant is the sum of $n!$ terms, in which each term (with its appropriate sign) is the product of n entries, with exactly one entry from each row and exactly one entry from each column. Therefore, an nth-order determinant is the algebraic sum of the products of n entries from different rows and columns.

(3) If the entries of a determinant are all numbers (polynomials), then its value is a number (polynomial).

Example 2.2.6.

(1) Evaluate the **anti-diagonal determinant**
$$
\begin{vmatrix}
0 & \cdots & \cdots & 0 & a_1 \\
\vdots & & & a_2 & 0 \\
\vdots & & \cdots & \cdots & \vdots \\
0 & a_{n-1} & & & \vdots \\
a_n & 0 & \cdots & \cdots & 0
\end{vmatrix}.
$$

Solution. Let $D =
\begin{vmatrix}
0 & \cdots & \cdots & 0 & a_1 \\
\vdots & & & a_2 & 0 \\
\vdots & & \cdots & \cdots & \vdots \\
0 & a_{n-1} & & & \vdots \\
a_n & 0 & \cdots & \cdots & 0
\end{vmatrix}.$

By definition, $D = \sum\limits_{j_1 j_2 \cdots j_n} (-1)^{\tau(j_1 j_2 \cdots j_n)} a_{1j_1} a_{2j_2} \cdots a_{nj_n}$. To compute D, it is enough to consider the terms $(-1)^{\tau(j_1 j_2 \cdots j_n)} a_{1j_1} a_{2j_2} \cdots a_{nj_n}$ with $j_1 = n$ in the expansion of D; similarly, it suffices to consider the terms $(-1)^{\tau(j_1 j_2 \cdots j_n)} a_{1j_1} a_{2j_2} \cdots a_{nj_n}$ with $j_2 = n - 1, \ldots, j_n = 1$. That is to say, in the expansion of D, except the term

$$(-1)^{\tau(n(n-1)\cdots 21)} a_{1n} a_{2,n-1} \cdots a_{n-1,2} a_{n1}$$

$$= (-1)^{\tau(n(n-1)\cdots 21)} a_1 a_2 \cdots a_{n-1} a_n,$$

the remaining terms are all zero. So $D = (-1)^{\frac{n(n-1)}{2}} a_1 a_2 \cdots a_n.$ $\qquad\square$

Similarly, we have

$$(2) \quad \begin{vmatrix} 0 & \cdots & \cdots & 0 & a_{1n} \\ \vdots & & \cdot^{\cdot^{\cdot}} & a_{2,n-1} & a_{2n} \\ \vdots & \cdot^{\cdot^{\cdot}} & \cdot^{\cdot^{\cdot}} & \vdots & \vdots \\ 0 & a_{n-1,2} & \cdots & a_{n-1,n-1} & a_{n-1,n} \\ a_{n1} & a_{n2} & \cdots & a_{n,n-1} & a_{nn} \end{vmatrix} = (-1)^{\frac{n(n-1)}{2}} a_{1n} a_{2,n-1} \cdots a_{n1}.$$

(3) The **lower triangular determinant**

$$\begin{vmatrix} a_{11} & 0 & \cdots & \cdots & 0 \\ a_{21} & a_{22} & \cdot^{\cdot^{\cdot}} & & \vdots \\ \vdots & \vdots & \ddots & \ddots & \vdots \\ a_{n-1,1} & a_{n-1,2} & \cdots & a_{n-1,n-1} & 0 \\ a_{n1} & a_{n2} & \cdots & a_{n,n-1} & a_{nn} \end{vmatrix} = a_{11} a_{22} \cdots a_{nn}.$$

(4) The **upper triangular determinant**

$$\begin{vmatrix} a_{11} & a_{12} & \cdots & a_{1,n-1} & a_{1n} \\ 0 & a_{22} & \cdots & a_{2,n-1} & a_{2n} \\ \vdots & \ddots & \ddots & \vdots & \vdots \\ \vdots & & \ddots & a_{n-1,n-1} & a_{n-1,n} \\ 0 & \cdots & \cdots & 0 & a_{nn} \end{vmatrix} = a_{11} a_{22} \cdots a_{nn}.$$

(5) The **diagonal determinant**

$$\begin{vmatrix} a_{11} & 0 & \cdots & & \cdots & 0 \\ 0 & a_{22} & \ddots & & & \vdots \\ \vdots & \ddots & \ddots & & \ddots & \vdots \\ \vdots & & & \ddots & a_{n-1,n-1} & 0 \\ 0 & \cdots & \cdots & & 0 & a_{nn} \end{vmatrix} = a_{11}a_{22}\cdots a_{nn}.$$

Example 2.2.7. Prove that

$$\begin{vmatrix} a_{11} & a_{12} & \cdots & a_{1,n-1} & a_{1n} \\ a_{21} & a_{22} & \cdots & a_{2,n-1} & a_{2n} \\ \vdots & \vdots & & \vdots & \vdots \\ a_{n-1,1} & a_{n-1,2} & \cdots & a_{n-1,n-1} & a_{n-1,n} \\ 0 & 0 & \cdots & 0 & 1 \end{vmatrix} = \begin{vmatrix} a_{11} & a_{12} & \cdots & a_{1,n-1} \\ a_{21} & a_{22} & \cdots & a_{2,n-1} \\ \vdots & \vdots & & \vdots \\ a_{n-1,1} & a_{n-1,2} & \cdots & a_{n-1,n-1} \end{vmatrix}.$$

Proof.

$$\begin{vmatrix} a_{11} & a_{12} & \cdots & a_{1,n-1} & a_{1n} \\ a_{21} & a_{22} & \cdots & a_{2,n-1} & a_{2n} \\ \vdots & \vdots & & \vdots & \vdots \\ a_{n-1,1} & a_{n-1,2} & \cdots & a_{n-1,n-1} & a_{n-1,n} \\ 0 & 0 & \cdots & 0 & 1 \end{vmatrix}$$

$$= \sum_{j_1 j_2 \cdots j_{n-1} j_n} (-1)^{\tau(j_1 j_2 \cdots j_{n-1} j_n)} a_{1j_1} a_{2j_2} \cdots a_{n-1,j_{n-1}} a_{nj_n}$$

$$= \sum_{j_1 j_2 \cdots j_{n-1} n} (-1)^{\tau(j_1 j_2 \cdots j_{n-1} n)} a_{1j_1} a_{2j_2} \cdots a_{n-1,j_{n-1}}$$

$$= \sum_{j_1 j_2 \cdots j_{n-1}} (-1)^{\tau(j_1 j_2 \cdots j_{n-1})} a_{1j_1} a_{2j_2} \cdots a_{n-1,j_{n-1}}$$

$$= \begin{vmatrix} a_{11} & a_{12} & \cdots & a_{1,n-1} \\ a_{21} & a_{22} & \cdots & a_{2,n-1} \\ \vdots & \vdots & & \vdots \\ a_{n-1,1} & a_{n-1,2} & \cdots & a_{n-1,n-1} \end{vmatrix}. \qquad \square$$

2.3 Properties of Determinants

The calculation of determinants is an important problem and is also a challenging problem. When n is large, it is almost impossible to calculate an nth-order determinant by definition. Therefore, we must discuss the properties of determinants. These properties can simplify the calculation of determinants.

In each term $(-1)^{\tau(j_1 j_2 \cdots j_n)} a_{1j_1} a_{2j_2} \cdots a_{nj_n}$ in the expansion of a determinant $|a_{ij}|_n$, the row indices form the natural permutation $1, 2, \ldots, n$. In fact, the row indices can be any permutation of $1, 2, \ldots, n$ as shown in the following.

Property 1. Let $l_1 l_2 \cdots l_n$ be a permutation of $1, 2, \ldots, n$. Then

$$\begin{vmatrix} a_{11} & a_{12} & \cdots & a_{1n} \\ a_{21} & a_{22} & \cdots & a_{2n} \\ \vdots & \vdots & & \vdots \\ a_{n1} & a_{n2} & \cdots & a_{nn} \end{vmatrix} = \sum_{k_1 k_2 \cdots k_n} (-1)^{\tau(l_1 l_2 \cdots l_n) + \tau(k_1 k_2 \cdots k_n)} a_{l_1 k_1} a_{l_2 k_2} \cdots a_{l_n k_n}.$$

Proof. By definition,

$$\begin{vmatrix} a_{11} & a_{12} & \cdots & a_{1n} \\ a_{21} & a_{22} & \cdots & a_{2n} \\ \vdots & \vdots & & \vdots \\ a_{n1} & a_{n2} & \cdots & a_{nn} \end{vmatrix} = \sum_{j_1 j_2 \cdots j_n} (-1)^{\tau(j_1 j_2 \cdots j_n)} a_{1j_1} a_{2j_2} \cdots a_{nj_n}.$$

In the product $a_{1j_1} a_{2j_2} \cdots a_{nj_n}$, after interchanging the positions of two elements s times, we have the product $a_{l_1 k_1} a_{l_2 k_2} \cdots a_{l_n k_n}$, then

$$a_{1j_1} a_{2j_2} \cdots a_{nj_n} = a_{l_1 k_1} a_{l_2 k_2} \cdots a_{l_n k_n},$$

where $l_1 l_2 \cdots l_n$ is obtained from $12 \cdots n$ after s transpositions, and $k_1 k_2 \cdots k_n$ is obtained from $j_1 j_2 \cdots j_n$ after the corresponding s transpositions. Thus

$$(-1)^{\tau(l_1 l_2 \cdots l_n)} = (-1)^s, \quad (-1)^{\tau(k_1 k_2 \cdots k_n)} = (-1)^{\tau(j_1 j_2 \cdots j_n)}(-1)^s,$$

and hence

$$(-1)^{\tau(j_1 j_2 \cdots j_n)} a_{1j_1} a_{2j_2} \cdots a_{nj_n}$$

$$= (-1)^{\tau(l_1 l_2 \cdots l_n) + \tau(k_1 k_2 \cdots k_n)} a_{l_1 k_1} a_{l_2 k_2} \cdots a_{l_n k_n}.$$

By Theorem 2.2.3 (3), when $j_1 j_2 \cdots j_n$ ranges over all permutations of $1, 2, \ldots, n$, $k_1 k_2 \cdots k_n$ also ranges over all permutations of $1, 2, \ldots, n$. So

$$\begin{vmatrix} a_{11} & a_{12} & \cdots & a_{1n} \\ a_{21} & a_{22} & \cdots & a_{2n} \\ \vdots & \vdots & & \vdots \\ a_{n1} & a_{n2} & \cdots & a_{nn} \end{vmatrix} = \sum_{k_1 k_2 \cdots k_n} (-1)^{\tau(l_1 l_2 \cdots l_n) + \tau(k_1 k_2 \cdots k_n)} a_{l_1 k_1} a_{l_2 k_2} \cdots a_{l_n k_n}.$$

$\square$

Similarly, we get the following.

Property 2. Let $k_1 k_2 \cdots k_n$ be a permutation of $1, 2, \ldots, n$. Then

$$\begin{vmatrix} a_{11} & a_{12} & \cdots & a_{1n} \\ a_{21} & a_{22} & \cdots & a_{2n} \\ \vdots & \vdots & & \vdots \\ a_{n1} & a_{n2} & \cdots & a_{nn} \end{vmatrix} = \sum_{l_1 l_2 \cdots l_n} (-1)^{\tau(l_1 l_2 \cdots l_n) + \tau(k_1 k_2 \cdots k_n)} a_{l_1 k_1} a_{l_2 k_2} \cdots a_{l_n k_n}.$$

In particular,

$$\begin{vmatrix} a_{11} & a_{12} & \cdots & a_{1n} \\ a_{21} & a_{22} & \cdots & a_{2n} \\ \vdots & \vdots & & \vdots \\ a_{n1} & a_{n2} & \cdots & a_{nn} \end{vmatrix} = \sum_{i_1 i_2 \cdots i_n} (-1)^{\tau(i_1 i_2 \cdots i_n)} a_{i_1 1} a_{i_2 2} \cdots a_{i_n n}.$$

Property 3. The determinant remains unchanged if we interchange the rows to columns and columns to rows, i.e.,

$$\begin{vmatrix} a_{11} & a_{12} & \cdots & a_{1n} \\ a_{21} & a_{22} & \cdots & a_{2n} \\ \vdots & \vdots & & \vdots \\ a_{n1} & a_{n2} & \cdots & a_{nn} \end{vmatrix} = \begin{vmatrix} a_{11} & a_{21} & \cdots & a_{n1} \\ a_{12} & a_{22} & \cdots & a_{n2} \\ \vdots & \vdots & & \vdots \\ a_{1n} & a_{2n} & \cdots & a_{nn} \end{vmatrix}.$$

In the above equality, the determinant on the right is called the **transpose** of the determinant on the left.

Proof. Let $b_{ij} = a_{ji}, i, j = 1, 2, \ldots, n$. Then

$$
\begin{vmatrix} a_{11} & a_{21} & \cdots & a_{n1} \\ a_{12} & a_{22} & \cdots & a_{n2} \\ \vdots & \vdots & & \vdots \\ a_{1n} & a_{2n} & \cdots & a_{nn} \end{vmatrix} = \begin{vmatrix} b_{11} & b_{12} & \cdots & b_{1n} \\ b_{21} & b_{22} & \cdots & b_{2n} \\ \vdots & \vdots & & \vdots \\ b_{n1} & b_{n2} & \cdots & b_{nn} \end{vmatrix}
$$

$$
= \sum_{i_1 i_2 \cdots i_n} (-1)^{\tau(i_1 i_2 \cdots i_n)} b_{i_1 1} b_{i_2 2} \cdots b_{i_n n}
$$

$$
= \sum_{i_1 i_2 \cdots i_n} (-1)^{\tau(i_1 i_2 \cdots i_n)} a_{1 i_1} a_{2 i_2} \cdots a_{n i_n}
$$

$$
= \begin{vmatrix} a_{11} & a_{12} & \cdots & a_{1n} \\ a_{21} & a_{22} & \cdots & a_{2n} \\ \vdots & \vdots & & \vdots \\ a_{n1} & a_{n2} & \cdots & a_{nn} \end{vmatrix}. \qquad \square
$$

From Properties 1, 2, and 3, we can see that, in a determinant, not only the positions of rows (columns) are equal but also the positions of rows and columns are equal. Therefore, any property valid for rows is also valid for columns.

Property 4. The determinant can be expanded along any row, i.e.,

$$
\begin{vmatrix} a_{11} & a_{12} & \cdots & a_{1n} \\ \vdots & \vdots & & \vdots \\ a_{i1} & a_{i2} & \cdots & a_{in} \\ \vdots & \vdots & & \vdots \\ a_{n1} & a_{n2} & \cdots & a_{nn} \end{vmatrix} = a_{i1} A_{i1} + a_{i2} A_{i2} + \cdots + a_{in} A_{in} \text{ for } 1 \leqslant i \leqslant n,
$$

where A_{ij} is the sum of all the remaining parts after taking out the common factor a_{ij} from the items containing a_{ij} in the expansion of $|a_{ij}|_n$, $j = 1, 2, \ldots, n$. Note that $A_{i1}, A_{i2}, \ldots, A_{in}$ are independent of the n entries of the ith row.

Proof. By definition, each term in the expansion of $|a_{ij}|_n$ contains exactly one entry from the n entries $a_{i1}, a_{i2}, \ldots, a_{in}$ of the ith row, where $1 \leqslant i \leqslant n$.

Divide the $n!$ terms in the expansion of $|a_{ij}|_n$ into n groups: The first group consists of items containing a_{i1}, the second group consists of items containing $a_{i2}, \ldots$, the nth group consists of items containing a_{in}.

Let A_{ij} be the sum of all the remaining parts after taking out the common factor a_{ij} from the items in the jth group, where $1 \leqslant j \leqslant n$. Then, we get the required equality. It is clear that $A_{i1}, A_{i2}, \ldots, A_{in}$ are independent of the n entries of the ith row. $\qquad\square$

Property 5. If all the entries of any row are multiplied by a scalar k, then the value of the determinant is also multiplied by k, i.e.,

$$
\begin{vmatrix}
a_{11} & a_{12} & \cdots & a_{1n} \\
\vdots & \vdots & & \vdots \\
ka_{i1} & ka_{i2} & \cdots & ka_{in} \\
\vdots & \vdots & & \vdots \\
a_{n1} & a_{n2} & \cdots & a_{nn}
\end{vmatrix}
= k
\begin{vmatrix}
a_{11} & a_{12} & \cdots & a_{1n} \\
\vdots & \vdots & & \vdots \\
a_{i1} & a_{i2} & \cdots & a_{in} \\
\vdots & \vdots & & \vdots \\
a_{n1} & a_{n2} & \cdots & a_{nn}
\end{vmatrix}
\quad \text{for } 1 \leqslant i \leqslant n.
$$

Proof. By Property 4,
$$
\begin{vmatrix}
a_{11} & a_{12} & \cdots & a_{1n} \\
\vdots & \vdots & & \vdots \\
ka_{i1} & ka_{i2} & \cdots & ka_{in} \\
\vdots & \vdots & & \vdots \\
a_{n1} & a_{n2} & \cdots & a_{nn}
\end{vmatrix}
$$

$$
= (ka_{i1})A_{i1} + (ka_{i2})A_{i2} + \cdots + (ka_{in})A_{in}
$$

$$
= k(a_{i1}A_{i1} + a_{i2}A_{i2} + \cdots + a_{in}A_{in})
$$

$$
= k
\begin{vmatrix}
a_{11} & a_{12} & \cdots & a_{1n} \\
\vdots & \vdots & & \vdots \\
a_{i1} & a_{i2} & \cdots & a_{in} \\
\vdots & \vdots & & \vdots \\
a_{n1} & a_{n2} & \cdots & a_{nn}
\end{vmatrix}.
\qquad\square
$$

Property 6. If all entries of any row are expressed as the sum of two terms, then the determinant can be expressed as the sum of two determinants, i.e.,

$$
\begin{vmatrix}
a_{11} & a_{12} & \cdots & a_{1n} \\
\vdots & \vdots & & \vdots \\
b_1 + c_1 & b_2 + c_2 & \cdots & b_n + c_n \\
\vdots & \vdots & & \vdots \\
a_{n1} & a_{n2} & \cdots & a_{nn}
\end{vmatrix} \leftarrow i\text{th row}
$$

$$
= \begin{vmatrix}
a_{11} & a_{12} & \cdots & a_{1n} \\
\vdots & \vdots & & \vdots \\
b_1 & b_2 & \cdots & b_n \\
\vdots & \vdots & & \vdots \\
a_{n1} & a_{n2} & \cdots & a_{nn}
\end{vmatrix}
+ \begin{vmatrix}
a_{11} & a_{12} & \cdots & a_{1n} \\
\vdots & \vdots & & \vdots \\
c_1 & c_2 & \cdots & c_n \\
\vdots & \vdots & & \vdots \\
a_{n1} & a_{n2} & \cdots & a_{nn}
\end{vmatrix}
\quad \text{for } 1 \leqslant i \leqslant n.
$$

Proof. By Property 4,

$$
\begin{vmatrix}
a_{11} & a_{12} & \cdots & a_{1n} \\
\vdots & \vdots & & \vdots \\
b_1 + c_1 & b_2 + c_2 & \cdots & b_n + c_n \\
\vdots & \vdots & & \vdots \\
a_{n1} & a_{n2} & \cdots & a_{nn}
\end{vmatrix}
$$

$$
= (b_1 + c_1)A_{i1} + (b_2 + c_2)A_{i2} + \cdots + (b_n + c_n)A_{in}
$$

$$
= (b_1 A_{i1} + b_2 A_{i2} + \cdots + b_n A_{in}) + (c_1 A_{i1} + c_2 A_{i2} + \cdots + c_n A_{in})
$$

$$
= \begin{vmatrix}
a_{11} & a_{12} & \cdots & a_{1n} \\
\vdots & \vdots & & \vdots \\
b_1 & b_2 & \cdots & b_n \\
\vdots & \vdots & & \vdots \\
a_{n1} & a_{n2} & \cdots & a_{nn}
\end{vmatrix}
+ \begin{vmatrix}
a_{11} & a_{12} & \cdots & a_{1n} \\
\vdots & \vdots & & \vdots \\
c_1 & c_2 & \cdots & c_n \\
\vdots & \vdots & & \vdots \\
a_{n1} & a_{n2} & \cdots & a_{nn}
\end{vmatrix}. \qquad \square
$$

Property 7. If any two rows are identical to each other, then the determinant is zero.

Proof. Suppose the ith row and the kth row of the determinant D are identical to each other, i.e.,

$$D = \begin{vmatrix} a_{11} & a_{12} & \cdots & a_{1n} \\ \vdots & \vdots & & \vdots \\ a_{i1} & a_{i2} & \cdots & a_{in} \\ \vdots & \vdots & & \vdots \\ a_{k1} & a_{k2} & \cdots & a_{kn} \\ \vdots & \vdots & & \vdots \\ a_{n1} & a_{n2} & \cdots & a_{nn} \end{vmatrix}, \text{ where } n \geqslant 2, \ a_{ij} = a_{kj}, j = 1, 2, \ldots, n, \ i \neq k.$$

By definition,

$$D = \sum_{j_1 j_2 \cdots j_n} (-1)^{\tau(j_1 j_2 \cdots j_i \cdots j_k \cdots j_n)} a_{1j_1} a_{2j_2} \cdots a_{ij_i} \cdots a_{kj_k} \cdots a_{nj_n}.$$

Suppose

$$(-1)^{\tau(j_1 j_2 \cdots j_i \cdots j_k \cdots j_n)} a_{1j_1} a_{2j_2} \cdots a_{ij_i} \cdots a_{kj_k} \cdots a_{nj_n}$$

is a term in the expansion of D, then

$$(-1)^{\tau(j_1 j_2 \cdots j_k \cdots j_i \cdots j_n)} a_{1j_1} a_{2j_2} \cdots a_{ij_k} \cdots a_{kj_i} \cdots a_{nj_n}$$

is also a term in the expansion of D. Since $a_{ij_i} = a_{kj_i}$ and $a_{kj_k} = a_{ij_k}$, the sum of the two terms above is 0. It follows that $D = 0$. $\square$

From Properties 5 and 7, we have the following.

Property 8. The determinant is zero if any two rows are proportional.

From Properties 6 and 8, we have the following.

Property 9. If a multiple of one row is added to another row, then the determinant is not changed, i.e.,

$$\begin{vmatrix} a_{11} & a_{12} & \cdots & a_{1n} \\ \vdots & \vdots & & \vdots \\ a_{i1} & a_{i2} & \cdots & a_{in} \\ \vdots & \vdots & & \vdots \\ a_{k1}+ca_{i1} & a_{k2}+ca_{i2} & \cdots & a_{kn}+ca_{in} \\ \vdots & \vdots & & \vdots \\ a_{n1} & a_{n2} & \cdots & a_{nn} \end{vmatrix} = \begin{vmatrix} a_{11} & a_{12} & \cdots & a_{1n} \\ \vdots & \vdots & & \vdots \\ a_{i1} & a_{i2} & \cdots & a_{in} \\ \vdots & \vdots & & \vdots \\ a_{k1} & a_{k2} & \cdots & a_{kn} \\ \vdots & \vdots & & \vdots \\ a_{n1} & a_{n2} & \cdots & a_{nn} \end{vmatrix},$$

where $n \geqslant 2, i \neq k, 1 \leqslant i, k \leqslant n$, c is a scalar.

Property 10. If two rows are interchanged, then the determinant is multiplied by -1, i.e.,

$$\begin{vmatrix} a_{11} & a_{12} & \cdots & a_{1n} \\ \vdots & \vdots & & \vdots \\ a_{i1} & a_{i2} & \cdots & a_{in} \\ \vdots & \vdots & & \vdots \\ a_{k1} & a_{k2} & \cdots & a_{kn} \\ \vdots & \vdots & & \vdots \\ a_{n1} & a_{n2} & \cdots & a_{nn} \end{vmatrix} = - \begin{vmatrix} a_{11} & a_{12} & \cdots & a_{1n} \\ \vdots & \vdots & & \vdots \\ a_{k1} & a_{k2} & \cdots & a_{kn} \\ \vdots & \vdots & & \vdots \\ a_{i1} & a_{i2} & \cdots & a_{in} \\ \vdots & \vdots & & \vdots \\ a_{n1} & a_{n2} & \cdots & a_{nn} \end{vmatrix}, \text{ where } n \geqslant 2, i \neq k.$$

Proof.

$$\begin{vmatrix} a_{11} & a_{12} & \cdots & a_{1n} \\ \vdots & \vdots & & \vdots \\ a_{i1} & a_{i2} & \cdots & a_{in} \\ \vdots & \vdots & & \vdots \\ a_{k1} & a_{k2} & \cdots & a_{kn} \\ \vdots & \vdots & & \vdots \\ a_{n1} & a_{n2} & \cdots & a_{nn} \end{vmatrix}$$

$$= \begin{vmatrix} a_{11} & a_{12} & \cdots & a_{1n} \\ \vdots & \vdots & & \vdots \\ a_{i1} & a_{i2} & \cdots & a_{in} \\ \vdots & \vdots & & \vdots \\ a_{k1}+a_{i1} & a_{k2}+a_{i2} & \cdots & a_{kn}+a_{in} \\ \vdots & \vdots & & \vdots \\ a_{n1} & a_{n2} & \cdots & a_{nn} \end{vmatrix}$$

$$= \begin{vmatrix} a_{11} & a_{12} & \cdots & a_{1n} \\ \vdots & \vdots & & \vdots \\ -a_{k1} & -a_{k2} & \cdots & -a_{kn} \\ \vdots & \vdots & & \vdots \\ a_{k1}+a_{i1} & a_{k2}+a_{i2} & \cdots & a_{kn}+a_{in} \\ \vdots & \vdots & & \vdots \\ a_{n1} & a_{n2} & \cdots & a_{nn} \end{vmatrix}$$

$$
= \begin{vmatrix} a_{11} & a_{12} & \cdots & a_{1n} \\ \vdots & \vdots & & \vdots \\ -a_{k1} & -a_{k2} & \cdots & -a_{kn} \\ \vdots & \vdots & & \vdots \\ a_{i1} & a_{i2} & \cdots & a_{in} \\ \vdots & \vdots & & \vdots \\ a_{n1} & a_{n2} & \cdots & a_{nn} \end{vmatrix} = - \begin{vmatrix} a_{11} & a_{12} & \cdots & a_{1n} \\ \vdots & \vdots & & \vdots \\ a_{k1} & a_{k2} & \cdots & a_{kn} \\ \vdots & \vdots & & \vdots \\ a_{i1} & a_{i2} & \cdots & a_{in} \\ \vdots & \vdots & & \vdots \\ a_{n1} & a_{n2} & \cdots & a_{nn} \end{vmatrix}. \qquad \square
$$

Example 2.3.1. Evaluate the determinant

$$
D = \begin{vmatrix} a & b & b & \cdots & \cdots & \cdots & b \\ b & a & b & & & & \vdots \\ b & b & a & \ddots & & & \vdots \\ \vdots & & \ddots & \ddots & \ddots & & \vdots \\ \vdots & & & \ddots & a & b & b \\ \vdots & & & & b & a & b \\ b & \cdots & \cdots & \cdots & b & b & a \end{vmatrix}.
$$

Solution. When $n = 1$, $D = a$. In the following, we assume that $n > 1$.

Adding the ith column to the first column for $i = 2, 3, \ldots, n$, we get

$$
D = \begin{vmatrix} a+(n-1)b & b & b & \cdots & \cdots & \cdots & b \\ a+(n-1)b & a & b & & & & \vdots \\ a+(n-1)b & b & a & \ddots & & & \vdots \\ \vdots & & \vdots & \ddots & \ddots & \ddots & \vdots \\ \vdots & & \vdots & & \ddots & a & b & b \\ \vdots & & \vdots & & & b & a & b \\ a+(n-1)b & b & \cdots & \cdots & & b & b & a \end{vmatrix}.
$$

Take $a + (n-1)b$ from the first column of the determinant above to obtain

$$D = (a + (n-1)b) \begin{vmatrix} 1 & b & b & \cdots & \cdots & \cdots & b \\ 1 & a & b & & & & \vdots \\ 1 & b & a & \ddots & & & \vdots \\ \vdots & \vdots & \ddots & \ddots & \ddots & & \vdots \\ \vdots & \vdots & & \ddots & a & b & b \\ \vdots & \vdots & & & b & a & b \\ 1 & b & \cdots & \cdots & b & b & a \end{vmatrix}.$$

By adding -1 times the first row to the ith row for $i = 2, 3, \ldots, n$, we have

$$D = (a + (n-1)b) \begin{vmatrix} 1 & b & b & \cdots & \cdots & \cdots & b \\ 0 & a-b & 0 & \cdots & \cdots & \cdots & 0 \\ 0 & 0 & a-b & \ddots & & & \vdots \\ \vdots & \vdots & \ddots & \ddots & \ddots & & \vdots \\ \vdots & \vdots & & \ddots & a-b & 0 & 0 \\ \vdots & \vdots & & & 0 & a-b & 0 \\ 0 & 0 & \cdots & \cdots & 0 & 0 & a-b \end{vmatrix}$$

$$= (a + (n-1)b)(a-b)^{n-1}.$$

When $n = 1$, the above equality still holds. So

$$D = (a + (n-1)b)(a-b)^{n-1}. \qquad \square$$

Example 2.3.2. Let $D = |a_{ij}|_n$. If $a_{ij} = -a_{ji}, i, j = 1, 2, \ldots, n$, then D is called a **skew-symmetric determinant**. Prove that the value of a skew-symmetric determinant of odd order is 0.

Solution. Let n be an odd number, and let $D = |a_{ij}|_n$ be a skew-symmetric determinant. Then

$$D = \begin{vmatrix} 0 & a_{12} & a_{13} & \cdots & a_{1n} \\ -a_{12} & 0 & a_{23} & \cdots & a_{2n} \\ -a_{13} & -a_{23} & 0 & \cdots & a_{3n} \\ \vdots & \vdots & \vdots & & \vdots \\ -a_{1n} & -a_{2n} & -a_{3n} & \cdots & 0 \end{vmatrix}$$

$$= (-1)^n \begin{vmatrix} 0 & -a_{12} & -a_{13} & \cdots & -a_{1n} \\ a_{12} & 0 & -a_{23} & \cdots & -a_{2n} \\ a_{13} & a_{23} & 0 & \cdots & -a_{3n} \\ \vdots & \vdots & \vdots & & \vdots \\ a_{1n} & a_{2n} & a_{3n} & \cdots & 0 \end{vmatrix}$$

$$= (-1)^n \begin{vmatrix} 0 & a_{12} & a_{13} & \cdots & a_{1n} \\ -a_{12} & 0 & a_{23} & \cdots & a_{2n} \\ -a_{13} & -a_{23} & 0 & \cdots & a_{3n} \\ \vdots & \vdots & \vdots & & \vdots \\ -a_{1n} & -a_{2n} & -a_{3n} & \cdots & 0 \end{vmatrix} = -D. \text{ So } D = 0. \qquad \square$$

Example 2.3.3. Let $n \geqslant 2$. Find $\displaystyle\sum_{j_1 j_2 \cdots j_n} \begin{vmatrix} a_{1j_1} & a_{1j_2} & \cdots & a_{1j_n} \\ a_{2j_1} & a_{2j_2} & \cdots & a_{2j_n} \\ \vdots & \vdots & & \vdots \\ a_{nj_1} & a_{nj_2} & \cdots & a_{nj_n} \end{vmatrix}$, where

$\displaystyle\sum_{j_1 j_2 \cdots j_n}$ is over all permutations $j_1 j_2 \cdots j_n$ of $1, 2, \ldots, n$.

Solution. According to Theorem 2.2.3 (2), any permutation $j_1 j_2 \cdots j_n$ can be changed to the natural permutation $12 \cdots n$ by a series of transpositions and vice versa; moreover, the number of transpositions has the same parity as the permutation. Thus, by Property 10, we have

$$\sum_{j_1 j_2 \cdots j_n} \begin{vmatrix} a_{1j_1} & a_{1j_2} & \cdots & a_{1j_n} \\ a_{2j_1} & a_{2j_2} & \cdots & a_{2j_n} \\ \vdots & \vdots & & \vdots \\ a_{nj_1} & a_{nj_2} & \cdots & a_{nj_n} \end{vmatrix}$$

$$= \sum_{j_1 j_2 \cdots j_n} (-1)^{\tau(j_1 j_2 \cdots j_n)} \begin{vmatrix} a_{11} & a_{12} & \cdots & a_{1n} \\ a_{21} & a_{22} & \cdots & a_{2n} \\ \vdots & \vdots & & \vdots \\ a_{n1} & a_{n2} & \cdots & a_{nn} \end{vmatrix} = 0. \qquad \square$$

2.4 Cofactor Expansions of Determinants

By Property 4 in the last section, we have

$$
\begin{vmatrix}
a_{11} & a_{12} & \cdots & a_{1n} \\
\vdots & \vdots & & \vdots \\
a_{i1} & a_{i2} & \cdots & a_{in} \\
\vdots & \vdots & & \vdots \\
a_{n1} & a_{n2} & \cdots & a_{nn}
\end{vmatrix}
= a_{i1}A_{i1} + a_{i2}A_{i2} + \cdots + a_{in}A_{in} \text{ for } 1 \leqslant i \leqslant n,
$$

where $A_{i1}, A_{i2}, \ldots, A_{in}$ are independent of the n entries of the ith row.

What exactly are $A_{i1}, A_{i2}, \ldots, A_{in}$?

Definition 2.4.1. Let $D = \begin{vmatrix} a_{11} & \cdots & a_{1j} & \cdots & a_{1n} \\ \vdots & & \vdots & & \vdots \\ a_{i1} & \cdots & a_{ij} & \cdots & a_{in} \\ \vdots & & \vdots & & \vdots \\ a_{n1} & \cdots & a_{nj} & \cdots & a_{nn} \end{vmatrix}$.

The $(n-1)$th-order determinant

$$
\begin{vmatrix}
a_{11} & \cdots & a_{1,j-1} & a_{1,j+1} & \cdots & a_{1n} \\
\vdots & & \vdots & \vdots & & \vdots \\
a_{i-1,1} & \cdots & a_{i-1,j-1} & a_{i-1,j+1} & \cdots & a_{i-1,n} \\
a_{i+1,1} & \cdots & a_{i+1,j-1} & a_{i+1,j+1} & \cdots & a_{i+1,n} \\
\vdots & & \vdots & \vdots & & \vdots \\
a_{n1} & \cdots & a_{n,j-1} & a_{n,j+1} & \cdots & a_{nn}
\end{vmatrix}
$$

obtained by deleting the ith row and jth column of D is called the **minor of** a_{ij}, denoted by M_{ij}, $i, j = 1, 2, \ldots, n$.

The following theorem tells us the exact meaning of A_{ij}, $j = 1, 2, \ldots, n$.

Theorem 2.4.2. $A_{ij} = (-1)^{i+j} M_{ij}$, $i, j = 1, 2, \ldots, n$.

Proof. First, we have

$$\begin{vmatrix} a_{11} & \cdots & a_{1j} & \cdots & a_{1n} \\ \vdots & & \vdots & & \vdots \\ a_{i1} & \cdots & a_{ij} & \cdots & a_{in} \\ \vdots & & \vdots & & \vdots \\ a_{n1} & \cdots & a_{nj} & \cdots & a_{nn} \end{vmatrix} = a_{i1}A_{i1} + \cdots + a_{ij}A_{ij} + \cdots + a_{in}A_{in}.$$

Taking $a_{ij} = 1$ and $a_{ik} = 0$ for $1 \leqslant k \leqslant n$ with $k \neq j$ in the equality above, we get

$$A_{ij} = \begin{vmatrix} a_{11} & \cdots & a_{1,j-1} & a_{1j} & a_{1,j+1} & \cdots & a_{1n} \\ \vdots & & \vdots & \vdots & \vdots & & \vdots \\ a_{i-1,1} & \cdots & a_{i-1,j-1} & a_{i-1,j} & a_{i-1,j+1} & \cdots & a_{i-1,n} \\ 0 & \cdots & 0 & 1 & 0 & \cdots & 0 \\ a_{i+1,1} & \cdots & a_{i+1,j-1} & a_{i+1,j} & a_{i+1,j+1} & \cdots & a_{i+1,n} \\ \vdots & & \vdots & \vdots & \vdots & & \vdots \\ a_{n1} & \cdots & a_{n,j-1} & a_{nj} & a_{n,j+1} & \cdots & a_{nn} \end{vmatrix}$$

$$= (-1)^{n-i} \begin{vmatrix} a_{11} & \cdots & a_{1,j-1} & a_{1j} & a_{1,j+1} & \cdots & a_{1n} \\ \vdots & & \vdots & \vdots & \vdots & & \vdots \\ a_{i-1,1} & \cdots & a_{i-1,j-1} & a_{i-1,j} & a_{i-1,j+1} & \cdots & a_{i-1,n} \\ a_{i+1,1} & \cdots & a_{i+1,j-1} & a_{i+1,j} & a_{i+1,j+1} & \cdots & a_{i+1,n} \\ \vdots & & \vdots & \vdots & \vdots & & \vdots \\ a_{n1} & \cdots & a_{n,j-1} & a_{nj} & a_{n,j+1} & \cdots & a_{nn} \\ 0 & \cdots & 0 & 1 & 0 & \cdots & 0 \end{vmatrix}$$

$$= (-1)^{n-i+n-j} \begin{vmatrix} a_{11} & \cdots & a_{1,j-1} & a_{1,j+1} & \cdots & a_{1n} & a_{1j} \\ \vdots & & \vdots & \vdots & & \vdots & \vdots \\ a_{i-1,1} & \cdots & a_{i-1,j-1} & a_{i-1,j+1} & \cdots & a_{i-1,n} & a_{i-1,j} \\ a_{i+1,1} & \cdots & a_{i+1,j-1} & a_{i+1,j+1} & \cdots & a_{i+1,n} & a_{i+1,j} \\ \vdots & & \vdots & \vdots & & \vdots & \vdots \\ a_{n1} & \cdots & a_{n,j-1} & a_{n,j+1} & \cdots & a_{nn} & a_{nj} \\ 0 & \cdots & 0 & 0 & \cdots & 0 & 1 \end{vmatrix}$$

$$\underset{\text{Example 2.2.7}}{=\!=\!=\!=\!=} (-1)^{i+j} M_{ij}. \qquad \qquad \square$$

Definition 2.4.3. Let $D = |a_{ij}|_n$. We usually call $A_{ij} = (-1)^{i+j} M_{ij}$ the **cofactor of** a_{ij} of D, $i, j = 1, 2, \ldots, n$.

Theorem 2.4.4. *Let $D = |a_{ij}|_n$. Then*

(1) *the sum of the products of the entries of any row and their cofactors is D, i.e.,*

$$a_{i1}A_{i1} + a_{i2}A_{i2} + \cdots + a_{in}A_{in} = D \text{ for } 1 \leqslant i \leqslant n;$$

(2) *the sum of the products of the entries of any row and the corresponding cofactors of the entries of another row is 0, i.e.,*

$$a_{k1}A_{i1} + a_{k2}A_{i2} + \cdots + a_{kn}A_{in} = 0 \text{ for } k \neq i \text{ with } 1 \leqslant k, i \leqslant n.$$

Proof. (1) follows from Property 4 in the last section.

(2) By hypothesis,
$$D = \begin{vmatrix} a_{11} & a_{12} & \cdots & a_{1n} \\ \vdots & \vdots & & \vdots \\ a_{i1} & a_{i2} & \cdots & a_{in} \\ \vdots & \vdots & & \vdots \\ a_{k1} & a_{k2} & \cdots & a_{kn} \\ \vdots & \vdots & & \vdots \\ a_{n1} & a_{n2} & \cdots & a_{nn} \end{vmatrix}, \text{ where } k \neq i.$$

Replacing the ith row of D with the kth row of D and leaving the other rows of D unchanged gives rise to the determinant

$$d = \begin{vmatrix} a_{11} & a_{12} & \cdots & a_{1n} \\ \vdots & \vdots & & \vdots \\ a_{k1} & a_{k2} & \cdots & a_{kn} \\ \vdots & \vdots & & \vdots \\ a_{k1} & a_{k2} & \cdots & a_{kn} \\ \vdots & \vdots & & \vdots \\ a_{n1} & a_{n2} & \cdots & a_{nn} \end{vmatrix}, \text{ then } d = 0.$$

Expanding d along the ith row, we have

$$a_{k1}A_{i1} + a_{k2}A_{i2} + \cdots + a_{kn}A_{in} = 0 \text{ for } k \neq i \text{ with } 1 \leqslant k, i \leqslant n. \quad \square$$

The above conclusion can be abbreviated as follows:

$$\sum_{s=1}^{n} a_{ks} A_{is} = D\delta_{ik} = \begin{cases} D, & \text{if } k = i, \\ 0, & \text{if } k \neq i. \end{cases}$$

By Property 3 in the last section, we get

$$\sum_{s=1}^{n} a_{sk} A_{si} = D\delta_{ik} = \begin{cases} D, & \text{if } k = i, \\ 0, & \text{if } k \neq i, \end{cases}$$

where $\delta_{ik} = \begin{cases} 1, \text{if } k = i, \\ 0, \text{if } k \neq i \end{cases}$ is the **Kronecker**[5] **delta**.

Remark 2.4.5.

(1) The above formulas are very important in theory; however, they are usually impractical in specific computations of determinants.

(2) Let $n = 3$ and $\alpha_i = (a_{i1}, a_{i2}, a_{i3})$, $a_{ij} \in \mathbb{R}$, $i, j = 1, 2, 3$. Then

$$\begin{vmatrix} a_{11} & a_{12} & a_{13} \\ a_{21} & a_{22} & a_{23} \\ a_{31} & a_{32} & a_{33} \end{vmatrix} = a_{11}A_{11} + a_{12}A_{12} + a_{13}A_{13} = \alpha_1 \cdot (\alpha_2 \times \alpha_3)$$

is the signed volume of a parallelepiped generated by α_1, α_2, and α_3, where $\alpha_1 \cdot (\alpha_2 \times \alpha_3)$ denotes the **mixed product** of the three vectors α_1, α_2, and α_3.

(3) Let $n = 2$ and $\alpha_i = (a_{i1}, a_{i2})$, $a_{ij} \in \mathbb{R}$, $i, j = 1, 2$. Then

$$\begin{vmatrix} a_{11} & a_{12} \\ a_{21} & a_{22} \end{vmatrix} = a_{11}a_{22} - a_{12}a_{21}$$

is the signed area of a parallelogram generated by α_1 and α_2.

Example 2.4.6. Let $n \geqslant 2$ be an integer. Prove that the **Vandermonde**[6] **determinant**

$$\begin{vmatrix} 1 & 1 & 1 & \cdots & 1 \\ a_1 & a_2 & a_3 & \cdots & a_n \\ a_1^2 & a_2^2 & a_3^2 & \cdots & a_n^2 \\ \vdots & \vdots & \vdots & & \vdots \\ a_1^{n-1} & a_2^{n-1} & a_3^{n-1} & \cdots & a_n^{n-1} \end{vmatrix} = \prod_{1 \leqslant j < i \leqslant n} (a_i - a_j).$$

---

[5] Leopold Kronecker, 1823–1891, German mathematician.

[6] Alexandre-Théophile Vandermonde, 1735–1796, French mathematician.

Proof. Let $V_n = \begin{vmatrix} 1 & 1 & 1 & \cdots & 1 \\ a_1 & a_2 & a_3 & \cdots & a_n \\ a_1^2 & a_2^2 & a_3^2 & \cdots & a_n^2 \\ \vdots & \vdots & \vdots & & \vdots \\ a_1^{n-1} & a_2^{n-1} & a_3^{n-1} & \cdots & a_n^{n-1} \end{vmatrix}$. We prove by induction on n.

When $n = 2$, $V_2 = \begin{vmatrix} 1 & 1 \\ a_1 & a_2 \end{vmatrix} = a_2 - a_1$, as desired.

For any integer $n > 2$, we assume that the result is true for a Vandermonde determinant of order $n - 1$. Now consider the case when the order of the Vandermonde determinant is n.

In V_n, subtracting a_1 times row $i - 1$ from row i for $i = n$, $n - 1, \ldots, 2$, we have

$$
V_n = \begin{vmatrix}
1 & 1 & 1 & \cdots & 1 \\
0 & a_2 - a_1 & a_3 - a_1 & \cdots & a_n - a_1 \\
0 & a_2^2 - a_1 a_2 & a_3^2 - a_1 a_3 & \cdots & a_n^2 - a_1 a_n \\
\vdots & \vdots & \vdots & & \vdots \\
0 & a_2^{n-1} - a_1 a_2^{n-2} & a_3^{n-1} - a_1 a_3^{n-2} & \cdots & a_n^{n-1} - a_1 a_n^{n-2}
\end{vmatrix}
$$

$$
= \begin{vmatrix}
a_2 - a_1 & a_3 - a_1 & \cdots & a_n - a_1 \\
a_2^2 - a_1 a_2 & a_3^2 - a_1 a_3 & \cdots & a_n^2 - a_1 a_n \\
\vdots & \vdots & & \vdots \\
a_2^{n-1} - a_1 a_2^{n-2} & a_3^{n-1} - a_1 a_3^{n-2} & \cdots & a_n^{n-1} - a_1 a_n^{n-2}
\end{vmatrix}
$$

$$
= (a_2 - a_1)(a_3 - a_1) \cdots (a_n - a_1) \begin{vmatrix}
1 & 1 & \cdots & 1 \\
a_2 & a_3 & \cdots & a_n \\
a_2^2 & a_3^2 & \cdots & a_n^2 \\
\vdots & \vdots & & \vdots \\
a_2^{n-2} & a_3^{n-2} & \cdots & a_n^{n-2}
\end{vmatrix}
$$

$$
= (a_2 - a_1)(a_3 - a_1) \cdots (a_n - a_1) \prod_{2 \leqslant j < i \leqslant n} (a_i - a_j)
$$

$$
= \prod_{1 \leqslant j < i \leqslant n} (a_i - a_j). \qquad \square
$$

Example 2.4.7. Prove that

$$
\begin{vmatrix}
a_{11} & \cdots & a_{1k} & 0 & \cdots & 0 \\
\vdots & & \vdots & \vdots & & \vdots \\
a_{k1} & \cdots & a_{kk} & 0 & \cdots & 0 \\
c_{11} & \cdots & c_{1k} & b_{11} & \cdots & b_{1r} \\
\vdots & & \vdots & \vdots & & \vdots \\
c_{r1} & \cdots & c_{rk} & b_{r1} & \cdots & b_{rr}
\end{vmatrix}
=
\begin{vmatrix}
a_{11} & \cdots & a_{1k} \\
\vdots & & \vdots \\
a_{k1} & \cdots & a_{kk}
\end{vmatrix}
\begin{vmatrix}
b_{11} & \cdots & b_{1r} \\
\vdots & & \vdots \\
b_{r1} & \cdots & b_{rr}
\end{vmatrix} .
$$

Proof.   The proof is by induction on k.

When $k = 1$, the conclusion is true by expanding the determinant on the left side along the first row.

For any integer $m > 1$, assume that the conclusion is true for $k = m - 1$. Now we consider the case for $k = m$.

Expanding the determinant on the left side along the first row, we have

$$
\begin{vmatrix}
a_{11} & \cdots & a_{1m} & 0 & \cdots & 0 \\
\vdots & & \vdots & \vdots & & \vdots \\
a_{m1} & \cdots & a_{mm} & 0 & \cdots & 0 \\
c_{11} & \cdots & c_{1m} & b_{11} & \cdots & b_{1r} \\
\vdots & & \vdots & \vdots & & \vdots \\
c_{r1} & \cdots & c_{rm} & b_{r1} & \cdots & b_{rr}
\end{vmatrix}
$$

$$
= a_{11}
\begin{vmatrix}
a_{22} & \cdots & a_{2m} & 0 & \cdots & 0 \\
\vdots & & \vdots & \vdots & & \vdots \\
a_{m2} & \cdots & a_{mm} & 0 & \cdots & 0 \\
c_{12} & \cdots & c_{1m} & b_{11} & \cdots & b_{1r} \\
\vdots & & \vdots & \vdots & & \vdots \\
c_{r2} & \cdots & c_{rm} & b_{r1} & \cdots & b_{rr}
\end{vmatrix}
+ \cdots
$$

$$+ (-1)^{1+i} a_{1i}
\begin{vmatrix}
a_{21} & \cdots & a_{2,i-1} & a_{2,i+1} & \cdots & a_{2m} & 0 & \cdots & 0 \\
\vdots & & \vdots & \vdots & & \vdots & \vdots & & \vdots \\
a_{m1} & \cdots & a_{m,i-1} & a_{m,i+1} & \cdots & a_{mm} & 0 & \cdots & 0 \\
c_{11} & \cdots & c_{1,i-1} & c_{1,i+1} & \cdots & c_{1m} & b_{11} & \cdots & b_{1r} \\
\vdots & & \vdots & \vdots & & \vdots & \vdots & & \vdots \\
c_{r1} & \cdots & c_{r,i-1} & c_{r,i+1} & \cdots & c_{rm} & b_{r1} & \cdots & b_{rr}
\end{vmatrix}$$

$$+ \cdots + (-1)^{1+m} a_{1m}
\begin{vmatrix}
a_{21} & \cdots & a_{2,m-1} & 0 & \cdots & 0 \\
\vdots & & \vdots & \vdots & & \vdots \\
a_{m1} & \cdots & a_{m,m-1} & 0 & \cdots & 0 \\
c_{11} & \cdots & c_{1,m-1} & b_{11} & \cdots & b_{1r} \\
\vdots & & \vdots & \vdots & & \vdots \\
c_{r1} & \cdots & c_{r,m-1} & b_{r1} & \cdots & b_{rr}
\end{vmatrix}$$

$$= \left(a_{11}
\begin{vmatrix}
a_{22} & \cdots & a_{2m} \\
\vdots & & \vdots \\
a_{m2} & \cdots & a_{mm}
\end{vmatrix} + \cdots \right.$$

$$+ (-1)^{1+i} a_{1i}
\begin{vmatrix}
a_{21} & \cdots & a_{2,i-1} & a_{2,i+1} & \cdots & a_{2m} \\
\vdots & & \vdots & \vdots & & \vdots \\
a_{m1} & \cdots & a_{m,i-1} & a_{m,i+1} & \cdots & a_{mm}
\end{vmatrix} + \cdots$$

$$\left. + (-1)^{1+m} a_{1m}
\begin{vmatrix}
a_{21} & \cdots & a_{2,m-1} \\
\vdots & & \vdots \\
a_{m1} & \cdots & a_{m,m-1}
\end{vmatrix} \right)
\begin{vmatrix}
b_{11} & \cdots & b_{1r} \\
\vdots & & \vdots \\
b_{r1} & \cdots & b_{rr}
\end{vmatrix}$$

$$=
\begin{vmatrix}
a_{11} & \cdots & a_{1m} \\
\vdots & & \vdots \\
a_{m1} & \cdots & a_{mm}
\end{vmatrix}
\begin{vmatrix}
b_{11} & \cdots & b_{1r} \\
\vdots & & \vdots \\
b_{r1} & \cdots & b_{rr}
\end{vmatrix}.$$

By the first form of induction, the equality to be proved is true.

$\square$

According to Theorem 2.4.4, determinants can be expanded along any row (column). The **Laplace**[7] **expansion** introduced in the following shows that determinants can be expanded along any k rows (columns).

First, we generalize the concepts of minors and cofactors.

Definition 2.4.8. Let $D = |a_{ij}|_n$ and $1 \leqslant k \leqslant n$. Choose any k rows and k columns of D. The kth-order determinant M obtained by keeping the entries in the intersections of the chosen k rows and k columns is called a **minor of order** k of D, and the $(n-k)$th-order determinant M' obtained from D by deleting the chosen k rows and k columns is called the **complementary minor** of the minor M. If the row indices and column indices of the minor M of order k are respectively $i_1, i_2, \ldots, i_k$ and $j_1, j_2, \ldots, j_k$, then $(-1)^{i_1+i_2+\cdots+i_k+j_1+j_2+\cdots+j_k} M'$ is called the **cofactor** of the minor M.

Remark 2.4.9. In Definition 2.4.8,

(1) when $k = n$, there is only one minor of order k, it is D itself;
(2) when $k = 1$, D has n^2 minors of order 1, they are n^2 entries of D, so each entry of D is a minor of order 1;
(3) if $M = a_{ij}$, a minor of order 1, then the complementary minor M' of M is exactly the minor M_{ij} of a_{ij} in Definition 2.4.1. Clearly, the complementary minor of M_{ij} is a_{ij}.

Theorem 2.4.10 (Laplace expansion). *Let D be a determinant of order n and $1 \leqslant k \leqslant n-1$, and choose any k rows of D. Then the sum of the products of the minors of order k obtained from the chosen k rows and their cofactors is equal to D, i.e.,*

$$D = M_1 A_1 + M_2 A_2 + \cdots + M_t A_t, \tag{2.1}$$

where $t = C_n^k$, $M_1, M_2, \ldots, M_t$ are all the minors of order k obtained from the chosen k rows, and $A_1, A_2, \ldots, A_t$ are the cofactors of $M_1, M_2, \ldots, M_t$, respectively.

---

[7]Pierre-Simon Laplace, 1749–1827, French mathematician.

Proof. By definition, $M_i A_i$ and $M_j A_j$ have no common terms when $i \neq j$. Since every M_i contains $k!$ terms, and every A_i contains $(n-k)!$ terms, the right side of (2.1) has $k!(n-k)!t = n!$ terms. Note that there are $n!$ terms in the expansion of D by definition.

We show that every term in $M_i A_i$ on the right side of (2.1) is a term in the expansion of D, so the equality (2.1) is true.

First, we let M be a minor of order k located at the upper left corner of D, i.e., $M = \begin{vmatrix} a_{11} & a_{12} & \cdots & a_{1k} \\ a_{21} & a_{22} & \cdots & a_{2k} \\ \vdots & \vdots & & \vdots \\ a_{k1} & a_{k2} & \cdots & a_{kk} \end{vmatrix}$, then its complementary

minor $M' = \begin{vmatrix} a_{k+1,k+1} & a_{k+1,k+2} & \cdots & a_{k+1,n} \\ a_{k+2,k+1} & a_{k+2,k+2} & \cdots & a_{k+2,n} \\ \vdots & \vdots & & \vdots \\ a_{n,k+1} & a_{n,k+2} & \cdots & a_{nn} \end{vmatrix}$. In this case, the cofactor

of M is $A = (-1)^{(1+2+\cdots+k)+(1+2+\cdots+k)} M' = M'$, and hence $MA = MM'$.

By definition, each term m in the expansion of M can be written as

$$m = (-1)^{\tau(\alpha_1 \alpha_2 \cdots \alpha_k)} a_{1\alpha_1} a_{2\alpha_2} \cdots a_{k\alpha_k},$$

where $\alpha_1 \alpha_2 \cdots \alpha_k$ is a permutation of $1, 2, \ldots, k$.

Let $b_{ij} = a_{k+i,k+j}$, $i, j = 1, 2, \ldots, n-k$. Then

$$M' = \begin{vmatrix} b_{11} & b_{12} & \cdots & b_{1,n-k} \\ b_{21} & b_{22} & \cdots & b_{2,n-k} \\ \vdots & \vdots & & \vdots \\ b_{n-k,1} & b_{n-k,2} & \cdots & b_{n-k,n-k} \end{vmatrix},$$

and hence each term m' in the expansion of M' can be written as

$$m' = (-1)^{\tau(\beta_1 \beta_2 \cdots \beta_{n-k})} b_{1\beta_1} b_{2\beta_2} \cdots b_{n-k,\beta_{n-k}},$$

where $\beta_1\beta_2\cdots\beta_{n-k}$ is a permutation of $1, 2, \ldots, n-k$. Thus,

$$mm' = (-1)^{\tau(\alpha_1\alpha_2\cdots\alpha_k)+\tau(\beta_1\beta_2\cdots\beta_{n-k})}a_{1\alpha_1}a_{2\alpha_2}$$

$$\cdots a_{k\alpha_k}b_{1\beta_1}b_{2\beta_2}\cdots b_{n-k,\beta_{n-k}}$$

$$= (-1)^{\tau(\alpha_1\alpha_2\cdots\alpha_k)+\tau(\beta_1\beta_2\cdots\beta_{n-k})}a_{1\alpha_1}a_{2\alpha_2}$$

$$\cdots a_{k\alpha_k}a_{k+1,k+\beta_1}a_{k+2,k+\beta_2}\cdots a_{n,k+\beta_{n-k}}.$$

Let $\alpha_{k+i} = k + \beta_i, i = 1, 2, \ldots, n-k$. Then $\alpha_{k+1}, \alpha_{k+2}, \ldots, \alpha_n$ is a permutation of $k+1, k+2, \ldots, n$, and

$$mm' = (-1)^{\tau(\alpha_1\alpha_2\cdots\alpha_k)+\tau(\beta_1\beta_2\cdots\beta_{n-k})}a_{1\alpha_1}a_{2\alpha_2}$$

$$\cdots a_{k\alpha_k}a_{k+1,\alpha_{k+1}}a_{k+2,\alpha_{k+2}}\cdots a_{n,\alpha_n}.$$

Since $\alpha_{k+i} > k$, $i = 1, 2, \ldots, n-k$, we have

$$\tau(\alpha_1\alpha_2\cdots\alpha_k\alpha_{k+1}\alpha_{k+2}\cdots\alpha_n)$$

$$= \tau(\alpha_1\alpha_2\cdots\alpha_k) + \tau(\alpha_{k+1}\alpha_{k+2}\cdots\alpha_n)$$

$$= \tau(\alpha_1\alpha_2\cdots\alpha_k) + \tau(\beta_1\beta_2\cdots\beta_{n-k}).$$

Obviously, $a_{1\alpha_1}a_{2\alpha_2}\cdots a_{k\alpha_k}a_{k+1,\alpha_{k+1}}a_{k+2,\alpha_{k+2}}\cdots a_{n,\alpha_n}$ is a product of n entries, with exactly one entry from each row and exactly one entry from each column of D. Thus,

$$mm' = (-1)^{\tau(\alpha_1\alpha_2\cdots\alpha_k\alpha_{k+1}\alpha_{k+2}\cdots\alpha_n)}a_{1\alpha_1}a_{2\alpha_2}$$

$$\cdots a_{k\alpha_k}a_{k+1,\alpha_{k+1}}a_{k+2,\alpha_{k+2}}\cdots a_{n,\alpha_n}$$

is a term in the expansion of D.

Next, we deal with the general situation. Suppose M is a minor of order k with row indices $i_1, i_2, \ldots, i_k$ and column indices $j_1, j_2, \ldots, j_k$, where $i_1 < i_2 < \cdots < i_k$; $j_1 < j_2 < \cdots < j_k$.

Change the order of rows and columns of D so that M is at the upper left corner of a new determinant. First, we move the i_1th row upwards by interchanging two adjacent rows until the i_1th row is the first row; this requires $i_1 - 1$ steps. Similarly, the i_2th row can be moved to the second row after $i_1 - 2$ steps, and so on. Therefore, the i_1th, i_2th, $\ldots$, i_kth rows of D can be moved to the first k rows after

$$(i_1 - 1) + (i_2 - 2) + \cdots + (i_k - k)$$

$$= (i_1 + i_2 + \cdots + i_k) - (1 + 2 + \cdots + k)$$

steps.

Similarly, the j_1th, j_2th, ..., j_kth columns of D can be moved to the first k columns after

$$(j_1 - 1) + (j_2 - 2) + \cdots + (j_k - k)$$
$$= (j_1 + j_2 + \cdots + j_k) - (1 + 2 + \cdots + k)$$

steps.

Let D_1 be the new determinant obtained after the interchanges above. Then

$$D = (-1)^{(i_1+i_2+\cdots+i_k)-(1+2+\cdots+k)+(j_1+j_2+\cdots+j_k)-(1+2+\cdots+k)} D_1$$
$$= (-1)^{(i_1+i_2+\cdots+i_k)+(j_1+j_2+\cdots+j_k)} D_1.$$

Now M is located at the upper left corner of D_1. So, each term in MM' is a term in the expansion of D_1 by the preceding proof.

Note the cofactor of M in D is

$$A = (-1)^{(i_1+i_2+\cdots+i_k)+(j_1+j_2+\cdots+j_k)} M'.$$

So, $MA = (-1)^{(i_1+i_2+\cdots+i_k)+(j_1+j_2+\cdots+j_k)} MM'$. It follows that each term in MA is a term in the expansion of $D = (-1)^{(i_1+i_2+\cdots+i_k)+(j_1+j_2+\cdots+j_k)} D_1$. $\qquad\square$

Using the Laplace expansion, it is easy to give another proof of Example 2.4.7.

Example 2.4.11. Evaluate the determinant

$$D_{2n} = \begin{vmatrix} a & 0 & \cdots & \cdots & \cdots & \cdots & \cdots & \cdots & 0 & b \\ 0 & a & \ddots & & & & & \ddots & b & 0 \\ \vdots & \ddots & \ddots & \ddots & & & \ddots & \ddots & \ddots & \vdots \\ \vdots & & \ddots & a & 0 & 0 & b & \ddots & & \vdots \\ \vdots & & & 0 & a & b & 0 & & & \vdots \\ \vdots & & & 0 & b & a & 0 & & & \vdots \\ \vdots & & \ddots & b & 0 & 0 & a & \ddots & & \vdots \\ \vdots & \ddots & \ddots & \ddots & & & \ddots & \ddots & \ddots & \vdots \\ 0 & b & \ddots & & & & & \ddots & a & 0 \\ b & 0 & \cdots & \cdots & \cdots & \cdots & \cdots & \cdots & 0 & a \end{vmatrix}.$$

Solution. Applying the Laplace expansion to the first and last rows in D_{2n} gives $D_{2n} = \begin{vmatrix} a & b \\ b & a \end{vmatrix} D_{2(n-1)} = (a^2 - b^2)D_{2(n-1)}$. Continuing this way, we have $D_{2n} = (a^2 - b^2)^2 D_{2(n-2)} = \cdots = (a^2 - b^2)^{n-1} D_2 = (a^2 - b^2)^n$. $\qquad\square$

Theorem 2.4.12 (Multiplication of determinants).

$$\begin{vmatrix} a_{11} & a_{12} & \cdots & a_{1n} \\ a_{21} & a_{22} & \cdots & a_{2n} \\ \vdots & \vdots & & \vdots \\ a_{n1} & a_{n2} & \cdots & a_{nn} \end{vmatrix} \begin{vmatrix} b_{11} & b_{12} & \cdots & b_{1n} \\ b_{21} & b_{22} & \cdots & b_{2n} \\ \vdots & \vdots & & \vdots \\ b_{n1} & b_{n2} & \cdots & b_{nn} \end{vmatrix} = \begin{vmatrix} c_{11} & c_{12} & \cdots & c_{1n} \\ c_{21} & c_{22} & \cdots & c_{2n} \\ \vdots & \vdots & & \vdots \\ c_{n1} & c_{n2} & \cdots & c_{nn} \end{vmatrix},$$

where $c_{ij} = a_{i1}b_{1j} + a_{i2}b_{2j} + \cdots + a_{in}b_{nj} = \sum_{k=1}^{n} a_{ik}b_{kj}, \ i, j = 1, 2, \ldots, n.$

Proof. Let

$$D = \begin{vmatrix} a_{11} & a_{12} & \cdots & a_{1,n-1} & a_{1n} & 0 & \cdots & \cdots & \cdots & 0 \\ a_{21} & a_{22} & \cdots & a_{2,n-1} & a_{2n} & \vdots & & & & \vdots \\ \vdots & \vdots & & \vdots & \vdots & \vdots & & & & \vdots \\ a_{n-1,1} & a_{n-1,2} & \cdots & a_{n-1,n-1} & a_{n-1,n} & \vdots & & & & \vdots \\ a_{n1} & a_{n2} & \cdots & a_{n,n-1} & a_{nn} & 0 & \cdots & \cdots & \cdots & 0 \\ -1 & 0 & \cdots & & \cdots & 0 & b_{11} & b_{12} & \cdots & b_{1,n-1} & b_{1n} \\ 0 & -1 & \ddots & & & \vdots & b_{21} & b_{22} & \cdots & b_{2,n-1} & b_{2n} \\ \vdots & & \ddots & \ddots & \ddots & \vdots & \vdots & \vdots & & \vdots & \vdots \\ \vdots & & & \ddots & -1 & 0 & b_{n-1,1} & b_{n-1,2} & \cdots & b_{n-1,n-1} & b_{n-1,n} \\ 0 & \cdots & & \cdots & 0 & -1 & b_{n1} & b_{n2} & \cdots & b_{n,n-1} & b_{nn} \end{vmatrix}.$$

Then, by Example 2.4.7 or the Laplace expansion, we get

$$D = \begin{vmatrix} a_{11} & a_{12} & \cdots & a_{1n} \\ a_{21} & a_{22} & \cdots & a_{2n} \\ \vdots & \vdots & & \vdots \\ a_{n1} & a_{n2} & \cdots & a_{nn} \end{vmatrix} \begin{vmatrix} b_{11} & b_{12} & \cdots & b_{1n} \\ b_{21} & b_{22} & \cdots & b_{2n} \\ \vdots & \vdots & & \vdots \\ b_{n1} & b_{n2} & \cdots & b_{nn} \end{vmatrix}.$$

On the other hand, adding b_{i1} times column i to column $(n+1)$ for $i = 1, 2, \ldots, n$, we have

$$
D = \begin{vmatrix}
a_{11} & a_{12} & \cdots & a_{1,n-1} & a_{1n} & c_{11} & 0 & \cdots & \cdots & 0 \\
a_{21} & a_{22} & \cdots & a_{2,n-1} & a_{2n} & c_{21} & \vdots & & & \vdots \\
\vdots & \vdots & & \vdots & \vdots & \vdots & \vdots & & & \vdots \\
a_{n-1,1} & a_{n-1,2} & \cdots & a_{n-1,n-1} & a_{n-1,n} & c_{n-1,1} & \vdots & & & \vdots \\
a_{n1} & a_{n2} & \cdots & a_{n,n-1} & a_{nn} & c_{n1} & 0 & \cdots & \cdots & 0 \\
-1 & 0 & \cdots & \cdots & 0 & 0 & b_{12} & \cdots & b_{1,n-1} & b_{1n} \\
0 & -1 & \ddots & & \vdots & 0 & b_{22} & \cdots & b_{2,n-1} & b_{2n} \\
\vdots & \ddots & \ddots & \ddots & \vdots & \vdots & \vdots & & \vdots & \vdots \\
\vdots & & \ddots & -1 & 0 & 0 & b_{n-1,2} & \cdots & b_{n-1,n-1} & b_{n-1,n} \\
0 & \cdots & \cdots & 0 & -1 & 0 & b_{n2} & \cdots & b_{n,n-1} & b_{nn}
\end{vmatrix}.
$$

Similarly, adding b_{ik} times column i to column $(n+k)$ for $i = 1, 2, \ldots, n$ and $k = 2, 3, \ldots, n$, we get

$$
D = \begin{vmatrix}
a_{11} & a_{12} & \cdots & a_{1,n-1} & a_{1n} & c_{11} & c_{12} & \cdots & c_{1,n-1} & c_{1n} \\
a_{21} & a_{22} & \cdots & a_{2,n-1} & a_{2n} & c_{21} & c_{22} & \cdots & c_{2,n-1} & c_{2n} \\
\vdots & \vdots & & \vdots & \vdots & \vdots & \vdots & & \vdots & \vdots \\
a_{n-1,1} & a_{n-1,2} & \cdots & a_{n-1,n-1} & a_{n-1,n} & c_{n-1,1} & c_{n-1,2} & \cdots & c_{n-1,n-1} & c_{n-1,n} \\
a_{n1} & a_{n2} & \cdots & a_{n,n-1} & a_{nn} & c_{n1} & c_{n2} & \cdots & c_{n,n-1} & c_{nn} \\
-1 & 0 & \cdots & \cdots & 0 & 0 & \cdots & \cdots & \cdots & 0 \\
0 & -1 & \ddots & & \vdots & \vdots & & & & \vdots \\
\vdots & \ddots & \ddots & \ddots & \vdots & \vdots & & & & \vdots \\
\vdots & & \ddots & -1 & 0 & \vdots & & & & \vdots \\
0 & \cdots & \cdots & 0 & -1 & 0 & \cdots & \cdots & \cdots & 0
\end{vmatrix}.
$$

Applying the Laplace expansion to the last n rows in the determinant above yields

$$D = (-1)^n (-1)^{(n+1)+(n+2)+\cdots+(n+n)+1+2+\cdots+n} \begin{vmatrix} c_{11} & c_{12} & \cdots & c_{1n} \\ c_{21} & c_{22} & \cdots & c_{2n} \\ \vdots & \vdots & & \vdots \\ c_{n1} & c_{n2} & \cdots & c_{nn} \end{vmatrix}$$

$$= (-1)^n (-1)^{n^2+n(n+1)} \begin{vmatrix} c_{11} & c_{12} & \cdots & c_{1n} \\ c_{21} & c_{22} & \cdots & c_{2n} \\ \vdots & \vdots & & \vdots \\ c_{n1} & c_{n2} & \cdots & c_{nn} \end{vmatrix} = \begin{vmatrix} c_{11} & c_{12} & \cdots & c_{1n} \\ c_{21} & c_{22} & \cdots & c_{2n} \\ \vdots & \vdots & & \vdots \\ c_{n1} & c_{n2} & \cdots & c_{nn} \end{vmatrix}.$$

So, the theorem is true. $\qquad\square$

Example 2.4.13. Evaluate $D_n = \begin{vmatrix} a_1+b_1 & a_1+b_2 & \cdots & a_1+b_n \\ a_2+b_1 & a_2+b_2 & \cdots & a_2+b_n \\ \vdots & \vdots & & \vdots \\ a_n+b_1 & a_n+b_2 & \cdots & a_n+b_n \end{vmatrix}.$

Solution. When $n=1$, $D_1 = a_1 + b_1$. Next, we let $n > 1$. Then

$$D_n = \begin{vmatrix} a_1 & 1 & 0 & \cdots & 0 \\ a_2 & 1 & 0 & \cdots & 0 \\ \vdots & \vdots & \vdots & & \vdots \\ a_n & 1 & 0 & \cdots & 0 \end{vmatrix} \begin{vmatrix} 1 & 1 & \cdots & 1 \\ b_1 & b_2 & \cdots & b_n \\ 0 & 0 & \cdots & 0 \\ \vdots & \vdots & & \vdots \\ 0 & 0 & \cdots & 0 \end{vmatrix} = \begin{cases} (a_1 - a_2)(b_2 - b_1), & \text{if } n = 2, \\ 0, & \text{if } n > 2. \end{cases}$$

So, $D_n = \begin{cases} a_1 + b_1, & n = 1, \\ (a_1 - a_2)(b_2 - b_1), & n = 2, \\ 0, & n > 2. \end{cases}$ $\qquad\square$

2.5 Cramer's Rule

Now, we use nth-order determinants to solve the problem of the solution of a system of n linear equations in n unknowns. Next, we get a conclusion similar to those for systems of binary and ternary linear equations.

Theorem 2.5.1 (Cramer's[8] rule). *Let*

$$\begin{cases} a_{11}x_1 + a_{12}x_2 + \cdots + a_{1n}x_n = b_1, \\ a_{21}x_1 + a_{22}x_2 + \cdots + a_{2n}x_n = b_2, \\ \quad\vdots \qquad\quad \vdots \qquad\qquad\quad \vdots \qquad \vdots \\ a_{n1}x_1 + a_{n2}x_2 + \cdots + a_{nn}x_n = b_n \end{cases} \tag{2.2}$$

be a system of n linear equations in n unknowns and let

$$d = \begin{vmatrix} a_{11} & a_{12} & \cdots & a_{1n} \\ a_{21} & a_{22} & \cdots & a_{2n} \\ \vdots & \vdots & & \vdots \\ a_{n1} & a_{n2} & \cdots & a_{nn} \end{vmatrix}$$

*be the **coefficient determinant**.*

If $d \neq 0$, then the system (2.2) has the unique solution:

$$x_1 = \frac{d_1}{d}, \ x_2 = \frac{d_2}{d}, \ \ldots, \ x_n = \frac{d_n}{d}, \tag{2.3}$$

where $d_j = \begin{vmatrix} a_{11} & \cdots & a_{1,j-1} & b_1 & a_{1,j+1} & \cdots & a_{1n} \\ a_{21} & \cdots & a_{2,j-1} & b_2 & a_{2,j+1} & \cdots & a_{2n} \\ \vdots & & \vdots & \vdots & \vdots & & \vdots \\ a_{n1} & \cdots & a_{n,j-1} & b_n & a_{n,j+1} & \cdots & a_{nn} \end{vmatrix}, \ j = 1, 2, \ldots, n.$

Proof. The linear system (2.2) can be abbreviated as

$$\sum_{j=1}^{n} a_{ij}x_j = b_i, \ i = 1, 2, \ldots, n. \tag{2.4}$$

We first show that (2.3) is a solution of (2.4).

Substituting $x_j = \frac{d_j}{d}$ ($1 \leqslant j \leqslant n$) into the ith equation in (2.4), we get

$$\sum_{j=1}^{n} a_{ij}x_j = \sum_{j=1}^{n} a_{ij}\frac{d_j}{d} = \frac{1}{d}\sum_{j=1}^{n} a_{ij}d_j, \quad i = 1, 2, \ldots, n.$$

---

[8]Gabriel Cramer, 1704–1752, Swiss mathematician.

Since $d_j = b_1 A_{1j} + b_2 A_{2j} + \cdots + b_n A_{nj} = \sum_{s=1}^{n} b_s A_{sj}$, $j = 1, 2, \ldots, n$,

$$\sum_{j=1}^{n} a_{ij} x_j = \frac{1}{d} \sum_{j=1}^{n} a_{ij} \sum_{s=1}^{n} b_s A_{sj} = \frac{1}{d} \sum_{s=1}^{n} \left(\sum_{j=1}^{n} a_{ij} A_{sj} \right) b_s = \frac{1}{d} db_i = b_i,$$

$i = 1, 2, \ldots, n$, showing that (2.3) is a solution of (2.4).

Suppose $x_1 = c_1, x_2 = c_2, \ldots, x_n = c_n$ is a solution of (2.4), then

$$\sum_{j=1}^{n} a_{ij} c_j = b_i, \quad i = 1, 2, \ldots, n. \tag{2.5}$$

We prove that $c_k = \frac{d_k}{d}$, $k = 1, 2, \ldots, n$.

Note that $d_k = \sum_{i=1}^{n} b_i A_{ik}$, $k = 1, 2, \ldots, n$.

Multiplying both sides of (2.5) by A_{ik}, we have

$$A_{ik} \sum_{j=1}^{n} a_{ij} c_j = b_i A_{ik}, \ i = 1, 2, \ldots, n.$$

Then

$$\sum_{i=1}^{n} A_{ik} \sum_{j=1}^{n} a_{ij} c_j = \sum_{i=1}^{n} b_i A_{ik}.$$

In the equality above, the left side is equal to $\sum_{j=1}^{n} \left(\sum_{i=1}^{n} a_{ij} A_{ik} \right) c_j = dc_k$, and the right side is equal to d_k. So $dc_k = d_k$, i.e., $c_k = \frac{d_k}{d}$, $k = 1, 2, \ldots, n$. $\qquad\square$

Remark 2.5.2. Cramer's rule only deals with cases where the coefficient determinant is not equal to zero; other cases are discussed in the following chapter. If $b_1 = b_2 = \cdots = b_n = 0$ in the linear system (2.2), then this system is called a system of **homogeneous linear equations** (see Section 3.2 for a general definition). Every system of homogeneous linear equations always has a solution because $x_1 = 0, x_2 = 0, \ldots, x_n = 0$ is a solution to such a system; it is called the **zero solution**. Any solution in which at least one unknown has a nonzero value is called a **nonzero solution**.

Theorem 2.5.3. *Let* $\begin{cases} a_{11}x_1 + a_{12}x_2 + \cdots + a_{1n}x_n = 0, \\ a_{21}x_1 + a_{22}x_2 + \cdots + a_{2n}x_n = 0, \\ \quad\vdots \qquad\quad \vdots \qquad\qquad\quad \vdots \qquad \vdots \\ a_{n1}x_1 + a_{n2}x_2 + \cdots + a_{nn}x_n = 0 \end{cases}$ *be a system of homogeneous linear equations. If its coefficient determinant does not equal zero, it only has the zero solution. In other words, if the system has a nonzero solution, its coefficient determinant equals zero.*

Proof. This result follows from Cramer's rule. $\qquad\square$

Example 2.5.4. Let $n \geqslant 2$ be an integer, and let $D = |a_{ij}|_n$, $\Delta = |A_{ij}|_n$, where A_{ij} is the cofactor of the (i,j) entry a_{ij} of D, $i,j = 1,2,\ldots,n$. Prove that $\Delta = D^{n-1}$.

Proof. Since

$$
D\Delta = \begin{vmatrix} a_{11} & a_{12} & \cdots & a_{1n} \\ a_{21} & a_{22} & \cdots & a_{2n} \\ \vdots & \vdots & & \vdots \\ a_{n1} & a_{n2} & \cdots & a_{nn} \end{vmatrix} \begin{vmatrix} A_{11} & A_{12} & \cdots & A_{1n} \\ A_{21} & A_{22} & \cdots & A_{2n} \\ \vdots & \vdots & & \vdots \\ A_{n1} & A_{n2} & \cdots & A_{nn} \end{vmatrix}
$$

$$
= \begin{vmatrix} a_{11} & a_{12} & \cdots & a_{1n} \\ a_{21} & a_{22} & \cdots & a_{2n} \\ \vdots & \vdots & & \vdots \\ a_{n1} & a_{n2} & \cdots & a_{nn} \end{vmatrix} \begin{vmatrix} A_{11} & A_{21} & \cdots & A_{n1} \\ A_{12} & A_{22} & \cdots & A_{n2} \\ \vdots & \vdots & & \vdots \\ A_{1n} & A_{2n} & \cdots & A_{nn} \end{vmatrix}
$$

$$
= \begin{vmatrix} D & 0 & \cdots\cdots & 0 \\ 0 & D & \ddots & \vdots \\ \vdots & \ddots & \ddots & \ddots & \vdots \\ \vdots & & \ddots & D & 0 \\ 0 & \cdots\cdots & & 0 & D \end{vmatrix} = D^n,
$$

i.e., $D\Delta = D^n$, $\Delta = D^{n-1}$ if $D \neq 0$.

In what follows, we assume that $D = 0$. We show that $\Delta = 0$.

Suppose $\Delta \neq 0$, then the linear system $\sum\limits_{j=1}^{n} A_{ij}x_j = 0$, $i = 1,2,\ldots,n$ only has the zero solution by Theorem 2.5.3. Since $D = 0$, by Theorem 2.4.4, we have

$$\sum_{j=1}^{n} A_{ij}a_{kj} = 0, \quad i = 1, 2, \ldots, n, \quad k = 1, 2, \ldots, n.$$

Therefore, $a_{kj} = 0$, $k, j = 1, 2, \ldots, n$, and hence $\Delta = 0$, a contradiction. So, $\Delta = D^{n-1}$. $\qquad\qquad\square$

2.6 Methods for Evaluating Determinants

This section mainly introduces some methods for evaluating determinants through examples.

I. Using triangular determinants

It is well known that evaluating an upper (a lower) triangular determinant is very easy. This method is to reduce the given determinant into a triangular one by the properties of determinants.

Example 2.6.1. Evaluate

$$D = \begin{vmatrix} 1 & 2 & 3 & \cdots\cdots & \cdots & n \\ 1 & -1 & 0 & \cdots\cdots & \cdots & 0 \\ 0 & 2 & -2 & \ddots & & \vdots \\ \vdots & \ddots & \ddots & \ddots & & \vdots \\ \vdots & & \ddots & \ddots & \ddots & \vdots \\ \vdots & & & \ddots & -(n-2) & 0 \\ 0 & \cdots\cdots & 0 & n-1 & -(n-1) \end{vmatrix}.$$

Solution.

$$D = \begin{vmatrix} \dfrac{n(n+1)}{2} & 2 & 3 & \cdots\cdots & \cdots & n \\ 0 & -1 & 0 & \cdots\cdots & \cdots & 0 \\ 0 & 2 & -2 & \ddots & & \vdots \\ \vdots & & \ddots & \ddots & & \vdots \\ \vdots & & \ddots & \ddots & \ddots & \vdots \\ \vdots & & & \ddots & -(n-2) & 0 \\ 0 & & \cdots\cdots & 0 & n-1 & -(n-1) \end{vmatrix}$$

$$= \frac{n(n+1)}{2} \begin{vmatrix} -1 & 0 & 0 & \cdots & & \cdots & & 0 \\ 2 & -2 & 0 & & & & & \vdots \\ 0 & 3 & -3 & \ddots & & & & \vdots \\ \vdots & \ddots & \ddots & \ddots & & \ddots & & \vdots \\ \vdots & & & \ddots & n-2 & -(n-2) & & 0 \\ 0 & \cdots & & \cdots & 0 & n-1 & & -(n-1) \end{vmatrix}$$

$$= \frac{(-1)^{n-1}}{2}(n+1)!. \qquad \square$$

Example 2.6.2. Evaluate the determinant D_{2n} in Example 2.4.11.

Solution. When $a = 0$,

$$D_{2n} = \begin{vmatrix} 0 & \cdots & & \cdots & 0 & b \\ \vdots & & & \cdot^{\cdot^{\cdot}} & b & 0 \\ \vdots & & \cdot^{\cdot^{\cdot}} & \cdot^{\cdot^{\cdot}} & \cdot^{\cdot^{\cdot}} & \vdots \\ 0 & b & \cdot^{\cdot^{\cdot}} & & & \vdots \\ b & 0 & \cdots & & \cdots & 0 \end{vmatrix} = (-1)^{\frac{2n(2n-1)}{2}} b^{2n} = (-b^2)^n.$$

When $a \neq 0$, adding $-\frac{b}{a}$ times row i to row $2n - (i-1)$ for $i = 1, 2, \ldots, n$, we get

$$D_{2n} = \begin{vmatrix} a & 0 & \cdots\cdots\cdots & \cdots & & \cdots & \cdots & 0 & b \\ 0 & a & \ddots & & & \cdot^{\cdot^{\cdot}} & b & 0 \\ \vdots & \ddots & \ddots & \ddots & & \cdot^{\cdot^{\cdot}} & \cdot^{\cdot^{\cdot}} & \cdot^{\cdot^{\cdot}} & \vdots \\ \vdots & & \ddots & a & 0 & 0 & b & \cdot^{\cdot^{\cdot}} & \vdots \\ \vdots & & & \ddots & a & b & 0 & & \vdots \\ \vdots & & & & \ddots & a - \frac{b^2}{a} & 0 & & \vdots \\ \vdots & & & & & \ddots & a - \frac{b^2}{a} & \ddots & \vdots \\ \vdots & & & & & & \ddots & \ddots & \ddots & \vdots \\ \vdots & & & & & & & \ddots & a - \frac{b^2}{a} & 0 \\ 0 & \cdots\cdots\cdots & \cdots & & \cdots & \cdots & & 0 & a - \frac{b^2}{a} \end{vmatrix}.$$

$$= a^n \left(a - \frac{b^2}{a} \right)^n = (a^2 - b^2)^n.$$

To sum up, $D_{2n} = (a^2 - b^2)^n$. $\square$

II. Using properties of polynomials

If the entries of a determinant are polynomials, then its value is a polynomial. This method is to evaluate some determinants with polynomial entries using the properties of polynomials.

Example 2.6.3. Evaluate $D = \begin{vmatrix} 1 & 2 & 3 & \cdots & n-1 & n \\ 1 & x+1 & 3 & \cdots & n-1 & n \\ 1 & 2 & x+1 & \cdots & n-1 & n \\ \vdots & \vdots & \vdots & & \vdots & \vdots \\ 1 & 2 & 3 & \cdots & x+1 & n \\ 1 & 2 & 3 & \cdots & n-1 & x+1 \end{vmatrix}$.

Solution. **Method 1** (Using properties of polynomials).

By the definition of a determinant, $D = D(x)$ is a monic polynomial of degree $n - 1$. Note that

$$D(1) = D(2) = \cdots = D(n - 1) = 0$$

by Property 7 of determinants, and hence

$$(x - 1)|D(x), (x - 2)|D(x), \ldots, (x - (n - 1))|D(x).$$

Since $x - 1, x - 2, \ldots, x - (n - 1)$ are pairwise coprime,

$$(x - 1)(x - 2) \cdots (x - (n - 1))|D(x).$$

So, $D(x) = (x - 1)(x - 2) \cdots (x - (n - 1))$.

Method 2 (Using triangular determinants).

Subtracting the first row from the ith row for $i = 2, 3, \ldots, n$, we have

$$D = \begin{vmatrix} 1 & 2 & 3 & \cdots & & & n \\ 0 & x-1 & 0 & \cdots & & & 0 \\ 0 & 0 & x-2 & \ddots & & & \vdots \\ \vdots & & & \ddots & \ddots & & \vdots \\ \vdots & & & & \ddots & x-(n-2) & 0 \\ 0 & \cdots & & \cdots & & 0 & x-(n-1) \end{vmatrix}$$

$$= (x - 1)(x - 2) \cdots (x - (n - 1)). \qquad \square$$

III. Using mathematical induction

This method is to use mathematical induction to prove some results on determinants.

Example 2.6.4. Let $n \geqslant 2$ be an integer. Prove that

$$
\begin{vmatrix}
x & 1 & 1 & \cdots & \cdots & 1 \\
1 & a_1 & 0 & \cdots & \cdots & 0 \\
1 & 0 & a_2 & \ddots & & \vdots \\
\vdots & \vdots & \ddots & \ddots & \ddots & \vdots \\
\vdots & \vdots & & \ddots & a_{n-1} & 0 \\
1 & 0 & \cdots & \cdots & 0 & a_n
\end{vmatrix}
= x a_1 a_2 \cdots a_n - \sum_{i=1}^{n} a_1 \cdots a_{i-1} a_{i+1} \cdots a_n.
$$

Proof. **Method 1.** Let $D_n = \begin{vmatrix}
x & 1 & 1 & \cdots & \cdots & 1 \\
1 & a_1 & 0 & \cdots & \cdots & 0 \\
1 & 0 & a_2 & \ddots & & \vdots \\
\vdots & \vdots & \ddots & \ddots & \ddots & \vdots \\
\vdots & \vdots & & \ddots & a_{n-1} & 0 \\
1 & 0 & \cdots & \cdots & 0 & a_n
\end{vmatrix}.$

We do an induction on n.

When $n = 2$, $D_2 = \begin{vmatrix} x & 1 & 1 \\ 1 & a_1 & 0 \\ 1 & 0 & a_2 \end{vmatrix} = a_2 \begin{vmatrix} x & 1 \\ 1 & a_1 \end{vmatrix} + 1 \times (-1)^{3+1} \begin{vmatrix} 1 & 1 \\ a_1 & 0 \end{vmatrix}$

$$
= a_2(x a_1 - 1) - a_1 = x a_1 a_2 - a_1 - a_2, \text{ as desired.}
$$

Let $n > 2$ be an integer. We assume the conclusion is true for the determinant D_{n-1}. Now consider the determinant D_n.

Expanding D_n along the last row, we get

$$
D_n = a_n D_{n-1} + (-1)^{n+1+1} \begin{vmatrix}
1 & 1 & \cdots & \cdots & 1 \\
a_1 & 0 & \cdots & \cdots & 0 \\
0 & a_2 & \ddots & & \vdots \\
\vdots & \ddots & \ddots & \ddots & \vdots \\
0 & \cdots & 0 & a_{n-1} & 0
\end{vmatrix}
$$

$$
= a_n D_{n-1} + (-1)^{n+1+1}(-1)^{1+n} a_1 a_2 \cdots a_{n-1}
$$

$$
= a_n D_{n-1} - a_1 a_2 \cdots a_{n-1}.
$$

By the inductive hypothesis,

$$D_{n-1} = xa_1a_2\cdots a_{n-1} - \sum_{i=1}^{n-1} a_1\cdots a_{i-1}a_{i+1}\cdots a_{n-1}.$$

Therefore,

$$D_n = a_n\left(xa_1a_2\cdots a_{n-1} - \sum_{i=1}^{n-1} a_1\cdots a_{i-1}a_{i+1}\cdots a_{n-1}\right)$$

$$- a_1a_2\cdots a_{n-1}$$

$$= xa_1a_2\cdots a_n - \sum_{i=1}^{n} a_1\cdots a_{i-1}a_{i+1}\cdots a_n.$$

By the first form of induction, the equality to be proved is true.

Method 2 (Using triangular determinants).

First, let $a_k \neq 0$, $k = 1, 2, \ldots, n$. Adding $-\frac{1}{a_{i-1}}$ times the ith column to the first column for $i = 2, 3, \ldots, n + 1$, we have

$$D_n = \begin{vmatrix} x - \sum_{i=1}^{n}\frac{1}{a_i} & 1 & 1 & \cdots & \cdots & 1 \\ 0 & a_1 & 0 & \cdots & \cdots & 0 \\ \vdots & & \ddots & \ddots & \ddots & \vdots \\ \vdots & & & \ddots & \ddots & \ddots & \vdots \\ \vdots & & & & \ddots & a_{n-1} & 0 \\ 0 & \cdots & \cdots & \cdots & 0 & a_n \end{vmatrix}$$

$$= a_1a_2\cdots a_n\left(x - \sum_{i=1}^{n}\frac{1}{a_i}\right)$$

$$= xa_1a_2\cdots a_n - \sum_{i=1}^{n} a_1\cdots a_{i-1}a_{i+1}\cdots a_n.$$

If $a_n = 0$, then $D_n = -a_1a_2\cdots a_{n-1}$ by expanding D_n along the last row (see Method 1).

If $a_i = 0$, $1 \leqslant i \leqslant n-1$, then

$$D_n = (-1)^{n-i}(-1)^{n-i} \begin{vmatrix} x & 1 & 1 & \cdots & \cdots & \cdots & \cdots & \cdots & 1 \\ 1 & a_1 & 0 & \cdots & \cdots & \cdots & \cdots & \cdots & 0 \\ 1 & 0 & a_2 & \ddots & & & & & \vdots \\ \vdots & \vdots & \ddots & \ddots & \ddots & & & & \vdots \\ \vdots & \vdots & & \ddots & a_{i-1} & & \ddots & & \vdots \\ \vdots & \vdots & & & \ddots & a_{i+1} & & \ddots & \vdots \\ \vdots & \vdots & & & & \ddots & \ddots & \ddots & \vdots \\ \vdots & \vdots & & & & & \ddots & a_n & 0 \\ 1 & 0 & \cdots & \cdots & \cdots & \cdots & \cdots & 0 & 0 \end{vmatrix}$$

$$= -a_1 \cdots a_{i-1} a_{i+1} \cdots a_n.$$

To sum up, $D_n = x a_1 a_2 \cdots a_n - \sum_{i=1}^{n} a_1 \cdots a_{i-1} a_{i+1} \cdots a_n.$ $\qquad \square$

Theorem 2.6.5. *Prove that*

$$\begin{vmatrix} \alpha+\beta & \alpha\beta & 0 & \cdots & \cdots & \cdots & \cdots & 0 \\ 1 & \alpha+\beta & \alpha\beta & \ddots & & & & \vdots \\ 0 & 1 & \alpha+\beta & \alpha\beta & \ddots & & & \vdots \\ \vdots & \ddots & 1 & \alpha+\beta & \ddots & \ddots & & \vdots \\ \vdots & & \ddots & \ddots & \ddots & \ddots & \ddots & \vdots \\ \vdots & & & \ddots & \ddots & \alpha+\beta & \alpha\beta & 0 \\ \vdots & & & & \ddots & 1 & \alpha+\beta & \alpha\beta \\ 0 & \cdots & \cdots & \cdots & \cdots & 0 & 1 & \alpha+\beta \end{vmatrix}_n = \frac{\alpha^{n+1} - \beta^{n+1}}{\alpha - \beta},$$

where $\alpha \neq \beta$.

Proof. Let D_n be the left side of the equality to be proved. We proceed by induction on n.

When $n = 1, 2$, it is easy to see the equality to be proved is true.

Let $n \geqslant 3$ be an integer. We assume the equality holds for determinants of order $< n$.

Now consider the case when the order of the determinant is n. Expanding D_n along the last row yields

$$D_n = (\alpha + \beta)D_{n-1} - \alpha\beta D_{n-2}.$$

By the inductive hypothesis,

$$D_{n-1} = \frac{\alpha^n - \beta^n}{\alpha - \beta}, \quad D_{n-2} = \frac{\alpha^{n-1} - \beta^{n-1}}{\alpha - \beta}.$$

Thus

$$D_n = (\alpha + \beta)\frac{\alpha^n - \beta^n}{\alpha - \beta} - \alpha\beta\frac{\alpha^{n-1} - \beta^{n-1}}{\alpha - \beta} = \frac{\alpha^{n+1} - \beta^{n+1}}{\alpha - \beta}.$$

By the second form of induction, the equality to be proved is true.

$\square$

IV. Using recurrence relation

Express a determinant D of order n in terms of the determinants of order $n - 1$ (or order $< n - 1$) of the same form. This expression is called a recurrence relation. This method is to find the general expression of D according to the recurrence relation.

Example 2.6.6. Evaluate

$$D_n = \begin{vmatrix} \alpha + \beta & \alpha\beta & 0 & \cdots & \cdots & \cdots & \cdots & 0 \\ 1 & \alpha + \beta & \alpha\beta & \ddots & & & & \vdots \\ 0 & 1 & \alpha + \beta & \alpha\beta & \ddots & & & \vdots \\ \vdots & \ddots & 1 & \alpha + \beta & \ddots & \ddots & & \vdots \\ & & \ddots & \ddots & \ddots & \ddots & \ddots & \vdots \\ & & & \ddots & \ddots & \alpha + \beta & \alpha\beta & 0 \\ \vdots & & & & \ddots & 1 & \alpha + \beta & \alpha\beta \\ 0 & \cdots & \cdots & \cdots & \cdots & 0 & 1 & \alpha + \beta \end{vmatrix}_n.$$

Solution. By Example 2.6.5, $D_n = (\alpha + \beta)D_{n-1} - \alpha\beta D_{n-2}$.

It follows that

$$D_n - \alpha D_{n-1} = \beta(D_{n-1} - \alpha D_{n-2})$$
$$= \beta\beta(D_{n-2} - \alpha D_{n-3})$$
$$= \beta^2(D_{n-2} - \alpha D_{n-3})$$
$$\vdots$$
$$= \beta^{n-2}(D_2 - \alpha D_1)$$
$$= \beta^{n-2}(\alpha^2 + \alpha\beta + \beta^2 - \alpha(\alpha + \beta)) = \beta^n,$$

i.e., $D_n - \alpha D_{n-1} = \beta^n$. Similarly, $D_n - \beta D_{n-1} = \alpha^n$.

If $\alpha \neq \beta$, then $D_n = \dfrac{\alpha^{n+1} - \beta^{n+1}}{\alpha - \beta} = \alpha^n + \alpha^{n-1}\beta + \cdots + \alpha\beta^{n-1} + \beta^n$.

If $\alpha = \beta$, then $D_n = \alpha D_{n-1} + \alpha^n$

$$= \alpha(\alpha D_{n-2} + \alpha^{n-1}) + \alpha^n$$
$$= \alpha^2 D_{n-2} + 2\alpha^n$$
$$\vdots$$
$$= \alpha^{n-1} D_1 + (n-1)\alpha^n$$
$$= \alpha^{n-1}(2\alpha) + (n-1)\alpha^n = (n+1)\alpha^n.$$

In summary, $D_n = \alpha^n + \alpha^{n-1}\beta + \cdots + \alpha\beta^{n-1} + \beta^n$. $\qquad\square$

V. Splitting rows (columns)

This method is to evaluate the given determinant D by splitting all entries of a row (column) of D into the sum of two or more terms.

Example 2.6.7. Evaluate

$$D_n = \begin{vmatrix} 1 & 1 & \cdots & 1 \\ x_1(x_1 - 1) & x_2(x_2 - 1) & \cdots & x_n(x_n - 1) \\ x_1^2(x_1 - 1) & x_2^2(x_2 - 1) & \cdots & x_n^2(x_n - 1) \\ \vdots & \vdots & & \vdots \\ x_1^{n-1}(x_1 - 1) & x_2^{n-1}(x_2 - 1) & \cdots & x_n^{n-1}(x_n - 1) \end{vmatrix}.$$

Solution. Note that $1 = x_i - (x_i - 1), i = 1, 2, \ldots, n$. Thus,

$$
D_n = \begin{vmatrix}
x_1 & x_2 & \cdots & x_n \\
x_1(x_1 - 1) & x_2(x_2 - 1) & \cdots & x_n(x_n - 1) \\
x_1^2(x_1 - 1) & x_2^2(x_2 - 1) & \cdots & x_n^2(x_n - 1) \\
\vdots & \vdots & & \vdots \\
x_1^{n-1}(x_1 - 1) & x_2^{n-1}(x_2 - 1) & \cdots & x_n^{n-1}(x_n - 1)
\end{vmatrix}
$$

$$
- \begin{vmatrix}
x_1 - 1 & x_2 - 1 & \cdots & x_n - 1 \\
x_1(x_1 - 1) & x_2(x_2 - 1) & \cdots & x_n(x_n - 1) \\
x_1^2(x_1 - 1) & x_2^2(x_2 - 1) & \cdots & x_n^2(x_n - 1) \\
\vdots & \vdots & & \vdots \\
x_1^{n-1}(x_1 - 1) & x_2^{n-1}(x_2 - 1) & \cdots & x_n^{n-1}(x_n - 1)
\end{vmatrix}
$$

$$
= \begin{vmatrix}
x_1 & x_2 & \cdots & x_n \\
x_1^2 & x_2^2 & \cdots & x_n^2 \\
x_1^3 & x_2^3 & \cdots & x_n^3 \\
\vdots & \vdots & & \vdots \\
x_1^n & x_2^n & \cdots & x_n^n
\end{vmatrix}
- \prod_{i=1}^{n}(x_i - 1)
\begin{vmatrix}
1 & 1 & \cdots & 1 \\
x_1 & x_2 & \cdots & x_n \\
x_1^2 & x_2^2 & \cdots & x_n^2 \\
\vdots & \vdots & & \vdots \\
x_1^{n-1} & x_2^{n-1} & \cdots & x_n^{n-1}
\end{vmatrix}
$$

$$
= \left(\prod_{i=1}^{n} x_i - \prod_{i=1}^{n}(x_i - 1) \right)
\begin{vmatrix}
1 & 1 & \cdots & 1 \\
x_1 & x_2 & \cdots & x_n \\
x_1^2 & x_2^2 & \cdots & x_n^2 \\
\vdots & \vdots & & \vdots \\
x_1^{n-1} & x_2^{n-1} & \cdots & x_n^{n-1}
\end{vmatrix}
$$

$$
= \left(\prod_{i=1}^{n} x_i - \prod_{i=1}^{n}(x_i - 1) \right) \prod_{1 \leqslant j < i \leqslant n} (x_i - x_j). \qquad \square
$$

VI. Adding edges

Let D be a determinant of order n. Then, expanding D along a row (column) of D reduces evaluating D to evaluating determinants of order $(n-1)$. However, sometimes, we may add a row and a column to D to get a determinant D_1 of order $n + 1$ such that $D = D_1$ and the evaluation of D_1 is easier. This method is usually called adding edges.

Example 2.6.8. Evaluate the determinant in Example 2.6.7 by adding edges.

Solution.
$$D_n = \begin{vmatrix} 1 & x_1 & x_2 & \cdots & x_n \\ 0 & 1 & 1 & \cdots & 1 \\ 0 & x_1(x_1-1) & x_2(x_2-1) & \cdots & x_n(x_n-1) \\ 0 & x_1^2(x_1-1) & x_2^2(x_2-1) & \cdots & x_n^2(x_n-1) \\ \vdots & \vdots & \vdots & & \vdots \\ 0 & x_1^{n-1}(x_1-1) & x_2^{n-1}(x_2-1) & \cdots & x_n^{n-1}(x_n-1) \end{vmatrix}$$

$$= \begin{vmatrix} 1 & x_1 & x_2 & \cdots & x_n \\ 0 & 1 & 1 & \cdots & 1 \\ 1 & x_1^2 & x_2^2 & \cdots & x_n^2 \\ 1 & x_1^3 & x_2^3 & \cdots & x_n^3 \\ \vdots & \vdots & \vdots & & \vdots \\ 1 & x_1^n & x_2^n & \cdots & x_n^n \end{vmatrix} = - \begin{vmatrix} 0 & 1 & 1 & \cdots & 1 \\ 1 & x_1 & x_2 & \cdots & x_n \\ 1 & x_1^2 & x_2^2 & \cdots & x_n^2 \\ 1 & x_1^3 & x_2^3 & \cdots & x_n^3 \\ \vdots & \vdots & \vdots & & \vdots \\ 1 & x_1^n & x_2^n & \cdots & x_n^n \end{vmatrix}$$

$$= - \left(\begin{vmatrix} 1 & 1 & 1 & \cdots & 1 \\ 1 & x_1 & x_2 & \cdots & x_n \\ 1 & x_1^2 & x_2^2 & \cdots & x_n^2 \\ 1 & x_1^3 & x_2^3 & \cdots & x_n^3 \\ \vdots & \vdots & \vdots & & \vdots \\ 1 & x_1^n & x_2^n & \cdots & x_n^n \end{vmatrix} + \begin{vmatrix} -1 & 0 & 0 & \cdots & 0 \\ 1 & x_1 & x_2 & \cdots & x_n \\ 1 & x_1^2 & x_2^2 & \cdots & x_n^2 \\ 1 & x_1^3 & x_2^3 & \cdots & x_n^3 \\ \vdots & \vdots & \vdots & & \vdots \\ 1 & x_1^n & x_2^n & \cdots & x_n^n \end{vmatrix} \right)$$

$$= \left(\prod_{i=1}^{n} x_i - \prod_{i=1}^{n}(x_i-1) \right) \prod_{1 \leqslant j < i \leqslant n} (x_i - x_j). \qquad \square$$

Example 2.6.9. Evaluate $D_n = \begin{vmatrix} a+x_1 & a+x_1^2 & \cdots & a+x_1^n \\ a+x_2 & a+x_2^2 & \cdots & a+x_2^n \\ \vdots & \vdots & & \vdots \\ a+x_n & a+x_n^2 & \cdots & a+x_n^n \end{vmatrix}.$

Solution. **Method 1 (Adding edges).**

$$D_n = \begin{vmatrix} 1 & a & a & \cdots & a \\ 0 & a+x_1 & a+x_1^2 & \cdots & a+x_1^n \\ 0 & a+x_2 & a+x_2^2 & \cdots & a+x_2^n \\ \vdots & \vdots & \vdots & & \vdots \\ 0 & a+x_n & a+x_n^2 & \cdots & a+x_n^n \end{vmatrix}$$

$$= \begin{vmatrix} 1 & a & a & \cdots & a \\ -1 & x_1 & x_1^2 & \cdots & x_1^n \\ -1 & x_2 & x_2^2 & \cdots & x_2^n \\ \vdots & \vdots & \vdots & & \vdots \\ -1 & x_n & x_n^2 & \cdots & x_n^n \end{vmatrix}$$

$$= \begin{vmatrix} 1+a-a & 0+a & 0+a & \cdots & 0+a \\ -1 & x_1 & x_1^2 & \cdots & x_1^n \\ -1 & x_2 & x_2^2 & \cdots & x_2^n \\ \vdots & \vdots & \vdots & & \vdots \\ -1 & x_n & x_n^2 & \cdots & x_n^n \end{vmatrix}$$

$$= \begin{vmatrix} 1+a & 0 & 0 & \cdots & 0 \\ -1 & x_1 & x_1^2 & \cdots & x_1^n \\ -1 & x_2 & x_2^2 & \cdots & x_2^n \\ \vdots & \vdots & \vdots & & \vdots \\ -1 & x_n & x_n^2 & \cdots & x_n^n \end{vmatrix} + \begin{vmatrix} -a & a & a & \cdots & a \\ -1 & x_1 & x_1^2 & \cdots & x_1^n \\ -1 & x_2 & x_2^2 & \cdots & x_2^n \\ \vdots & \vdots & \vdots & & \vdots \\ -1 & x_n & x_n^2 & \cdots & x_n^n \end{vmatrix}$$

$$= (1+a) \begin{vmatrix} x_1 & x_1^2 & \cdots & x_1^n \\ x_2 & x_2^2 & \cdots & x_2^n \\ \vdots & \vdots & & \vdots \\ x_n & x_n^2 & \cdots & x_n^n \end{vmatrix} - a \begin{vmatrix} 1 & 1 & 1 & \cdots & 1 \\ 1 & x_1 & x_1^2 & \cdots & x_1^n \\ 1 & x_2 & x_2^2 & \cdots & x_2^n \\ \vdots & \vdots & \vdots & & \vdots \\ 1 & x_n & x_n^2 & \cdots & x_n^n \end{vmatrix}$$

$$= \left((1+a) \prod_{i=1}^{n} x_i - a \prod_{i=1}^{n} (x_i - 1) \right) \prod_{1 \leqslant j < i \leqslant n} (x_i - x_j).$$

Method 2 (Using the result of Example 2.6.7).

Subtract column $i-1$ from column i for $i = n, n-1, \ldots, 2$ to obtain

$$D_n = \begin{vmatrix} a+x_1 & x_1(x_1-1) & \cdots & x_1^{n-1}(x_1-1) \\ a+x_2 & x_2(x_2-1) & \cdots & x_2^{n-1}(x_2-1) \\ \vdots & \vdots & & \vdots \\ a+x_n & x_n(x_n-1) & \cdots & x_n^{n-1}(x_n-1) \end{vmatrix}$$

$$= \begin{vmatrix} a & x_1(x_1-1) & \cdots & x_1^{n-1}(x_1-1) \\ a & x_2(x_2-1) & \cdots & x_2^{n-1}(x_2-1) \\ \vdots & \vdots & & \vdots \\ a & x_n(x_n-1) & \cdots & x_n^{n-1}(x_n-1) \end{vmatrix} + \begin{vmatrix} x_1 & x_1^2 & \cdots & x_1^n \\ x_2 & x_2^2 & \cdots & x_2^n \\ \vdots & \vdots & & \vdots \\ x_n & x_n^2 & \cdots & x_n^n \end{vmatrix}$$

$$= \begin{vmatrix} a & x_1(x_1-1) & \cdots & x_1^{n-1}(x_1-1) \\ a & x_2(x_2-1) & \cdots & x_2^{n-1}(x_2-1) \\ \vdots & \vdots & & \vdots \\ a & x_n(x_n-1) & \cdots & x_n^{n-1}(x_n-1) \end{vmatrix} + \prod_{i=1}^{n} x_i \prod_{1 \leqslant j < i \leqslant n} (x_i - x_j)$$

$$= a \begin{vmatrix} 1 & x_1(x_1-1) & \cdots & x_1^{n-1}(x_1-1) \\ 1 & x_2(x_2-1) & \cdots & x_2^{n-1}(x_2-1) \\ \vdots & \vdots & & \vdots \\ 1 & x_n(x_n-1) & \cdots & x_n^{n-1}(x_n-1) \end{vmatrix} + \prod_{i=1}^{n} x_i \prod_{1 \leqslant j < i \leqslant n} (x_i - x_j)$$

$$\overset{\underline{\text{Example 2.6.7}}}{=\!=\!=\!=} a \left(\prod_{i=1}^{n} x_i - \prod_{i=1}^{n}(x_i - 1) \right) \prod_{1 \leqslant j < i \leqslant n} (x_i - x_j)$$

$$+ \prod_{i=1}^{n} x_i \prod_{1 \leqslant j < i \leqslant n} (x_i - x_j)$$

$$= \left((1+a) \prod_{i=1}^{n} x_i - a \prod_{i=1}^{n}(x_i - 1) \right) \prod_{1 \leqslant j < i \leqslant n} (x_i - x_j). \qquad \square$$

VII. Using the Vandermonde determinant

This method is to evaluate a determinant by using the Vandermonde determinant.

Example 2.6.10. Evaluate $D_n = \begin{vmatrix} 1 & 2 & 3 & \cdots & n \\ 1 & 2^3 & 3^3 & \cdots & n^3 \\ 1 & 2^5 & 3^5 & \cdots & n^5 \\ \vdots & \vdots & \vdots & & \vdots \\ 1 & 2^{2n-1} & 3^{2n-1} & \cdots & n^{2n-1} \end{vmatrix}$.

Solution.
$$D_n = (n!) \begin{vmatrix} 1 & 1 & 1 & \cdots & 1 \\ 1 & 2^2 & 3^2 & \cdots & n^2 \\ 1 & (2^2)^2 & (3^2)^2 & \cdots & (n^2)^2 \\ \vdots & \vdots & \vdots & & \vdots \\ 1 & (2^2)^{n-1} & (3^2)^{n-1} & \cdots & (n^2)^{n-1} \end{vmatrix}$$

$$= (n!) \prod_{1 \leqslant j < i \leqslant n} (i^2 - j^2)$$

$$= (n!) \prod_{1 \leqslant j < i \leqslant n} (i + j) \prod_{1 \leqslant j < i \leqslant n} (i - j)$$

$$= 1!3!5! \cdots (2n - 1)!. \qquad \square$$

VIII. Using the Laplace expansion

This method is to evaluate a determinant by using the Laplace expansion.

Example 2.6.11. Evaluate $D = \begin{vmatrix} 1 & 1 & 0 & 0 & 0 & 1 \\ x_1 & x_2 & 0 & 0 & 0 & x_3 \\ a_1 & b_1 & 1 & 1 & 1 & c_1 \\ a_2 & b_2 & x_1 & x_2 & x_3 & c_2 \\ a_3 & b_3 & x_1^2 & x_2^2 & x_3^2 & c_3 \\ x_1^2 & x_2^2 & 0 & 0 & 0 & x_3^2 \end{vmatrix}$.

Solution. **Method 1 (Using the Laplace expansion).**

$$D = \begin{vmatrix} 1 & 1 & 1 \\ x_1 & x_2 & x_3 \\ x_1^2 & x_2^2 & x_3^2 \end{vmatrix} \times (-1)^{1+2+6+1+2+6} \begin{vmatrix} 1 & 1 & 1 \\ x_1 & x_2 & x_3 \\ x_1^2 & x_2^2 & x_3^2 \end{vmatrix}$$

$$= (x_2 - x_1)^2 (x_3 - x_1)^2 (x_3 - x_2)^2.$$

Method 2 (Using properties of determinants).

$$D = (-1)^3 \begin{vmatrix} 1 & 1 & 0 & 0 & 0 & 1 \\ x_1 & x_2 & 0 & 0 & 0 & x_3 \\ x_1^2 & x_2^2 & 0 & 0 & 0 & x_3^2 \\ a_1 & b_1 & 1 & 1 & 1 & c_1 \\ a_2 & b_2 & x_1 & x_2 & x_3 & c_2 \\ a_3 & b_3 & x_1^2 & x_2^2 & x_3^2 & c_3 \end{vmatrix}$$

$$= (-1)^3 (-1)^3 \begin{vmatrix} 1 & 1 & 1 & 0 & 0 & 0 \\ x_1 & x_2 & x_3 & 0 & 0 & 0 \\ x_1^2 & x_2^2 & x_3^2 & 0 & 0 & 0 \\ a_1 & b_1 & c_1 & 1 & 1 & 1 \\ a_2 & b_2 & c_2 & x_1 & x_2 & x_3 \\ a_3 & b_3 & c_3 & x_1^2 & x_2^2 & x_3^2 \end{vmatrix}$$

$$= \begin{vmatrix} 1 & 1 & 1 \\ x_1 & x_2 & x_3 \\ x_1^2 & x_2^2 & x_3^2 \end{vmatrix}^2 = (x_2 - x_1)^2 (x_3 - x_1)^2 (x_3 - x_2)^2. \qquad \square$$

IX. Using multiplication of determinants

This method is to prove some results using the multiplication of determinants or express the given determinant as a product of two determinants that are easier to evaluate.

Example 2.6.12. Let $n \geqslant 2$ be an integer. Evaluate

$$D = \begin{vmatrix} 1 + x_1 y_1 & 1 + x_1 y_2 & \cdots & 1 + x_1 y_n \\ 1 + x_2 y_1 & 1 + x_2 y_2 & \cdots & 1 + x_2 y_n \\ \vdots & \vdots & & \vdots \\ 1 + x_n y_1 & 1 + x_n y_2 & \cdots & 1 + x_n y_n \end{vmatrix}.$$

Solution. **Method 1 (Using multiplication of determinants).**

$$D = \begin{vmatrix} 1 & x_1 & 0 & \cdots & 0 \\ 1 & x_2 & 0 & \cdots & 0 \\ \vdots & \vdots & \vdots & & \vdots \\ 1 & x_n & 0 & \cdots & 0 \end{vmatrix} \begin{vmatrix} 1 & 1 & \cdots & 1 \\ y_1 & y_2 & \cdots & y_n \\ 0 & 0 & \cdots & 0 \\ \vdots & \vdots & & \vdots \\ 0 & 0 & \cdots & 0 \end{vmatrix}$$

$$= \begin{cases} (x_2 - x_1)(y_2 - y_1), & \text{if } n = 2, \\ 0, & \text{if } n > 2. \end{cases}$$

Method 2 (Using properties of determinants).

$$
D = \begin{vmatrix} 1 & 1+x_1y_2 & \cdots & 1+x_1y_n \\ 1 & 1+x_2y_2 & \cdots & 1+x_2y_n \\ \vdots & \vdots & & \vdots \\ 1 & 1+x_ny_2 & \cdots & 1+x_ny_n \end{vmatrix} + y_1 \begin{vmatrix} x_1 & 1+x_1y_2 & \cdots & 1+x_1y_n \\ x_2 & 1+x_2y_2 & \cdots & 1+x_2y_n \\ \vdots & \vdots & & \vdots \\ x_n & 1+x_ny_2 & \cdots & 1+x_ny_n \end{vmatrix}
$$

$$
= \begin{vmatrix} 1 & x_1y_2 & \cdots & x_1y_n \\ 1 & x_2y_2 & \cdots & x_2y_n \\ \vdots & \vdots & & \vdots \\ 1 & x_ny_2 & \cdots & x_ny_n \end{vmatrix} + y_1 \begin{vmatrix} x_1 & 1 & \cdots & 1 \\ x_2 & 1 & \cdots & 1 \\ \vdots & \vdots & & \vdots \\ x_n & 1 & \cdots & 1 \end{vmatrix}
$$

$$
= y_2 \cdots y_n \begin{vmatrix} 1 & x_1 & \cdots & x_1 \\ 1 & x_2 & \cdots & x_2 \\ \vdots & \vdots & & \vdots \\ 1 & x_n & \cdots & x_n \end{vmatrix} + y_1 \begin{vmatrix} x_1 & 1 & \cdots & 1 \\ x_2 & 1 & \cdots & 1 \\ \vdots & \vdots & & \vdots \\ x_n & 1 & \cdots & 1 \end{vmatrix}
$$

$$
= \begin{cases} (x_2 - x_1)(y_2 - y_1), & \text{if } n = 2, \\ 0, & \text{if } n > 2. \end{cases} \qquad \square
$$

Example 2.6.13. Let $s_k = x_1^k + x_2^k + \cdots + x_n^k, k = 1, 2, \ldots.$ Evaluate

$$
D = \begin{vmatrix} s_1 & s_2 & s_3 & \cdots & s_n \\ s_2 & s_3 & s_4 & \cdots & s_{n+1} \\ s_3 & s_4 & s_5 & \cdots & s_{n+2} \\ \vdots & \vdots & \vdots & & \vdots \\ s_n & s_{n+1} & s_{n+2} & \cdots & s_{2n-1} \end{vmatrix}.
$$

Solution.

$$
D = \begin{vmatrix} x_1 & x_2 & x_3 & \cdots & x_n \\ x_1^2 & x_2^2 & x_3^2 & \cdots & x_n^2 \\ x_1^3 & x_2^3 & x_3^3 & \cdots & x_n^3 \\ \vdots & \vdots & \vdots & & \vdots \\ x_1^n & x_2^n & x_3^n & \cdots & x_n^n \end{vmatrix} \begin{vmatrix} 1 & x_1 & x_1^2 & \cdots & x_1^{n-1} \\ 1 & x_2 & x_2^2 & \cdots & x_2^{n-1} \\ 1 & x_3 & x_3^2 & \cdots & x_3^{n-1} \\ \vdots & \vdots & \vdots & & \vdots \\ 1 & x_n & x_n^2 & \cdots & x_n^{n-1} \end{vmatrix}
$$

$$= \left(\prod_{i=1}^{n} x_i \right) \begin{vmatrix} 1 & 1 & 1 & \cdots & 1 \\ x_1 & x_2 & x_3 & \cdots & x_n \\ x_1^2 & x_2^2 & x_3^2 & \cdots & x_n^2 \\ \vdots & \vdots & \vdots & & \vdots \\ x_1^{n-1} & x_2^{n-1} & x_3^{n-1} & \cdots & x_n^{n-1} \end{vmatrix} \begin{vmatrix} 1 & x_1 & x_1^2 & \cdots & x_1^{n-1} \\ 1 & x_2 & x_2^2 & \cdots & x_2^{n-1} \\ 1 & x_3 & x_3^2 & \cdots & x_3^{n-1} \\ \vdots & \vdots & \vdots & & \vdots \\ 1 & x_n & x_n^2 & \cdots & x_n^{n-1} \end{vmatrix}$$

$$= \prod_{i=1}^{n} x_i \prod_{1 \leqslant j < i \leqslant n} (x_i - x_j)^2. \qquad \square$$

Example 2.6.14. Let

$$D = \begin{vmatrix} a_0 & a_1 & a_2 & \cdots\cdots & a_{n-3} & a_{n-2} & a_{n-1} \\ a_{n-1} & a_0 & a_1 & \ddots \quad \ddots & \ddots & a_{n-3} & a_{n-2} \\ a_{n-2} & a_{n-1} & a_0 & \ddots \quad \ddots \quad \ddots & & \ddots & a_{n-3} \\ \vdots & & \ddots & \ddots \; \ddots \; \ddots & \ddots & \ddots & \vdots \\ \vdots & & \ddots & \ddots \; \ddots \; \ddots & \ddots & \ddots & \vdots \\ a_3 & & \ddots & \ddots \quad \ddots \quad \ddots & a_0 & a_1 & a_2 \\ a_2 & a_3 & & \ddots \quad \ddots & a_{n-1} & a_0 & a_1 \\ a_1 & a_2 & a_3 & \cdots\cdots & a_{n-2} & a_{n-1} & a_0 \end{vmatrix}.$$

Prove that $D = f(\zeta_0)f(\zeta_1)\cdots f(\zeta_{n-1})$, where $\zeta_0,\ \zeta_1,\ \ldots,\ \zeta_{n-1}$ are n roots of $x^n - 1$ (i.e., the n nth roots of unity, see Remark 1.4.9), $f(x) = a_0 + a_1 x + \cdots + a_{n-1}x^{n-1}$. Usually, D is called a **circulant determinant**.

Solution. Let $V = \begin{vmatrix} 1 & 1 & \cdots & 1 \\ \zeta_0 & \zeta_1 & \cdots & \zeta_{n-1} \\ \vdots & \vdots & & \vdots \\ \zeta_0^{n-1} & \zeta_1^{n-1} & \cdots & \zeta_{n-1}^{n-1} \end{vmatrix}$. Then

$$DV = \begin{vmatrix} f(\zeta_0) & f(\zeta_1) & \cdots & f(\zeta_{n-1}) \\ \zeta_0 f(\zeta_0) & \zeta_1 f(\zeta_1) & \cdots & \zeta_{n-1} f(\zeta_{n-1}) \\ \vdots & \vdots & & \vdots \\ \zeta_0^{n-1} f(\zeta_0) & \zeta_1^{n-1} f(\zeta_1) & \cdots & \zeta_{n-1}^{n-1} f(\zeta_{n-1}) \end{vmatrix}$$

$$= f(\zeta_0)f(\zeta_1)\cdots f(\zeta_{n-1})V.$$

Since ζ_0, ζ_1, $\ldots$, ζ_{n-1} are different from each other, $V \neq 0$. It follows that $D = f(\zeta_0)f(\zeta_1)\cdots f(\zeta_{n-1})$. $\qquad\qquad\square$

2.7 Definition and Operations of Matrices

The concept of matrices was gradually formed in the 19th century. Arthur Cayley is recognized as the founder of matrix theory. When he began to study matrices as an independent mathematical object, many properties related to matrices were found in the study of determinants, making Cayley think that the introduction of matrices is very natural. The term "matrix" was coined by Sylvester[9] in 1850, who understood a matrix as an object giving rise to many determinants today called minors, that is to say, determinants of smaller matrices that derive from the original one by removing columns and rows. In an 1851 paper, Sylvester explains: I have in previous papers defined a "Matrix" as a rectangular array of terms, out of which different systems of determinants may be engendered as from the womb of a common parent.

Definition 2.7.1. Let m and n be positive integers. For any given mn numbers or polynomials $a_{ij}, i = 1, 2, \ldots, m; j = 1, 2, \ldots, n$, the rectangular array with parentheses

$$\begin{pmatrix} a_{11} & a_{12} & \cdots & a_{1n} \\ a_{21} & a_{22} & \cdots & a_{2n} \\ \vdots & \vdots & & \vdots \\ a_{m1} & a_{m2} & \cdots & a_{mn} \end{pmatrix}$$

is called an $m \times n$ **matrix**, denoted by $(a_{ij})_{m \times n}$, where a_{ij} is called the (i, j) **entry** of this matrix. The index i is called the **row index**, and j is the **column index** of a_{ij}.

Remark 2.7.2.

(1) Unlike a determinant, an $m \times n$ matrix is only a rectangular array of mn numbers (polynomials) with m rows and n columns.

---

[9]James Joseph Sylvester, 1814–1897, English mathematician.

(2) In what follows, we often use uppercase English letters to represent matrices, for example, $A = (a_{ij})_{m \times n}$, $B = (b_{ij})_{s \times t}, \ldots$, or write an $m \times n$ matrix as $A_{m \times n}, B_{m \times n}$, and so on.

(3) Let $A = (a_{ij})_{m \times n}$ and $B = (b_{ij})_{s \times t}$. A and B are **equal**, denoted by $A = B$, if $m = s, n = t$, and $a_{ij} = b_{ij}$ for $i = 1, 2, \ldots, m$ and $j = 1, 2, \ldots, n$.

(4) Any $m \times n$ matrix all of whose entries are 0 is called a **zero matrix** and is denoted by $0_{m \times n}$, or, simply, 0. Thus $(a_{ij})_{m \times n} = 0$ if and only if $a_{ij} = 0$ for $i = 1, 2, \ldots, m$ and $j = 1, 2, \ldots, n$.

(5) A $1 \times n$ matrix $(a_1, a_2, \ldots, a_n)$ is called an n-dimensional **row vector** (in this case, the entries of the matrix are often separated by commas); and an $n \times 1$ matrix $\begin{pmatrix} a_1 \\ a_2 \\ \vdots \\ a_n \end{pmatrix}$ is called an n-dimensional **column vector**, a_i is called the ith **component** of the n-dimensional row (column) vector, $i = 1, 2, \ldots, n$.

(6) An $n \times n$ matrix is called a **square matrix** of order n. In particular, a 1×1 matrix (a_{11}) is denoted by a_{11}, that is, a 1×1 matrix and its entry are indistinguishable.

(7) Let F be a number field and $A = (a_{ij})_{m \times n}$. If all $a_{ij} \in F$, then A is a matrix over F; if all $a_{ij} \in F[x]$, then A is a polynomial matrix over F or λ-matrix (see Chapter 6).

(8) In this chapter, from now on, the matrices involved are all matrices over a number field F unless otherwise stated. The set of all $m \times n$ matrices over F is denoted by $\mathrm{M}_{m \times n}(F)$. If $m = n$, then $\mathrm{M}_{m \times n}(F)$ is abbreviated as $\mathrm{M}_n(F)$; if $n = 1$, then $\mathrm{M}_{m \times 1}(F)$ is denoted by $F^{m \times 1}$; if $m = 1$, then $\mathrm{M}_{1 \times n}(F)$ is denoted by $F^{1 \times n}$.

Definition 2.7.3. Let $A = (a_{ij})_{m \times n}, B = (b_{ij})_{m \times n} \in \mathrm{M}_{m \times n}(F)$. The **sum** $A + B$ of A and B is defined by $A + B = (a_{ij} + b_{ij})_{m \times n}$. If $k \in F$, then the **scalar multiplication** kA of A by k is defined by $kA = (ka_{ij})_{m \times n}$.

It should be noted that the sum of the matrices A and B is defined only when A and B have the same number of rows and the same number of columns. To obtain the sum of A and B, we merely add corresponding entries. The scalar multiplication kA of A by k is obtained by multiplying each entry of A by k. Given a matrix A, we call $(-1)A$

the **negative** of A, denoted by $-A$, i.e., $(-1)A = -A$. The **difference** $A - B$ between A and B is defined by $A - B = A + (-1)B$.

Remark 2.7.4. Let $A, B, C \in \mathrm{M}_{m \times n}(F)$ and $k, l \in F$. By definition, it is not difficult to verify that the addition and scalar multiplication of matrices meet the following operation rules:

(1) **Commutative law of addition:** $A + B = B + A$.
(2) **Associative law of addition:** $(A + B) + C = A + (B + C)$.
(3) Zero matrix is the **identity for addition:** $A + 0 = A$.
(4) Negation produces the **additive inverse:** $A + (-A) = 0$.
(5) **Associative law of scalar multiplication:** $(kl)A = k(lA)$.
(6) **Unitary law:** $1A = A$.
(7) **Distributive law of scalars over a matrix:**

$$(k + l)A = kA + lA.$$

(8) **Distributive law of a scalar over matrices:**

$$k(A + B) = kA + kB.$$

Since m-dimensional column vectors over a number field F are special matrices, the addition, scalar multiplication, and the corresponding operation rules of matrices are also valid for m-dimensional column vectors. So we have the following.

Definition 2.7.5. The set $F^{m \times 1}$ of m-dimensional column vectors over a number field F together with the addition and scalar multiplication is called the m-dimensional **column vector space** over F.

Similarly, we may define the n-dimensional **row vector space** $F^{1 \times n}$ over F.

Remark 2.7.6. Let F be a number field.

$$F^{1 \times n} = \{(a_1, a_2, \ldots, a_n) \mid a_i \in F, \ i = 1, 2, \ldots, n\}$$

is the n-dimensional row vector space over F, and

$$F^{n \times 1} = \left\{ \begin{pmatrix} a_1 \\ a_2 \\ \vdots \\ a_n \end{pmatrix} \ \middle| \ a_i \in F, \ i = 1, 2, \ldots, n \right\}$$

is the n-dimensional column vector space over F. There is no essential difference between the two spaces, but the elements are written differently, so they are abbreviated as F^n, collectively called the n-dimensional **vector space** over F. In the following, we refer to an n-dimensional row (column) vector over F as an n-dimensional **vector** over F unless otherwise stated. We can see whether we are talking about row or column vectors from the context on specific occasions. When $F = \mathbb{R}$ ($\mathbb{C}$), F^n is called the n-dimensional real (complex) vector space.

We have just discussed matrix addition and scalar multiplication. Next, we deal with matrix multiplication.

Definition 2.7.7. Let $A = (a_{ij})_{s \times n}$ and $B = (b_{ij})_{n \times m}$. The **product** AB of A and B is defined by $AB = (c_{ij})_{s \times m}$, where $c_{ij} = \sum_{k=1}^{n} a_{ik}b_{kj}$, $i = 1, 2, \ldots, s, j = 1, 2, \ldots, m$.

For matrix multiplication, the number of columns of the first matrix must be equal to the number of rows of the second matrix. The product of two matrices has the number of rows of the first matrix and the number of columns of the second matrix. The (i, j) entry of the product is equal to the sum of the products of the entries of the ith row of the first matrix and the corresponding entries of the jth column of the second matrix.

Example 2.7.8. The system of linear equations

$$\begin{cases} a_{11}x_1 + a_{12}x_2 + \cdots + a_{1n}x_n = b_1, \\ a_{21}x_1 + a_{22}x_2 + \cdots + a_{2n}x_n = b_2, \\ \quad \vdots \qquad \vdots \qquad\qquad \vdots \qquad \vdots \\ a_{m1}x_1 + a_{m2}x_2 + \cdots + a_{mn}x_n = b_m \end{cases}$$

can be expressed as the matrix equation

$$\begin{pmatrix} a_{11} & a_{12} & \cdots & a_{1n} \\ a_{21} & a_{22} & \cdots & a_{2n} \\ \vdots & \vdots & & \vdots \\ a_{m1} & a_{m2} & \cdots & a_{mn} \end{pmatrix} \begin{pmatrix} x_1 \\ x_2 \\ \vdots \\ x_n \end{pmatrix} = \begin{pmatrix} b_1 \\ b_2 \\ \vdots \\ b_m \end{pmatrix}.$$

Example 2.7.9.

(1) Let $A = \begin{pmatrix} 1 & 2 & 3 \\ 0 & 1 & 2 \end{pmatrix}$, $B = \begin{pmatrix} 1 & 1 & 0 & 1 \\ 0 & 1 & 0 & 0 \\ 0 & 0 & 1 & 0 \end{pmatrix}$.

Then $AB = \begin{pmatrix} 1 & 3 & 3 & 1 \\ 0 & 1 & 2 & 0 \end{pmatrix}$.

(2) Let $A = (a_1, a_2, \ldots, a_n)$, $B = \begin{pmatrix} b_1 \\ b_2 \\ \vdots \\ b_n \end{pmatrix}$.

Then $AB = a_1 b_1 + a_2 b_2 + \cdots + a_n b_n$, $BA = \begin{pmatrix} b_1 a_1 & b_1 a_2 & \cdots & b_1 a_n \\ b_2 a_1 & b_2 a_2 & \cdots & b_2 a_n \\ \vdots & \vdots & & \vdots \\ b_n a_1 & b_n a_2 & \cdots & b_n a_n \end{pmatrix}$.

(3) Let $A = \begin{pmatrix} 1 & 0 \\ 1 & 0 \end{pmatrix}$, $B = \begin{pmatrix} 0 & 0 \\ 1 & 1 \end{pmatrix}$, $C = \begin{pmatrix} 0 & 0 \\ 0 & 0 \end{pmatrix}$.

Then $AB = AC = \begin{pmatrix} 0 & 0 \\ 0 & 0 \end{pmatrix}$, $BA = \begin{pmatrix} 0 & 0 \\ 2 & 0 \end{pmatrix}$.

Remark 2.7.10. From Example 2.7.9 we know the following:

(1) Matrix multiplication does not satisfy the commutative law:

    (1.1) When AB is defined, BA may not be defined (see Example 2.7.9 (1)).

    (1.2) Even though both AB and BA are defined, their rows or columns may not be the same (see Example 2.7.9 (2)).

    (1.3) Even though both AB and BA are square matrices of order n, AB may not be equal to BA (see Example 2.7.9 (3)).

(2) Matrix multiplication does not satisfy the cancellation law, that is, $AB = AC$ and $A \neq 0$ do not imply $B = C$ in general (see Example 2.7.9 (3)).

As pointed out before, the multiplication of matrices does not satisfy the commutative law, but for two specific matrices A and B, $AB = BA$ may hold. So, we have the following.

Definition 2.7.11. Two matrices A and B are said to be **commuting** if $AB = BA$.

It can be seen from the definition that two commuting matrices must be square matrices of the same order.

Example 2.7.12. Let $A = \begin{pmatrix} 2 & 3 \\ 0 & 2 \end{pmatrix}$ and $B = \begin{pmatrix} 5 & -7 \\ 0 & 5 \end{pmatrix}$. Then $AB = BA = \begin{pmatrix} 10 & 1 \\ 0 & 10 \end{pmatrix}$. So, A and B are commuting.

Next, we introduce some special types of matrices.

Definition 2.7.13.

(1) The $n \times n$ matrix of the form

$$\begin{pmatrix} a_{11} & a_{12} & \cdots & a_{1,n-1} & a_{1n} \\ 0 & a_{22} & \cdots & a_{2,n-1} & a_{2n} \\ \vdots & \ddots & \ddots & \vdots & \vdots \\ \vdots & & \ddots & a_{n-1,n-1} & a_{n-1,n} \\ 0 & \cdots & \cdots & 0 & a_{nn} \end{pmatrix}$$

is called an **upper triangular matrix**.

(2) The $n \times n$ matrix of the form

$$\begin{pmatrix} a_{11} & 0 & \cdots & \cdots & 0 \\ a_{21} & a_{22} & \ddots & & \vdots \\ \vdots & \vdots & \ddots & \ddots & \vdots \\ a_{n-1,1} & a_{n-1,2} & \cdots & a_{n-1,n-1} & 0 \\ a_{n1} & a_{n2} & \cdots & a_{n,n-1} & a_{nn} \end{pmatrix}$$

is called a **lower triangular matrix**.

(3) The $n \times n$ matrix of the form
$$\begin{pmatrix} a_1 & 0 & \cdots & \cdots & 0 \\ 0 & a_2 & \ddots & & \vdots \\ \vdots & \ddots & \ddots & \ddots & \vdots \\ \vdots & & \ddots & a_{n-1} & 0 \\ 0 & \cdots & \cdots & 0 & a_n \end{pmatrix}$$
is

called a **diagonal matrix**, abbreviated as $\begin{pmatrix} a_1 & & & \\ & a_2 & & \\ & & \ddots & \\ & & & a_n \end{pmatrix}$, or

$\mathrm{diag}(a_1, a_2, \ldots, a_n)$.

(4) The $n \times n$ matrix of the form $\begin{pmatrix} 1 & & & \\ & 1 & & \\ & & \ddots & \\ & & & 1 \end{pmatrix}$ is called the **iden-**

tity matrix, denoted by I_n, or abbreviated as I (without causing confusion).

(5) The $n \times n$ matrix of the form $kI_n = \begin{pmatrix} k & & & \\ & k & & \\ & & \ddots & \\ & & & k \end{pmatrix}$ is called a

scalar matrix, where k is a number.

Remark 2.7.14. It is not difficult to verify that the following operation rules are true:

(1) **Associative law of matrix multiplication:** Let $A \in \mathrm{M}_{s \times n}(F)$, $B \in \mathrm{M}_{n \times m}(F)$, $C \in \mathrm{M}_{m \times t}(F)$. Then

$$(AB)C = A(BC).$$

(2) **Left distributive law of matrix multiplication with respect to matrix addition:** Let $C \in \mathrm{M}_{s \times n}(F)$, $A, B \in \mathrm{M}_{n \times m}(F)$. Then

$$C(A + B) = CA + CB.$$

(3) **Right distributive law of matrix multiplication with respect to matrix addition:** Let $A, B \in \mathrm{M}_{n \times m}(F), D \in \mathrm{M}_{m \times t}(F)$. Then

$$(A + B)D = AD + BD.$$

(4) **Associative law of matrix multiplication and scalar multiplication:** Let $A \in \mathrm{M}_{s \times n}(F), B \in \mathrm{M}_{n \times m}(F), k \in F$. Then

$$k(AB) = (kA)B = A(kB).$$

(5) The identity matrix is the **identity for multiplication:** Let $A \in \mathrm{M}_{s \times n}(F)$. Then $AI_n = A, I_s A = A$.

Let $A \in \mathrm{M}_n(F)$ and $k \in F$. Then $kA = (kI_n)A = A(kI_n)$ by Remark 2.7.14 (4) (5). Therefore, scalar matrices of order n commute with all matrices of order n. In fact, if a matrix A of order n commutes with all matrices of order n, then A must be a scalar matrix (see Exercise 25).

Definition 2.7.15. Let $A \in \mathrm{M}_n(F)$. The **powers of** A are defined as follows: $A^0 = I_n$, $A^1 = A$, $A^{k+1} = A^k A$, $k \geqslant 1$.
  Let $f(x) = a_0 x^m + a_1 x^{m-1} + \cdots + a_{m-1} x + a_m \in F[x]$. Then

$$f(A) = a_0 A^m + a_1 A^{m-1} + \cdots + a_{m-1} A + a_m I_n$$

is called a **polynomial in the matrix** A.

Remark 2.7.16. Let $A, B \in \mathrm{M}_n(F)$.

(1) For any two natural numbers k and l, we always have

$$A^k A^l = A^l A^k = A^{k+l}, (A^k)^l = A^{kl}.$$

(2) Let $f(x), g(x) \in F[x]$ and $h(x) = f(x)g(x)$. Then $h(A) = f(A)g(A)$. Since $h(x) = g(x)f(x)$, it follows that $h(A) = g(A)f(A)$, whence $f(A)g(A) = g(A)f(A)$. Therefore, any two polynomials in the matrix A are commuting.

(3) If $AB = BA$, then

$$A^2 - B^2 = (A + B)(A - B),$$
$$(A+B)^m = A^m + C_m^1 A^{m-1}B + \cdots + C_m^{m-1}AB^{m-1} + B^m, \ m \geqslant 1,$$
$$(AB)^k = A^k B^k, \ k \geqslant 1.$$

(4) The equalities in (3) may not hold in general.

Definition 2.7.17. Let $A = \begin{pmatrix} a_{11} & a_{12} & \cdots & a_{1n} \\ a_{21} & a_{22} & \cdots & a_{2n} \\ \vdots & \vdots & & \vdots \\ a_{m1} & a_{m2} & \cdots & a_{mn} \end{pmatrix} \in M_{m \times n}(F)$. The

transpose A' of A is defined by $A' = \begin{pmatrix} a_{11} & a_{21} & \cdots & a_{m1} \\ a_{12} & a_{22} & \cdots & a_{m2} \\ \vdots & \vdots & & \vdots \\ a_{1n} & a_{2n} & \cdots & a_{mn} \end{pmatrix}$.

Obviously, the transpose of an $m \times n$ matrix is an $n \times m$ matrix. The transpose A' of A is often denoted by A^t or A^T.

Direct verification shows that the transposition of matrices meets the following rules:

(1) Let $A \in M_{m \times n}(F)$. Then $(A')' = A$.
(2) Let $A, B \in M_{m \times n}(F)$. Then $(A + B)' = A' + B'$.
(3) Let $A \in M_{m \times n}(F)$ and $k \in F$. Then $(kA)' = kA'$.
(4) Let $A \in M_{m \times n}(F)$ and $B \in M_{n \times s}(F)$. Then $(AB)' = B'A'$.

Definition 2.7.18. Let $A \in M_n(F)$. Then A is called a **symmetric matrix** if $A' = A$. A is called a **skew-symmetric matrix** if $A' = -A$.

Remark 2.7.19.

(1) The product of two symmetric matrices is not necessarily a symmetric matrix. Please give counterexamples.
(2) Let $A, B \in M_n(F)$. If A and B are symmetric, then AB is symmetric if and only if A and B are commuting. The proof is left to the reader.
(3) Any square matrix can be expressed as the sum of a symmetric matrix and a skew-symmetric matrix (see Exercise 30).

As mentioned earlier, matrices and determinants are different! However, if A is a square matrix over a number field F, we get a determinant, which can be denoted directly in terms of the entries of

A by writing vertical lines instead of parentheses. So the determinant can be regarded as a function on $\mathrm{M}_n(F)$.

Definition 2.7.20. Let $A = \begin{pmatrix} a_{11} & a_{12} & \cdots & a_{1n} \\ a_{21} & a_{22} & \cdots & a_{2n} \\ \vdots & \vdots & & \vdots \\ a_{n1} & a_{n2} & \cdots & a_{nn} \end{pmatrix} \in \mathrm{M}_n(F)$. The

determinant $|A|$ of A is defined by $|A| = \begin{vmatrix} a_{11} & a_{12} & \cdots & a_{1n} \\ a_{21} & a_{22} & \cdots & a_{2n} \\ \vdots & \vdots & & \vdots \\ a_{n1} & a_{n2} & \cdots & a_{nn} \end{vmatrix}$.

The determinant $|A|$ of A is often denoted by $\det(A)$. According to Theorem 2.4.12, the determinant of a product of two square matrices is the product of their determinants, i.e., we have the following.

Theorem 2.7.21. *If* $A, B \in \mathrm{M}_n(F)$, *then* $|AB| = |A||B|$.

We have dealt with three operations: addition, subtraction, and multiplication of matrices. The concept of invertible matrices is introduced in the following.

Definition 2.7.22. Let $A \in \mathrm{M}_n(F)$. If there exists $B \in \mathrm{M}_n(F)$ such that $AB = BA = I_n$, then we say that A is **invertible**.

Remark 2.7.23.

(1) Let A be an $n \times n$ invertible matrix. Then, the matrix satisfying $AB = BA = I_n$ is unique. In fact, if $AB_1 = B_1 A = I_n, AB_2 = B_2 A = I_n$, then

$$B_1 = B_1 I_n = B_1(AB_2) = (B_1 A)B_2 = I_n B_2 = B_2.$$

Therefore, if B satisfies $AB = BA = I_n$, then B is called the **inverse** of A and is denoted by A^{-1}.

(2) The set of all $n \times n$ invertible matrices over a number field F is often denoted by $\mathrm{GL}_n(F)$.

(3) An invertible matrix is also called a **nondegenerate** matrix, or a **nonsingular** matrix.

Definition 2.7.24. Let $A = \begin{pmatrix} a_{11} & a_{12} & \cdots & a_{1n} \\ a_{21} & a_{22} & \cdots & a_{2n} \\ \vdots & \vdots & & \vdots \\ a_{n1} & a_{n2} & \cdots & a_{nn} \end{pmatrix} \in \mathrm{M}_n(F)$. The

adjoint A^* of A is defined by $A^* = \begin{pmatrix} A_{11} & A_{21} & \cdots & A_{n1} \\ A_{12} & A_{22} & \cdots & A_{n2} \\ \vdots & \vdots & & \vdots \\ A_{1n} & A_{2n} & \cdots & A_{nn} \end{pmatrix}$, where

A_{ij} is the cofactor of a_{ij} of $|A|$, $i, j = 1, 2, \ldots, n$.

By matrix multiplication and Theorem 2.4.4, we always have

$$AA^* = A^*A = |A|I_n, \tag{2.6}$$

for any $A \in \mathrm{M}_n(F)$.

Theorem 2.7.25. *Let $A \in \mathrm{M}_n(F)$. Then $A \in \mathrm{GL}_n(F)$ if and only if $|A| \neq 0$.*

Proof. Necessity. Since $A \in \mathrm{GL}_n(F)$, $AA^{-1} = I_n$. Thus $|A| \, |A^{-1}| = 1$, and so $|A| \neq 0$.

Sufficiency. Suppose $|A| \neq 0$, then

$$A\left(\frac{1}{|A|}A^*\right) = \left(\frac{1}{|A|}A^*\right)A = I_n.$$

It follows that $A \in \mathrm{GL}_n(F)$, and $A^{-1} = \frac{1}{|A|}A^*$. $\qquad\qquad\square$

Theorem 2.7.26.

(1) *If $A \in \mathrm{GL}_n(F)$, then $|A^{-1}| = \frac{1}{|A|}$.*
(2) *If $A \in \mathrm{GL}_n(F)$, then $A^{-1} \in \mathrm{GL}_n(F)$ and $(A^{-1})^{-1} = A$. That is, A and A^{-1} are inverses of each other.*
(3) *Let $A, B \in \mathrm{M}_n(F)$. If $AB = I_n$ or $BA = I_n$, then $A, B \in \mathrm{GL}_n(F)$. It follows that A and B are inverses of each other, i.e., $B = A^{-1}$.*
(4) *If $A, B \in \mathrm{GL}_n(F)$, then $AB \in \mathrm{GL}_n(F)$ and $(AB)^{-1} = B^{-1}A^{-1}$.*
(5) *If $A \in \mathrm{GL}_n(F)$, then $A' \in \mathrm{GL}_n(F)$ and $(A')^{-1} = (A^{-1})'$.*

Proof. (1) Since $A \in \mathrm{GL}_n(F)$, $AA^{-1} = I_n$, whence $|A||A^{-1}| = 1$. So $|A^{-1}| = \frac{1}{|A|}$.

(2) Since $A^{-1}A = AA^{-1} = I_n$, $A^{-1} \in \mathrm{GL}_n(F)$ and $(A^{-1})^{-1} = A$ by definition.

(3) Since $AB = I_n$, $|A||B| = 1$. Thus $|A| \neq 0, |B| \neq 0$, and so $A, B \in \mathrm{GL}_n(F)$. Left multiplication by A^{-1} on both sides of $AB = I_n$ gives $B = A^{-1}$. Similarly, $A = B^{-1}$.

(4) Since $(AB)(B^{-1}A^{-1}) = A(BB^{-1})A^{-1} = AI_nA^{-1} = AA^{-1} = I_n$, $AB \in \mathrm{GL}_n(F)$ and $(AB)^{-1} = B^{-1}A^{-1}$.

(5) Since $A'(A^{-1})' = (A^{-1}A)' = I_n' = I_n$, $A' \in \mathrm{GL}_n(F)$ and $(A')^{-1} = (A^{-1})'$. $\square$

Corollary 2.7.27. *If $A_1, A_2, \ldots, A_t \in \mathrm{GL}_n(F)$, then $A_1 A_2 \cdots A_t \in \mathrm{GL}_n(F)$, and $(A_1 A_2 \cdots A_t)^{-1} = A_t^{-1} \cdots A_2^{-1} A_1^{-1}$.*

Example 2.7.28. Let $A, B \in \mathrm{M}_n(F)$ and $AB = A + B$. Prove that $AB = BA$.

Proof. Since $AB = A + B$, $A(B - I) = B$, where $I = I_n$. It follows that $A(B - I) = B - I + I$, and so $(A - I)(B - I) = I$. Thus, $A - I$ and $B - I$ are inverses of each other. Therefore, $(B - I)(A - I) = I$, and hence $BA = A + B = AB$. $\square$

Example 2.7.29 (Jacobson's[10] lemma). Let $A, B \in \mathrm{M}_n(F)$ and $I = I_n$. If $I - AB \in \mathrm{GL}_n(F)$, prove that $I - BA \in \mathrm{GL}_n(F)$, and find $(I - BA)^{-1}$.

Proof. It suffices to find an $X \in \mathrm{M}_n(F)$ such that $(I - BA)X = I$. In fact,

$$
\begin{aligned}
I &= I - BA + BA \\
&= I - BA + BIA \\
&= I - BA + B(I - AB)(I - AB)^{-1}A \\
&= I - BA + (B - BAB)(I - AB)^{-1}A \\
&= I - BA + (I - BA)B(I - AB)^{-1}A \\
&= (I - BA)(I + B(I - AB)^{-1}A).
\end{aligned}
$$

Thus, $I - BA \in \mathrm{GL}_n(F)$ and $(I - BA)^{-1} = I + B(I - AB)^{-1}A$. $\square$

---

[10]Nathan Jacobson, 1910–1999, American mathematician.

Remark 2.7.30. Readers may ask: How did you think of this proof? Why is it called Jacobson's lemma?

(1) It is well known that $1 - x^2 = (1 + x)(1 - x)$. Generally, we have

$$1 - x^n = (1 + x + \cdots + x^{n-1})(1 - x).$$

If $|x| < 1$, then $(1 - x)^{-1} = \frac{1}{1-x} = 1 + x + x^2 + x^3 + x^4 + \cdots$.

Now we pretend not to know that the above expression for $(1 - x)^{-1}$ requires conditions. Replacing 1 by I and x by BA in the above expression for $(1 - x)^{-1}$ gives

$$(I - BA)^{-1} = I + BA + BABA + BABABA$$
$$+ BABABABA + \cdots.$$

It follows (it doesn't really, but it's fun to keep "pretending") that

$$(I - BA)^{-1} = I + B(I + AB + ABAB + ABABAB + \cdots)A$$
$$= I + B(I - AB)^{-1}A.$$

Now stop "pretense"! Although the "derivation" is unlawful, we can still verify the formula above works! In fact,

$$(I - BA)(I + B(I - AB)^{-1}A)$$
$$= I - BA + (I - BA)B(I - AB)^{-1}A$$
$$= I - BA + (B - BAB)(I - AB)^{-1}A$$
$$= I - BA + B(I - AB)(I - AB)^{-1}A$$
$$= I - BA + BA$$
$$= I.$$

Therefore,

$$(I - BA)^{-1} = I + B(I - AB)^{-1}A. \tag{2.7}$$

Once the statement is put in this way, its proof becomes a matter of perfectly legal mechanical computation.

Halmos[11] attributed the previous trick to Jacobson, although this trick cannot be directly used to prove this lemma! Interested readers can refer to [15, 23].

(2) Kaplansky[12] once remarked that, using this trick, you can always reproduce formula (2.7) without fail "even if you are thrown up on a desert island, with all your books and papers lost" [23, p. 4]. Resonating with Kaplansky's witty comment, (2.7) is sometimes fondly referred to as the "Desert Island Formula". For details, see [23–25].

(3) The foregoing remarks tell us that sometimes "unreliable" ideas may bring a "turnaround" or a "breakthrough".

Example 2.7.31. Let $A, B, AB - I \in \mathrm{GL}_n(F)$, where $I = I_n$.

(1) Prove that $A - B^{-1} \in \mathrm{GL}_n(F)$, and find $(A - B^{-1})^{-1}$.

(2) Prove that $(A - B^{-1})^{-1} - A^{-1} \in \mathrm{GL}_n(F)$, and find

$$((A - B^{-1})^{-1} - A^{-1})^{-1}.$$

Proof. (1) Since $A - B^{-1} = ABB^{-1} - IB^{-1} = (AB - I)B^{-1}$, $A - B^{-1} \in \mathrm{GL}_n(F)$ and $(A - B^{-1})^{-1} = B(AB - I)^{-1}$.

(2) Since

$$
\begin{aligned}
(A - B^{-1})^{-1} - A^{-1} &= B(AB - I)^{-1} - A^{-1}I \\
&= B(AB - I)^{-1} - A^{-1}(AB - I)(AB - I)^{-1} \\
&= B(AB - I)^{-1} - (B - A^{-1})(AB - I)^{-1} \\
&= (B - (B - A^{-1}))(AB - I)^{-1} \\
&= A^{-1}(AB - I)^{-1},
\end{aligned}
$$

$(A - B^{-1})^{-1} - A^{-1} \in \mathrm{GL}_n(F)$ and

$$((A - B^{-1})^{-1} - A^{-1})^{-1} = ABA - A, \tag{2.8}$$

which is called **Hua's**[13] **identity.** $\qquad\square$

---

[11] Paul Richard Halmos, 1916–2006, Hungarian-born American mathematician.

[12] Irving Kaplansky, 1917–2006, American mathematician.

[13] Luogeng Hua, 1910–1985, Chinese mathematician.

Remark 2.7.32. Hua's identity is the first identity named after Chinese people we met after entering the university. This identity only involves matrix operations, but the first person finding this identity is fantastic!

2.8 The Rank of a Matrix

According to Theorem 2.7.25, an $n \times n$ matrix A is invertible if and only if $|A| \neq 0$. For a general $m \times n$ matrix A, how can we use determinant to study A? In this section, we introduce the concept of the rank of a matrix by using its minors.

Definition 2.8.1. Let F be a number field and $A \in M_{m \times n}(F)$, and let $1 \leqslant k \leqslant \min\{m, n\}$. Choose any k rows and k columns of A. The kth-order determinant obtained by keeping the entries in the intersections of the chosen k rows and k columns is called a **minor of order** k of A.

Obviously, for a matrix as above, there are a total of $C_m^k C_n^k$ minors of order k.

Definition 2.8.2. Let $A \in M_{m \times n}(F)$. The **rank** of A is r if there exists at least one nonzero minor of order r of A and every minor of order $r + 1$ (if any) of A is 0. The **rank** of A is defined as 0 if $A = 0$. Usually, the rank of A is denoted by rank$A$ or rank(A).

Remark 2.8.3. We have the following conclusions from the definition of the rank of a matrix and the properties of determinants:

(1) If $A \in M_{m \times n}(F)$, then rank$A \leqslant \min\{m, n\}$.
(2) For any matrix A, rank$A = $ rankA'.
(3) Let $A \in M_n(F)$. Then $A \in GL_n(F)$ if and only if rank$A = n$. So, an invertible matrix is also called a **full rank matrix**.
(4) Let $0 \neq A \in M_{m \times n}(F)$. Then

    (4.1) rank$A = r$ if and only if there exists at least one nonzero minor of order r of A, and every minor of order k with $k > r$ of A is 0. So, the rank of A equals the highest order of nonzero minors of A.

    (4.2) rank$A \leqslant r$ if and only if every minor of order $r + 1$ (if any) of A is 0.

(4.3) rank$A \geqslant r$ if and only if there exists at least one nonzero minor of order r of A.

To find the rank of a matrix, we need to evaluate the values of many determinants, which is usually difficult! Therefore, it is necessary to find a better method to find the rank of a matrix.

Our idea is as follows: If we want to find the rank of the matrix A, we hope to transform A into a new matrix B so that they have the same rank, and the rank of B is easy to find, it's better to know at a glance. According to the discussion for determinants, Properties $5, 9$, and 10 of determinants can simplify the evaluation of determinants. Inspired by this, we define the operations described in the following, which play the role we hope.

Definition 2.8.4. Elementary operations on a matrix over a number field F are the following six types of operations:

(1) Multiply a row by a nonzero number k in F.
(2) Add b times a row to another row, where $b \in F$.
(3) Interchange two rows.
(4) Multiply a column by a nonzero number k in F.
(5) Add b times a column to another column, where $b \in F$.
(6) Interchange two columns.

The first three types of operations are **elementary row operations**, and the last three are **elementary column operations**.

In general, one matrix is transformed into another matrix through elementary operations. We often use the notation $A \to B$ to mean that A is transformed to B through a finite sequence of elementary operations. For the convenience of later description, we use the following notations:

$A \xrightarrow{k \times r_1} B$ means that A is transformed to B by multiplying the first row of A by a nonzero k; $A \xrightarrow{k \times c_1} B$ means that A is transformed to B by multiplying the first column of A by a nonzero k.

$A \xrightarrow{b \times r_1 + r_2} B$ means that A is transformed to B by adding b times the first row of A to the second row of A; $A \xrightarrow{b \times c_1 + c_2} B$ means that A is transformed to B by adding b times the first column of A to the second column of A.

$A \xrightarrow{(r_1, r_2)} B$ means that A is transformed to B by interchanging the first and the second rows of A; $A \xrightarrow{(c_1, c_2)} B$ means that A is transformed to B by interchanging the first and the second columns of A.

The annotations can be located above or below the arrow $\longrightarrow$. For example, $A \xrightarrow[k \times r_1]{} B$ also means that A is transformed to B by multiplying the first row of A by a nonzero k.

Note that elementary operations are reversible. For example, when a matrix A is transformed to a matrix B through a finite sequence of elementary operations, B can be converted to A through a finite sequence of the corresponding elementary operations.

Theorem 2.8.5. *Elementary operations do not change the rank of a matrix.*

Proof. It is enough to show that $\operatorname{rank}A = \operatorname{rank}B$ when a matrix A becomes a matrix B through one single elementary operation.

Suppose $\operatorname{rank}A = r$, then every minor of order $r + 1$ of A is 0.

Let M be a minor of order $r + 1$ of B. Then, by Properties $5, 9$, and 10 of determinants, M is either a nonzero multiple of a minor of order $r + 1$ of A or a sum of a minor of order $r + 1$ of A and a multiple of another minor of order $r + 1$ of A. Thus $M = 0$, and hence $\operatorname{rank}B \leqslant \operatorname{rank}A$.

Note that B can be transformed to A through an elementary operation. It follows that $\operatorname{rank}A \leqslant \operatorname{rank}B$, and so $\operatorname{rank}A = \operatorname{rank}B$. $\qquad \square$

For convenience, a row consisting of entirely zero in a matrix is called a **zero row**, and it is called a **nonzero row** otherwise.

Definition 2.8.6. A matrix A is said to be a **row echelon matrix** if A satisfies the following properties:

(1) All rows below a zero row (if any) are zero rows.
(2) The column index of the first (counting from left to right) nonzero entry of a nonzero row is greater than the column index of the first nonzero entry of the row above it (if any).

A **reduced row echelon matrix** is a row echelon matrix in which the first (counting from left to right) nonzero entry of each nonzero row is 1 (called a leading 1) and each column containing a leading 1 has zeros in all its other entries.

A matrix A is said to be a (**reduced**) **column echelon matrix** if its transpose A' is a (reduced) row echelon matrix.

Only (reduced) row echelon matrices are considered in the remainder of this book. For simplicity, a (reduced) row echelon matrix is called a (**reduced**) **echelon matrix** hereafter.

Example 2.8.7.
$$\begin{pmatrix} 1 & 2 & 3 & 4 \\ 0 & 2 & 2 & 3 \\ 0 & 0 & 3 & 1 \\ 0 & 0 & 0 & 0 \end{pmatrix}, \begin{pmatrix} 1 & 2 & 3 & 4 \\ 0 & 0 & 2 & 3 \\ 0 & 0 & 0 & 5 \\ 0 & 0 & 0 & 0 \end{pmatrix}, \text{ and } \begin{pmatrix} 0 & 2 & 3 & 4 \\ 0 & 0 & 2 & 3 \\ 0 & 0 & 0 & 4 \end{pmatrix}$$

are echelon matrices, but $\begin{pmatrix} 1 & 2 & 3 & 4 \\ 0 & 0 & 0 & 0 \\ 0 & 0 & 1 & 0 \\ 0 & 0 & 0 & 0 \end{pmatrix}$ and $\begin{pmatrix} 1 & 0 & 0 & 0 \\ 0 & 0 & 2 & 0 \\ 0 & 0 & 1 & 1 \\ 0 & 0 & 0 & 0 \end{pmatrix}$ are not echelon matrices.

Example 2.8.8.
$$\begin{pmatrix} 1 & 0 & 0 & 0 & -2 & 1 \\ 0 & 1 & 0 & 0 & 3 & 5 \\ 0 & 0 & 0 & 1 & 7 & 4 \\ 0 & 0 & 0 & 0 & 0 & 0 \end{pmatrix}, \begin{pmatrix} 1 & 2 & 0 & 0 \\ 0 & 0 & 1 & 0 \\ 0 & 0 & 0 & 1 \\ 0 & 0 & 0 & 0 \end{pmatrix}, \text{ and } \begin{pmatrix} 1 & 0 & 0 & 0 \\ 0 & 1 & 0 & 0 \\ 0 & 0 & 1 & 0 \\ 0 & 0 & 0 & 1 \end{pmatrix}$$

are reduced echelon matrices, but $\begin{pmatrix} 1 & 2 & 3 & 4 \\ 0 & 1 & 0 & 6 \\ 0 & 0 & 1 & 2 \\ 0 & 0 & 0 & 0 \\ 0 & 0 & 0 & 0 \end{pmatrix}$ and $\begin{pmatrix} 1 & 0 & 0 & 0 \\ 0 & 2 & 1 & 0 \\ 0 & 0 & 0 & 1 \\ 0 & 0 & 0 & 0 \end{pmatrix}$

are not reduced echelon matrices.

Theorem 2.8.9. *Any matrix over a number field F can be transformed to a (reduced) echelon matrix through a finite sequence of elementary row operations.*

Proof. Let $A = (a_{ij}) \in M_{s\times n}(F)$. If $A = 0$, then A is already an echelon matrix. Now we assume that $A \neq 0$. We look in matrix A for the first column with a nonzero entry; say this is column k. Suppose $a_{ik} \neq 0$. Now interchange, if necessary, rows 1 and i, getting a matrix $B = (b_{ij})_{s\times n}$. Thus $b_{1k} \neq 0$. We add suitable multiples of row 1 of B to row i of B ($2 \leqslant i \leqslant s$) such that all entries in column k and rows $2, 3, \ldots, s$ are zero. Denote the resulting matrix by C. Note that we have used only elementary row operations.

Next, consider the $(s-1) \times n$ matrix A_1 obtained by deleting the first row of C. We now repeat the procedure above with matrix A_1 instead of A. Continuing this way, we get an echelon matrix D.

Multiplying each row of D by a suitable nonzero number in F, we obtain an echelon matrix E in which the first (counting from left to right) nonzero entry of each nonzero row is 1. Suppose that the number of nonzero rows of E is r and the leading 1 occurs in row i and column c_i of E, $i = 1, 2, \ldots, r$. Then $c_1 < c_2 < \cdots < c_r$. Now subtract suitable multiples of row i of E to make all entries in column c_i and rows $i-1, i-2, \ldots, 1$ of E equal to zero for $i = 2, 3, \ldots, r$. The resulting matrix is a reduced echelon matrix. $\qquad\square$

Theorem 2.8.10. *The rank of an echelon matrix equals the number of nonzero rows of the echelon matrix.*

Proof. Let A be an echelon matrix and r the number of nonzero rows of A. Since elementary operations do not change the rank of a matrix, by exchanging the order of the columns of A, we may assume

$$
A = \begin{pmatrix}
a_{11} & a_{12} & \cdots & a_{1,r-1} & a_{1r} & a_{1,r+1} & \cdots & a_{1n} \\
0 & a_{22} & \cdots & a_{2,r-1} & a_{2r} & a_{2,r+1} & \cdots & a_{2n} \\
\vdots & \ddots & \ddots & \vdots & \vdots & \vdots & & \vdots \\
\vdots & & \ddots & a_{r-1,r-1} & a_{r-1,r} & a_{r-1,r+1} & \cdots & a_{r-1,n} \\
0 & \cdots & \cdots & 0 & a_{rr} & a_{r,r+1} & \cdots & a_{rn} \\
0 & \cdots & \cdots & 0 & 0 & \cdots & \cdots & 0 \\
\vdots & & & \vdots & \vdots & & & \vdots \\
0 & \cdots & \cdots & 0 & 0 & \cdots & \cdots & 0
\end{pmatrix},
$$

where $a_{ii} \neq 0$, $i = 1, 2, \ldots, r$.

Clearly, A has a minor

$$
\begin{vmatrix}
a_{11} & a_{12} & \cdots & a_{1,r-1} & a_{1r} \\
0 & a_{22} & \cdots & a_{2,r-1} & a_{2r} \\
\vdots & \ddots & \ddots & \vdots & \vdots \\
\vdots & & \ddots & a_{r-1,r-1} & a_{r-1,r} \\
0 & \cdots & \cdots & 0 & a_{rr}
\end{vmatrix} = a_{11} a_{22} \cdots a_{rr} \neq 0,
$$

but all minors of order $r+1$ of A are 0. So rank$A = r$. $\qquad\square$

Example 2.8.11. Let $n > 1$ be an integer, and let

$$A = \begin{pmatrix} 1 & a & \cdots & \cdots & a \\ a & 1 & \ddots & & \vdots \\ \vdots & \ddots & \ddots & \ddots & \vdots \\ \vdots & & \ddots & 1 & a \\ a & \cdots & \cdots & a & 1 \end{pmatrix} \in \mathrm{M}_n(F).$$

Find rank$A$.

Solution. $|A| = \begin{vmatrix} 1 & a & \cdots\cdots & a \\ a & 1 & \ddots & \vdots \\ \vdots & \ddots & \ddots & \vdots \\ \vdots & & \ddots & 1 & a \\ a & \cdots\cdots & a & 1 \end{vmatrix} \overset{\text{Example 2.3.1}}{=\!=\!=\!=\!=} (1+(n-1)a)(1-a)^{n-1}.$

If $a \neq 1$ and $a \neq \frac{1}{1-n}$, then $|A| \neq 0$. So rank$A = n$.

If $a = 1$, then it is clear that rank$A = 1$.

If $a = \frac{1}{1-n}$, then

$$A \xrightarrow{(1-n)\times r_i,\ i=1,2,\ldots,n} \begin{pmatrix} 1-n & 1 & 1 & \cdots & \cdots & 1 \\ 1 & 1-n & 1 & & & \vdots \\ 1 & 1 & 1-n & \ddots & & \vdots \\ \vdots & \vdots & & \ddots & \ddots & \vdots \\ \vdots & \vdots & & & \ddots\, 1-n & 1 \\ 1 & 1 & \cdots & \cdots & 1 & 1-n \end{pmatrix}$$

$$\xrightarrow{c_i+c_1,\ i=2,3,\ldots,n} \begin{pmatrix} 0 & 1 & 1 & \cdots & \cdots & 1 \\ 0 & 1-n & 1 & & & \vdots \\ 0 & 1 & 1-n & \ddots & & \vdots \\ \vdots & \vdots & & \ddots & \ddots & \vdots \\ \vdots & \vdots & & & \ddots\, 1-n & 1 \\ 0 & 1 & \cdots & \cdots & 1 & 1-n \end{pmatrix}$$

$$\xrightarrow{\ (-1)\times r_1+r_i,\ \ i=2,3,\ldots,n\ } \begin{pmatrix} 0 & 1 & 1 & \cdots & \cdots & 1 \\ 0 & -n & 0 & \cdots & \cdots & 0 \\ 0 & 0 & -n & \ddots & & \vdots \\ \vdots & \vdots & & \ddots & \ddots & \vdots \\ \vdots & \vdots & & & \ddots & -n & 0 \\ 0 & 0 & \cdots & \cdots & 0 & -n \end{pmatrix} = B.$$

Note that $\mathrm{rank}B = n - 1$, and hence $\mathrm{rank}A = n - 1$.

$$\text{So } \mathrm{rank}A = \begin{cases} 1, & \text{if } a = 1, \\ n-1, & \text{if } a = \dfrac{1}{1-n}, \\ n, & \text{if } a \neq 1 \text{ and } a \neq \dfrac{1}{1-n}. \end{cases} \qquad \square$$

2.9   Equivalence of Matrices

According to the discussion in Section 2.8, a matrix can be transformed into an echelon matrix using only elementary row operations. In fact, if elementary row and column operations are used simultaneously, a matrix can be transformed into a "standard echelon matrix". For this reason, we introduce the concept of equivalent matrices.

Definition 2.9.1. Let F be a number field and $A, B \in M_{s\times n}(F)$. A is said to be **equivalent** to B, denoted by $A \sim B$, if B can be obtained by applying a finite sequence of elementary operations to A.

It is easy to see that the equivalence of matrices is an equivalence relation which partitions the set $M_{s\times n}(F)$ into disjoint equivalence classes.

Theorem 2.9.2. *Every matrix $A = (a_{ij})_{s \times n}$ over a number field is equivalent to an $s \times n$ matrix of the form*

$$
\begin{pmatrix}
1 & 0 & \cdots & \cdots & 0 & 0 & \cdots & 0 \\
0 & 1 & \ddots & & \vdots & \vdots & & \vdots \\
\vdots & \ddots & \ddots & \ddots & \vdots & \vdots & & \vdots \\
\vdots & & \ddots & 1 & 0 & \vdots & & \vdots \\
0 & \cdots & \cdots & 0 & 1 & 0 & \cdots & 0 \\
0 & \cdots & \cdots & \cdots & 0 & 0 & \cdots & 0 \\
\vdots & & & & \vdots & \vdots & & \vdots \\
0 & \cdots & \cdots & \cdots & 0 & 0 & \cdots & 0
\end{pmatrix},
$$

*which is called the **canonical form** of A, where the number of 1 on the main diagonal is equal to the rank of A (the number of 1 may be 0).*

Proof. If $A = 0$, then A is already a canonical form.

Let $A \neq 0$. Then A can be transformed to a matrix whose upper left corner entry is not equal to 0 by elementary operations. So we assume that $a_{11} \neq 0$. Subtracting $a_{11}^{-1} a_{i1}$ times row 1 from row i $(2 \leqslant i \leqslant s)$ and $a_{11}^{-1} a_{1j}$ times column 1 from column j $(2 \leqslant j \leqslant n)$, and multiplying row 1 by a_{11}^{-1}, we get

$$
B = \begin{pmatrix}
1 & 0 & \cdots & 0 \\
0 & b_{22} & \cdots & b_{2n} \\
\vdots & \vdots & & \vdots \\
0 & b_{s2} & \cdots & a_{sn}
\end{pmatrix}.
$$

Next, consider the $(s-1) \times (n-1)$ matrix A_1 obtained by deleting the first row and column of B. We now repeat the procedure above with matrix A_1 instead of A. Continuing this way, we get the desired canonical form. Clearly, the rank of the canonical form of A is the number of 1 on the main diagonal. Since elementary operations do not change the rank of a matrix, the number of 1 on the main diagonal is the rank of A. $\qquad\square$

Corollary 2.9.3. *Let $A, B \in \mathrm{M}_{s \times n}(F)$. Then $A \sim B$ if and only if* $\mathrm{rank} A = \mathrm{rank} B$.

To further reveal the relationship between equivalent matrices, we introduce the concept of elementary matrices, which links elementary operations and matrix multiplication!

Definition 2.9.4. An **elementary matrix** is a matrix obtained from the identity matrix by performing one single elementary operation.

By definition, each elementary operation corresponds to an elementary matrix, so there are only three types of elementary matrices:

(1) Multiplying row (or column) i of the identity matrix by a nonzero number k yields the elementary matrix of the form

$$
E(i(k)) = \begin{pmatrix} 1 & & & & & & \\ & \ddots & & & & & \\ & & 1 & & & & \\ & & & k & & & \\ & & & & 1 & & \\ & & & & & \ddots & \\ & & & & & & 1 \end{pmatrix} \leftarrow \text{ row } i.
$$

(2) Adding b times row j to row i (or b times column i to column j) of the identity matrix yields the elementary matrix of the form

$$
E(i, j(b)) = \begin{pmatrix} 1 & & & & & & \\ & \ddots & & & & & \\ & & 1 & & b & & \\ & & & \ddots & & & \\ & & & & 1 & & \\ & & & & & \ddots & \\ & & & & & & 1 \end{pmatrix} \begin{matrix} \\ \\ \leftarrow \text{ row } i \\ \\ \leftarrow \text{ row } j \\ \\ \end{matrix} , \ i < j,
$$

$$\text{or } E(i,j(b)) = \begin{pmatrix} 1 & & & & & & & \\ & \ddots & & & & & & \\ & & 1 & & & & & \\ & & & \ddots & & & & \\ & & b & & 1 & & & \\ & & & & & \ddots & & \\ & & & & & & 1 \end{pmatrix} \begin{matrix} \\ \\ \leftarrow \text{ row } j \\ \\ \leftarrow \text{ row } i \\ \\ \\ \end{matrix} \;, \quad i > j.$$

(3) Interchanging rows (or columns) i and j of the identity matrix yields the elementary matrix of the form

$$E(i,j) = \begin{pmatrix} 1 & & & & & & & & & & \\ & \ddots & & & & & & & & & \\ & & 1 & & & & & & & & \\ & & & 0 & & & & 1 & & & \\ & & & & 1 & & & & & & \\ & & & & & \ddots & & & & & \\ & & & & & & 1 & & & & \\ & & & 1 & & & & 0 & & & \\ & & & & & & & & 1 & & \\ & & & & & & & & & \ddots & \\ & & & & & & & & & & 1 \end{pmatrix}.$$

Theorem 2.9.5. *Let $A \in \mathrm{M}_{s \times n}(F)$.*

(1) *Applying an elementary row operation to A is equivalent to multiplying A on the left by the corresponding elementary matrix of order s.*

(2) *Applying an elementary column operation to A is equivalent to multiplying A on the right by the corresponding elementary matrix of order n.*

Proof. We only consider the case of elementary row operations. A completely analogous proof can be given for the case of elementary column operations. Let $A = \begin{pmatrix} a_{11} & a_{12} & \cdots & a_{1n} \\ a_{21} & a_{22} & \cdots & a_{2n} \\ \vdots & \vdots & & \vdots \\ a_{s1} & a_{s2} & \cdots & a_{sn} \end{pmatrix} \in \mathrm{M}_{s\times n}(F)$. By matrix multiplication, we have

$$E(i(k))A = \begin{pmatrix} a_{11} & a_{12} & \cdots & a_{1n} \\ \vdots & \vdots & & \vdots \\ ka_{i1} & ka_{i2} & \cdots & ka_{in} \\ \vdots & \vdots & & \vdots \\ a_{s1} & a_{s2} & \cdots & a_{sn} \end{pmatrix},$$

$$E(i,j(b))A = \begin{pmatrix} a_{11} & a_{12} & \cdots & a_{1n} \\ \vdots & \vdots & & \vdots \\ a_{i1}+ba_{j1} & a_{i2}+ba_{j2} & \cdots & a_{in}+ba_{jn} \\ \vdots & \vdots & & \vdots \\ a_{j1} & a_{j2} & \cdots & a_{jn} \\ \vdots & \vdots & & \vdots \\ a_{s1} & a_{s2} & \cdots & a_{sn} \end{pmatrix}, \quad i < j,$$

$$E(i,j)A = \begin{pmatrix} a_{11} & a_{12} & \cdots & a_{1n} \\ \vdots & \vdots & & \vdots \\ a_{j1} & a_{j2} & \cdots & a_{jn} \\ \vdots & \vdots & & \vdots \\ a_{i1} & a_{i2} & \cdots & a_{in} \\ \vdots & \vdots & & \vdots \\ a_{s1} & a_{s2} & \cdots & a_{sn} \end{pmatrix}, \quad i < j.$$

$\square$

Remark 2.9.6.

(1) From the theorem above, we get

$$E(i(k^{-1}))E(i(k)) = I_n, \quad E(i,j(-b))E(i,j(b)) = I_n,$$
$$E(i,j)E(i,j) = I_n.$$

Therefore, elementary matrices are invertible, and

$$E(i(k))^{-1} = E(i(k^{-1})), \quad E(i,j(b))^{-1} = E(i,j(-b)),$$

$$E(i,j)^{-1} = E(i,j).$$

It follows that the inverse of an elementary matrix is also an elementary matrix; moreover, the elementary matrix and its inverse are of the same type.

(2) It is clear that

$$E(i(k))' = E(i(k)), \quad E(i,j(b))' = E(j,i(b)), \quad E(i,j)' = E(i,j).$$

Thus, $E(i(k))'AE(i(k))$ represents the matrix obtained by multiplying row i of A by a nonzero number k to get a matrix A_1 and then multiplying column i of A_1 by the same number k; $E(i,j(b))'AE(i,j(b))$ represents the matrix obtained by adding b times row i of A to row j of A to get a matrix A_2 and then adding b times column i of A_2 to column j of A_2; $E(i,j)'AE(i,j)$ represents the matrix obtained by interchanging rows i and j of A to get a matrix A_3 and then interchanging columns i and j of A_3.

By Theorem 2.9.5, applying an elementary row (column) operation to a matrix is equivalent to multiplying the matrix on the left (right) by the corresponding elementary matrix, so we have the following.

Corollary 2.9.7. *Let* $A, B \in M_{s \times n}(F)$. *Then* $A \sim B$ *if and only if there exist elementary matrices* $P_1, P_2, \ldots, P_l$ *of order* s *and elementary matrices* $Q_1, Q_2, \ldots, Q_t$ *of order* n *such that* $A = P_l \cdots P_2 P_1 B Q_1 Q_2 \cdots Q_t$.

Corollary 2.9.8. *Let* $A \in M_n(F)$. *Then the following statements are equivalent:*

(1) $A \in \mathrm{GL}_n(F)$.
(2) $\mathrm{rank} A = n$.
(3) $A \sim I_n$.
(4) *A is a product of elementary matrices.*

Corollary 2.9.9. *Let* $A, B \in M_{s \times n}(F)$. *Then* $A \sim B$ *if and only if there exist* $P \in \mathrm{GL}_s(F)$ *and* $Q \in \mathrm{GL}_n(F)$ *such that* $A = PBQ$.

Corollary 2.9.10. *Let $A \in \mathrm{M}_{s \times n}(F)$, $P \in \mathrm{GL}_s(F)$ and $Q \in \mathrm{GL}_n(F)$. Then $\mathrm{rank} A = \mathrm{rank}(PA) = \mathrm{rank}(AQ) = \mathrm{rank}(PAQ)$.*

Corollary 2.9.11. *If $A \in \mathrm{GL}_n(F)$, then A can be transformed to the identity matrix only through a finite sequence of elementary row (or column) operations.*

Proof. Since A is invertible, A is a product of elementary matrices:

$$A = Q_1 Q_2 \cdots Q_m,$$

where each Q_j is an elementary matrix, $j = 1, 2, \ldots, m$. Therefore,

$$Q_m^{-1} \cdots Q_2^{-1} Q_1^{-1} A = I_n \ (\text{or } A Q_m^{-1} \cdots Q_2^{-1} Q_1^{-1} = I_n).$$

Since the inverse of an elementary matrix is also an elementary matrix, A can be transformed to the identity matrix only through a finite sequence of elementary row (or column) operations. $\qquad \square$

The result above provides a method of finding the inverse of an invertible matrix by elementary operations.

Let A be an invertible matrix. Then there exist elementary matrices $P_1, P_2, \ldots, P_m$ such that

$$P_m \cdots P_2 P_1 A = I_n.$$

Multiplying both sides of this equation on the right by A^{-1}, we get

$$P_m \cdots P_2 P_1 I_n = A^{-1}.$$

Therefore, if A is transformed to the identity matrix I_n by a finite sequence of elementary row operations, the same sequence of operations, when applied to I_n, yields A^{-1}.

Let $A \in \mathrm{GL}_n(F)$. To find A^{-1}, we form the $n \times 2n$ matrix $(A | I_n)$ and perform elementary row operations to reduce A on the left side of $(A | I_n)$ to I_n, then I_n on the right side of $(A | I_n)$ becomes A^{-1}, i.e.,

$$(A | I_n) \xrightarrow{\text{elementary row operations}} (I_n | A^{-1}). \qquad (2.9)$$

In the same way, we can obtain the inverse A^{-1} by elementary column operations. We form the $2n \times n$ matrix $\left(\dfrac{A}{I_n} \right)$ and perform elementary column operations to reduce A at the top of $\left(\dfrac{A}{I_n}\right)$ to I_n, then I_n at the bottom of $\left(\dfrac{A}{I_n}\right)$ becomes A^{-1}, i.e.,

$$\left(\frac{A}{I_n}\right) \xrightarrow{\text{elementary column operations}} \left(\frac{I_n}{A^{-1}}\right). \qquad (2.10)$$

Remark 2.9.12. Let $A \in \mathrm{GL}_n(F)$.

(1) If we replace I_n on the left side of (2.9) by an $n \times m$ matrix B, then A^{-1} on the right side of (2.9) becomes $A^{-1}B$, which is the solution of the matrix equation $AX = B$.
(2) If we replace I_n on the left side of (2.10) by an $m \times n$ matrix B, then A^{-1} on the right side of (2.10) becomes BA^{-1}, which is the solution of the matrix equation $XA = B$.

Example 2.9.13. Let $A = \begin{pmatrix} 1 & 1 & 1 & 1 \\ 0 & 1 & 1 & 1 \\ 0 & 0 & 1 & 1 \\ 0 & 0 & 0 & 1 \end{pmatrix}$. Find A^{-1}.

Solution. Note that

$$(A \mid I_4) = \left(\begin{array}{cccc|cccc} 1 & 1 & 1 & 1 & 1 & 0 & 0 & 0 \\ 0 & 1 & 1 & 1 & 0 & 1 & 0 & 0 \\ 0 & 0 & 1 & 1 & 0 & 0 & 1 & 0 \\ 0 & 0 & 0 & 1 & 0 & 0 & 0 & 1 \end{array}\right)$$

$$\xrightarrow{(-1)\times r_4 + r_i,\ i=1,2,3} \left(\begin{array}{cccc|cccc} 1 & 1 & 1 & 0 & 1 & 0 & 0 & -1 \\ 0 & 1 & 1 & 0 & 0 & 1 & 0 & -1 \\ 0 & 0 & 1 & 0 & 0 & 0 & 1 & -1 \\ 0 & 0 & 0 & 1 & 0 & 0 & 0 & 1 \end{array}\right)$$

$$\xrightarrow{(-1)\times r_3 + r_i,\ i=1,2} \left(\begin{array}{cccc|cccc} 1 & 1 & 0 & 0 & 1 & 0 & -1 & 0 \\ 0 & 1 & 0 & 0 & 0 & 1 & -1 & 0 \\ 0 & 0 & 1 & 0 & 0 & 0 & 1 & -1 \\ 0 & 0 & 0 & 1 & 0 & 0 & 0 & 1 \end{array}\right)$$

$$\xrightarrow{(-1)\times r_2 + r_1} \left(\begin{array}{cccc|cccc} 1 & 0 & 0 & 0 & 1 & -1 & 0 & 0 \\ 0 & 1 & 0 & 0 & 0 & 1 & -1 & 0 \\ 0 & 0 & 1 & 0 & 0 & 0 & 1 & -1 \\ 0 & 0 & 0 & 1 & 0 & 0 & 0 & 1 \end{array}\right).$$

Thus $A^{-1} = \begin{pmatrix} 1 & -1 & 0 & 0 \\ 0 & 1 & -1 & 0 \\ 0 & 0 & 1 & -1 \\ 0 & 0 & 0 & 1 \end{pmatrix}$. $\qquad\qquad\square$

2.10 Block Matrices

Generally, when dealing with a large matrix, we divide it into several small blocks and treat each small block as a small matrix. The large matrix can be regarded as a matrix composed of small matrices, just as a matrix is composed of entries. In particular, these small matrices can be treated as entries in matrix operations.

Definition 2.10.1. A **partitioning of a matrix** A is to split A up into a number of smaller rectangular blocks, each of which is called a **submatrix** of A. The partitioning is usually indicated by dashed horizontal lines between rows and vertical lines between columns. The matrix formed by taking these submatrices as entries is called a **block matrix** or **partitioned matrix**.

By definition, any matrix may be interpreted as a block matrix in one or more ways, with each interpretation defined by how its rows and columns are partitioned.

For example, partition the matrix

$$\left(\begin{array}{ccccc|ccc} 1 & 0 & \cdots & \cdots & 0 & 0 & \cdots & 0 \\ 0 & 1 & \ddots & & \vdots & \vdots & & \vdots \\ \vdots & \ddots & \ddots & \ddots & \vdots & \vdots & & \vdots \\ \vdots & & \ddots & 1 & 0 & \vdots & & \vdots \\ 0 & \cdots & \cdots & 0 & 1 & 0 & \cdots & 0 \\ \hline 0 & \cdots & \cdots & \cdots & 0 & 0 & \cdots & 0 \\ \vdots & & & & \vdots & \vdots & & \vdots \\ 0 & \cdots & \cdots & \cdots & 0 & 0 & \cdots & 0 \end{array}\right)$$

in Theorem 2.9.2 to obtain a 2×2 block matrix $\begin{pmatrix} I_r & 0 \\ 0 & 0 \end{pmatrix}$. This form is concise and highlights the characteristics of the matrix.

For another example, partitioning a matrix by rows (columns) is also common.

Let $A = \begin{pmatrix} a_{11} & a_{12} & \cdots & a_{1n} \\ a_{21} & a_{22} & \cdots & a_{2n} \\ \vdots & \vdots & & \vdots \\ a_{m1} & a_{m2} & \cdots & a_{mn} \end{pmatrix}$ be an $m \times n$ matrix.

Partition A by rows to obtain the block matrix $\begin{pmatrix} \alpha_1 \\ \alpha_2 \\ \vdots \\ \alpha_m \end{pmatrix}$, where

$\alpha_1, \alpha_2, \ldots, \alpha_m$ are the first, second, $\ldots$, mth rows of A, respectively. Usually, $\alpha_1, \alpha_2, \ldots, \alpha_m$ are called the **row vectors** of A.

Partition A by columns to obtain the block matrix $(\beta_1, \beta_2, \ldots, \beta_n)$, where $\beta_1, \beta_2, \ldots, \beta_n$ are the first, second, $\ldots$, nth columns of A, respectively. Usually, $\beta_1, \beta_2, \ldots, \beta_n$ are called the **column vectors** of A.

Operations with block matrices are similar to those with ordinary matrices.

Theorem 2.10.2.

(1) *Let $A, B \in \mathrm{M}_{m\times n}(F)$. If A and B are partitioned in the same way, i.e.,*

$$A = \begin{pmatrix} A_{11} & A_{12} & \cdots & A_{1t} \\ A_{21} & A_{22} & \cdots & A_{2t} \\ \vdots & \vdots & & \vdots \\ A_{s1} & A_{s2} & \cdots & A_{st} \end{pmatrix}, \quad B = \begin{pmatrix} B_{11} & B_{12} & \cdots & B_{1t} \\ B_{21} & B_{22} & \cdots & B_{2t} \\ \vdots & \vdots & & \vdots \\ B_{s1} & B_{s2} & \cdots & B_{st} \end{pmatrix},$$

where $A_{ij}, B_{ij} \in \mathrm{M}_{p_i \times q_j}(F)$, $i = 1, 2, \ldots, s$; $j = 1, 2, \ldots, t$, then

$$A + B = \begin{pmatrix} A_{11} + B_{11} & A_{12} + B_{12} & \cdots & A_{1t} + B_{1t} \\ A_{21} + B_{21} & A_{22} + B_{22} & \cdots & A_{2t} + B_{2t} \\ \vdots & \vdots & & \vdots \\ A_{s1} + B_{s1} & A_{s2} + B_{s2} & \cdots & A_{st} + B_{st} \end{pmatrix},$$

$$kA = \begin{pmatrix} kA_{11} & kA_{12} & \cdots & kA_{1t} \\ kA_{21} & kA_{22} & \cdots & kA_{2t} \\ \vdots & \vdots & & \vdots \\ kA_{s1} & kA_{s2} & \cdots & kA_{st} \end{pmatrix}, \ \text{where } k \in F,$$

$$A' = \begin{pmatrix} A'_{11} & A'_{21} & \cdots & A'_{s1} \\ A'_{12} & A'_{22} & \cdots & A'_{s2} \\ \vdots & \vdots & & \vdots \\ A'_{1t} & A'_{2t} & \cdots & A'_{st} \end{pmatrix}.$$

(2) *Let $A \in \mathrm{M}_{s \times n}(F)$ and $B \in \mathrm{M}_{n \times m}(F)$. If the column partition of A matches the row partition of B, i.e.,*

$$A = \begin{pmatrix} A_{11} & A_{12} & \cdots & A_{1t} \\ A_{21} & A_{22} & \cdots & A_{2t} \\ \vdots & \vdots & & \vdots \\ A_{r1} & A_{r2} & \cdots & A_{rt} \end{pmatrix}, \quad B = \begin{pmatrix} B_{11} & B_{12} & \cdots & B_{1p} \\ B_{21} & B_{22} & \cdots & B_{2p} \\ \vdots & \vdots & & \vdots \\ B_{t1} & B_{t2} & \cdots & B_{tp} \end{pmatrix},$$

where $A_{ik} \in \mathrm{M}_{u_i \times v_k}(F)$, $B_{kj} \in \mathrm{M}_{v_k \times w_j}(F)$, $i = 1, 2, \ldots, r$; $k = 1, 2, \ldots, t$; $j = 1, 2, \ldots, p$, then

$$AB = (C_{ij})_{r \times p} = \begin{pmatrix} C_{11} & C_{12} & \cdots & C_{1p} \\ C_{21} & C_{22} & \cdots & C_{2p} \\ \vdots & \vdots & & \vdots \\ C_{r1} & C_{r2} & \cdots & C_{rp} \end{pmatrix},$$

where $C_{ij} = \sum\limits_{k=1}^{t} A_{ik} B_{kj}$; $i = 1, 2, \ldots, r$, $j = 1, 2, \ldots, p$.

Similar to upper (lower) triangular matrices and diagonal matrices, we have the following.

Definition 2.10.3.

(1) The block matrix of the form

$$\begin{pmatrix} A_{11} & A_{12} & \cdots & A_{1,n-1} & A_{1n} \\ 0 & A_{22} & \cdots & A_{2,n-1} & A_{2n} \\ \vdots & \ddots & \ddots & \vdots & \vdots \\ \vdots & & \ddots & A_{n-1,n-1} & A_{n-1,n} \\ 0 & \cdots & \cdots & 0 & A_{nn} \end{pmatrix}$$

is called a **block upper triangular matrix**, where each A_{ii} is a square matrix, $i = 1, 2, \ldots, n$.

(2) The block matrix of the form

$$\begin{pmatrix} A_{11} & 0 & \cdots & \cdots & 0 \\ A_{21} & A_{22} & \ddots & & \vdots \\ \vdots & \vdots & \ddots & \ddots & \vdots \\ A_{n-1,1} & A_{n-1,2} & \cdots & A_{n-1,n-1} & 0 \\ A_{n1} & A_{n2} & \cdots & A_{n,n-1} & A_{nn} \end{pmatrix}$$

is called a **block lower triangular matrix**, where each A_{ii} is a square matrix, $i = 1, 2, \ldots, n$.

(3) The block matrix of the form

$$\begin{pmatrix} A_1 & 0 & \cdots & \cdots & 0 \\ 0 & A_2 & \ddots & & \vdots \\ \vdots & \ddots & \ddots & \ddots & \vdots \\ \vdots & & \ddots & A_{n-1} & 0 \\ 0 & \cdots & \cdots & 0 & A_n \end{pmatrix}$$

is called a **block diagonal matrix**, often denoted by

$$\begin{pmatrix} A_1 & & & \\ & A_2 & & \\ & & \ddots & \\ & & & A_n \end{pmatrix} \quad \text{or} \quad \operatorname{diag}(A_1, A_2, \ldots, A_n),$$

where each A_i is a square matrix, $i = 1, 2, \ldots, n$.

Remark 2.10.4. Let $A = \begin{pmatrix} A_1 & & & \\ & A_2 & & \\ & & \ddots & \\ & & & A_n \end{pmatrix}$ and $B = \begin{pmatrix} B_1 & & & \\ & B_2 & & \\ & & \ddots & \\ & & & B_n \end{pmatrix}$

be block diagonal matrices partitioned in the same way. Then

$$(1) \quad A + B = \begin{pmatrix} A_1 + B_1 & & & \\ & A_2 + B_2 & & \\ & & \ddots & \\ & & & A_n + B_n \end{pmatrix},$$

$$(2) \quad AB = \begin{pmatrix} A_1 B_1 & & & \\ & A_2 B_2 & & \\ & & \ddots & \\ & & & A_n B_n \end{pmatrix},$$

$$(3) \quad A^{-1} = \begin{pmatrix} A_1^{-1} & & & \\ & A_2^{-1} & & \\ & & \ddots & \\ & & & A_n^{-1} \end{pmatrix} \quad \text{if } A_1, A_2, \ldots, A_n \text{ are invertible.}$$

Example 2.10.5. Let

$$A = \begin{pmatrix} a_{11} & a_{12} & \cdots & a_{1n} \\ a_{21} & a_{22} & \cdots & a_{2n} \\ \vdots & \vdots & & \vdots \\ a_{s1} & a_{s2} & \cdots & a_{sn} \end{pmatrix} \quad \text{and} \quad B = \begin{pmatrix} b_{11} & b_{12} & \cdots & b_{1m} \\ b_{21} & b_{22} & \cdots & b_{2m} \\ \vdots & \vdots & & \vdots \\ b_{n1} & b_{n2} & \cdots & b_{nm} \end{pmatrix}.$$

Express AB using the row (column) vectors of A or B.

Solution. Note that $A = \begin{pmatrix} \alpha_1 \\ \alpha_2 \\ \vdots \\ \alpha_s \end{pmatrix} = (\beta_1, \beta_2, \ldots, \beta_n)$, $B = \begin{pmatrix} \gamma_1 \\ \gamma_2 \\ \vdots \\ \gamma_n \end{pmatrix} = (\delta_1, \delta_2, \ldots, \delta_m)$. Thus,

$$AB = A(\delta_1, \delta_2, \ldots, \delta_m) = (A\delta_1, A\delta_2, \ldots, A\delta_m),$$

$$AB = \begin{pmatrix} \alpha_1 \\ \alpha_2 \\ \vdots \\ \alpha_s \end{pmatrix} B = \begin{pmatrix} \alpha_1 B \\ \alpha_2 B \\ \vdots \\ \alpha_s B \end{pmatrix},$$

$$AB = (\beta_1, \beta_2, \ldots, \beta_n) \begin{pmatrix} \gamma_1 \\ \gamma_2 \\ \vdots \\ \gamma_n \end{pmatrix} = \beta_1\gamma_1 + \beta_2\gamma_2 + \cdots + \beta_n\gamma_n,$$

$$AB = \begin{pmatrix} a_{11} & a_{12} & \cdots & a_{1n} \\ a_{21} & a_{22} & \cdots & a_{2n} \\ \vdots & \vdots & & \vdots \\ a_{s1} & a_{s2} & \cdots & a_{sn} \end{pmatrix} \begin{pmatrix} \gamma_1 \\ \gamma_2 \\ \vdots \\ \gamma_n \end{pmatrix} = \begin{pmatrix} a_{11}\gamma_1 + a_{12}\gamma_2 + \cdots + a_{1n}\gamma_n \\ a_{21}\gamma_1 + a_{22}\gamma_2 + \cdots + a_{2n}\gamma_n \\ \vdots \\ a_{s1}\gamma_1 + a_{s2}\gamma_2 + \cdots + a_{sn}\gamma_n \end{pmatrix},$$

$$AB = (\beta_1, \beta_2, \ldots, \beta_n) \begin{pmatrix} b_{11} & b_{12} & \cdots & b_{1m} \\ b_{21} & b_{22} & \cdots & b_{2m} \\ \vdots & \vdots & & \vdots \\ b_{n1} & b_{n2} & \cdots & b_{nm} \end{pmatrix}$$

$$= (b_{11}\beta_1 + b_{21}\beta_2 + \cdots + b_{n1}\beta_n, \ldots,$$
$$b_{1m}\beta_1 + b_{2m}\beta_2 + \cdots + b_{nm}\beta_n). \qquad \square$$

Definition 2.10.6. Elementary operations on a block matrix are the following six types of operations:

(1) Multiply a row on the left by an invertible matrix.
(2) Add a multiple of a row on the left by a matrix to another row.
(3) Interchange two rows.
(4) Multiply a column on the right by an invertible matrix.
(5) Add a multiple of a column on the right by a matrix to another column.
(6) Interchange two columns.

The first three types of operations are **elementary row operations**, and the last three are **elementary column operations**.

Definition 2.10.7. A block diagonal matrix of the form
$$\begin{pmatrix} I_{n_1} & & & \\ & I_{n_2} & & \\ & & \ddots & \\ & & & I_{n_t} \end{pmatrix}$$
is called a **block identity matrix**.

For example, $\begin{pmatrix} I_m & 0 \\ 0 & I_n \end{pmatrix}$ is a 2×2 block identity matrix.

Definition 2.10.8. An **elementary block matrix** is a block matrix obtained from a block identity matrix by performing one single elementary operation.

Let $I = \begin{pmatrix} I_m & 0 \\ 0 & I_n \end{pmatrix}$. Multiply the first row (respectively, column) of I on the left (respectively, right) by an invertible matrix P to obtain the elementary block matrix $\begin{pmatrix} P & 0 \\ 0 & I_n \end{pmatrix}$. Add $Q_{n\times m}$ times the first row on the left to the second row of I (or add $Q_{n\times m}$ times the second column on the right to the first column of I) to obtain the elementary block matrix $\begin{pmatrix} I_m & 0 \\ Q & I_n \end{pmatrix}$. Interchange two rows of I to obtain the elementary block matrix $\begin{pmatrix} 0 & I_n \\ I_m & 0 \end{pmatrix}$. Interchange two columns of I to obtain the elementary block matrix $\begin{pmatrix} 0 & I_m \\ I_n & 0 \end{pmatrix}$.

Theorem 2.10.9. *Let* $M = \begin{pmatrix} A & B \\ C & D \end{pmatrix}$ *be a* 2×2 *block matrix.*

(1) *Applying an elementary row operation to* M *is equivalent to multiplying* M *on the left by the corresponding elementary block matrix.*

(2) *Applying an elementary column operation to* M *is equivalent to multiplying* M *on the right by the corresponding elementary block matrix.*

Proof. First, we prove (1).

$$\begin{pmatrix} P & 0 \\ 0 & I_n \end{pmatrix}\begin{pmatrix} A & B \\ C & D \end{pmatrix} = \begin{pmatrix} PA & PB \\ C & D \end{pmatrix},$$

$$\begin{pmatrix} I_m & 0 \\ Q & I_n \end{pmatrix}\begin{pmatrix} A & B \\ C & D \end{pmatrix} = \begin{pmatrix} A & B \\ C+QA & D+QB \end{pmatrix}, \qquad (2.11)$$

$$\begin{pmatrix} 0 & I_n \\ I_m & 0 \end{pmatrix}\begin{pmatrix} A & B \\ C & D \end{pmatrix} = \begin{pmatrix} C & D \\ A & B \end{pmatrix}.$$

The proof of (2) is similar to that of (1). $\qquad\qquad\square$

If A is invertible and $Q = -CA^{-1}$ in (2.11), then

$$\begin{pmatrix} I_m & 0 \\ -CA^{-1} & I_n \end{pmatrix} \begin{pmatrix} A & B \\ C & D \end{pmatrix} = \begin{pmatrix} A & B \\ 0 & D - CA^{-1}B \end{pmatrix}.$$

Remark 2.10.10. Elementary block matrices are invertible, and elementary operations on a block matrix do not change the rank of the block matrix.

Proposition 2.10.11. *Let A, B, P, and Q be matrices.*

(1) *If A and B have the same number of rows, then*

$$\operatorname{rank}A \leqslant \operatorname{rank}(A, B).$$

(2) *If P and Q have the same number of columns, then*

$$\operatorname{rank}P \leqslant \operatorname{rank}\begin{pmatrix} P \\ Q \end{pmatrix}.$$

Proof. Let $\operatorname{rank}A = r$. Then A has a nonzero minor of order r, which is also a nonzero minor of order r of (A, B). So $\operatorname{rank}(A, B) \geqslant r = \operatorname{rank}A$.

Thus, $\operatorname{rank}P = \operatorname{rank}P' \leqslant \operatorname{rank}(P', Q') = \operatorname{rank}\begin{pmatrix} P \\ Q \end{pmatrix}' = \operatorname{rank}\begin{pmatrix} P \\ Q \end{pmatrix}.$ $\qquad\square$

Proposition 2.10.12. *Let $A \in \mathrm{M}_{s\times n}(F)$ and $B \in \mathrm{M}_{t\times m}(F)$. Then*

$$\operatorname{rank}\begin{pmatrix} A & 0 \\ 0 & B \end{pmatrix} = \operatorname{rank}A + \operatorname{rank}B.$$

Proof. Let $\operatorname{rank}A = r$ and $\operatorname{rank}B = s$. Then A has a nonzero minor $|A_1|$ of order r and B has a nonzero minor $|B_1|$ of order s. Therefore, $\begin{pmatrix} A & 0 \\ 0 & B \end{pmatrix}$ has a nonzero minor $\begin{vmatrix} A_1 & 0 \\ 0 & B_1 \end{vmatrix} = |A_1||B_1| \neq 0.$

Note that every minor of order $r + 1$ of A is 0, and every minor of order $s + 1$ of B is 0. Thus every minor of order $r + s + 1$ of $\begin{pmatrix} A & 0 \\ 0 & B \end{pmatrix}$ is 0. So

$$\operatorname{rank}\begin{pmatrix} A & 0 \\ 0 & B \end{pmatrix} = r + s = \operatorname{rank}A + \operatorname{rank}B. \qquad\square$$

Proposition 2.10.13. *Let A, B, P, and Q be matrices.*

(1) *If A and B have the same number of rows, then*

$$\mathrm{rank}(A, B) \leqslant \mathrm{rank}A + \mathrm{rank}B.$$

(2) *If P and Q have the same number of columns, then*

$$\mathrm{rank}\begin{pmatrix} P \\ Q \end{pmatrix} \leqslant \mathrm{rank}P + \mathrm{rank}Q.$$

Proof. By Propositions 2.10.11 and 2.10.12, we have

$$\mathrm{rank}(A, B) \leqslant \mathrm{rank}\begin{pmatrix} A & B \\ 0 & B \end{pmatrix} = \mathrm{rank}\begin{pmatrix} A & 0 \\ 0 & B \end{pmatrix} = \mathrm{rank}A + \mathrm{rank}B.$$

Similarly, $\mathrm{rank}\begin{pmatrix} P \\ Q \end{pmatrix} \leqslant \mathrm{rank}P + \mathrm{rank}Q$. $\qquad\square$

Proposition 2.10.14. *Let $A \in \mathrm{M}_{s\times n}(F)$, $B \in \mathrm{M}_{t\times m}(F)$, and $C \in \mathrm{M}_{t\times n}(F)$. Then*

$$\mathrm{rank}A + \mathrm{rank}B \leqslant \mathrm{rank}\begin{pmatrix} A & 0 \\ C & B \end{pmatrix} \leqslant \mathrm{rank}A + \mathrm{rank}B + \mathrm{rank}C.$$

Proof. On one hand, by Proposition 2.10.13, we get

$$\mathrm{rank}\begin{pmatrix} A & 0 \\ C & B \end{pmatrix} \leqslant \mathrm{rank}\begin{pmatrix} A \\ C \end{pmatrix} + \mathrm{rank}\begin{pmatrix} 0 \\ B \end{pmatrix}$$

$$\leqslant \mathrm{rank}A + \mathrm{rank}B + \mathrm{rank}C.$$

On the other hand, let $\mathrm{rank}A = r$ and $\mathrm{rank}B = s$. Then

$$\begin{pmatrix} A & 0 \\ C & B \end{pmatrix} \to \begin{pmatrix} I_r & 0 & 0 & 0 \\ 0 & 0 & 0 & 0 \\ C_1 & C_2 & I_s & 0 \\ C_3 & C_4 & 0 & 0 \end{pmatrix} \to \begin{pmatrix} I_r & 0 & 0 & 0 \\ 0 & 0 & 0 & 0 \\ 0 & 0 & I_s & 0 \\ 0 & C_4 & 0 & 0 \end{pmatrix} \to \begin{pmatrix} I_r & 0 & 0 & 0 \\ 0 & I_s & 0 & 0 \\ 0 & 0 & C_4 & 0 \\ 0 & 0 & 0 & 0 \end{pmatrix}.$$

Thus, by Proposition 2.10.12, we have

$$\mathrm{rank}\begin{pmatrix} A & 0 \\ C & B \end{pmatrix} = r + s + \mathrm{rank}C_4 \geqslant r + s = \mathrm{rank}A + \mathrm{rank}B. \qquad\square$$

Theorem 2.10.15. *Let $A \in \mathrm{M}_{s \times n}(F)$ and $B \in \mathrm{M}_{n \times m}(F)$. Then*

$$\mathrm{rank}A + \mathrm{rank}B - n \leqslant \mathrm{rank}(AB) \leqslant \min\{\mathrm{rank}A, \mathrm{rank}B\}.$$

In particular, if $AB = 0$, then $\mathrm{rank}A + \mathrm{rank}B \leqslant n$.

Proof. Let $\mathrm{rank}A = r$. Then there exist invertible matrices P of order s and Q of order n such that $A = P \begin{pmatrix} I_r & 0 \\ 0 & 0 \end{pmatrix} Q$, and so

$$AB = P \begin{pmatrix} I_r & 0 \\ 0 & 0 \end{pmatrix} QB.$$

Let $QB = \begin{pmatrix} G \\ H \end{pmatrix}$, where $G \in \mathrm{M}_{r \times m}(F)$, $H \in \mathrm{M}_{(n-r) \times m}(F)$. Then

$$AB = P \begin{pmatrix} I_r & 0 \\ 0 & 0 \end{pmatrix} \begin{pmatrix} G \\ H \end{pmatrix} = P \begin{pmatrix} G \\ 0 \end{pmatrix}.$$

Since P is invertible, $\mathrm{rank}(AB) = \mathrm{rank} \begin{pmatrix} G \\ 0 \end{pmatrix} = \mathrm{rank}G$. Note that G is an $r \times m$ matrix, and hence $\mathrm{rank}(AB) = \mathrm{rank}G \leqslant r = \mathrm{rank}A$. Since Q is invertible, $\mathrm{rank}(AB) = \mathrm{rank}G \leqslant \mathrm{rank}(QB) = \mathrm{rank}B$. So

$$\mathrm{rank}(AB) \leqslant \min\{\mathrm{rank}A, \mathrm{rank}B\}.$$

On the other hand, since H is an $(n - r) \times m$ matrix,

$$\begin{aligned}
\mathrm{rank}B &= \mathrm{rank}(QB) \\
&\leqslant \mathrm{rank}G + \mathrm{rank}H \\
&= \mathrm{rank}(AB) + \mathrm{rank}H \\
&\leqslant \mathrm{rank}(AB) + n - r.
\end{aligned}$$

Thus, $\mathrm{rank}(AB) \geqslant \mathrm{rank}B + r - n = \mathrm{rank}A + \mathrm{rank}B - n$. $\qquad\square$

Remark 2.10.16. The inequality $\mathrm{rank}(AB) \geqslant \mathrm{rank}A + \mathrm{rank}B - n$ is often called **Sylvester's rank inequality**.

Corollary 2.10.17. *Let $A, B \in \mathrm{M}_{s \times n}(F)$. Then*

$$\mathrm{rank}(A + B) \leqslant \mathrm{rank}A + \mathrm{rank}B.$$

Proof. Since $A + B = (A, B) \begin{pmatrix} I_n \\ I_n \end{pmatrix}$, we have

$$\operatorname{rank}(A + B) \leqslant \operatorname{rank}(A, B) \leqslant \operatorname{rank}A + \operatorname{rank}B$$

by Theorem 2.10.15 and Proposition 2.10.13. □

Example 2.10.18. Let $A, B \in \mathrm{M}_n(F)$ and $AB = BA$. Prove that

$$\operatorname{rank}(A + B) + \operatorname{rank}(AB) \leqslant \operatorname{rank}A + \operatorname{rank}B.$$

Proof. It is easy to check that

$$\begin{pmatrix} A + B & 0 \\ A & AB \end{pmatrix} = \begin{pmatrix} I_n & I_n \\ I_n & 0 \end{pmatrix} \begin{pmatrix} A & 0 \\ 0 & B \end{pmatrix} \begin{pmatrix} I_n & B \\ I_n & -A \end{pmatrix}.$$

Thus,

$$\begin{aligned}
\operatorname{rank}(A + B) + \operatorname{rank}(AB) &\leqslant \operatorname{rank} \begin{pmatrix} A + B & 0 \\ A & AB \end{pmatrix} \\
&\leqslant \operatorname{rank} \begin{pmatrix} A & 0 \\ 0 & B \end{pmatrix} \\
&= \operatorname{rank}A + \operatorname{rank}B
\end{aligned}$$

by Proposition 2.10.14, Theorem 2.10.15, and Proposition 2.10.12. □

Example 2.10.19. Let $A \in \mathrm{GL}_m(F)$ and $B \in \mathrm{GL}_n(F)$. If $T = \begin{pmatrix} A & 0 \\ C & B \end{pmatrix}$, prove that $T \in \mathrm{GL}_{m+n}(F)$ and find T^{-1}.

Proof. Since $A \in \mathrm{GL}_m(F)$ and $B \in \mathrm{GL}_n(F)$, $|A| \neq 0$ and $|B| \neq 0$. It follows that $|T| = |A||B| \neq 0$, and hence $T \in \mathrm{GL}_{m+n}(F)$.

Next, we find T^{-1}. Note that $\begin{pmatrix} I_m & 0 \\ -CA^{-1} & I_n \end{pmatrix} \begin{pmatrix} A & 0 \\ C & B \end{pmatrix} = \begin{pmatrix} A & 0 \\ 0 & B \end{pmatrix}$

and $\begin{pmatrix} A & 0 \\ 0 & B \end{pmatrix}^{-1} = \begin{pmatrix} A^{-1} & 0 \\ 0 & B^{-1} \end{pmatrix}$. Thus

$$T^{-1} = \begin{pmatrix} A^{-1} & 0 \\ 0 & B^{-1} \end{pmatrix} \begin{pmatrix} I_m & 0 \\ -CA^{-1} & I_n \end{pmatrix} = \begin{pmatrix} A^{-1} & 0 \\ -B^{-1}CA^{-1} & B^{-1} \end{pmatrix}. \quad \square$$

Example 2.10.20. Let $A, B \in M_n(F)$. Prove that $|AB| = |A||B|$.

Proof. Consider the $\begin{pmatrix} A & 0 \\ -I & B \end{pmatrix}$, where $I = I_n$. Then

$$\begin{pmatrix} I & A \\ 0 & I \end{pmatrix} \begin{pmatrix} A & 0 \\ -I & B \end{pmatrix} = \begin{pmatrix} 0 & AB \\ -I & B \end{pmatrix}.$$

Taking determinants of both sides of the equality above yields

$$\left| \begin{pmatrix} I & A \\ 0 & I \end{pmatrix} \begin{pmatrix} A & 0 \\ -I & B \end{pmatrix} \right| = \left| \begin{matrix} 0 & AB \\ -I & B \end{matrix} \right|.$$

Note that $\left| \begin{matrix} 0 & AB \\ -I & B \end{matrix} \right| = (-1)^n \left| \begin{matrix} AB & 0 \\ B & -I \end{matrix} \right| = (-1)^n |AB||-I| = |AB|.$

Next, we show that $\left| \begin{pmatrix} I & A \\ 0 & I \end{pmatrix} \begin{pmatrix} A & 0 \\ -I & B \end{pmatrix} \right| = |A||B|.$

Let $A = (a_{ij})_{n \times n}$. Then $A = \sum_{i,j=1}^{n} a_{ij} E_{ij}$, where E_{ij} denotes the

$n \times n$ matrix in which the (i,j) entry is equal to 1, and all the other entries are zero (see Exercise 25), $i, j = 1, 2, \ldots, n$.

Note that

$$\begin{pmatrix} I & X \\ 0 & I \end{pmatrix} \begin{pmatrix} I & Y \\ 0 & I \end{pmatrix} = \begin{pmatrix} I & X+Y \\ 0 & I \end{pmatrix}$$

for all $n \times n$ matrices X and Y. Thus

$$\begin{pmatrix} I & A \\ 0 & I \end{pmatrix} = \begin{pmatrix} I & \sum_{i,j=1}^{n} a_{ij} E_{ij} \\ 0 & I \end{pmatrix} = \prod_{i,j=1}^{n} \begin{pmatrix} I & a_{ij} E_{ij} \\ 0 & I \end{pmatrix}.$$

Let $P_{ij} = \begin{pmatrix} I & a_{ij} E_{ij} \\ 0 & I \end{pmatrix}$, where $i, j = 1, 2, \ldots, n$. Then $\begin{pmatrix} I & A \\ 0 & I \end{pmatrix} = \prod_{i,j=1}^{n} P_{ij}.$

Note that each P_{ij} is an elementary matrix, and the corresponding elementary operation is a multiple of one row added to another row, which does not change the value of the determinant of the matrix. Thus

$$\left| \begin{pmatrix} I & A \\ 0 & I \end{pmatrix} \begin{pmatrix} A & 0 \\ -I & B \end{pmatrix} \right| = \left| \left(\prod_{i,j=1}^{n} P_{ij} \right) \begin{pmatrix} A & 0 \\ -I & B \end{pmatrix} \right| = \left| \begin{matrix} A & 0 \\ -I & B \end{matrix} \right| = |A||B|.$$

In conclusion, $|AB| = |A||B|$. $\hspace{3cm}$ $\square$

Example 2.10.21. Let $A \in M_{n \times m}(F)$ and $B \in M_{m \times n}(F)$. Prove that

$$\lambda^m |\lambda I_n - AB| = \lambda^n |\lambda I_m - BA|$$

for any $\lambda \in F$.

Proof. Consider the block matrix $\begin{pmatrix} \lambda I_m & B \\ A & I_n \end{pmatrix}$. Then

$$\begin{pmatrix} I_m & 0 \\ 0 & \lambda I_n \end{pmatrix} \begin{pmatrix} I_m & -B \\ 0 & I_n \end{pmatrix} \begin{pmatrix} \lambda I_m & B \\ A & I_n \end{pmatrix} = \begin{pmatrix} \lambda I_m - BA & 0 \\ \lambda A & \lambda I_n \end{pmatrix},$$

$$\begin{pmatrix} \lambda I_m & B \\ A & I_n \end{pmatrix} \begin{pmatrix} I_m & 0 \\ 0 & \lambda I_n \end{pmatrix} \begin{pmatrix} I_m & -B \\ 0 & I_n \end{pmatrix} = \begin{pmatrix} \lambda I_m & 0 \\ A & \lambda I_n - AB \end{pmatrix}.$$

Taking determinants of both sides of the equalities above, we have

$$\begin{vmatrix} I_m & 0 \\ 0 & \lambda I_n \end{vmatrix} \begin{vmatrix} I_m & -B \\ 0 & I_n \end{vmatrix} \begin{vmatrix} \lambda I_m & B \\ A & I_n \end{vmatrix} = \begin{vmatrix} \lambda I_m - BA & 0 \\ \lambda A & \lambda I_n \end{vmatrix},$$

$$\begin{vmatrix} \lambda I_m & B \\ A & I_n \end{vmatrix} \begin{vmatrix} I_m & 0 \\ 0 & \lambda I_n \end{vmatrix} \begin{vmatrix} I_m & -B \\ 0 & I_n \end{vmatrix} = \begin{vmatrix} \lambda I_m & 0 \\ A & \lambda I_n - AB \end{vmatrix}.$$

Thus, $\lambda^n |\lambda I_m - BA| = \lambda^n \begin{vmatrix} \lambda I_m & B \\ A & I_n \end{vmatrix} = \lambda^m |\lambda I_n - AB|.$ $\hspace{1.5cm}$ $\square$

Remark 2.10.22. Let A and B be the same as in Example 2.10.21.

(1) Taking $\lambda = 1$ in Example 2.10.21 gives $|I_n - AB| = |I_m - BA|$.

(2) Taking $\lambda = -1$ yields $|I_n + AB| = |I_m + BA|$. This equality is called **Sylvester's determinant identity**.

(3) If $m = n$, then $|\lambda I_n - AB| = |\lambda I_n - BA|$.

In fact, if $\lambda = 0$, then $|\lambda I_n - AB| = |-AB| = (-1)^n |A||B|$,

$$|\lambda I_n - BA| = |-BA| = (-1)^n |B||A|,$$

so $|\lambda I_n - AB| = |\lambda I_n - BA|$.

If $\lambda \neq 0$, it is clear that $|\lambda I_n - AB| = |\lambda I_n - BA|$.

Therefore, if $m = n$, we always have $|\lambda I_n - AB| = |\lambda I_n - BA|$.

2.11* Cauchy–Binet Formula

If A and B are square matrices, then $|AB| = |A||B|$ by Theorem 2.7.21 (or Example 2.10.20). Let's consider that A and B are not square matrices, but AB is a square matrix. What is $|AB|$ at this time? The Cauchy–Binet[14] formula introduced in the following answers this question.

For convenience, we introduce the following notation. Suppose that $A \in \mathrm{M}_{m \times n}(F)$ and $1 \leqslant s \leqslant \min\{m, n\}$. Choose any s rows $i_1, i_2, \ldots, i_s$ and s columns $j_1, j_2, \ldots, j_s$ of A. We use $A \begin{pmatrix} i_1, \ i_2, \ \ldots, \ i_s \\ j_1, \ j_2, \ \ldots, \ j_s \end{pmatrix}$ to denote the minor of order s whose entries are located at the intersections of the chosen s rows $i_1, i_2, \ldots, i_s$ and s columns $j_1, j_2, \ldots, j_s$. Usually, the minor of the form $A \begin{pmatrix} i_1, \ i_2, \ \ldots, \ i_s \\ i_1, \ i_2, \ \ldots, \ i_s \end{pmatrix}$ is called a **principal minor** of order s.

Theorem 2.11.1 (Cauchy–Binet formula). *Let* $A \in \mathrm{M}_{m \times n}(F)$ *and* $B \in \mathrm{M}_{n \times m}(F)$.

(1) *If* $m > n$, *then* $|AB| = 0$.

(2) *If* $m \leqslant n$, *then* $|AB|$ *is equal to the sum of the products of all minors of order* m *of* A *and the corresponding minors of order* m *of* B, *that is,*

$$|AB| = \sum_{1 \leqslant i_1 < \cdots < i_m \leqslant n} A \begin{pmatrix} 1, \ 2, \ \ldots, \ m \\ i_1, \ i_2, \ \ldots, \ i_m \end{pmatrix} B \begin{pmatrix} i_1, \ i_2, \ \ldots, \ i_m \\ 1, \ 2, \ \ldots, \ m \end{pmatrix}.$$

---

[14] Jacques Philippe Marie Binet, 1786–1856, French mathematician.

Proof. To evaluate $|AB|$, inspired by the proof of Example 2.10.20, we consider the block matrix $M = \begin{pmatrix} A & 0 \\ -I_n & B \end{pmatrix}$.

Since $\begin{pmatrix} I_m & A \\ 0 & I_n \end{pmatrix} \begin{pmatrix} A & 0 \\ -I_n & B \end{pmatrix} = \begin{pmatrix} 0 & AB \\ -I_n & B \end{pmatrix}$, we have

$$|M| = \begin{vmatrix} 0 & AB \\ -I_n & B \end{vmatrix} = (-1)^{mn} \begin{vmatrix} AB & 0 \\ B & -I_n \end{vmatrix} = (-1)^{mn+n}|AB|,$$

that is, $|AB| = (-1)^{mn+n}|M|$.

Next, we apply the Laplace expansion to the first m rows of $|M|$.

(1) If $m > n$, then all minors of order m obtained from the first m rows of $|M|$ are 0, and hence $|M| = 0$. Thus $|AB| = (-1)^{mn+n}|M| = 0$.

(2) If $m \leqslant n$, then there are possibly C_n^m nonzero minors of order m obtained from the first m rows of $|M|$. Each of these minors is of the form

$$A \begin{pmatrix} 1, & 2, & \ldots, & m \\ i_1, & i_2, & \ldots, & i_m \end{pmatrix},$$

where $1 \leqslant i_1 < i_2 < \cdots < i_m \leqslant n$, and its cofactor is

$$(-1)^{1+2+\cdots+m+i_1+i_2+\cdots+i_m}|E, B|,$$

where E denotes the $n \times (n - m)$ matrix obtained from $-I_n$ by deleting columns $i_1, i_2, \ldots, i_m$. Thus

$$|M| = \sum_{1 \leqslant i_1 < i_2 < \cdots < i_m \leqslant n} A \begin{pmatrix} 1, & 2, & \ldots, & m \\ i_1, & i_2, & \ldots, & i_m \end{pmatrix} \times k_1 |E, B|,$$

where $k_1 = (-1)^{1+2+\cdots+m+i_1+i_2+\cdots+i_m}$.

Now, let's evaluate $|E, B|$. Note that rows $i_1, i_2, \ldots, i_m$ of E are all zero. Choose rows $i_1, i_2, \ldots, i_m$ of the determinant $|E, B|$. Then the only possible nonzero minor of order m obtained from the chosen m rows of $|E, B|$ is $B \begin{pmatrix} i_1, & i_2, & \ldots, & i_m \\ 1, & 2, & \ldots, & m \end{pmatrix}$, and its cofactor is

$$k_2 = (-1)^{i_1+i_2+\cdots+i_m+(n-m+1)+(n-m+2)+\cdots+(n-m+m)}|-I_{n-m}|$$

$$= (-1)^{i_1+i_2+\cdots+i_m+(n-m+1)+(n-m+2)+\cdots+(n-m+m)}(-1)^{n-m}.$$

Thus, $|E, B| = k_2 B \begin{pmatrix} i_1, i_2, \ldots, i_m \\ 1, 2, \ldots, m \end{pmatrix}$ by the Laplace expansion, and hence

$$|M| = \sum_{1 \leqslant i_1 < i_2 < \cdots < i_m \leqslant n} A \begin{pmatrix} 1, 2, \ldots, m \\ i_1, i_2, \ldots, i_m \end{pmatrix} \times k_1 k_2 B \begin{pmatrix} i_1, i_2, \ldots, i_m \\ 1, 2, \ldots, m \end{pmatrix}.$$

It is easy to see that $k_1 k_2 = (-1)^{mn+n}$. Therefore,

$$|AB| = (-1)^{mn+n} |M|$$
$$= \sum_{1 \leqslant i_1 < i_2 < \cdots < i_m \leqslant n} A \begin{pmatrix} 1, 2, \ldots, m \\ i_1, i_2, \ldots, i_m \end{pmatrix} B \begin{pmatrix} i_1, i_2, \ldots, i_m \\ 1, 2, \ldots, m \end{pmatrix}. \qquad \square$$

Remark 2.11.2. When $m = n$, the Cauchy–Binet formula is consistent with Theorem 2.7.21 (or Example 2.10.20).

Corollary 2.11.3. *Let $A \in M_{s \times n}(F)$, $B \in M_{n \times m}(F)$, $C = AB$, and $1 \leqslant r \leqslant min\{s, m\}$.*

(1) *If $r > n$, then any minor of order r of C is equal to 0.*

(2) *If $r \leqslant n$, then any minor $C \begin{pmatrix} i_1, i_2, \ldots, i_r \\ j_1, j_2, \ldots, j_r \end{pmatrix}$ of order r of C is*

equal to

$$\sum_{1 \leqslant k_1 < k_2 < \cdots < k_r \leqslant n} A \begin{pmatrix} i_1, i_2, \ldots, i_r \\ k_1, k_2, \ldots, k_r \end{pmatrix} B \begin{pmatrix} k_1, k_2, \ldots, k_r \\ j_1, j_2, \ldots, j_r \end{pmatrix}.$$

Proof. By the definition of matrix multiplication, we have

$$C \begin{pmatrix} i_1, i_2, \ldots, i_r \\ j_1, j_2, \ldots, j_r \end{pmatrix} = \left| \begin{pmatrix} a_{i_1 1} & a_{i_1 2} & \cdots & a_{i_1 n} \\ a_{i_2 1} & a_{i_2 2} & \cdots & a_{i_2 n} \\ \vdots & \vdots & & \vdots \\ a_{i_r 1} & a_{i_r 2} & \cdots & a_{i_r n} \end{pmatrix} \begin{pmatrix} b_{1 j_1} & b_{1 j_2} & \cdots & b_{1 j_r} \\ b_{2 j_1} & b_{2 j_2} & \cdots & b_{2 j_r} \\ \vdots & \vdots & & \vdots \\ b_{n j_1} & b_{n j_2} & \cdots & b_{n j_r} \end{pmatrix} \right|.$$

Therefore, the conclusion follows from the Cauchy–Binet formula.

$$\square$$

Remark 2.11.4. Let $A \in M_{s \times n}(F)$ and $B \in M_{n \times m}(F)$. Then, by Theorem 2.10.15, $\text{rank}(AB) \leqslant min\{\text{rank}A, \text{rank}B\}$, which can be proved easily using Corollary 2.11.3.

The following examples illustrate applications of the Cauchy–Binet formula.

Example 2.11.5. Let $A \in \mathrm{M}_{s \times n}(\mathbb{R})$. Then any principal minor of AA' is nonnegative.

Proof. Let $1 \leqslant r \leqslant s$. By Corollary 2.11.3, if $r > n$, then any minor of order r of AA' is equal to 0. If $r \leqslant n$, then any principal minor $(AA') \begin{pmatrix} i_1, i_2, \ldots, i_r \\ i_1, i_2, \ldots, i_r \end{pmatrix}$ of order r of AA' is equal to

$$\sum_{1 \leqslant k_1 < k_2 < \cdots < k_r \leqslant n} A \begin{pmatrix} i_1, i_2, \ldots, i_r \\ k_1, k_2, \ldots, k_r \end{pmatrix} A' \begin{pmatrix} k_1, k_2, \ldots, k_r \\ i_1, i_2, \ldots, i_r \end{pmatrix}$$

$$= \sum_{1 \leqslant k_1 < k_2 < \cdots < k_r \leqslant n} \left(A \begin{pmatrix} i_1, i_2, \ldots, i_r \\ k_1, k_2, \ldots, k_r \end{pmatrix} \right)^2 \geqslant 0. \qquad \square$$

Example 2.11.6. Let $a_1, a_2, \ldots, a_n$; $b_1, b_2, \ldots, b_n$ be all real numbers. Prove the **Cauchy–Bunyakovsky[15]–Schwarz[16] inequality:**

$$\left(\sum_{i=1}^{n} a_i^2 \right) \left(\sum_{i=1}^{n} b_i^2 \right) \geqslant \left(\sum_{i=1}^{n} a_i b_i \right)^2.$$

Proof.

$$\left(\sum_{i=1}^{n} a_i^2 \right) \left(\sum_{i=1}^{n} b_i^2 \right) - \left(\sum_{i=1}^{n} a_i b_i \right)^2$$

$$= \begin{vmatrix} \sum_{i=1}^{n} a_i^2 & \sum_{i=1}^{n} a_i b_i \\ \sum_{i=1}^{n} a_i b_i & \sum_{i=1}^{n} b_i^2 \end{vmatrix}$$

$$= \left| \begin{pmatrix} a_1 & a_2 & \cdots & a_n \\ b_1 & b_2 & \cdots & b_n \end{pmatrix} \begin{pmatrix} a_1 & b_1 \\ a_2 & b_2 \\ \vdots & \vdots \\ a_n & b_n \end{pmatrix} \right|$$

---

[15] Viktor Yakovlevich Bunyakovsky, 1804–1889, Russian mathematician.

[16] Karl Hermann Amandus Schwarz, 1843–1921, German mathematician.

$$= \sum_{1 \leqslant j < k \leqslant n} \begin{vmatrix} a_j & a_k \\ b_j & b_k \end{vmatrix} \begin{vmatrix} a_j & b_j \\ a_k & b_k \end{vmatrix}$$

$$= \sum_{1 \leqslant j < k \leqslant n} \begin{vmatrix} a_j & a_k \\ b_j & b_k \end{vmatrix}^2 \geqslant 0. \qquad \square$$

Exercises

1. Find the inversion numbers of the following permutations and determine whether each of the permutations is even or odd:

   (1) $1746\,2538$; (2) $2\,8967\,5341$; (3) $8675\,3421$.

2. Choose i and k such that

   (1) $817i25k49$ is an even permutation;
   (2) $1i25k4897$ is an odd permutation.

3. Suppose the inversion number of $i_1 i_2 \cdots i_{n-1} i_n$ is k, what is the inversion number of $i_n i_{n-1} \cdots i_2 i_1$?

4. Let n and k be positive integers with $1 \leqslant k \leqslant n$. Among all permutations of $1, 2, \ldots, n$,

   (1) if the number 1 is located in the kth position of a permutation, how many inversions are there?
   (2) if the number n is located in the kth position of a permutation, how many inversions are there?

5. In the expansion of a 6th-order determinant, what is the sign associated with each of the following two terms?

   (1) $a_{23}a_{31}a_{42}a_{56}a_{14}a_{65}$; (2) $a_{32}a_{43}a_{14}a_{51}a_{66}a_{25}$.

6. Write down all terms with a minus sign and containing the factor a_{23} in the expansion of a 4th-order determinant.

7. Let $n \geqslant 2$ be an integer. Prove that there are $\frac{n!}{2}$ even and $\frac{n!}{2}$ odd permutations by using $\begin{vmatrix} 1 & 1 & \cdots & 1 \\ 1 & 1 & \cdots & 1 \\ \vdots & \vdots & & \vdots \\ 1 & 1 & \cdots & 1 \end{vmatrix} = 0.$

8. Prove that

$$\begin{vmatrix} a_{11} & a_{12} & a_{13} & a_{14} & a_{15} \\ a_{21} & a_{22} & a_{23} & a_{24} & a_{25} \\ a_{31} & a_{32} & 0 & 0 & 0 \\ a_{41} & a_{42} & 0 & 0 & 0 \\ a_{51} & a_{52} & 0 & 0 & 0 \end{vmatrix} = 0$$

using the definition of determinants.

9. Let $f(x) = \begin{vmatrix} 2x & x & 1 & 2 \\ 1 & x & 1 & 1 \\ 3 & 2 & x & 1 \\ 1 & 1 & 1 & x \end{vmatrix}$. Use the definition of determinants to

find the coefficients of x^4 and x^3 and the constant of $f(x)$.

10. Evaluate the following determinants by definition:

(1) $\begin{vmatrix} 0 & 1 & 0 & \cdots\cdots & \cdots & & 0 \\ 0 & 0 & 2 & \ddots & & & \vdots \\ \vdots & \vdots & & \ddots & \ddots & \ddots & \vdots \\ \vdots & \vdots & & & \ddots & \ddots & \ddots & \vdots \\ \vdots & \vdots & & & & \ddots & n-2 & 0 \\ 0 & 0 & \cdots\cdots\cdots & & 0 & n-1 \\ n & 0 & \cdots\cdots\cdots & & 0 & 0 \end{vmatrix}$;

(2) $\begin{vmatrix} 0 & & \cdots & \cdots\cdots & 0 & 1 & 0 \\ \vdots & & & & \ddots & 2 & 0 & 0 \\ \vdots & & & \ddots & \ddots & \ddots & \vdots & \vdots \\ \vdots & & \ddots & \ddots & \ddots & & \vdots & \vdots \\ 0 & n-2 & \ddots & & & & \vdots & \vdots \\ n-1 & 0 & \cdots\cdots\cdots & & & 0 & 0 \\ 0 & 0 & \cdots\cdots\cdots & & & 0 & n \end{vmatrix}$.

11. Evaluate the following determinants:

(1) $\begin{vmatrix} 103 & 100 & 204 \\ 199 & 200 & 395 \\ 301 & 300 & 600 \end{vmatrix}$;

(2) $\begin{vmatrix} 246 & 427 & 327 \\ 1014 & 543 & 443 \\ -342 & 721 & 621 \end{vmatrix}$;

(3) $\begin{vmatrix} x & y & x+y \\ y & x+y & x \\ x+y & x & y \end{vmatrix}$;

(4) $\begin{vmatrix} 1 & 2 & 0 & 1 \\ 1 & 3 & 5 & 0 \\ 0 & 1 & 5 & 6 \\ 1 & 2 & 3 & 4 \end{vmatrix}$;

(5) $\begin{vmatrix} 2 & 1 & 7 & 6 \\ 8 & 3 & 0 & 7 \\ 1 & 0 & 4 & 2 \\ 3 & -1 & -2 & -5 \end{vmatrix}$;

(6) $\begin{vmatrix} x_1 & 0 & y_1 & 0 \\ 0 & u_1 & 0 & v_1 \\ x_2 & 0 & y_2 & 0 \\ 0 & u_2 & 0 & v_2 \end{vmatrix}$.

12. Find the roots of the following polynomials:

$$(1) \quad f(x) = \begin{vmatrix} 1 & 1 & 2 \\ 1 & 1 & x^2 - 2 \\ 2 & x^2 + 1 & 1 \end{vmatrix};$$

$$(2) \quad g(x) = \begin{vmatrix} x - 2 & x - 1 & x - 2 & x - 3 \\ 2x - 2 & 2x - 1 & 2x - 2 & 2x - 3 \\ 3x - 3 & 3x - 2 & 4x - 5 & 3x - 5 \\ 4x & 4x - 3 & 5x - 7 & 4x - 3 \end{vmatrix}.$$

13. Evaluate the following determinants of order n:

$$(1) \quad \begin{vmatrix} 1 & 2 & 3 & 4 & \cdots & n \\ 2 & 2 & 3 & 4 & \cdots & n \\ 3 & 3 & 3 & 4 & \cdots & n \\ \vdots & \vdots & \vdots & \vdots & & \vdots \\ n & n & n & n & \cdots & n \end{vmatrix};$$

$$(2) \quad \begin{vmatrix} x_1 - y_1 & x_1 - y_2 & \cdots & x_1 - y_n \\ x_2 - y_1 & x_2 - y_2 & \cdots & x_2 - y_n \\ \vdots & \vdots & & \vdots \\ x_n - y_1 & x_n - y_2 & \cdots & x_n - y_n \end{vmatrix};$$

$$(3) \quad \begin{vmatrix} x & y & 0 & \cdots & & & & 0 \\ 0 & x & y & \ddots & & & & \vdots \\ 0 & 0 & x & \ddots & \ddots & & & \vdots \\ \vdots & \vdots & \ddots & \ddots & \ddots & \ddots & & \vdots \\ \vdots & \vdots & & \ddots & x & y & 0 \\ 0 & 0 & \cdots & \cdots & 0 & x & y \\ y & 0 & \cdots & \cdots & 0 & 0 & x \end{vmatrix};$$

$$(4) \quad \begin{vmatrix} 1 & 1 & 1 & \cdots & \cdots & 1 \\ 1 & 2 & 1 & & & \vdots \\ 1 & 1 & 3 & \ddots & & \vdots \\ \vdots & & \ddots & \ddots & \ddots & \vdots \\ \vdots & & & \ddots & n-1 & 1 \\ 1 & \cdots & \cdots & \cdots & 1 & n \end{vmatrix};$$

$$(5) \quad \begin{vmatrix} 1 + a_1 & 1 & 1 & \cdots & \cdots & 1 \\ 1 & 1 + a_2 & 1 & & & \vdots \\ 1 & 1 & 1 + a_3 & \ddots & & \vdots \\ \vdots & & & \ddots & \ddots & \vdots \\ \vdots & & & & 1 + a_{n-1} & 1 \\ 1 & \cdots & \cdots & \cdots & 1 & 1 + a_n \end{vmatrix}.$$

14. Let n be a positive integer. Prove the following equalities:

(1)
$$\begin{vmatrix} x & 0 & 0 & \cdots\cdots & 0 & a_0 \\ -1 & x & 0 & \cdots\cdots & 0 & a_1 \\ 0 & -1 & x & \ddots & 0 & a_2 \\ \vdots & \ddots & \ddots & \ddots & \vdots & \vdots \\ \vdots & & \ddots & x & 0 & a_{n-3} \\ \vdots & & & -1 & x & a_{n-2} \\ 0 & \cdots\cdots & & 0 & -1 & x+a_{n-1} \end{vmatrix} = x^n + a_{n-1}x^{n-1} + \cdots + a_1 x + a_0, \text{ where } n \geqslant 2;$$

(2)
$$\begin{vmatrix} \cos\alpha & 1 & 0 & \cdots & \cdots & \cdots & 0 \\ 1 & 2\cos\alpha & 1 & \ddots & & & \vdots \\ 0 & 1 & 2\cos\alpha & \ddots & \ddots & & \vdots \\ \vdots & \ddots & \ddots & \ddots & \ddots & \ddots & \vdots \\ \vdots & & \ddots & \ddots & 2\cos\alpha & 1 & 0 \\ \vdots & & & \ddots & 1 & 2\cos\alpha & 1 \\ 0 & \cdots & \cdots & \cdots & 0 & 1 & 2\cos\alpha \end{vmatrix}_n = \cos n\alpha.$$

15. Use the Laplace expansion to find the values of the following determinants and evaluate them by row (column) expansion to verify whether the two results are consistent.

(1) Expand $\begin{vmatrix} 2 & 3 & 5 & 8 \\ -1 & 0 & 2 & 3 \\ 0 & 1 & 7 & 4 \\ 4 & 1 & -2 & 1 \end{vmatrix}$ along rows 2 and 4;

(2) Expand $\begin{vmatrix} -1 & -5 & 2 & 3 \\ 1 & 2 & 0 & 0 \\ 0 & 1 & 0 & 3 \\ 4 & 2 & 5 & 7 \end{vmatrix}$ along rows 1 and 3.

16. Prove the following:

$$(1) \quad \begin{vmatrix} a_{11}+x & a_{12}+x & \cdots & a_{1n}+x \\ a_{21}+x & a_{22}+x & \cdots & a_{2n}+x \\ \vdots & \vdots & & \vdots \\ a_{n1}+x & a_{n2}+x & \cdots & a_{nn}+x \end{vmatrix} = \begin{vmatrix} a_{11} & a_{12} & \cdots & a_{1n} \\ a_{21} & a_{22} & \cdots & a_{2n} \\ \vdots & \vdots & & \vdots \\ a_{n1} & a_{n2} & \cdots & a_{nn} \end{vmatrix} + x \sum_{i=1}^{n} \sum_{j=1}^{n} A_{ij},$$

where A_{ij} is the cofactor of a_{ij}, $i, j = 1, 2, \ldots n$;

$$(2) \quad \sum_{i=1}^{n} \sum_{j=1}^{n} A_{ij} = \begin{vmatrix} a_{11}-a_{12} & a_{12}-a_{13} & \cdots & a_{1,n-1}-a_{1n} & 1 \\ a_{21}-a_{22} & a_{22}-a_{23} & \cdots & a_{2,n-1}-a_{2n} & 1 \\ \vdots & \vdots & & \vdots & \vdots \\ a_{n1}-a_{n2} & a_{n2}-a_{n3} & \cdots & a_{n,n-1}-a_{nn} & 1 \end{vmatrix}.$$

17. Use Cramer's rule to solve the following systems of linear equations:

$$(1) \quad \begin{cases} 4x_1 - x_2 + 3x_3 + 2x_4 = 8, \\ 2x_1 - 3x_2 + 3x_3 + 2x_4 = 4, \\ 3x_1 - x_2 - x_3 + 2x_4 = 3, \\ 2x_1 - x_2 + 3x_3 - x_4 = 3; \end{cases}$$

$$(2) \quad \begin{cases} 2x_1 - x_2 + 3x_3 + 2x_4 = 6, \\ 3x_1 - 3x_2 + 3x_3 + 2x_4 = 5, \\ 3x_1 - x_2 - x_3 + 2x_4 = 3, \\ 3x_1 - x_2 + 3x_3 - x_4 = 4; \end{cases}$$

$$(3) \quad \begin{cases} x_1 + 2x_2 + 3x_3 - 2x_4 = 6, \\ 2x_1 - x_2 - 2x_3 - 3x_4 = 8, \\ 3x_1 + 2x_2 - x_3 + 2x_4 = 4, \\ 2x_1 - 3x_2 + 2x_3 + x_4 = -8. \end{cases}$$

18. Let F be a number field, and let $a_1, a_2, \ldots, a_n \in F$ be different from each other. Suppose $b_1, b_2, \ldots, b_n \in F$. Use Cramer's rule to prove that there is a unique $f(x) = c_0 x^{n-1} + c_1 x^{n-2} + \cdots + c_{n-1} \in F[x]$ such that $f(a_i) = b_i$, $i = 1, 2, \ldots, n$.

 Advanced Algebra

19. Evaluate the following determinants of order n:

$$(1)\quad \begin{vmatrix} \lambda & x & x & x & \cdots & \cdots & \cdots & x \\ y & \alpha & \beta & \beta & \cdots & \cdots & \cdots & \beta \\ y & \beta & \alpha & \beta & & & & \vdots \\ y & \beta & \beta & \alpha & \ddots & & & \vdots \\ \vdots & \vdots & & \ddots & \ddots & \ddots & & \vdots \\ \vdots & \vdots & & & \ddots & \alpha & \beta & \beta \\ \vdots & \vdots & & & & \beta & \alpha & \beta \\ y & \beta & \cdots & \cdots & \cdots & \beta & \beta & \alpha \end{vmatrix} ; \quad (2)\quad \begin{vmatrix} x & a & a & \cdots\cdots\cdots & a \\ -a & x & a & & \vdots \\ -a & -a & x & \ddots & \vdots \\ \vdots & & \ddots & \ddots\ddots\ddots & \vdots \\ \vdots & & & \ddots & x & a & a \\ \vdots & & & & -a & x & a \\ -a & \cdots\cdots\cdots & -a & -a & x \end{vmatrix} ;$$

$$(3)\quad \begin{vmatrix} x & y & y & \cdots & \cdots & \cdots & y \\ z & x & y & & & & \vdots \\ z & z & x & \ddots & & & \vdots \\ \vdots & & \ddots & \ddots & \ddots & & \vdots \\ \vdots & & & \ddots & x & y & y \\ \vdots & & & & z & x & y \\ z & \cdots & \cdots & \cdots & z & z & x \end{vmatrix} ; \quad (4)\quad \begin{vmatrix} 1 & 1 & \cdots & 1 \\ x_1 & x_2 & \cdots & x_n \\ \vdots & \vdots & & \vdots \\ x_1^{n-2} & x_2^{n-2} & \cdots & x_n^{n-2} \\ x_1^n & x_2^n & \cdots & x_n^n \end{vmatrix} ;$$

$$(5)\quad \begin{vmatrix} a & b & 0 & \cdots & \cdots & \cdots & \cdots & 0 \\ c & a & b & \ddots & & & & \vdots \\ 0 & c & a & b & \ddots & & & \vdots \\ \vdots & \ddots & c & a & \ddots & \ddots & & \vdots \\ \vdots & & \ddots & \ddots & \ddots & \ddots & \ddots & \vdots \\ \vdots & & & \ddots & \ddots & a & b & 0 \\ \vdots & & & & \ddots & c & a & b \\ 0 & \cdots & \cdots & \cdots & \cdots & 0 & c & a \end{vmatrix} .$$

20. Let $a_{ij}(t)$ be polynomials, $i, j = 1, 2, \ldots, n$. Prove that

$$\left(\begin{vmatrix} a_{11}(t) & a_{12}(t) & \cdots & a_{1n}(t) \\ a_{21}(t) & a_{22}(t) & \cdots & a_{2n}(t) \\ \vdots & \vdots & & \vdots \\ a_{n1}(t) & a_{n2}(t) & \cdots & a_{nn}(t) \end{vmatrix} \right)' = \sum_{j=1}^{n} \begin{vmatrix} a_{11}(t) & \cdots & a_{1j}'(t) & \cdots & a_{1n}(t) \\ a_{21}(t) & \cdots & a_{2j}'(t) & \cdots & a_{2n}(t) \\ \vdots & & \vdots & & \vdots \\ a_{n1}(t) & \cdots & a_{nj}'(t) & \cdots & a_{nn}(t) \end{vmatrix} .$$

21.* Let x be an indeterminate and m a positive integer, and let $m' = 2m - 1$. Prove that

$$\begin{vmatrix} 1 & 0 & \cdots & 0 & 1 & 0 & \cdots & 0 \\ x^2 & x^2 & \cdots & x^2 & 1 & 1 & \cdots & 1 \\ x^4 & 2x^4 & \cdots & 2^{m-1}x^4 & 1 & 2 & \cdots & 2^{m-1} \\ x^6 & 3x^6 & \cdots & 3^{m-1}x^6 & 1 & 3 & \cdots & 3^{m-1} \\ \vdots & \vdots & & \vdots & \vdots & \vdots & & \vdots \\ x^{2m'} & m'x^{2m'} & \cdots & (m')^{m-1}x^{2m'} & 1 & m' & \cdots & (m')^{m-1} \end{vmatrix}$$

$$= \left(\prod_{i=1}^{m-1} i! \right)^2 x^{m^2-m} \left(1 - x^2\right)^{m^2}.$$

22. Let $A = \begin{pmatrix} 1 & 0 & 1 & -1 \\ 2 & -1 & 0 & -2 \\ -3 & 1 & -2 & 4 \\ 1 & -1 & 0 & -1 \end{pmatrix}$, $B = \begin{pmatrix} 7 & -2 & 4 & -8 \\ 4 & -3 & 0 & -4 \\ -2 & 2 & 1 & 2 \\ 4 & 0 & 4 & -5 \end{pmatrix}$, and

$C = \begin{pmatrix} 2 & 1 & 1 & 0 \\ 0 & 1 & 0 & -2 \\ 1 & -1 & 0 & 1 \\ 3 & 0 & 1 & 1 \end{pmatrix}$. Compute $AB, BC, (AB)C$, and $A(BC)$.

23. Let $A = (a_{ij})_{n \times n} \in M_n(F)$. The **trace** of A, $\mathrm{Tr}(A)$, is defined as the sum of all entries on the main diagonal of A, $\mathrm{Tr}(A) = \sum_{i=1}^{n} a_{ii}$. Prove that $\mathrm{Tr}(BC) = \mathrm{Tr}(CB)$ for $B \in M_{m \times n}(F)$ and $C \in M_{n \times m}(F)$.

24. Let $A = \begin{pmatrix} a_1 & & & \\ & a_2 & & \\ & & \ddots & \\ & & & a_n \end{pmatrix}$, where $a_1, a_2, \ldots, a_n$ are different from

each other. Prove that matrices commuting with A can only be diagonal matrices.

25. Let $A = (a_{ij})_{n \times n} \in M_n(F)$, and let E_{ij} denote the $n \times n$ matrix in which the (i, j) entry is equal to 1, and all the other entries are zero. Usually, E_{ij} is called a **matrix unit**. Prove the following:

   (1) If $AE_{12} = E_{12}A$, then $a_{k1} = 0$ when $k \neq 1$; $a_{2k} = 0$ when $k \neq 2$;

(2) If $AE_{ij} = E_{ij}A$, then $a_{ki} = 0$ when $k \neq i$; and $a_{jk} = 0$ when $k \neq j$; moreover $a_{ii} = a_{jj}$;

(3) If A commutes with all matrices of order n, then A must be a scalar matrix.

26. Let n be a positive integer. Compute the following:

$$(1)\ \begin{pmatrix} 1 & 1 \\ 0 & 1 \end{pmatrix}^n; \qquad\qquad (2)\ \begin{pmatrix} \cos\theta & -\sin\theta \\ \sin\theta & \cos\theta \end{pmatrix}^n.$$

27. Let $A = \begin{pmatrix} \lambda & 1 & 0 & \cdots & \cdots & \cdots & 0 \\ 0 & \lambda & 1 & \ddots & & & \vdots \\ 0 & 0 & \lambda & \ddots & \ddots & & \vdots \\ \vdots & & \ddots & \ddots & \ddots & \ddots & \vdots \\ \vdots & & & \ddots & \lambda & 1 & 0 \\ \vdots & & & & 0 & \lambda & 1 \\ 0 & \cdots & \cdots & \cdots & 0 & 0 & \lambda \end{pmatrix}$ be a square matrix of

order n. Compute A^n.

28. Let

$$(1)\ f(x) = x^2 - x - 1, \quad A = \begin{pmatrix} 2 & 1 & 1 \\ 3 & 1 & 2 \\ 1 & -1 & 0 \end{pmatrix};$$

$$(2)\ f(x) = x^2 - 5x + 3, \quad A = \begin{pmatrix} 2 & -1 \\ -3 & 3 \end{pmatrix}.$$

Compute $f(A)$.

29. Let $A \in M_n(F)$. If $AX = 0$ for all n-dimensional column vectors $X = (x_1, x_2, \ldots, x_n)'$, prove that $A = 0$.

30. Prove that any square matrix can be expressed as the sum of a symmetric matrix and a skew-symmetric matrix.

31. Prove the following:

(1) If A is an invertible (skew-)symmetric matrix, then so is A^{-1};

(2) There is no invertible skew-symmetric matrices of odd order.

32. Prove the following:

(1) The product of two upper (lower) triangular matrices is still an upper (lower) triangular matrix;

(2) The inverse of an invertible upper (lower) triangular matrix is still an upper (lower) triangular matrix.

33. Let $A \in \mathrm{M}_n(F)$. Then A is called **nilpotent** if $A^k = 0$ for some positive integer k. Let $A \in \mathrm{M}_n(F)$ be a nilpotent matrix. Prove that $I_n - A$ is invertible and find $(I_n - A)^{-1}$.

34. Let $A \in \mathrm{M}_n(F)$ be a nilpotent matrix. If $B \in \mathrm{GL}_n(F)$ and $AB = BA$, prove that $A + B, A - B \in \mathrm{GL}_n(F)$.

35. Prove Hua's identity by Jacobson's lemma.

36.* Let $A, B \in \mathrm{M}_n(F)$ and $n \geqslant 2$ be an integer. Prove that $(AB)^* = B^*A^*$.

37. Let $A \in \mathrm{M}_n(F)$ and $n \geqslant 2$ be an integer. If $k \in F$, prove that $(kA)^* = k^{n-1}A^*$.

38. Find the rank of each of the following matrices:

(1) $\begin{pmatrix} 1 & 2 & 3 \\ 2 & 3 & -5 \\ 4 & 7 & 1 \end{pmatrix}$;

(2) $\begin{pmatrix} 0 & 1 & 1 & -1 & 2 \\ 0 & 2 & -2 & -2 & 0 \\ 0 & -1 & -1 & 1 & 1 \\ 1 & 1 & 0 & 1 & -1 \end{pmatrix}$;

(3) $\begin{pmatrix} 1 & -2 & -1 & 0 & 2 \\ -2 & 4 & 2 & 6 & -6 \\ 2 & -1 & 0 & 2 & 3 \\ 3 & 3 & 3 & 3 & 4 \end{pmatrix}$;

(4) $\begin{pmatrix} 1 & 0 & 1 & 0 & 0 \\ 1 & 1 & 0 & 0 & 0 \\ 0 & 1 & 1 & 0 & 0 \\ 0 & 0 & 1 & 1 & 0 \\ 0 & 1 & 0 & 1 & 1 \end{pmatrix}$;

(5) $\begin{pmatrix} 0 & 1 & 0 & 1 & 0 \\ 0 & 2 & 0 & -2 & 0 \\ 0 & 3 & 0 & -1 & 0 \\ 1 & 0 & 0 & 1 & 2 \end{pmatrix}$;

(6) $\begin{pmatrix} 1 & -1 & 2 & 1 & 0 \\ 2 & -2 & 4 & -2 & 0 \\ 3 & 0 & 6 & -1 & 1 \\ 0 & 3 & 0 & 0 & 1 \end{pmatrix}$.

39. Find the inverse of each of the following matrices:

(1) $\begin{pmatrix} a & b \\ c & d \end{pmatrix}$, where $ad - cb = 1$;

(2) $\begin{pmatrix} 1 & 2 & 3 \\ 2 & 2 & 1 \\ 3 & 4 & 3 \end{pmatrix}$;

(3) $\begin{pmatrix} 1 & 1 & -1 \\ 2 & 1 & 0 \\ 1 & -1 & 0 \end{pmatrix}$;

(4) $\begin{pmatrix} 1 & 1 & 1 & 1 \\ 1 & 1 & -1 & -1 \\ 1 & -1 & 1 & -1 \\ 1 & -1 & -1 & 1 \end{pmatrix}$;

(5) $\begin{pmatrix} 1 & 3 & -5 & 7 \\ 0 & 1 & 2 & -3 \\ 0 & 0 & 1 & 2 \\ 0 & 0 & 0 & 1 \end{pmatrix}$;

(6) $\begin{pmatrix} 2 & 1 & 0 & 0 & 0 \\ 0 & 2 & 1 & 0 & 0 \\ 0 & 0 & 2 & 1 & 0 \\ 0 & 0 & 0 & 2 & 1 \\ 0 & 0 & 0 & 0 & 2 \end{pmatrix}$.

40. Solve the following matrix equations:

$$(1) \quad \begin{pmatrix} 1 & 1 & -1 \\ 0 & 2 & 2 \\ 1 & -1 & 0 \end{pmatrix} X = \begin{pmatrix} 1 & -1 & 1 \\ 1 & 1 & 0 \\ 2 & 1 & 1 \end{pmatrix};$$

$$(2) \quad X \begin{pmatrix} 1 & 1 & -1 \\ 0 & 2 & 2 \\ 1 & -1 & 0 \end{pmatrix} = \begin{pmatrix} 1 & -1 & 1 \\ 1 & 1 & 0 \\ 2 & 1 & 1 \end{pmatrix}.$$

41. Let $A \in M_2(F)$ and $|A| = 1$. Prove that A can be expressed as the product of matrices of the form $\begin{pmatrix} 1 & x \\ 0 & 1 \end{pmatrix}$ and $\begin{pmatrix} 1 & 0 \\ x & 1 \end{pmatrix}$.

42. Let $n \geqslant 2$ be an integer, and let $A \in M_n(F)$ with $|A| = 1$. Prove that A can be expressed as the product of finitely many elementary matrices of the form $E(i, j(b))$, where $b \in F$.

43. Let

$$A = \begin{pmatrix} a_1 I_{n_1} & & & \\ & a_2 I_{n_2} & & \\ & & \ddots & \\ & & & a_t I_{n_t} \end{pmatrix},$$

where $a_1, a_2, \ldots, a_t$ are different from each other. Prove that matrices commuting with A can only be block diagonal matrices.

44. Let $A \in M_n(F)$ with $\mathrm{rank}A = r$. Prove that there exists $P \in GL_n(F)$ such that the last $n - r$ rows of PAP^{-1} are all zero.

45. A matrix A is called a **row (column) full rank matrix** if $\mathrm{rank}A$ is equal to the number of rows (columns) of A. Let $A \in M_{m \times r}(F)$. Prove the following:

    (1) A is a column full rank matrix if and only if there exists $P \in GL_m(F)$ such that $A = P \begin{pmatrix} I_r \\ 0 \end{pmatrix}$;

    (2) A is a row full rank matrix if and only if there exists $Q \in GL_r(F)$ such that $A = (I_m, 0)Q$.

46. Let $A \in M_{m \times n}(F)$. Prove that $\mathrm{rank}A = r$ if and only if there exists a column full rank matrix $B \in M_{m \times r}(F)$ and a row full rank matrix $C \in M_{r \times n}(F)$ such that $A = BC$.

47. Let $B \in M_r(F)$ and $C \in M_{r \times n}(F)$ with $\mathrm{rank} C = r$. Prove the following:

   (1) If $BC = 0$, then $B = 0$;

   (2) If $BC = C$, then $B = I_r$.

48. Let $A \in M_n(F)$. Prove that there exists $R \in \mathrm{GL}_n(F)$ such that AR is a symmetric matrix.

49. Let

$$A = \left(\begin{array}{cc|cc} 1 & 1 & 1 & 1 \\ 1 & -1 & 1 & -1 \\ \hline 1 & 1 & -1 & -1 \\ 1 & -1 & -1 & 1 \end{array} \right)$$

be a 4×4 matrix. Find the inverse of A by using

   (1) elementary operations of a matrix, or

   (2) elementary operations of a block matrix according to the partition in A.

50.* Let $A, B, C, D \in M_n(F)$.

   (1) If $|A| \neq 0$ and $AC = CA$, prove that $\begin{vmatrix} A & B \\ C & D \end{vmatrix} = |AD - CB|$.

   (2) If $|A| = 0$ and $AC = CA$, whether the conclusion above holds?

   (3) If $AC \neq CA$, what is the conclusion?

51. Let $A, B \in M_n(F)$. Prove the following:

   (1) $\begin{vmatrix} A & B \\ B & A \end{vmatrix} = |A + B||A - B|$;

   (2) $\begin{vmatrix} A & I_n \\ I_n & B \end{vmatrix} = |AB - I_n|$.

52. Let $A \in M_{s \times m}(F)$, $B \in M_{m \times n}(F)$, and $C \in M_{n \times t}(F)$. Prove **Frobenius'[17] rank inequality:**

$$\mathrm{rank}(ABC) \geqslant \mathrm{rank}(AB) + \mathrm{rank}(BC) - \mathrm{rank} B.$$

53.* Let $A \in M_n(F)$. Prove that $\mathrm{rank} A^n = \mathrm{rank} A^{n+1}$.

---

[17]Ferdinand Georg Frobenius, 1849–1917, German mathematician.

54. Let $A \in M_n(F)$ with $\text{rank}A = 1$. Prove the following:

(1) $A = \begin{pmatrix} a_1 \\ a_2 \\ \vdots \\ a_n \end{pmatrix} (b_1, b_2, \ldots, b_n)$;

(2) $A^2 = kA$, where $k \in F$.

55.* Let $A \in M_2(F)$. If there is an integer $l \geq 2$ such that $A^l = 0$, prove that $A^2 = 0$. Suppose $A \in M_n(F)$ and there is an integer $l \geq n$ such that $A^l = 0$, whether $A^n = 0$?

56. Let $n \geq 2$ be an integer and $A \in M_n(F)$. Prove the following:

(1) $|A^*| = |A|^{n-1}$.

(2) $\text{rank}A^* = \begin{cases} n, & \text{if } \text{rank}A = n, \\ 1, & \text{if } \text{rank}A = n - 1, \\ 0, & \text{if } \text{rank}A < n - 1. \end{cases}$

(3) $(A^*)^* = |A|^{n-2}A$.

57. Let $A \in M_n(F)$. Then A is called **idempotent** if $A^2 = A$. Prove that the following statements are equivalent:

(1) A is an idempotent matrix.

(2) $\text{rank}A + \text{rank}(A - I_n) = n$.

(3) There exists an invertible matrix P with $A = P \begin{pmatrix} I_r & 0 \\ 0 & 0 \end{pmatrix} P^{-1}$.

58. Let $A \in M_n(F)$. Then A is called **involutory** if $A^2 = I_n$. Prove that the following statements are equivalent:

(1) A an involutory matrix.

(2) $\text{rank}(A + I_n) + \text{rank}(A - I_n) = n$.

(3) There exists an invertible matrix P with $A = P \begin{pmatrix} I_r & 0 \\ 0 & -I_{n-r} \end{pmatrix} P^{-1}$.

59. Let $A = (a_{ij})_{n \times m}, B = (b_{ij})_{n \times m} \in M_{n \times m}(F)$. The matrix

$$(a_{ij}b_{ij})_{n \times m} = \begin{pmatrix} a_{11}b_{11} & a_{12}b_{12} & \cdots & a_{1m}b_{1m} \\ a_{21}b_{21} & a_{22}b_{22} & \cdots & a_{2m}b_{2m} \\ \vdots & \vdots & & \vdots \\ a_{n1}b_{n1} & a_{n2}b_{n2} & \cdots & a_{nm}b_{nm} \end{pmatrix}$$

is called the **Hadamard**[18] **product** of A and B, denoted by $A * B$. Prove that the following statements hold:

(1) $A * B = B * A$.

(2) If $A, B_1, B_2, \ldots, B_s \in \mathrm{M}_{n \times m}(F)$, then $A * \left(\sum\limits_{i=1}^{s} B_i \right) = \sum\limits_{i=1}^{s} A * B_i$.

(3) If $\alpha, \beta \in F^n$ and $\gamma, \delta \in F^m$ are all column vectors, then

$$(\alpha * \beta)(\gamma * \delta)' = (\alpha \gamma') * (\beta \delta').$$

(4) $\mathrm{rank}(A * B) \leqslant \mathrm{rank}A \times \mathrm{rank}B$.

60. Let $A = (a_{ij})_{n \times n} \in \mathrm{M}_n(F)$, $B \in \mathrm{M}_m(F)$. The matrix

$$\begin{pmatrix} a_{11}B & a_{12}B & \cdots & a_{1n}B \\ a_{21}B & a_{22}B & \cdots & a_{2n}B \\ \vdots & \vdots & & \vdots \\ a_{n1}B & a_{n2}B & \cdots & a_{nn}B \end{pmatrix}$$

is called the **tensor product** or **Kronecker product** of A and B, denoted by $A \otimes B$. Prove that the following statements hold:

(1) If $A, B, C \in \mathrm{M}_n(F)$, then

$$A \otimes (B+C) = A \otimes B + A \otimes C; \quad (B+C) \otimes A = B \otimes A + C \otimes A.$$

(2) If $A, C \in \mathrm{M}_n(F)$ and $B, D \in \mathrm{M}_m(F)$, then

$$(A \otimes B)(C \otimes D) = (AC) \otimes (BD).$$

(3) If $A \in \mathrm{M}_n(F)$, $B \in \mathrm{M}_m(F)$ and $C \in \mathrm{M}_q(F)$, then

$$(A \otimes B) \otimes C = A \otimes (B \otimes C).$$

(4) $\mathrm{rank}(A \otimes B) = \mathrm{rank}A \times \mathrm{rank}B$.

(5) $|A \otimes B| = |A|^m |B|^n$.

(6) If A and B are invertible, then $A \otimes B$ is invertible and

$$(A \otimes B)^{-1} = A^{-1} \otimes B^{-1}.$$

61.[*] Let $A = (a_{ij})_{n \times n}$, where $a_{ij} = (i, j)$ is the positive greatest common divisor of i and j, $1 \leqslant i, j \leqslant n$. Find $|A|$.

---

[18] Jacques Hadamard, 1865–1963, French mathematician.

62. Use the Cauchy–Binet formula to evaluate the determinant in Example 2.6.12.

63. Let $n \geqslant 2$ be an integer. Prove the **Binet–Cauchy identity**:

$$\left(\sum_{i=1}^{n} a_i c_i \right) \left(\sum_{i=1}^{n} b_i d_i \right) - \left(\sum_{i=1}^{n} a_i d_i \right) \left(\sum_{i=1}^{n} b_i c_i \right)$$
$$= \sum_{1 \leqslant j < k \leqslant n} (a_j b_k - a_k b_j)(c_j d_k - c_k d_j).$$

Chapter 3

Linear Systems

3.1 Introduction

A system of linear equations (or just a linear system) is a natural generalization of linear equations in one or two unknowns, which were studied in primary and middle schools. Historically, linear equations and the related solutions appeared in the chapter named "Equations" of the Chinese mathematical book "The Nine Chapters on the Mathematical Art", which is around 1700 years ahead of Europe. The commonly used form today originated in Europe, mainly with the introduction of analytic geometry by René Descartes in 1637. In fact, lines and planes in analytic geometry are described by linear equations, and calculating the intersection of these geometric objects is equivalent to solving linear equations. The first systematic method to solve linear equations in modern times was to use determinants introduced by Takakazu Seki and Gottfried Wilhelm Leibniz (see the introduction at the beginning of Chapter 2). In 1750, Gabriel Cramer gave Cramer's rule introduced in the previous chapter. Later, the famous mathematician Carl Friedrich Gauss presented the widely used elimination method, which is indeed the same method used in "The Nine Chapters on the Mathematical Art".

This chapter aims to theoretically summarize and improve the well-known knowledge about linear equation(s), giving a criterion for the existence of solutions of a system of linear equations. We provide a neat expression of all solutions if there is a solution to a system of linear equations. We see that the elementary operations of matrices are the primary means to solve linear systems. Therefore, we start with elimination methods and elementary operations.

3.2 Elimination Methods and Elementary Operations

Let F be a number field. We call the following equations

$$\begin{cases} a_{11}x_1 + a_{12}x_2 + \cdots + a_{1n}x_n = b_1, \\ a_{21}x_1 + a_{22}x_2 + \cdots + a_{2n}x_n = b_2, \\ \quad\vdots \qquad\quad \vdots \qquad\qquad\quad \vdots \qquad \vdots \\ a_{m1}x_1 + a_{m2}x_2 + \cdots + a_{mn}x_n = b_m \end{cases} \tag{3.1}$$

a **system of linear equations** in n unknowns, or just a **linear system**, where $x_1, x_2, \ldots, x_n$ are the **unknowns**, m is the number of linear equations, $a_{ij} \in F$ ($1 \leqslant i \leqslant m$, $1 \leqslant j \leqslant n$) are the **coefficients** of this linear system, and $b_1, b_2, \ldots, b_m \in F$ are the **constant terms**. An n-dimensional column vector $(k_1, k_2, \ldots, k_n)' \in F^n$ is called a **solution** of the linear system (3.1) if each equation in (3.1) becomes an identity when $x_1 = k_1, x_2 = k_2, \ldots, x_n = k_n$ are substituted in (3.1). The set of all solutions of (3.1) is called the **solution set** of (3.1). If two linear systems have the same solution set, we call them **linear systems with identical solutions**.

Moreover, let

$$A = \begin{pmatrix} a_{11} & a_{12} & \ldots & a_{1n} \\ a_{21} & a_{22} & \ldots & a_{2n} \\ \vdots & \vdots & & \vdots \\ a_{m1} & a_{m2} & \ldots & a_{mn} \end{pmatrix}, \quad X = \begin{pmatrix} x_1 \\ x_2 \\ \vdots \\ x_n \end{pmatrix}, \quad \beta = \begin{pmatrix} b_1 \\ b_2 \\ \vdots \\ b_m \end{pmatrix}.$$

We call A, (A, β), X, and β the **coefficient matrix**, the **augmented matrix**, the **unknown vector**, and the **constant vector** of (3.1), respectively. It is easy to see that the system of linear equations (3.1) and its augmented matrix are mutually uniquely determined without considering the notation of the unknowns. If the augmented matrix of (3.1) is a (reduced) echelon matrix, then the linear system (3.1) is called a **linear system of (reduced) echelon form** or just a **(reduced) echelon system**.

By matrix multiplication, the linear system (3.1) can be written as[1]

$$AX = \beta. \tag{3.2}$$

If $\beta = 0$, then $AX = 0$ is called a **system of homogeneous linear equations** or a **homogeneous linear system**; if $\beta \neq 0$, then $AX = \beta$ is called a **system of nonhomogeneous linear equations** or a **nonhomogeneous linear system**.

Remark 3.2.1. In this book, we always assume that the coefficient matrix of a linear system has no zero columns; otherwise, this linear system will have fewer unknowns.

Given a linear system, there are two natural tasks:

(1) Determine whether the system has a solution or not.
(2) Describe the solution set if this system has solutions.

Let's look at a simple example first.

Example 3.2.2. Use the elimination method to solve the following linear system:

$$\begin{cases} x_1 + x_2 + x_3 = 1, \\ 2x_1 + 2x_2 + 2x_3 = 2, \\ x_1 + x_2 \qquad\;\; = 3, \\ x_1 - x_2 \qquad\;\; = 1. \end{cases} \tag{3.3}$$

Solution. The augmented matrix of this linear system is
$$\begin{pmatrix} 1 & 1 & 1 & 1 \\ 2 & 2 & 2 & 2 \\ 1 & 1 & 0 & 3 \\ 1 & -1 & 0 & 1 \end{pmatrix}.$$

---

[1]Linear system (3.1) can also be written as $X'A' = \beta'$. Therefore, $AX = \beta$ and $X'A' = \beta'$ are both called linear systems.

                    Advanced Algebra

The process of the elimination method corresponds to the elementary row operations of the augmented matrix:

$$\text{Eliminations} \qquad\qquad \text{Elementary row operations}$$

$$\begin{cases} x_1 + x_2 + x_3 = 1, \\ 2x_1 + 2x_2 + 2x_3 = 2, \\ x_1 + x_2 \phantom{{}+x_3} = 3, \\ x_1 - x_2 \phantom{{}+x_3} = 1. \end{cases} \qquad \begin{pmatrix} 1 & 1 & 1 & 1 \\ 2 & 2 & 2 & 2 \\ 1 & 1 & 0 & 3 \\ 1 & -1 & 0 & 1 \end{pmatrix}$$

$$\downarrow \qquad\qquad\qquad \downarrow$$

$$\begin{cases} x_1 + x_2 + x_3 = 1, \\ x_1 + x_2 \phantom{{}+x_3} = 3, \\ x_1 - x_2 \phantom{{}+x_3} = 1, \\ 2x_1 + 2x_2 + 2x_3 = 2. \end{cases} \qquad \begin{pmatrix} 1 & 1 & 1 & 1 \\ 1 & 1 & 0 & 3 \\ 1 & -1 & 0 & 1 \\ 2 & 2 & 2 & 2 \end{pmatrix}$$

$$\downarrow \qquad\qquad\qquad \downarrow$$

$$\begin{cases} x_1 + x_2 + x_3 = 1, \\ - x_3 = 2, \\ - 2x_2 - x_3 = 0, \\ 0 = 0. \end{cases} \qquad \begin{pmatrix} 1 & 1 & 1 & 1 \\ 0 & 0 & -1 & 2 \\ 0 & -2 & -1 & 0 \\ 0 & 0 & 0 & 0 \end{pmatrix}$$

$$\downarrow \qquad\qquad\qquad \downarrow$$

$$\begin{cases} x_1 + x_2 + x_3 = 1, \\ - 2x_2 - x_3 = 0, \\ - x_3 = 2, \\ 0 = 0. \end{cases} \qquad \begin{pmatrix} 1 & 1 & 1 & 1 \\ 0 & -2 & -1 & 0 \\ 0 & 0 & -1 & 2 \\ 0 & 0 & 0 & 0 \end{pmatrix}$$

$$\downarrow \qquad\qquad\qquad \downarrow$$

$$\begin{cases} x_1 + x_2 + x_3 = 1, \\ x_2 + \tfrac{1}{2}x_3 = 0, \\ x_3 = -2, \\ 0 = 0. \end{cases} \qquad \begin{pmatrix} 1 & 1 & 1 & 1 \\ 0 & 1 & \tfrac{1}{2} & 0 \\ 0 & 0 & 1 & -2 \\ 0 & 0 & 0 & 0 \end{pmatrix}$$

$$\downarrow \qquad\qquad\qquad \downarrow$$

$$\begin{cases} x_1 \phantom{{}+ x_2 + x_3} = 2, \\ x_2 \phantom{{}+ x_3} = 1, \\ x_3 = -2, \\ 0 = 0. \end{cases} \qquad \begin{pmatrix} 1 & 0 & 0 & 2 \\ 0 & 1 & 0 & 1 \\ 0 & 0 & 1 & -2 \\ 0 & 0 & 0 & 0 \end{pmatrix}$$

From this, we find that $x_1 = 2, x_2 = 1, x_3 = -2$. $\qquad\qquad\square$

After careful investigation, we find that the elimination method repeats the following three kinds of operations on the equations:

(1) Multiply an equation by a nonzero number $k \in F$.
(2) Add b times an equation to another equation, where $b \in F$.
(3) Interchange the positions of two equations.

We call these three kinds of operations the **elementary operations** of a linear system. It is easy to see that applying elementary operations to a linear system is equivalent to applying elementary row operations to its augmented matrix. Elementary operations will turn a linear system into another one with identical solutions. Since any matrix can be transformed into a (reduced) echelon matrix by elementary row operations, any linear system can be transformed into a (reduced) echelon system with identical solutions. It can be seen that elementary operations are beneficial to simplify the linear system without changing its solution set.

Example 3.2.3. Solve the following linear systems:

$$(1) \quad \begin{cases} x_1 + x_2 + x_3 = 1, \\ 2x_1 + 3x_2 + 5x_3 = 7, \\ 3x_1 + 4x_2 + 6x_3 = 9; \end{cases}$$

$$(2) \quad \begin{cases} 2x_1 - x_2 + 3x_3 = 1, \\ 4x_1 - 2x_2 + 7x_3 = -1, \\ 2x_1 - x_2 + 4x_3 = -2. \end{cases}$$

Solution. (1) Applying elementary operations, we have

$$\begin{cases} x_1 \quad\quad - 2x_3 = 0, \\ \quad x_2 + 3x_3 = 0, \\ \quad\quad\quad 0 = 1. \end{cases}$$

This system has no solutions; thus, the original system has no solutions.

(2) Applying elementary operations, we have

$$\begin{cases} x_1 - \dfrac{1}{2}x_2 = \dfrac{10}{2}, \\ \qquad\qquad x_3 = -3, \\ \qquad\qquad\quad 0 = 0. \end{cases}$$

From this, we know that the original linear system has infinitely

many solutions:
$$\begin{cases} x_1 = \dfrac{1}{2}(10 + k), \\ x_2 = k, \\ x_3 = -3. \end{cases} \qquad \text{for } k \in F. \qquad\qquad \square$$

From Examples 3.2.2 and 3.2.3, we know that there are at least three possibilities about the solutions of linear systems: no solution, unique solution, and infinitely many solutions. The solutions of general linear systems are just these three possibilities, and we have the following theorem.

Theorem 3.2.4. *Let $A \in \mathrm{M}_{m\times n}(F)$ and $\beta \in F^m$.*

(1) *$AX = \beta$ has a solution if and only if $\mathrm{rank}A = \mathrm{rank}(A, \beta)$. Moreover,*

     (1.1) *$AX = \beta$ has a unique solution if and only if $\mathrm{rank}A = \mathrm{rank}(A, \beta) = n$;*

     (1.2) *$AX = \beta$ has infinitely many solutions if and only if $\mathrm{rank}A = \mathrm{rank}(A, \beta) < n$.*

(2) *The homogeneous system $AX = 0$ always has the zero solution and*

     (2.1) *$AX = 0$ only has the zero solution if and only if $\mathrm{rank}A = n$;*

     (2.2) *$AX = 0$ has nonzero solutions if and only if $\mathrm{rank}A < n$.*

In particular, for $A \in \mathrm{M}_n(F)$, $AX = 0$ only has the zero solution if and only if $|A| \neq 0$.

Proof. Since (2) is a special case of (1), we only need to prove (1). By Theorem 2.8.9, the matrix (A, β) can be transformed into a reduced echelon matrix (B, γ) by elementary row operations. Thus, $AX = \beta$ and $BX = \gamma$ are linear systems with identical solutions.

By Theorem 2.8.5, $\mathrm{rank}A = \mathrm{rank}B$ and $\mathrm{rank}(A, \beta) = \mathrm{rank}(B, \gamma)$. Therefore, we may assume that (A, β) is just a reduced echelon matrix. Let $\mathrm{rank}A = r \geqslant 1$. By Theorem 2.8.10 or Proposition 2.10.13, $\mathrm{rank}(A, \beta) = r$ or $r + 1$.

If $\mathrm{rank}A \neq \mathrm{rank}(A, \beta)$, then $\mathrm{rank}(A, \beta) = r + 1$. Thus $AX = \beta$ has no solutions since the $(r + 1)$th equation is $0 = 1$ now. So the "only if" part follows. Next, we prove the "if" part.

If $\mathrm{rank}(A, \beta) = r = n$, then $(A, \beta) = \begin{pmatrix} 1 & & & d_1 \\ & \ddots & & \vdots \\ & & 1 & d_n \\ 0 & \cdots & 0 & 0 \\ \vdots & & \vdots & \vdots \\ 0 & \cdots & 0 & 0 \end{pmatrix}$. Clearly $AX = \beta$ has a unique solution $X = (d_1, d_2, \ldots, d_n)'$.

If $\mathrm{rank}(A, \beta) = r < n$, then we denote the column indices of the first nonzero elements 1 of the r nonzero rows by $i_1, \ldots, i_r$ with $i_1 < \cdots < i_r$ and the remaining $n - r$ column indices by $i_{r+1}, \ldots, i_n$ with $i_{r+1} < \cdots < i_n$.

Now $AX = \beta$ has the same solution set as

$$\begin{cases} x_{i_1} \qquad\quad + c_{1i_{r+1}}x_{i_{r+1}} + \cdots + c_{1i_n}x_{i_n} = d_1, \\ \quad \ddots \qquad\qquad\qquad\quad \vdots \\ \quad\; x_{i_r} + c_{ri_{r+1}}x_{i_{r+1}} + \cdots + c_{ri_n}x_{i_n} = d_r. \end{cases} \tag{3.4}$$

Therefore, for any given $k_{i_{r+1}}, \ldots, k_{i_n} \in F$, we get the following solution of $AX = \beta$:

$$\begin{cases} x_{i_1} \;\; = d_1 - c_{1i_{r+1}}k_{i_{r+1}} - \cdots - c_{1i_n}k_{i_n}, \\ \;\; \vdots \qquad \vdots \\ x_{i_r} \;\; = d_r - c_{ri_{r+1}}k_{i_{r+1}} - \cdots - c_{ri_n}k_{i_n}, \\ x_{i_{r+1}} = k_{i_{r+1}}, \\ \;\; \vdots \qquad \vdots \\ x_{i_n} \;\; = k_{i_n}. \end{cases}$$

Since $n - r \geqslant 1$ and F is a number field, $AX = \beta$ has infinitely many solutions. $\qquad\square$

By the linear system (3.4), $x_{i_1}, \ldots, x_{i_r}$ are represented by $x_{i_{r+1}}, \ldots, x_{i_n}$. Therefore, we call $x_{i_{r+1}}, \ldots, x_{i_n}$ the **free unknowns** of $AX = \beta$.

Corollary 3.2.5. *Assume that* $A \in \mathrm{M}_{m \times n}(F)$. *If* $m < n$, *then* $AX = 0$ *has nonzero solutions.*

Proof. Since $\mathrm{rank} A \leqslant m < n$, $AX = 0$ has nonzero solutions by (2.2) of Theorem 3.2.4. $\qquad\square$

3.3 Linear Dependence

The solution set of the linear system (3.1) is a subset of F^n. To study the structure of the solutions of a linear system, we need to examine the relations among vectors in the vector space F^n.

Assume that $A = (\beta_1, \beta_2, \ldots, \beta_n) \in \mathrm{M}_{m \times n}(F)$, that is, $\beta_1, \beta_2, \ldots, \beta_n \in F^m$ are the column vectors of A. By the multiplication of block matrices, the linear system (3.2) can be written as

$$x_1 \beta_1 + x_2 \beta_2 + \cdots + x_n \beta_n = \beta. \tag{3.5}$$

It follows that a column vector $\alpha = (k_1, k_2, \ldots, k_n)' \in F^n$ is a solution of (3.2) if and only if

$$k_1 \beta_1 + k_2 \beta_2 + \cdots + k_n \beta_n = \beta. \tag{3.6}$$

Definition 3.3.1. Let $\alpha_1, \alpha_2, \ldots, \alpha_s, \beta_1, \beta_2, \ldots, \beta_t, \alpha$ be vectors in F^n.

(1) For any $k_1, k_2, \ldots, k_s \in F$, the vector $k_1 \alpha_1 + k_2 \alpha_2 + \cdots + k_s \alpha_s$ is called a **linear combination** of $\alpha_1, \alpha_2, \ldots, \alpha_s$.
(2) The vector α is said to be **linearly expressed** by the sequence $\alpha_1, \alpha_2, \ldots, \alpha_s$ if there are $k_1, k_2, \ldots, k_s \in F$ such that

$$\alpha = k_1 \alpha_1 + k_2 \alpha_2 + \cdots + k_s \alpha_s.$$

(3) The sequence $\alpha_1, \alpha_2, \ldots, \alpha_s$ is said to be linearly expressed by the sequence $\beta_1, \beta_2, \ldots, \beta_t$ if each α_i can be linearly expressed by the sequence $\beta_1, \beta_2, \ldots, \beta_t$ for any $i = 1, 2, \ldots, s$.
(4) If two sequences $\alpha_1, \alpha_2, \ldots, \alpha_s$ and $\beta_1, \beta_2, \ldots, \beta_t$ can be linearly expressed by each other, then they are called **linearly equivalent** (or simply, **equivalent**).

(5) The sequence $\alpha_1, \alpha_2, \ldots, \alpha_s$ is called **linearly dependent** if there are $k_1, k_2, \ldots, k_s \in F$, not all zero, such that $k_1\alpha_1 + k_2\alpha_2 + \cdots + k_s\alpha_s = 0$, and it is called **linearly independent** otherwise.

Remark 3.3.2. Unless otherwise specified, a sequence of vectors involved in this book only contains finitely many vectors. We sometimes say that $\alpha_1, \alpha_2, \ldots, \alpha_s$ are linearly dependent (or linearly independent) instead of saying that the sequence $\alpha_1, \alpha_2, \ldots, \alpha_s$ is linearly dependent (or linearly independent).

Theorem 3.3.3. *Let $A = (\beta_1, \beta_2, \ldots, \beta_n) \in \mathrm{M}_{m \times n}(F)$ and $\beta \in F^m$.*

(1) *The linear system $AX = \beta$ has a solution if and only if β can be linearly expressed by $\beta_1, \beta_2, \ldots, \beta_n$.*
(2) *The homogeneous linear system $AX = 0$ has a nonzero solution if and only if $\beta_1, \beta_2, \ldots, \beta_n$ are linearly dependent.*
(3) *The homogeneous linear system $AX = 0$ only has the zero solution if and only if $\beta_1, \beta_2, \ldots, \beta_n$ are linearly independent.*

Proof. This follows from (3.6) and the definitions of a linear expression, linear dependence, and linear independence. $\qquad\square$

The theorem above shows that the problem of whether a linear system has a solution or not can be transformed into a related problem of a linear expression, linear dependence, and linear independence.

Example 3.3.4. Let n be a positive integer. For $1 \leqslant i \leqslant n$, define $\varepsilon_i \in F^n$ such that the ith component is 1 and the others are zero. That is,

$$\varepsilon_1 = \begin{pmatrix} 1 \\ 0 \\ 0 \\ \vdots \\ 0 \\ 0 \end{pmatrix}, \varepsilon_2 = \begin{pmatrix} 0 \\ 1 \\ 0 \\ \vdots \\ 0 \\ 0 \end{pmatrix}, \ldots, \varepsilon_n = \begin{pmatrix} 0 \\ 0 \\ 0 \\ \vdots \\ 0 \\ 1 \end{pmatrix} \in F^n.$$

Then for any $\alpha = (k_1, k_2, \ldots, k_n)' \in F^n$, one has $\alpha = k_1\varepsilon_1 + k_2\varepsilon_2 + \cdots + k_n\varepsilon_n$. Thus $k_1\varepsilon_1 + k_2\varepsilon_2 + \cdots + k_n\varepsilon_n = 0$ if and only if $k_1 = k_2 = \cdots = k_n = 0$. It follows that $\varepsilon_1, \varepsilon_2, \ldots, \varepsilon_n$ are linearly independent and any $\alpha \in F^n$ can be linearly expressed by $\varepsilon_1, \varepsilon_2, \ldots, \varepsilon_n$.

Example 3.3.5.

(1) Let $A \in M_{n \times m}(F)$ and $B \in M_{m \times s}(F)$. Then the column vectors of AB can be linearly expressed by the column vectors of A, and the row vectors of AB can be linearly expressed by the row vectors of B.

(2) Let $\alpha_1, \alpha_2, \ldots, \alpha_m, \gamma_1, \gamma_2, \ldots, \gamma_s \in F^n$ be column vectors. Then $\gamma_1, \gamma_2, \ldots, \gamma_s$ can be linearly expressed by $\alpha_1, \alpha_2, \ldots, \alpha_m$ if and only if there exists $B \in M_{m \times s}(F)$ such that

$$(\gamma_1, \gamma_2, \ldots, \gamma_s) = (\alpha_1, \alpha_2, \ldots, \alpha_m)B.$$

Proof. (1) follows from Example 2.10.5.

(2) Sufficiency. Let $A = (\alpha_1, \alpha_2, \ldots, \alpha_m)$, one gets the result from (1).

Necessity. Assume that $\gamma_1, \gamma_2, \ldots, \gamma_s$ can be linearly expressed by $\alpha_1, \alpha_2, \ldots, \alpha_m$, then for $j = 1, 2, \ldots, s$, there are $b_{1j}, b_{2j}, \ldots, b_{mj} \in F$ such that

$$\gamma_j = b_{1j}\alpha_1 + b_{2j}\alpha_2 + \cdots + b_{mj}\alpha_m = (\alpha_1, \alpha_2, \ldots, \alpha_m)\begin{pmatrix} b_{1j} \\ b_{2j} \\ \vdots \\ b_{mj} \end{pmatrix}.$$

Now let $B = (b_{ij})_{m \times s}$, then it is obvious that $(\gamma_1, \gamma_2, \ldots, \gamma_s) = (\alpha_1, \alpha_2, \ldots, \alpha_m)B$. $\qquad\square$

Proposition 3.3.6. *The linear equivalence is an equivalence relation.*

Proof. See Exercise 3. $\qquad\square$

Proposition 3.3.7. *Let F be a number field, and let n, m, p, r be positive integers.*

(1) *Any sequence of vectors containing the zero vector is linearly dependent.*

(2) *Let $\alpha \in F^n$. Then α is linearly dependent if and only if $\alpha = 0$, and α is linearly independent if and only if $\alpha \neq 0$.*

(3) *For nonzero vectors*

$$\alpha = (a_1, a_2, \ldots, a_n) \text{ and } \beta = (b_1, b_2, \ldots, b_n) \in F^n,$$

we have

(3.1) *α, β are linearly dependent if and only if α is proportional to β, that is, there exists a nonzero number λ such that $a_i = \lambda b_i$ for $i = 1, 2, \ldots, n$;*

(3.2) *α, β are linearly independent if and only if there are i and j with $1 \leqslant i \neq j \leqslant n$ such that $a_i b_j - a_j b_i \neq 0$.*

(4) *Assume that $m \geqslant 2$, then*

(4.1) *$\alpha_1, \alpha_2, \ldots, \alpha_m \in F^n$ are linearly dependent if and only if there exists $i \in \{1, 2, \ldots, m\}$ such that α_i can be linearly expressed by the other vectors;*

(4.2) *$\alpha_1, \alpha_2, \ldots, \alpha_m \in F^n$ are linearly independent if and only if α_i can not be linearly expressed by the other vectors for any $i \in \{1, 2, \ldots, m\}$.*

(5) *Let $\{i_1, i_2, \ldots, i_r\}$ be a subset of $\{1, 2, \ldots, m\}$.*

(5.1) *If $\alpha_{i_1}, \ldots, \alpha_{i_r}$ are linearly dependent, then $\alpha_1, \alpha_2, \ldots, \alpha_m$ are linearly dependent;*

(5.2) *If $\alpha_1, \alpha_2, \ldots, \alpha_m$ are linearly independent, then $\alpha_{i_1}, \ldots, \alpha_{i_r}$ are linearly independent.*

(6) *Assume that $\alpha_1, \alpha_2, \ldots, \alpha_m \in F^n$ and $\beta_1, \beta_2, \ldots, \beta_m \in F^p$ are column vectors. Let*

$$\gamma_1 = \begin{pmatrix} \alpha_1 \\ \beta_1 \end{pmatrix}, \ \gamma_2 = \begin{pmatrix} \alpha_2 \\ \beta_2 \end{pmatrix}, \ \ldots, \ \gamma_m = \begin{pmatrix} \alpha_m \\ \beta_m \end{pmatrix} \in F^{n+p}.$$

(6.1) *If $\alpha_1, \alpha_2, \ldots, \alpha_m$ are linearly independent, then so are $\gamma_1, \gamma_2, \ldots, \gamma_m$;*

(6.2) *If $\gamma_1, \gamma_2, \ldots, \gamma_m$ are linearly dependent, then so are $\alpha_1, \alpha_2, \ldots, \alpha_m$.*

Proof. All of these statements are direct consequences of definitions. $\qquad\square$

Theorem 3.3.8. *Let $\alpha_1, \alpha_2, \ldots, \alpha_m, \beta \in F^n$. Assume that $\alpha_1, \alpha_2, \ldots, \alpha_m$ are linearly independent and $\alpha_1, \alpha_2, \ldots, \alpha_m, \beta$ are linearly dependent, then β can be linearly expressed by $\alpha_1, \alpha_2, \ldots, \alpha_m$ in a unique way.*

Proof. We first prove the existence. We must show that β can be linearly expressed by $\alpha_1, \ldots, \alpha_m$. In fact, since $\alpha_1, \alpha_2, \ldots, \alpha_m, \beta$ are linearly dependent, there are $k_1, k_2, \ldots, k_m, k \in F$, not all zero, such that

$$k_1\alpha_1 + k_2\alpha_2 + \cdots + k_m\alpha_m + k\beta = 0. \tag{3.7}$$

We claim that $k \neq 0$. Otherwise, if $k = 0$, then we have $k_1\alpha_1 + k_2\alpha_2 + \cdots + k_m\alpha_m = 0$. Note that $\alpha_1, \alpha_2, \ldots, \alpha_m$ are linearly independent, and so $k_1 = k_2 = \cdots = k_m = 0$. This contradicts the fact that $k_1, k_2, \ldots, k_m, k$ are not all zero. Therefore, $k \neq 0$. By (3.7), $\beta = -\frac{k_1}{k}\alpha_1 - \frac{k_2}{k}\alpha_2 - \cdots - \frac{k_m}{k}\alpha_m$, and hence β can be linearly expressed by $\alpha_1, \alpha_2, \ldots, \alpha_m$.

For the uniqueness, suppose there are $k_1, k_2, \ldots, k_m, l_1, l_2, \ldots, l_m \in F$ such that

$$k_1\alpha_1 + k_2\alpha_2 + \cdots + k_m\alpha_m = \beta,$$
$$l_1\alpha_1 + l_2\alpha_2 + \cdots + l_m\alpha_m = \beta.$$

Then $(k_1 - l_1)\alpha_1 + (k_2 - l_2)\alpha_2 + \cdots + (k_m - l_m)\alpha_m = 0$. Since $\alpha_1, \alpha_2, \ldots, \alpha_m$ are linearly independent, $k_i - l_i = 0$, i.e., $k_i = l_i$ for $i = 1, 2, \ldots, m$. $\square$

Lemma 3.3.9 (Steinitz[2] exchange lemma). *Assume that $\alpha_1, \ldots, \alpha_r, \beta_1, \ldots, \beta_m \in F^n$, and $\alpha_1, \ldots, \alpha_r$ can be linearly expressed by $\beta_1, \ldots, \beta_m$. If $\alpha_1, \ldots, \alpha_r$ are linearly independent, then $r \leqslant m$ and we can use $\alpha_1, \ldots, \alpha_r$ to replace r vectors, say $\beta_1, \ldots, \beta_r$, in $\beta_1, \ldots, \beta_m$ such that $\alpha_1, \ldots, \alpha_r, \beta_{r+1}, \ldots, \beta_m$ and $\beta_1, \ldots, \beta_r, \beta_{r+1}, \ldots, \beta_m$ are linearly equivalent.*

Proof. We proceed by induction on r. When $r = 1$, clearly $r \leqslant m$. Since α_1 can be linearly expressed by $\beta_1, \ldots, \beta_m$, there exist $k_1, k_2, \ldots, k_m \in F$ such that

$$\alpha_1 = k_1\beta_1 + k_2\beta_2 + \cdots + k_m\beta_m. \tag{3.8}$$

Note that α_1 is linearly independent, so $\alpha_1 \neq 0$, whence $k_1, k_2, \ldots, k_m$ are not all 0. Without loss of generality, we may assume that $k_1 \neq 0$.

---

[2]Ernst Steinitz, 1871–1928, German mathematician.

Hence

$$\beta_1 = \frac{1}{k_1}\alpha_1 - \frac{k_2}{k_1}\beta_2 - \cdots - \frac{k_m}{k_1}\beta_m.$$

It follows that $\beta_1, \beta_2, \ldots, \beta_m$ can be linearly expressed by $\alpha_1, \beta_2, \ldots, \beta_m$. By (3.8), $\alpha_1, \beta_2, \ldots, \beta_m$ can be linearly expressed by $\beta_1, \beta_2, \ldots, \beta_m$. Therefore, $\alpha_1, \beta_2, \ldots, \beta_m$ and $\beta_1, \beta_2, \ldots, \beta_m$ are linearly equivalent.

Now assume that $r \geqslant 2$ and the theorem is true for the case $r - 1$. Next, we prove the theorem for the case r. By assumption, $\alpha_1, \ldots, \alpha_{r-1}$ can be linearly expressed by $\beta_1, \ldots, \beta_m$. By induction, $r - 1 \leqslant m$ and one can replace $r - 1$ vectors, say $\beta_1, \ldots, \beta_{r-1}$, in $\beta_1, \ldots, \beta_m$ by $\alpha_1, \ldots, \alpha_{r-1}$ such that

$$\text{the sequences } \alpha_1, \ldots, \alpha_{r-1}, \beta_r, \ldots, \beta_m \text{ and}$$
$$\beta_1, \ldots, \beta_{r-1}, \beta_r, \ldots, \beta_m \text{ are linearly equivalent.} \tag{3.9}$$

Since α_r can be linearly expressed by $\beta_1, \ldots, \beta_{r-1}, \beta_r, \ldots, \beta_m$, α_r can be linearly expressed by $\alpha_1, \ldots, \alpha_{r-1}, \beta_r, \ldots, \beta_m$, that is, there are $k_1, \ldots, k_m \in F$ such that

$$\alpha_r = k_1\alpha_1 + \cdots + k_{r-1}\alpha_{r-1} + k_r\beta_r + \cdots + k_m\beta_m. \tag{3.10}$$

If $r - 1 = m$ or $k_r = \cdots = k_m = 0$, then (3.10) shows that α_r can be linearly expressed by $\alpha_1, \ldots, \alpha_{r-1}$, contradicting the fact that $\alpha_1, \ldots, \alpha_{r-1}, \alpha_r$ are linearly independent. Thus $r - 1 < m$, that is, $r \leqslant m$, and $k_r, \ldots, k_m$ are not all zero, say $k_r \neq 0$. By (3.10), we have

$$\beta_r = -\frac{k_1}{k_r}\alpha_1 - \cdots - \frac{k_{r-1}}{k_r}\alpha_{r-1} + \frac{1}{k_r}\alpha_r - \frac{k_{r+1}}{k_r}\beta_{r+1} - \cdots - \frac{k_m}{k_r}\beta_m,$$

and hence $\alpha_1, \ldots, \alpha_{r-1}, \beta_r, \beta_{r+1} \ldots, \beta_m$ can be linearly expressed by $\alpha_1, \ldots, \alpha_{r-1}, \alpha_r, \beta_{r+1}, \ldots, \beta_m$. Due to (3.9), $\beta_1, \ldots, \beta_r, \beta_{r+1}, \ldots, \beta_m$ are linearly expressed by $\alpha_1, \ldots, \alpha_r, \beta_{r+1}, \ldots, \beta_m$. By assumption, $\alpha_1, \ldots, \alpha_r, \beta_{r+1}, \ldots, \beta_m$ can be linearly expressed by $\beta_1, \ldots, \beta_r, \beta_{r+1}, \ldots, \beta_m$. Therefore, $\alpha_1, \ldots, \alpha_r, \beta_{r+1}, \ldots, \beta_m$ and $\beta_1, \ldots, \beta_r, \beta_{r+1}, \ldots, \beta_m$ are linearly equivalent. $\square$

Corollary 3.3.10. *Let F be a number field and n, m, r be positive integers.*

(1) *Assume that $\alpha_1, \ldots, \alpha_r, \beta_1, \ldots, \beta_m \in F^n$ and $\alpha_1, \ldots, \alpha_r$ can be linearly expressed by $\beta_1, \ldots, \beta_m$. If $r > m$, then $\alpha_1, \ldots, \alpha_r$ are linearly dependent.*

(2) *Any sequence of $n + 1$ vectors in F^n is linearly dependent.*

(3) *In F^n, any two linearly equivalent sequences of linearly independent vectors must have the same number of vectors.*

Proof. (1) If $\alpha_1, \ldots, \alpha_r$ are linearly independent, then by the Steinitz exchange lemma, we have $r \leqslant m$, which contradicts $r > m$.

(2) By Example 3.3.4, any sequence of $n + 1$ vectors in F^n always can be linearly expressed by $\varepsilon_1, \varepsilon_2, \ldots, \varepsilon_n$. Then, by (1), we get what we want.

(3) follows from the Steinitz exchange lemma. $\qquad\qquad\square$

Definition 3.3.11. Assume that $\alpha_1, \alpha_2, \ldots, \alpha_m \in F^n$ are not all zero. A subsequence $\alpha_{i_1}, \ldots, \alpha_{i_r}$ of the sequence $\alpha_1, \alpha_2, \ldots, \alpha_m$ is called a **maximal linearly independent set** if it satisfies

(1) $\alpha_{i_1}, \ldots, \alpha_{i_r}$ are linearly independent and

(2) $\alpha_j, \alpha_{i_1}, \ldots, \alpha_{i_r}$ are linearly dependent for any $j = 1, 2, \ldots, m$.

Remark 3.3.12. By Theorem 3.3.8, the condition (2) in Definition 3.3.11 can be replaced with

    (2′) α_j can be linearly expressed by $\alpha_{i_1}, \ldots, \alpha_{i_r}$ for any $j = 1, 2, \ldots, m$.

Theorem 3.3.13. *Let S be a sequence of vectors $\alpha_1, \alpha_2, \ldots, \alpha_m$ in F^n which are not all zero. Then*

(1) *S has a maximal linearly independent set;*

(2) *S is linearly equivalent to any maximal linearly independent set in S;*

(3) *any two maximal linearly independent sets in S are linearly equivalent;*

(4) *any two maximal linearly independent sets in S have the same number of vectors, and this number is less than or equal to n.*

Proof. (1) We use induction on m to prove the result.

Let $m = 1$. Since $\alpha_1 \neq 0$, α_1 is linearly independent, and we get the conclusion.

Assume that $m \geqslant 2$ and the statement is true for the case $m - 1$. Now we consider the sequence of vectors $\alpha_1, \alpha_2, \ldots, \alpha_m$, which are not all zero. The statement is clearly true if $\alpha_1, \alpha_2, \ldots, \alpha_m$ are linearly independent. If $\alpha_1, \alpha_2, \ldots, \alpha_m$ are linearly dependent, then by Proposition 3.3.7 (2), there exists a vector, say α_m, such that α_m can be linearly expressed by $\alpha_1, \ldots, \alpha_{m-1}$. By induction, the sequence $\alpha_1, \ldots, \alpha_{m-1}$ has a maximal linearly independent set $\alpha_{i_1}, \ldots, \alpha_{i_r}$. By transitivity of linear expressions, α_m can be linearly expressed by $\alpha_{i_1}, \ldots, \alpha_{i_r}$. Therefore, by definition, the subsequence $\alpha_{i_1}, \ldots, \alpha_{i_r}$ is a maximal linearly independent set of the sequence $\alpha_1, \alpha_2, \ldots, \alpha_m$. By the first form of induction, (1) is true.

(2) follows from the definition of a maximal linearly independent set.

(3) comes from (2) and the transitivity of linear equivalence.

(4) is clear by (3) and Corollary 3.3.10. $\qquad\square$

Definition 3.3.14. Let $\alpha_1, \ldots, \alpha_m$ be a sequence of vectors in F^n.

(1) If the sequence $\alpha_1, \ldots, \alpha_m$ consists of only zero vectors, then we define its **rank** to be zero, denoted by $\operatorname{rank}(\alpha_1, \ldots, \alpha_m) = 0$.

(2) If the sequence $\alpha_1, \ldots, \alpha_m$ contains a nonzero vector, then we define its **rank** to be the number of vectors in a maximal linearly independent set in the sequence $\alpha_1, \ldots, \alpha_m$ and denote this number by $\operatorname{rank}(\alpha_1, \ldots, \alpha_m)$.

Remark 3.3.15. By Theorem 3.3.13 (4), $\operatorname{rank}(\alpha_1, \ldots, \alpha_m)$ is well defined, and the rank of a sequence of vectors in F^n is less than or equal to n.

Proposition 3.3.16. *Let* $\alpha_1, \ldots, \alpha_r, \beta_1, \ldots, \beta_m \in F^n$, $s \in \mathbb{N}^*$. *Then*

(1) $\operatorname{rank}(\alpha_1, \ldots, \alpha_r) \leqslant r$;
(2) *the sequence* $\alpha_1, \ldots, \alpha_r$ *is linearly independent if and only if* $\operatorname{rank}(\alpha_1, \ldots, \alpha_r) = r$;
(3) *if the rank of a sequence of vectors is s, then any set of s linearly independent vectors in this sequence must be a maximal linearly independent set;*
(4) *if the sequence* $\alpha_1, \ldots, \alpha_r$ *can be linearly expressed by the sequence* $\beta_1, \ldots, \beta_m$, *then* $\operatorname{rank}(\alpha_1, \ldots, \alpha_r) \leqslant \operatorname{rank}(\beta_1, \ldots, \beta_m) \leqslant m$;
(5) *linearly equivalent sequences of vectors have the same rank.*

Proof. (1), (2), and (3) follow directly from the definition of the rank of a sequence of vectors.

(4) Assume that $\operatorname{rank}(\alpha_1, \ldots, \alpha_r) = s$ and $\operatorname{rank}(\beta_1, \ldots, \beta_m) = t$, then clearly we have $t \leqslant m$. Let $\alpha_{i_1}, \ldots, \alpha_{i_s}$ and $\beta_{j_1}, \ldots, \beta_{j_t}$ be maximal linearly independent sets of the sequences $\alpha_1, \ldots, \alpha_r$ and $\beta_1, \ldots, \beta_m$, respectively. By Theorem 3.3.13 (2), $\alpha_1, \ldots, \alpha_r$ and $\alpha_{i_1}, \ldots, \alpha_{i_s}$ are linearly equivalent, and $\beta_1, \ldots, \beta_m$ and $\beta_{j_1}, \ldots, \beta_{j_t}$ are linearly equivalent. Then, by the transitivity of linear expressions and the fact that $\alpha_1, \ldots, \alpha_r$ can be linearly expressed by $\beta_1, \ldots, \beta_m$, the sequence $\alpha_{i_1}, \ldots, \alpha_{i_s}$ can be linearly expressed by the sequence $\beta_{j_1}, \ldots, \beta_{j_t}$. By the Steinitz exchange lemma, $\operatorname{rank}(\alpha_1, \ldots, \alpha_r) = s \leqslant t = \operatorname{rank}(\beta_1, \ldots, \beta_m) \leqslant m$.

(5) follows from (4). $\qquad\square$

Once the definition of the rank of a sequence of vectors in F^n is given, we have the following concepts naturally.

Definition 3.3.17. Let $A \in \mathrm{M}_{n \times m}(F)$. The rank of the row vectors of A is called the **row rank** of A, and similarly the rank of the column vectors of A is called the **column rank** of A.

Theorem 3.3.18. *Let $A \in \mathrm{M}_{n \times m}(F)$.*

(1) *Elementary row (respectively, column) operations don't change the row (respectively, column) rank of A.*

(2) *The row rank of A and the column rank of A are equal to the rank of A, and thus elementary operations don't change the row and column ranks of A.*

Proof. (1) This result follows from the definitions of elementary row (respectively, column) operations.

(2) Assume that A was transformed into an echelon matrix B after elementary row operations. Then, by (1), the row rank of A is equal to that of B, which is just the number of nonzero row vectors of B. By Theorem 2.8.10, the rank of A also equals the number of nonzero row vectors of B, and so the row rank of A equals the rank of A. Similarly, the column rank of A also equals the rank of A. This fact can also be proved in the following way: By the previous result, the row rank of A' equals the rank of A', and thus the column rank of A equals the rank of A. Therefore, the row rank of A and the column rank of A equal the rank of A. By Theorem 2.8.5, elementary operations don't change the row and column ranks of A. $\qquad\square$

Example 3.3.19. Assume that $A \in M_{n \times m}(F)$ and $\mathrm{rank}A = r \geqslant 1$. Then the minor of order r whose entries are located at the intersections of any r linearly independent row vectors and r linearly independent column vectors of A is nonzero.

Proof. Without loss of generality, we may assume that the first r rows of A are linearly independent, the first r columns of A are linearly independent, and $A = \begin{pmatrix} A_1 & A_2 \\ A_3 & A_4 \end{pmatrix}$, where A_1 is a square matrix of order r. Since the rank of A equals the column rank of A, the last $n - r$ columns of A can be linearly expressed by the first r columns of A. Therefore, the column vectors of A_2 can be linearly expressed by the column vectors of A_1, so the column rank of the block matrix (A_1, A_2) equals that of A_1. Thus

$$\mathrm{rank}A_1 = \mathrm{rank}(A_1, A_2) = r.$$

So $|A_1| \neq 0$. $\qquad\qquad\square$

In the previous section, we use elementary operations to prove Theorem 3.2.4. Next, we use linear expressions of vectors to give another proof of this theorem.

Let $A = (\beta_1, \beta_2, \ldots, \beta_n)$. Then

the linear system $AX = \beta$ has a solution

$\Longleftrightarrow x_1\beta_1 + x_2\beta_2 + \cdots + x_n\beta_n = \beta$ has a solution

$\Longleftrightarrow \beta$ can be linearly expressed by $\beta_1, \beta_2, \ldots, \beta_n$

$\Longleftrightarrow$ the sequences $\beta_1, \beta_2, \ldots, \beta_n$ and $\beta_1, \beta_2, \ldots, \beta_n, \beta$ are linearly equivalent

$\Longleftrightarrow \mathrm{rank}(\beta_1, \beta_2, \ldots, \beta_n) = \mathrm{rank}(\beta_1, \beta_2, \ldots, \beta_n, \beta)$

$\Longleftrightarrow \mathrm{rank}A = \mathrm{rank}(A, \beta)$.

When $\mathrm{rank}A = \mathrm{rank}(A, \beta) = n$, we have $\mathrm{rank}(\beta_1, \beta_2, \ldots, \beta_n) = \mathrm{rank}(\beta_1, \beta_2, \ldots, \beta_n, \beta) = n$. Thus $\beta_1, \beta_2, \ldots, \beta_n$ are linearly independent, and $\beta_1, \beta_2, \ldots, \beta_n, \beta$ are linearly dependent. By Theorem 3.3.8, β is linearly expressed by $\beta_1, \beta_2, \ldots, \beta_n$ in a unique way. That is, the linear system $AX = \beta$ has a unique solution.

When $\mathrm{rank}A = \mathrm{rank}(A, \beta) = r < n$,

$$\mathrm{rank}(\beta_1, \beta_2, \ldots, \beta_n) = \mathrm{rank}(\beta_1, \beta_2, \ldots, \beta_n, \beta) = r < n.$$

It is harmless to assume that the subsequence $\beta_1, \ldots, \beta_r$ is a maximal linearly independent set of the sequence $\beta_1, \beta_2, \ldots, \beta_n$. Then the subsequence $\beta_1, \ldots, \beta_r$ is also a maximal linearly independent set of $\beta_1, \beta_2, \ldots, \beta_n, \beta$. So $\beta_{r+1}, \ldots, \beta_n, \beta$ can be linearly expressed by $\beta_1, \ldots, \beta_r$. Therefore, for any $k_{r+1}, \ldots, k_n \in F$, by Theorem 3.3.8, there are unique $k_1, \ldots, k_r \in F$ such that

$$k_1\beta_1 + \cdots + k_r\beta_r = \beta - k_{r+1}\beta_{r+1} - \cdots - k_n\beta_n,$$

that is, $k_1\beta_1 + \cdots + k_r\beta_r + k_{r+1}\beta_{r+1} + \cdots + k_n\beta_n = \beta$. It follows that $X = (k_1, k_2, \ldots, k_n)'$ is a solution of the system

$$x_1\beta_1 + x_2\beta_2 + \cdots + x_n\beta_n = \beta,$$

which is just $AX = \beta$.

From the above discussion, we know that for any $k_{r+1}, \ldots, k_n \in F$, the linear system $AX = \beta$ has a solution

$$X = (k_1, \ldots, k_r, k_{r+1}, \ldots, k_n)'.$$

Since $r < n$ and F contains infinitely many elements, $AX = \beta$ has infinitely many solutions. $\square$

Corollary 3.3.20. *Let $A \in \mathrm{M}_{m \times n}(F)$ and $B \in \mathrm{M}_{m \times r}(F)$. Then the matrix equation $AX = B$ has a solution if and only if $\mathrm{rank}(A, B) = \mathrm{rank}A$.*

Proof.   The matrix equation $AX = B$ has a solution
     $\Longleftrightarrow$ the column vectors of B are linearly expressed by those of A
     $\Longleftrightarrow$ the column vectors of (A, B) are linearly equivalent to those of A
     $\Longleftrightarrow$ $\mathrm{rank}(A, B) = \mathrm{rank}A$. $\square$

Theorem 3.2.4 only gives the criteria for the existence of solutions of linear equations and does not give all solutions when there are infinitely many solutions. However, from the above proof, when $AX = \beta$ has infinitely many solutions, finding a maximal linearly independent set of the column vectors of the coefficient matrix A is necessary. At the end of this section, we give a method to find a maximal linearly independent set of a sequence of vectors and use this set to express the other vectors.

Theorem 3.3.21. *Let $A = (\alpha_1, \alpha_2, \ldots, \alpha_m) \in \mathrm{M}_{n \times m}(F)$, and assume that A is transformed to $B = (\beta_1, \beta_2, \ldots, \beta_m)$ after elementary row operations, and $1 \leqslant i_1 < i_2 < \cdots < i_r \leqslant m$.*

(1) $\alpha_{i_1}, \alpha_{i_2}, \ldots, \alpha_{i_r}$ *are linearly dependent if and only if* $\beta_{i_1}, \beta_{i_2}, \ldots, \beta_{i_r}$ *are linearly dependent.*

(2) $\alpha_{i_1}, \alpha_{i_2}, \ldots, \alpha_{i_r}$ *are linearly independent if and only if* $\beta_{i_1}, \beta_{i_2}, \ldots, \beta_{i_r}$ *are linearly independent.*

(3) *if* $\alpha_{i_1}, \alpha_{i_2}, \ldots, \alpha_{i_r}$ *are linearly independent and* $1 \leqslant j \leqslant m$, *then* $\alpha_j = k_{i_1} \alpha_{i_1} + k_{i_2} \alpha_{i_2} + \cdots + k_{i_r} \alpha_{i_r}$ *if and only if* $\beta_j = k_{i_1} \beta_{i_1} + k_{i_2} \beta_{i_2} + \cdots + k_{i_r} \beta_{i_r}$.

(4) *The subsequence* $\alpha_{i_1}, \alpha_{i_2}, \ldots, \alpha_{i_r}$ *is a maximal linearly independent set of the sequence* $\alpha_1, \alpha_2, \ldots, \alpha_m$ *if and only if the subsequence* $\beta_{i_1}, \beta_{i_2}, \ldots, \beta_{i_r}$ *is a maximal linearly independent set of the sequence* $\beta_1, \beta_2, \ldots, \beta_m$.

(5) *If*

$$
B = (\beta_1, \beta_2, \ldots, \beta_m) =
\begin{pmatrix}
1 & 0 & \cdots & & 0 & c_{1,t+1} & \cdots & c_{1m} \\
0 & 1 & \ddots & & \vdots & c_{2,t+1} & \cdots & c_{2m} \\
\vdots & \ddots & \ddots & \ddots & \vdots & \vdots & & \\
\vdots & & \ddots & \ddots & 0 & c_{t-1,t+1} & \cdots & c_{t-1,m} \\
0 & \cdots & \cdots & 0 & 1 & c_{t,t+1} & \cdots & c_{tm} \\
0 & \cdots & \cdots & 0 & 0 & \cdots & & 0 \\
\vdots & & & \vdots & \vdots & & & \\
0 & \cdots & \cdots & 0 & 0 & \cdots & & 0
\end{pmatrix},
$$

then the subsequence $\alpha_1, \ldots, \alpha_t$ *is a maximal linearly independent set of the column vectors* $\alpha_1, \ldots, \alpha_m$ *of* A *and*

$$
\alpha_j = c_{1j}\alpha_1 + \cdots + c_{tj}\alpha_t, \quad j = t+1, \ldots, m.
$$

Proof. Since elementary row operations don't change any combination coefficients of column vectors, we get (1)–(4) directly.

(5) Clearly, the subsequence $\beta_1, \ldots, \beta_t$ is a maximal linearly independent set of the column vectors $\beta_1, \ldots, \beta_m$ of B and

$$
\beta_j = c_{1j}\beta_1 + \cdots + c_{tj}\beta_t, \quad j = t+1, \ldots, m.
$$

So (5) follows from (3) and (4). $\square$

Question. Can we remove the linear independence condition in Theorem 3.3.21 (3)?

Remark 3.3.22. Assume that the matrix $B = (\beta_1, \beta_2, \ldots, \beta_m)$ in Theorem 3.3.21 is of reduced echelon form with t nonzero rows and the column indices of the first nonzero elements 1 of the nonzero rows are $i_1, \ldots, i_t$, that is, the kth component of β_{i_k} is 1 and the others are 0 for $1 \leqslant k \leqslant t$. Since B is a reduced echelon matrix, the other $m - t$ column vectors of B can be written as $\beta_{i_k} = (c_{1i_k}, \ldots, c_{ti_k}, 0, \ldots, 0)', k = t+1, \ldots, m$. Therefore, the subsequence $\alpha_{i_1}, \ldots, \alpha_{i_t}$ is a maximal linearly independent set of the column vectors $\alpha_1, \ldots, \alpha_m$ of A and $\alpha_{i_k} = c_{1i_k}\alpha_{i_1} + c_{2i_k}\alpha_{i_2} + \cdots + c_{ti_k}\alpha_{i_t}$, $k = t+1, \ldots, m$.

Example 3.3.23. Suppose that $\alpha_1 = (1, -1, 1, 5)$, $\alpha_2 = (2, -2, 2, 10)$, $\alpha_3 = (1, 0, 2, 5)$, $\alpha_4 = (1, 3, 5, 5)$, $\alpha_5 = (2, -3, 2, 13)$, $\alpha_6 = (0, -1, 2, 9)$. Find a maximal linearly independent set of the sequence $\alpha_1, \alpha_2, \alpha_3, \alpha_4, \alpha_5, \alpha_6$ and express the other vectors by the vectors in this maximal set.

Solution. At first, transpose the vectors $\alpha_1, \alpha_2, \alpha_3, \alpha_4, \alpha_5, \alpha_6$ to get the following matrix:

$$(\alpha_1', \alpha_2', \alpha_3', \alpha_4', \alpha_5', \alpha_6') = \begin{pmatrix} 1 & 2 & 1 & 1 & 2 & 0 \\ -1 & -2 & 0 & 3 & -3 & -1 \\ 1 & 2 & 2 & 5 & 2 & 2 \\ 5 & 10 & 5 & 5 & 13 & 9 \end{pmatrix}.$$

Applying elementary row operations, we obtain

$$\begin{pmatrix} 1 & 2 & 1 & 1 & 2 & 0 \\ -1 & -2 & 0 & 3 & -3 & -1 \\ 1 & 2 & 2 & 5 & 2 & 2 \\ 5 & 10 & 5 & 5 & 13 & 9 \end{pmatrix}$$

$$\xrightarrow[\substack{(-1) \times r_1 + r_3 \\ (-5) \times r_1 + r_4}]{1 \times r_1 + r_2} \begin{pmatrix} 1 & 2 & 1 & 1 & 2 & 0 \\ 0 & 0 & 1 & 4 & -1 & -1 \\ 0 & 0 & 1 & 4 & 0 & 2 \\ 0 & 0 & 0 & 0 & 3 & 9 \end{pmatrix}$$

$$\xrightarrow{(-1) \times r_2 + r_3} \begin{pmatrix} 1 & 2 & 1 & 1 & 2 & 0 \\ 0 & 0 & 1 & 4 & -1 & -1 \\ 0 & 0 & 0 & 0 & 1 & 3 \\ 0 & 0 & 0 & 0 & 3 & 9 \end{pmatrix}$$

$$
\xrightarrow[\quad]{(-3)\times r_3 + r_4}
\begin{pmatrix}
1 & 2 & 1 & 1 & 2 & 0 \\
0 & 0 & 1 & 4 & -1 & -1 \\
0 & 0 & 0 & 0 & 1 & 3 \\
0 & 0 & 0 & 0 & 0 & 0
\end{pmatrix}
$$

$$
\xrightarrow[(-2)\times r_3 + r_1]{1\times r_3 + r_2}
\begin{pmatrix}
1 & 2 & 1 & 1 & 0 & -6 \\
0 & 0 & 1 & 4 & 0 & 2 \\
0 & 0 & 0 & 0 & 1 & 3 \\
0 & 0 & 0 & 0 & 0 & 0
\end{pmatrix}
$$

$$
\xrightarrow{(-1)\times r_2 + r_1}
\begin{pmatrix}
1 & 2 & 0 & -3 & 0 & -8 \\
0 & 0 & 1 & 4 & 0 & 2 \\
0 & 0 & 0 & 0 & 1 & 3 \\
0 & 0 & 0 & 0 & 0 & 0
\end{pmatrix}
= (\beta_1, \beta_2, \beta_3, \beta_4, \beta_5, \beta_6).
$$

Obviously, the subsequence $\beta_1, \beta_3, \beta_5$ is a maximal linearly independent set of the sequence $\beta_1, \beta_2, \beta_3, \beta_4, \beta_5, \beta_6$ and

$$
\beta_2 = 2\beta_1, \quad \beta_4 = -3\beta_1 + 4\beta_3, \quad \beta_6 = -8\beta_1 + 2\beta_3 + 3\beta_5.
$$

Therefore, the subsequence $\alpha_1, \alpha_3, \alpha_5$ is a maximal linearly independent set of the sequence $\alpha_1, \alpha_2, \alpha_3, \alpha_4, \alpha_5, \alpha_6$ and

$$
\alpha_2 = 2\alpha_1, \quad \alpha_4 = -3\alpha_1 + 4\alpha_3, \quad \alpha_6 = -8\alpha_1 + 2\alpha_3 + 3\alpha_5. \qquad \square
$$

3.4 Structure of Solutions

Theorem 3.2.4 solves the problem of the existence of solutions of a linear system. In this section, we give a quantitative analysis of a linear system and provide all solutions if this linear system has a solution. We start with homogeneous linear systems first.

Lemma 3.4.1. *Let $A \in M_{m\times n}(F)$ and $\alpha_1, \ldots, \alpha_s \in F^n$ be solutions of the homogeneous linear system $AX = 0$. Then for any $k_1, \ldots, k_s \in F$, $k_1\alpha_1 + \cdots + k_s\alpha_s$ is also a solution of $AX = 0$.*

Proof. In fact, $A(k_1\alpha_1 + \cdots + k_s\alpha_s) = k_1 A\alpha_1 + \cdots + k_s A\alpha_s = 0$ and thus $k_1\alpha_1 + \cdots + k_s\alpha_s$ is a solution of $AX = 0$. $\qquad \square$

Theorem 3.4.2 (Structure of solutions of a homogeneous system). *Let $A \in \mathrm{M}_{m\times n}(F)$.*

(1) *If* $\mathrm{rank}A = n$, *then* $AX = 0$ *only has the zero solution.*

(2) *If* $\mathrm{rank}A = r < n$, *then* $AX = 0$ *has* $n - r$ *linearly independent solutions* $\alpha_1, \dots, \alpha_{n-r}$ *such that any solution of* $AX = 0$ *can be linearly expressed by* $\alpha_1, \dots, \alpha_{n-r}$, *that is, the solution set of* $AX = 0$ *is*

$$\{k_1\alpha_1 + \cdots + k_{n-r}\alpha_{n-r} \mid k_1, \dots, k_{n-r} \in F\}.$$

Usually, we call $k_1\alpha_1 + \cdots + k_{n-r}\alpha_{n-r}$ *the **general solution** of* $AX = 0$, *where* $k_1, \dots, k_{n-r} \in F$.

Proof. By Theorem 3.2.4 (2.1), we have (1). So, it suffices to prove (2). Let $A = (\beta_1, \dots, \beta_n)$. Then $\mathrm{rank}(\beta_1, \dots, \beta_n) = \mathrm{rank}A = r$. We may assume that the subsequence $\beta_1, \dots, \beta_r$ is a maximal linearly independent set of the sequence $\beta_1, \dots, \beta_n$, and hence $\beta_{r+1}, \dots, \beta_n$ can be linearly expressed by $\beta_1, \dots, \beta_r$. Thus the vector $-k_{r+1}\beta_{r+1} - \cdots - k_n\beta_n$ can be linearly expressed by $\beta_1, \dots, \beta_r$ for any $k_{r+1}, \dots, k_n \in F$. It follows that there are $k_1, \dots, k_r \in F$ such that

$$k_1\beta_1 + \cdots + k_r\beta_r = -k_{r+1}\beta_{r+1} - \cdots - k_n\beta_n.$$

So $\alpha = (k_1, \dots, k_r, k_{r+1}, \dots, k_n)'$ is a solution of

$$x_1\beta_1 + \cdots + x_r\beta_r + x_{r+1}\beta_{r+1} + \cdots + x_n\beta_n = 0$$

which is just $AX = 0$.

Now letting $(k_{r+1}, \dots, k_n)$ be $(1, 0, \dots, 0), \dots, (0, \dots, 0, 1)$ respectively, we get the following solutions of $AX = 0$:

$$
\begin{aligned}
\alpha_1 \;&=\; (k_{11}, \quad \dots, \quad k_{r1}, \; 1, 0, 0, \dots, 0, 0)', \\
\alpha_2 \;&=\; (k_{12}, \quad \dots, \quad k_{r2}, \; 0, 1, 0, \dots, 0, 0)', \\
&\;\;\vdots \qquad\qquad\quad \vdots \\
\alpha_{n-r} \;&=\; (k_{1,n-r}, \; \dots, \; k_{r,n-r}, \; 0, 0, 0, \dots, 0, 1)'.
\end{aligned}
$$

Since $(1, 0, \dots, 0), \dots, (0, \dots, 0, 1)$ are linearly independent, so are $\alpha_1, \dots, \alpha_{n-r}$.

By Lemma 3.4.1, any linear combination of $\alpha_1, \ldots, \alpha_{n-r}$ is still a solution of $AX = 0$. Now let $\alpha = (k_1, \ldots, k_r, k_{r+1}, \ldots, k_n)'$ be a solution of $AX = 0$, by Lemma 3.4.1 again,

$$\alpha - k_{r+1}\alpha_1 - k_{r+2}\alpha_2 - \cdots - k_n\alpha_{n-r} = (l_1, \ldots, l_r, 0, \ldots, 0)'$$

is a solution of $AX = 0$. Therefore, $l_1\beta_1 + l_2\beta_2 + \cdots + l_r\beta_r = 0$. Since $\beta_1, \beta_2, \ldots, \beta_r$ are linearly independent, $l_1 = l_2 = \cdots = l_r = 0$. Therefore, $\alpha - k_{r+1}\alpha_1 - k_{r+2}\alpha_2 - \cdots - k_n\alpha_{n-r} = 0$, that is,

$$\alpha = k_{r+1}\alpha_1 + k_{r+2}\alpha_2 + \cdots + k_n\alpha_{n-r}.$$

It follows that the solution set of $AX = 0$ is

$$\{k_1\alpha_1 + \cdots + k_{n-r}\alpha_{n-r} \mid k_1, \ldots, k_{n-r} \in F\}. \qquad \square$$

Definition 3.4.3. A sequence of solutions $\alpha_1, \ldots, \alpha_t$ of a linear system $AX = 0$ is called a **basis for the solution set** of $AX = 0$ if it satisfies

(1) $\alpha_1, \ldots, \alpha_t$ are linearly independent and
(2) any solution of $AX = 0$ can be linearly expressed by $\alpha_1, \ldots, \alpha_t$.

Remark 3.4.4. Let $A \in \mathrm{M}_{m\times n}(F)$. If $\mathrm{rank}A = r < n$, by Theorem 3.4.2, we know that $AX = 0$ always has a basis for the solution set. Any sequence of $n - r$ linearly independent solutions of $AX = 0$ is a basis for the solution set (see Exercise 19).

Theorem 3.4.5 (Structure of solutions of a nonhomogeneous system). *Let $A \in \mathrm{M}_{m\times n}(F)$ and $\beta \in F^m$, and let $\gamma_0 \in F^n$ be a solution of $AX = \beta$, which is called a **particular solution** of $AX = \beta$.*

(1) *Assume that γ is a solution of $AX = \beta$, then $\gamma - \gamma_0$ is solution of $AX = 0$.*
(2) *Assume that α is a solution of $AX = 0$, then $\gamma = \gamma_0 + \alpha$ is a solution of $AX = \beta$.*
(3) *If $\mathrm{rank}A = n$, then $AX = \beta$ only has the particular solution γ_0.*
(4) *Assume that $\mathrm{rank}A = r < n$ and the sequence $\alpha_1, \ldots, \alpha_{n-r}$ is a basis for the solution set of $AX = 0$. Then, the solution set of $AX = \beta$ is*

$$\{\gamma_0 + k_1\alpha_1 + \cdots + k_{n-r}\alpha_{n-r} \mid k_1, \ldots, k_{n-r} \in F\}.$$

*Usually, we call $\gamma_0 + k_1\alpha_1 + \cdots + k_{n-r}\alpha_{n-r}$ the **general solution** of $AX = \beta$, where $k_1, \ldots, k_{n-r} \in F$.*

Proof. (1) Since $A(\gamma - \gamma_0) = A\gamma - A\gamma_0 = \beta - \beta = 0$, $\gamma - \gamma_0$ is a solution of $AX = 0$.

(2) Since $A\gamma = A(\gamma_0 + \alpha) = A\gamma_0 + A\alpha = \beta + 0 = \beta$, γ is a solution of $AX = \beta$.

(3) follows from Theorem 3.2.4 (1.1).

(4) comes from (1), (2) and Theorem 3.4.2. $\square$

For convenience, we call $AX = 0$ the **associated homogeneous system** of $AX = \beta$.

Given a linear system $AX = \beta$, how do we determine whether it has a solution and find its general solutions if there is a solution? The general steps are as follows:

Step 1: Simplify the augmented matrix. Apply the elementary row operations to make the augmented matrix (A, β) into an echelon matrix:

$$\begin{pmatrix}
a'_{1i_1} & \cdots & a'_{1i_2} & \cdots & a'_{1i_r} & \cdots & a'_{1i_n} & b'_1 \\
0 & \cdots & a'_{2i_2} & \cdots & a'_{2i_r} & \cdots & a'_{2i_n} & b'_2 \\
\vdots & & & & \vdots & & \vdots & \vdots \\
0 & \cdots & 0 & \cdots & a'_{ri_r} & \cdots & a'_{ri_n} & b'_r \\
0 & \cdots & 0 & \cdots & 0 & \cdots & 0 & b'_{r+1} \\
0 & \cdots & 0 & \cdots & 0 & \cdots & 0 & 0 \\
\vdots & & \vdots & & \vdots & & \vdots & \vdots \\
0 & \cdots & 0 & \cdots & 0 & \cdots & 0 & 0
\end{pmatrix}_{m \times n},$$

where $a'_{ji_j} \neq 0$, $j = 1, 2, \ldots, r$.

Step 2: Determine whether there is a solution. If $b'_{r+1} \neq 0$, then $\text{rank}(A, \beta) = r + 1 > r = \text{rank}A$. In this case, $AX = \beta$ has no solutions.

If $r = m$ or $b'_{r+1} = 0$, then $\text{rank}(A, \beta) = \text{rank}A$. In these cases, $AX = \beta$ has solutions.

Step 3: Find a particular solution. Let $\{i_{r+1}, i_{r+2} \ldots, i_n\}$ be the complementary set of $\{i_1, i_2, \ldots, i_r\}$ in $\{1, 2, \ldots, n\}$ and assume that $i_{r+1} < i_{r+2} < \cdots < i_n$. Letting $(x_{i_{r+1}}, x_{i_{r+2}}, \ldots, x_{i_n}) = (0, 0, \ldots, 0)$ yields a particular solution γ_0 of $AX = \beta$.

Step 4: Find general solutions. Letting $(x_{i_{r+1}}, x_{i_{r+2}}, \ldots, x_{i_n})$ be $(1, 0, \ldots, 0)$, $\ldots$, $(0, 0, \ldots, 1)$ respectively, we get a basis $\alpha_1, \alpha_2, \ldots, \alpha_{n-r}$ for the solution set of the associated homogeneous system $AX = 0$. Therefore, the general solutions of $AX = \beta$ are

$$X = \gamma_0 + k_1 \alpha_1 + \cdots + k_{n-r} \alpha_{n-r},$$

where $k_1, \ldots, k_{n-r} \in F$.

Example 3.4.6. Determine the value of the real number λ such that the linear system

$$\begin{cases} (2-\lambda)x_1 + & 2x_2 - & 2x_3 = 1, \\ 2x_1 + (5-\lambda)x_2 - & 4x_3 = 2, \\ 2x_1 + & 4x_2 - (5-\lambda)x_3 = \lambda + 1 \end{cases}$$

(1) has no solutions and explain the reason, (2) has a unique solution and find this solution, and (3) has infinitely many solutions and find its general solutions.

Solution. Applying elementary row operations, we have

$$\begin{pmatrix} 2-\lambda & 2 & -2 & 1 \\ 2 & 5-\lambda & -4 & 2 \\ 2 & 4 & \lambda-5 & \lambda+1 \end{pmatrix}$$

$$\xrightarrow{(r_1,r_3)} \begin{pmatrix} 2 & 4 & \lambda-5 & \lambda+1 \\ 2 & 5-\lambda & -4 & 2 \\ 2-\lambda & 2 & -2 & 1 \end{pmatrix}$$

$$\xrightarrow[\frac{\lambda-2}{2} \times r_1 + r_3]{(-1) \times r_1 + r_2} \begin{pmatrix} 2 & 4 & \lambda-5 & \lambda+1 \\ 0 & 1-\lambda & 1-\lambda & 1-\lambda \\ 0 & -2+2\lambda & \frac{1}{2}\lambda^2 - \frac{7}{2}\lambda+3 & \frac{1}{2}(\lambda^2-\lambda) \end{pmatrix}$$

$$\xrightarrow{2 \times r_2 + r_3} \begin{pmatrix} 2 & 4 & \lambda-5 & \lambda+1 \\ 0 & 1-\lambda & 1-\lambda & 1-\lambda \\ 0 & 0 & \frac{1}{2}(\lambda-1)(\lambda-10) & \frac{1}{2}(\lambda-1)(\lambda-4) \end{pmatrix}.$$

(1) When $\lambda = 10$, the rank of the coefficient matrix is 2 while the rank of the augmented matrix is 3, and thus the linear system has no solutions.

(2) When $\lambda \neq 1$ and $\lambda \neq 10$, the ranks of the coefficient matrix and augmented matrix are both 3 and thus the linear system has a unique solution: $x_1 = \frac{-3}{\lambda-10}$, $x_2 = \frac{-6}{\lambda-10}$, $x_3 = \frac{\lambda-4}{\lambda-10}$.

(3) When $\lambda = 1$, the ranks of the coefficient matrix and augmented matrix are both 1, which is smaller than 3 and thus, the linear system has infinitely many solutions. In such a case, this system has the same solution set as the equation $2x_1 + 4x_2 - 4x_3 = 2$, that is, $x_1 = 1 - 2x_2 + 2x_3$.

Letting $x_2 = x_3 = 0$, we get a particular solution $\gamma_0 = (1, 0, 0)'$. Then letting (x_2, x_3) be $(1, 0)$ and $(0, 1)$ respectively, we have a basis $\eta_1 = (-2, 1, 0)'$, $\eta_2 = (2, 0, 1)'$ for the solution set of its associated homogeneous system. Therefore, the general solutions of the original linear system are $X = \gamma_0 + k_1\eta_1 + k_2\eta_2$, where k_1, k_2 are arbitrary real numbers. $\qquad\square$

Example 3.4.7. Let $1 \leqslant r < n$. Prove that any r linearly independent vectors in F^n must be a basis for the solution set of a homogeneous linear system.

Proof. Assume that the vectors $\beta_1, \beta_2, \ldots, \beta_r \in F^n$ are linearly independent. Let $B = (\beta_1, \beta_2, \ldots, \beta_r) \in M_{n \times r}(F)$. Then $\mathrm{rank}B' = \mathrm{rank}B = r$. Suppose that the sequence of column vectors $\alpha_1, \alpha_2, \ldots, \alpha_{n-r}$ is a basis for the solution set of $B'X = 0$. Let $A = (\alpha_1, \alpha_2, \ldots, \alpha_{n-r})' \in M_{(n-r) \times n}(F)$. Then $\mathrm{rank}A = n - r$ and $B'A' = 0$. Thus $AB = 0$ and $\beta_1, \beta_2, \ldots, \beta_r$ are linearly independent solutions of $AX = 0$. Since $r = n - (n-r) = n - \mathrm{rank}A$, the sequence $\beta_1, \beta_2, \ldots, \beta_r$ is a basis for the solution set of $AX = 0$. $\qquad\square$

Example 3.4.8. Assume that $A \in M_{s \times n}(F)$, $B \in M_{t \times n}(F)$ and all solutions of $AX = 0$ are solutions of $BX = 0$. Prove that

(1) the row vectors of B can be linearly expressed by the row vectors of A, that is, there exists $C \in M_{t \times s}(F)$ such that $B = CA$;
(2) $\mathrm{rank}B \leqslant \mathrm{rank}A$, and the equality holds if and only if $AX = 0$ and $BX = 0$ have the same solution set.

Proof. (1) **Method 1.** By assumption, $AX = 0$ and $\begin{pmatrix} A \\ B \end{pmatrix} X = 0$ have the same solution set. It follows that either they only have the zero solution or have the same basis for their solution sets. Thus, we always have

$$n - \operatorname{rank}A = n - \operatorname{rank}\begin{pmatrix} A \\ B \end{pmatrix},$$

that is, $\operatorname{rank}A = \operatorname{rank}\begin{pmatrix} A \\ B \end{pmatrix}$. Thus a maximal linearly independent set of the row vectors of A is also a maximal linearly independent set of the row vectors of $\begin{pmatrix} A \\ B \end{pmatrix}$, and hence the row vectors of A and $\begin{pmatrix} A \\ B \end{pmatrix}$ are linearly equivalent. Therefore, the row vectors of B can be linearly expressed by those of A.

Method 2. If $\operatorname{rank}A = n$, then by Theorem 3.3.8 and Corollary 3.3.10 (2), the row vectors of B can be linearly expressed by any maximal linearly independent set of the row vectors of A and hence by the row vectors of A. Assume that $\operatorname{rank}A = r < n$ and the sequence $\gamma_1, \gamma_2, \ldots, \gamma_{n-r}$ in F^n is a basis for the solution set of $AX = 0$. Let $D = (\gamma_1, \gamma_2, \ldots, \gamma_{n-r})$. Then $D \in M_{n \times (n-r)}(F)$, $\operatorname{rank}D = n - r$, and $AD = 0$. By assumption, $BD = 0$. Note that $D'A' = 0$ and $D'B' = 0$; the column vectors of A' and B' are both solutions of $D'X = 0$. Assume that the sequence $\alpha_1, \alpha_2, \ldots, \alpha_r$ is a maximal linearly independent set of the column vectors of A'. Since $r = n - (n-r) = n - \operatorname{rank}D'$, the sequence $\alpha_1, \alpha_2, \ldots, \alpha_r$ is a basis for the solution set of $D'X = 0$. Thus, the column vectors of B' can be linearly expressed by $\alpha_1, \alpha_2, \ldots, \alpha_r$ and hence by the column vectors of A'. So, the row vectors of B can be linearly expressed by those of A.

(2) By (1), $B = CA$ and thus $\operatorname{rank}B \leqslant \operatorname{rank}A$. If $\operatorname{rank}B = \operatorname{rank}A = n$, then both $AX = 0$ and $BX = 0$ only have the zero solution, so they have the same solution set. Suppose $\operatorname{rank}B = \operatorname{rank}A < n$. Since all solutions of $AX = 0$ are also solutions of $BX = 0$, any basis for the solution set of $AX = 0$ is also a basis for the solution set of $BX = 0$. Therefore, $AX = 0$ and $BX = 0$ have the same solution set. Conversely, let $AX = 0$ and $BX = 0$ have the same solution set. If they only have the zero solution, then $\operatorname{rank}B = n = \operatorname{rank}A$.

If they have a nonzero solution, then $n - \mathrm{rank}A = n - \mathrm{rank}B$ and thus $\mathrm{rank}A = \mathrm{rank}B$. $\qquad\qquad\qquad\qquad\qquad\qquad\qquad\square$

3.5*   Resultants and Systems of Higher Degree Binary Equations

Now, let's consider systems of equations of higher degrees. The basic idea of solving systems of equations of higher degrees is also elimination, reducing the number of unknowns. This idea is practical for the system of higher-degree equations in two unknowns.

Lemma 3.5.1. *Let $f(x), g(x) \in F[x]$ and assume that $\deg(f(x)) = n > 0$, $\deg(g(x)) = m > 0$. Then $f(x)$ and $g(x)$ have a nonconstant common factor if and only if there are $u(x)$ and $v(x)$ in $F[x]$ such that $u(x)f(x) = v(x)g(x)$ with $\deg(u(x)) < m$ and $\deg(v(x)) < n$.*

Proof.   Necessity. Assume that $f(x)$ and $g(x)$ have a nonconstant common factor $d(x)$. Then $f(x) = f_1(x)d(x)$, $g(x) = g_1(x)d(x)$, where $\deg(f_1(x)) < n$ and $\deg(g_1(x)) < m$. Let $u(x) = g_1(x)$ and $v(x) = f_1(x)$, then we have $u(x)f(x) = d(x)f_1(x)g_1(x) = v(x)g(x)$.

Sufficiency. Assume that there are $u(x), v(x) \in F[x]$ such that $u(x)f(x) = v(x)g(x)$ with $\deg(v(x)) < n$ and $\deg(u(x)) < m$. Let $(f(x), v(x)) = d(x)$, then there exist $f_1(x), v_1(x) \in F[x]$ such that $f(x) = f_1(x)d(x)$, $v(x) = v_1(x)d(x)$. Therefore, $d(x)u(x)f_1(x) = d(x)v_1(x)g(x)$ and thus $u(x)f_1(x) = v_1(x)g(x)$. Since $d(x)|v(x)$, the degree of $d(x)$ is smaller than n, and so the degree of $f_1(x)$ is greater than zero. Since $(f_1(x), v_1(x)) = 1$, $f_1(x)|v_1(x)g(x)$ implies that $f_1(x)|g(x)$. Thus, $f(x)$ and $g(x)$ have a nonconstant common factor $f_1(x)$. $\qquad\qquad\square$

Now, we convert the condition in the above lemma to the case that linear algebra can handle. To this end, let

$$
\begin{aligned}
f(x) &= a_0 x^n &&+ a_1 x^{n-1} &&+ \cdots + && a_n, \\
g(x) &= b_0 x^m &&+ b_1 x^{m-1} &&+ \cdots + && b_m, \\
u(x) &= u_0 x^{m-1} &&+ u_1 x^{m-2} &&+ \cdots + && u_{m-1}, \\
v(x) &= v_0 x^{n-1} &&+ v_1 x^{n-2} &&+ \cdots + && v_{n-1}.
\end{aligned}
$$

By the condition $u(x)f(x) = v(x)g(x)$ in the above lemma, we have the following relation between coefficients:

$$
\begin{cases}
a_0 u_0 & = b_0 v_0, \\
a_1 u_0 + a_0 u_1 & = b_1 v_0 + b_0 v_1, \\
a_2 u_0 + a_1 u_1 + a_0 u_2 & = b_2 v_0 + b_1 v_1 + b_0 v_2, \\
\quad \vdots & \qquad \vdots \\
a_n u_{m-2} + a_{n-1} u_{m-1} & = b_m v_{n-2} + b_{m-1} v_{n-1}, \\
a_n u_{m-1} & = b_m v_{n-1}.
\end{cases}
\tag{3.11}
$$

The relation (3.11) can be regarded as a homogeneous linear system in the unknowns $u_0, u_1, \ldots, u_{m-1}, v_0, v_1, \ldots, v_{n-1}$. This linear system has exactly $m + n$ equations and $m + n$ unknowns. The condition "There are $u(x)$ and $v(x)$ in $F[x]$ such that $u(x)f(x) = v(x)g(x)$ with $\deg(u(x)) < m$ and $\deg(v(x)) < n$" in Lemma 3.5.1 is equivalent to that the homogeneous linear system has nonzero solutions, which is equivalent to that the determinant of the coefficient matrix of the homogeneous linear system is zero.

Swapping the rows and columns of the coefficient matrix of this linear system and reversing the signs of the last n rows, we get the following determinant:

$$
\begin{vmatrix}
a_0 & a_1 & a_2 & \ldots & a_n & & & & \\
 & a_0 & a_1 & a_2 & \ldots & a_n & & & \\
 & & \ddots & \ddots & \ddots & & \ddots & & \\
 & & & & a_0 & a_1 & a_2 & \ldots & a_n \\
b_0 & b_1 & b_2 & \ldots & b_m & & & & \\
 & b_0 & b_1 & b_2 & \ldots & b_m & & & \\
 & & \ddots & \ddots & \ddots & & \ddots & & \\
 & & & b_0 & b_1 & b_2 & \ldots & b_m &
\end{vmatrix}_{m+n}.
$$

In this determinant, the row vector $(a_0, a_1, \ldots, a_n)$ appears m times and the row vector $(b_0, b_1, \ldots, b_m)$ appears n times.

Definition 3.5.2. The above determinant is called the **resultant** of the polynomials

$$f(x) = a_0 x^n + a_1 x^{n-1} + \cdots + a_n \quad \text{and}$$

$$g(x) = b_0 x^m + b_1 x^{m-1} + \cdots + b_m,$$

and it is denoted by $R(f, g)$.

To sum up, we draw the following conclusion.

Theorem 3.5.3. *Let*

$$f(x) = a_0 x^n + a_1 x^{n-1} + \cdots + a_n,$$

$$g(x) = b_0 x^m + b_1 x^{m-1} + \cdots + b_m$$

be two polynomials in $F[x]$, where $m, n > 0$. Then $R(f, g) = 0$ if and only if either $f(x)$ and $g(x)$ have a nonconstant common factor or their leading coefficients a_0, b_0 are both zero.

Proof. Sufficiency. If a_0, b_0 are both zero or one of $f(x)$ and $g(x)$ is zero, then obviously $R(f, g) = 0$. If $f(x)$ and $g(x)$ are not zero and have a nonconstant common factor, then by Lemma 3.5.1, there are $u(x)$ and $v(x)$ in $F[x]$ such that $u(x)f(x) = v(x)g(x)$ with $\deg(u(x)) < m$ and $\deg(v(x)) < n$. Thus the linear system (3.11) has a nonzero solution and so $R(f, g) = 0$.

Necessity. Assume that $R(f, g) = 0$. The conclusion is true if one of $f(x)$ and $g(x)$ is zero. Suppose both $f(x)$ and $g(x)$ are not zero and a_0, b_0 are not both zero, by $R(f, g) = 0$, we know that the linear system (3.11) has a nonzero solution, and thus there are polynomials

$$u(x) = u_0 x^{m-1} + u_1 x^{m-2} + \cdots + u_{m-1} \quad \text{and}$$

$$v(x) = v_0 x^{n-1} + v_1 x^{n-2} + \cdots + v_{n-1},$$

which are not both zero, such that $u(x)f(x) = v(x)g(x)$. Note that both $f(x)$ and $g(x)$ are not zero, so both $u(x)$ and $v(x)$ are not zero. Since $\deg(u(x)) < m$ and $\deg(v(x)) < n$, Lemma 3.5.1 implies that $f(x)$ and $g(x)$ have a nonconstant common factor. The rest is the case $a_0 = 0$ and $b_0 = 0$. $\qquad\square$

Example 3.5.4. Find the resultant of the polynomials $f(x) = x^4 + 11x^3 + 13x^2 - 44x + 9$ and $g(x) = x^3 + 15x^2 + 52x - 18$.

Solution. By Definition 3.5.2,

$$
R(f,g) = \begin{vmatrix}
1 & 11 & 13 & -44 & 9 & & \\
 & 1 & 11 & 13 & -44 & 9 & \\
 & & 1 & 11 & 13 & -44 & 9 \\
1 & 15 & 52 & -18 & & & \\
 & 1 & 15 & 52 & -18 & & \\
 & & 1 & 15 & 52 & -18 & \\
 & & & 1 & 15 & 52 & -18
\end{vmatrix} = 0. \qquad \square
$$

By Theorem 3.5.3, $f(x)$ and $g(x)$ have a nonconstant common factor. In fact, by Euclidean algorithm, one can get

$$(f(x), g(x)) = x + 9.$$

Let $f(x,y), g(x,y)$ be two polynomials with complex coefficients in two indeterminates. We aim to find all solutions to the following system of equations:

$$
\begin{cases}
f(x,y) = 0, \\
g(x,y) = 0.
\end{cases}
\tag{3.12}
$$

These two polynomials can be written as

$$f(x,y) = a_0(y)x^n + a_1(y)x^{n-1} + \cdots + a_n(y),$$

$$g(x,y) = b_0(y)x^m + b_1(y)x^{m-1} + \cdots + b_m(y),$$

where $a_i(y), b_j(y)$ $(0 \leqslant i \leqslant n,\ 0 \leqslant j \leqslant m)$ are polynomials in the indeterminate y. In such a way, we can regard $f(x,y),\ g(x,y)$ as polynomials in the indeterminate x. Let

$$
R_x(f,g) = \begin{vmatrix}
a_0(y) & a_1(y) & a_2(y) & \cdots & a_n(y) & & & \\
 & a_0(y) & a_1(y) & a_2(y) & \cdots & a_n(y) & & \\
 & & \ddots & \ddots & \ddots & & \ddots & \\
 & & & a_0(y) & a_1(y) & a_2(y) & \cdots & a_n(y) \\
b_0(y) & b_1(y) & b_2(y) & \cdots & b_m(y) & & & \\
 & b_0(y) & b_1(y) & b_2(y) & \cdots & b_m(y) & & \\
 & & \ddots & \ddots & \ddots & & \ddots & \\
 & & & b_0(y) & b_1(y) & b_2(y) & \cdots & b_m(y)
\end{vmatrix}.
$$

Then, $R_x(f, g)$ is a polynomial with complex coefficients in the indeterminate y. By Theorem 3.5.3, we have the following.

Theorem 3.5.5.

(1) *If (x_0, y_0) is a complex solution of (3.12), then y_0 is a root of $R_x(f, g)$.*

(2) *If y_0 is a root of $R_x(f, g)$, then either $a_0(y_0) = b_0(y_0) = 0$ or there exists a complex number x_0 such that (x_0, y_0) is a solution of (3.12).*

By Theorem 3.5.5, to find all solutions of (3.12), we can solve the polynomial equation $R_x(f, g) = 0$ at first and then substitute the roots of $R_x(f, g) = 0$ into (3.12) to find the values of x. In such a way, one can get all solutions of (3.12).

Example 3.5.6. Find all solutions of the system of equations

$$\begin{cases} x^3 + y^3 = 7x + 7y, \\ x^2 + y^2 = 13. \end{cases}$$

Solution. We first rewrite this system of equations as follows:

$$\begin{cases} x^3 - 7x + y^3 - 7y = 0, \\ \qquad\quad x^2 + y^2 - 13 = 0. \end{cases} \tag{3.13}$$

Thus

$$R_x(f, g) = \begin{vmatrix} 1 & 0 & -7 & y^3 - 7y & 0 \\ 0 & 1 & 0 & -7 & y^3 - 7y \\ 1 & 0 & y^2 - 13 & 0 & 0 \\ 0 & 1 & 0 & y^2 - 13 & 0 \\ 0 & 0 & 1 & 0 & y^2 - 13 \end{vmatrix}$$

$$= 2y^6 - 39y^4 + 241y^2 - 468.$$

So the resultant $R_x(f, g)$ has six roots $y = \pm 2, \pm 3, \pm\frac{\sqrt{26}}{2}$. Substituting them into (3.13) respectively, we get six solutions of the original system of equations:

$$(2, 3), \ (-2, -3), \ (3, 2), \ (-3, -2), \ \left(\frac{\sqrt{26}}{2}, -\frac{\sqrt{26}}{2}\right), \ \left(-\frac{\sqrt{26}}{2}, \frac{\sqrt{26}}{2}\right).$$

$\square$

Remark 3.5.7. By the symmetry of x and y, we can also find solutions of $R_y(f, g) = 0$ at first and then substitute these roots into (3.12) to get the values of y. In this way, one can also get all solutions of (3.12).

Exercises

1. Use the elimination method to solve the following linear systems:

(1)
$$\begin{cases} x_1 + 3x_2 + 5x_3 - 4x_4 \qquad\quad = 1, \\ x_1 + 3x_2 + 2x_3 - 2x_4 + x_5 = -1, \\ x_1 - 2x_2 + x_3 - x_4 - x_5 = 3, \\ x_1 - 4x_2 + x_3 + x_4 - x_5 = 3, \\ x_1 + 2x_2 + x_3 - x_4 + x_5 = -1; \end{cases}$$

(2)
$$\begin{cases} x_1 + 2x_2 - 3x_3 + x_4 - 3x_5 = 1, \\ 2x_1 - 3x_2 + 4x_3 - 5x_4 + 2x_5 = 7, \\ x_1 - x_2 - 3x_3 + x_4 - 3x_5 = 2, \\ 9x_1 - 9x_2 + 6x_3 - 16x_4 + 2x_5 = 25; \end{cases}$$

(3)
$$\begin{cases} x_1 + x_2 + x_3 + x_4 + x_5 = 1, \\ x_1 + 3x_2 + 2x_3 - 2x_4 + x_5 = -1, \\ 2x_1 + 4x_2 + 3x_3 - x_4 + 2x_5 = 0, \\ 2x_2 + x_3 - 3x_4 = -1, \\ x_1 + 2x_2 + x_3 - x_4 + x_5 = -1; \end{cases}$$

(4)
$$\begin{cases} x_1 + 2x_2 + 3x_3 - x_4 = 1, \\ 3x_1 + 2x_2 + x_3 - x_4 = 1, \\ 2x_1 + 3x_2 + x_3 + x_4 = 1, \\ 2x_1 + 2x_2 + 2x_3 - x_4 = 1, \\ 5x_1 + 5x_2 + 2x_3 = 2; \end{cases}$$

(5)
$$\begin{cases} x_1 + x_2 + x_3 + x_4 + x_5 = 0, \\ x_1 + 2x_2 + 2x_3 + 2x_4 + 2x_5 = 0, \\ 2x_1 + 3x_2 + 3x_3 + 3x_4 + 3x_5 = 0, \\ x_1 - 4x_2 + x_3 + x_4 - x_5 = 0, \\ x_1 + 2x_2 + x_3 - x_4 + x_5 = 0; \end{cases}$$

$$(6) \quad \begin{cases} 2x_1 + x_2 - x_3 + x_4 = 0, \\ 3x_1 - 2x_2 + 2x_3 - 3x_4 = 0, \\ 2x_1 - x_2 + x_3 - 3x_4 = 0, \\ 5x_1 + x_2 - x_3 + 2x_4 = 0. \end{cases}$$

2. Express the vector β as a linear combination of $\alpha_1, \alpha_2, \alpha_3, \alpha_4$:

   (1) $\alpha_1 = (1,1,1,1), \quad \alpha_2 = (1,1,-1,-1), \quad \alpha_3 = (1,-1,1,-1),$
       $\alpha_4 = (1,-1,-1,1), \; \beta = (1,2,1,1);$

   (2) $\alpha_1 = (1,1,1,1), \quad \alpha_2 = (0,1,1,1), \quad \alpha_3 = (0,0,1,1),$
       $\alpha_4 = (0,0,0,1), \; \beta = (2,1,3,4).$

3. Prove that the linear equivalence defined for vectors in F^n is an equivalence relation.

4. Let $\alpha_i = (a_{i1}, a_{i2}, \ldots, a_{in})$, $1 \leqslant i \leqslant n$. Prove that the sequence $\alpha_1, \alpha_2, \ldots, \alpha_n$ is linearly independent if and only if the determinant $|a_{ij}|_n \neq 0$.

5. Let $1 \leqslant r \leqslant n$ and $t_1, t_2, \ldots, t_r$ be r numbers different from each other. If $\alpha_i = (1, t_i, \ldots, t_i^{n-1})$ for $1 \leqslant i \leqslant r$, prove that the sequence $\alpha_1, \alpha_2, \ldots, \alpha_r$ is linearly independent.

6. Let $\alpha_1, \alpha_2, \ldots, \alpha_r$ be a sequence of linearly independent vectors in F^n and $\beta_i = \sum_{j=1}^{r} a_{ij}\alpha_j$, $1 \leqslant i \leqslant r$. Prove that the sequence $\beta_1, \beta_2, \ldots, \beta_r$ is linearly independent if and only if the determinant $|a_{ij}|_r \neq 0$.

7. Prove that the linear system

$$\begin{cases} a_{11}x_1 + a_{12}x_2 + \cdots + a_{1n}x_n = b_1, \\ a_{21}x_1 + a_{22}x_2 + \cdots + a_{2n}x_n = b_2, \\ \;\;\vdots \qquad\quad \vdots \qquad\qquad\quad \vdots \qquad\quad \vdots \\ a_{n1}x_1 + a_{n2}x_2 + \cdots + a_{nn}x_n = b_n \end{cases}$$

always has a solution for any $b_1, b_2, \ldots, b_n$ if and only if the determinant $|a_{ij}|_n$ of the coefficient matrix of the linear system is not zero.

8. Let $\alpha_1, \alpha_2, \ldots, \alpha_n$ be a sequence of vectors in F^n. Prove that the sequence $\alpha_1, \alpha_2, \ldots, \alpha_n$ is linearly independent if and only if any vector in F^n can be linearly expressed by this sequence of vectors.

9. Let $\alpha_1, \alpha_2, \ldots, \alpha_n$ be a sequence of linearly independent vectors in F^n. If every α_i $(1 \leqslant i \leqslant m)$ can be linearly expressed by the

sequence $\beta_1, \beta_2, \ldots, \beta_s$ of vectors in F^n, prove that there exists β_j $(1 \leqslant j \leqslant s)$ such that the sequence $\beta_j, \alpha_2, \ldots, \alpha_m$ is linearly independent.

10. Let $\alpha_1 = (1, -1, 2, 4)$, $\alpha_2 = (0, 3, 1, 2)$, $\alpha_3 = (3, 0, 7, 14)$, $\alpha_4 = (1, -1, 2, 0)$, $\alpha_5 = (2, 1, 5, 6)$. Prove that the sequence α_1, α_2 is linearly independent and extend it to a maximal linearly independent set of the sequence α_1, α_2, α_3, α_4, α_5.

11. Let S be a sequence of vectors in F^n. Prove that any linearly independent subsequence of S can be extended to a maximal linearly independent set of S.

12. Let S be a sequence of vectors $\alpha_1, \alpha_2, \ldots, \alpha_m$ in F^n, and let $\operatorname{rank}(\alpha_1, \alpha_2, \ldots, \alpha_m) = r$. If $\alpha_{i_1}, \alpha_{i_2}, \ldots, \alpha_{i_r}$ are r vectors in S, prove that the subsequence $\alpha_{i_1}, \alpha_{i_2}, \ldots, \alpha_{i_r}$ of S is a maximal linearly independent subset of S if and only if any vector in S can be linearly expressed by $\alpha_{i_1}, \alpha_{i_2}, \ldots, \alpha_{i_r}$.

13. Assume that the ranks of the two sequences $\alpha_1, \alpha_2, \ldots, \alpha_r$ and $\beta_1, \beta_2, \ldots, \beta_s$ of vectors in F^n are equal. If α_1, α_2, $\ldots$, α_r can be linearly expressed by $\beta_1, \beta_2, \ldots, \beta_s$, prove that the two sequences $\alpha_1, \alpha_2, \ldots, \alpha_r$ and $\beta_1, \beta_2, \ldots, \beta_s$ are linearly equivalent.

14. Let $\alpha_1, \alpha_2, \ldots, \alpha_s$ be a sequence of vectors with $\operatorname{rank}(\alpha_1, \alpha_2, \ldots, \alpha_s) = r$. Now take any m vectors $\alpha_{i_1}, \alpha_{i_2}, \ldots, \alpha_{i_m}$ from this sequence. Prove that $\operatorname{rank}(\alpha_{i_1}, \alpha_{i_2}, \ldots, \alpha_{i_m}) \geqslant r + m - s$.

15. Let A be a symmetric matrix of order n whose rank is assumed to be $r > 0$. Prove that A has a nonzero principal minor of order r (see Section 2.11 for the definition of a principal minor).

16. Let $\alpha_1, \alpha_2, \ldots, \alpha_r$ be a sequence of linearly independent vectors in F^n, and let $(\beta_1, \beta_2, \ldots, \beta_m) = (\alpha_1, \alpha_2, \ldots, \alpha_r)C$, where $C = (\gamma_1, \gamma_2, \ldots, \gamma_m) \in \mathrm{M}_{r \times m}(F)$. Prove the following:

    (1) For any $1 \leqslant i_1 < i_2 < \cdots < i_s \leqslant m$, $\beta_{i_1}, \beta_{i_2}, \ldots, \beta_{i_s}$ are linearly dependent if and only if $\gamma_{i_1}, \gamma_{i_2}, \ldots, \gamma_{i_s}$ are linearly dependent; $\beta_{i_1}, \beta_{i_2}, \ldots, \beta_{i_s}$ are linearly independent if and only if $\gamma_{i_1}, \gamma_{i_2}, \ldots, \gamma_{i_s}$ are linearly independent.

    (2) $\operatorname{rank}(\beta_1, \beta_2, \ldots, \beta_m) = \operatorname{rank} C$.

17. Let $A \in \mathrm{M}_{n \times (n+1)}(F)$. Prove that the matrix equation $AX = I_n$ has a solution if and only if $\operatorname{rank} A = n$.

18. Let $\alpha_1 = (-1, 2, 0, 4)$, $\alpha_2 = (5, 0, 3, 1)$, $\alpha_3 = (3, -1, 4, -2)$, $\alpha_4 = (-2, 4, -5, 9)$, $\alpha_5 = (1, 3, -1, 7)$.

(1) Find $\operatorname{rank}(\alpha_1, \alpha_2, \alpha_3, \alpha_4, \alpha_5)$.

(2) Find a maximal linearly independent subset of the sequence $\alpha_1, \alpha_2, \alpha_3, \alpha_4, \alpha_5$.

(3) Express the other vectors in the sequence $\alpha_1, \alpha_2, \alpha_3, \alpha_4, \alpha_5$ as linear combinations of the vectors in the maximal linearly independent subset found in (2).

19. Assume that the rank of the coefficient matrix of a homogeneous linear system in n unknowns is r. Prove that any sequence of $n - r$ linearly independent solutions of this linear system is a basis for the solution set.

20. Find general solutions of the following homogeneous linear systems:

(1)
$$\begin{cases} x_1 + x_2 + x_3 + x_4 + x_5 = 0, \\ 3x_1 + 2x_2 + x_3 + x_4 - 3x_5 = 0, \\ 5x_1 + 4x_2 + 3x_3 + 3x_4 - x_5 = 0; \end{cases}$$

(2)
$$\begin{cases} x_1 + x_2 - 3x_4 - x_5 = 0, \\ x_1 - x_2 + 2x_3 - x_4 = 0, \\ 4x_1 - 2x_2 + 6x_3 + 3x_4 - 4x_5 = 0, \\ 2x_1 + 4x_2 - 2x_3 + 4x_4 - 7x_5 = 0; \end{cases}$$

(3)
$$\begin{cases} x_1 - 2x_2 + x_3 + x_4 - x_5 = 0, \\ 2x_1 + x_2 - x_3 - x_4 - x_5 = 0, \\ x_1 + 7x_2 - 5x_3 - 5x_4 + 5x_5 = 0, \\ 3x_1 - x_2 - 2x_3 + x_4 - x_5 = 0; \end{cases}$$

(4)
$$\begin{cases} x_1 - 2x_2 + x_3 - x_4 + x_5 = 0, \\ 2x_1 + x_2 - x_3 + 2x_4 - 3x_5 = 0, \\ 3x_1 - 2x_2 - x_3 + x_4 - 2x_5 = 0, \\ 2x_1 - 5x_2 + x_3 - 2x_4 + 2x_5 = 0. \end{cases}$$

21. Given a linear system

$$\begin{cases} a_{11}x_1 + a_{12}x_2 + \cdots + a_{1n}x_n = 0, \\ a_{21}x_1 + a_{22}x_2 + \cdots + a_{2n}x_n = 0, \\ \quad\vdots \qquad\qquad \vdots \qquad\qquad\quad \vdots \qquad \vdots \\ a_{n-1,1}x_1 + a_{n-1,2}x_2 + \cdots + a_{n-1,n}x_n = 0, \end{cases}$$

and let M_i be the determinant of the square matrix of order $n-1$ obtained from the coefficient matrix $A = (a_{ij})_{(n-1)\times n}$ by deleting the ith column for $1 \leqslant i \leqslant n$.

(1) Prove that $\alpha = (M_1, \ldots, (-1)^{i-1}M_i, \ldots, (-1)^{n-1}M_n)'$ is a solution of the given linear system.

(2) If the rank of A is $n-1$, prove that any solution of the given linear system must be a multiple of α.

22. Let $0 \neq A \in M_{n\times m}(F)$ and $\beta \in F^n$. Prove that $AX = \beta$ has a solution if and only if all solutions of $A'Y = 0$ are solutions of $Y'\beta = 0$.

23. Let $A \in M_{n\times m}(\mathbb{R})$. Prove that $\mathrm{rank}(A'A) = \mathrm{rank}A$.

24. Let $A, B, C \in M_n(\mathbb{R})$. If $A'AB = A'AC$, prove that $AB = AC$.

25. Prove that the matrix equation $A'AX = A'\beta$ has a solution for any $A \in M_{n\times m}(\mathbb{R})$ and $\beta \in \mathbb{R}^n$.

26. Let $A \in M_{m\times n}(F)$, $B \in M_{n\times s}(F)$ and assume that $\mathrm{rank}(AB) = \mathrm{rank}B$. Prove that $\mathrm{rank}(ABC) = \mathrm{rank}(BC)$ for any $C \in M_{s\times t}(F)$.

27. Use the theory on linear systems to prove Exercise 53 of Chapter 2.

28. Prove that if two linear systems have the same solution set, then the sequences of row vectors of their augmented matrices are linearly equivalent.

29. Determine whether the following equations have solutions, if any, then find their general solutions:

(1)
$$\begin{cases} 2x_1 + x_2 - 2x_3 = 1, \\ x_1 + x_2 + x_3 = 3, \\ x_1 + 2x_2 - 3x_3 = 1; \end{cases}$$

(2)
$$\begin{cases} x_1 - x_2 + 3x_3 - x_4 = 1, \\ 2x_1 - x_2 - x_3 + 4x_4 = 2, \\ 3x_1 - 2x_2 + 2x_3 + 3x_4 = 3, \\ x_1 \qquad - 4x_3 + 5x_4 = -1; \end{cases}$$

(3)
$$\begin{cases} x_1 - x_2 + x_3 = 1, \\ 3x_1 - 2x_2 - x_3 = -1, \\ 4x_1 - 2x_2 + 2x_3 = 5, \\ 8x_1 - 5x_2 + 2x_3 = 5. \end{cases}$$

30. Solve the following equations which contain a parameter λ:

$$(1) \quad \begin{cases} 2x_1 - x_2 + x_3 + x_4 = 1, \\ x_1 + 2x_2 - x_3 + 4x_4 = 2, \\ x_1 + 7x_2 - 4x_3 + 11x_4 = \lambda; \end{cases}$$

$$(2) \quad \begin{cases} \lambda x_1 + x_2 + x_3 + \lambda x_4 = 1, \\ x_1 + \lambda x_2 + x_3 + \lambda x_4 = 1, \\ x_1 + x_2 + \lambda x_3 + \lambda x_4 = 1. \end{cases}$$

31. Let $A \in \mathrm{M}_{m \times n}(F)$, $0 \neq \beta \in F^m$, $\mathrm{rank}(A, \beta) = \mathrm{rank} A = r < n$, and $s = n - r$. Prove that the linear system $AX = \beta$ has $s + 1$ linearly independent solutions $\gamma_0, \gamma_1, \ldots, \gamma_s$ and the solution set of $AX = \beta$ is

$$\left\{ \gamma \in F^n \;\middle|\; \gamma = \sum_{i=0}^{s} k_i \gamma_i, \ k_0, k_1, \ldots, k_s \in F \text{ and } \sum_{i=0}^{s} k_i = 1 \right\}.$$

32. Let $f(x), g(x) \in F[x]$ with $(f(x), g(x)) = 1$, and let $N \in \mathrm{M}_n(F)$, $A = f(N)$ and $B = g(N)$. Prove that any solution γ of $ABX = 0$ has a decomposition $\gamma = \alpha + \beta$, where α is a solution of $AX = 0$ and β is a solution of $BX = 0$.

33.* Let $A \in \mathrm{M}_{s \times n}(\mathbb{R})$ with $\mathrm{rank} A = s < n$. Prove that there exists $B \in \mathrm{M}_{(n-s) \times n}(\mathbb{R})$ such that $\begin{pmatrix} A \\ B \end{pmatrix}$ is an invertible matrix and $AB' = 0$.

34.* Let $\alpha = (1, 2, 3)$, $\beta = (0, 1, 2) \in \mathbb{R}^3$. Determine the set

$$\left\{ \gamma \in \mathbb{R}^3 \;\middle|\; \begin{array}{l} \text{there exists } A \in \mathrm{M}_3(\mathbb{R}) \text{ such that} \\ |A| = 0, \ \alpha A = \beta, \ \beta A = \gamma \text{ and} \\ \gamma A = \alpha \end{array} \right\}.$$

35.* Let $A, B \in \mathrm{M}_n(F)$ and let $C = \begin{pmatrix} A \\ B \end{pmatrix}$. Prove that if $AB = BA$, then

$$\mathrm{rank} A + \mathrm{rank} B \geqslant \mathrm{rank} C + \mathrm{rank}(AB).$$

36.[*] (Minkowski[3]) Let $n \geqslant 2$ be an integer and $A = (a_{ij})_{n \times n} \in M_n(\mathbb{R})$. Prove the following:

(1) If $|a_{ii}| > \sum_{j \neq i} |a_{ij}|$, $1 \leqslant i \leqslant n$, then $|A| \neq 0$.

(2) If $a_{ii} > \sum_{j \neq i} |a_{ij}|$, $1 \leqslant i \leqslant n$, then $|A| > 0$.

We remark that the matrix satisfying the condition stated in (1) is called a **diagonally dominant matrix** while the matrix satisfying the condition stated in (2) is called a **strictly diagonally dominant matrix**.

37.[*] Let $n \geqslant 3$ be an integer and $A = (a_{ij})_{n \times n} \in M_n(\mathbb{R})$. If

$$|a_{ii}a_{jj}| > \left(\sum_{k \neq i} |a_{ik}| \right) \left(\sum_{l \neq j} |a_{jl}| \right)$$

for any $1 \leqslant i \neq j \leqslant n$, prove that $|A| \neq 0$.

38. Find the resultants of the following polynomials:

(1) $x^3 + x^2 + x + 1$ and $x^5 + x^4 + x^3 + x^2 + x + 1$;
(2) $x^n + x + 1$ and $x^2 - 3x + 2$;
(3) $x^n + 1$ and $(x - 1)^n$.

39. Solve the following equations:

(1) $\begin{cases} 5y^2 - 6xy + 5x^2 - 16 = 0, \\ y^2 - xy + 2x^2 - y - x - 4 = 0; \end{cases}$

(2) $\begin{cases} x^2 + y^2 + 4x - 2y + 3 = 0, \\ x^2 + 4xy - y^2 + 10y - 9 = 0. \end{cases}$

---

[3]Hermann Minkowski, 1864–1909, Russian mathematician.

Chapter 4

Linear Spaces

4.1 Introduction

A linear space is one of the most important mathematical concepts in linear algebra, and it is also the first abstract mathematical concept we have encountered so far. Historically, the idea of a linear space originated from the analytical geometry created by René Descartes in the 17th century. Descartes is the founder of analytic geometry. One of his most important contributions is introducing the coordinate system, which lays the foundation for the emergence of the concept of a vector. Surprisingly, vectors appeared relatively late. It was not until the 19th century that the concept of a vector was given by Bellavitis[1] and was called a "bipoint", that is, a directed segment from one point to another point. This "bipoint" is the vector we learned in middle school. This concept was quickly used to express the complex numbers, which prompted Hamilton's[2] discovery of quaternions. In 1857, Arthur Cayley formally introduced the concept of a matrix, which provides the most direct help for the emergence of a linear space. The concept of a linear space we now see was given by Giuseppe Peano in 1888.

Originally, linear spaces are defined over the field of real numbers and are always finite-dimensional. Such linear spaces are actually

---

[1]Giusto Bellavitis, 1803–1880, Italian mathematician.

[2]William Rowan Hamilton, 1805–1865, Irish mathematician.

Euclidean spaces, which are discussed in Chapter 8, and their elements are the vectors we are familiar with. With the establishment of calculus and the need for the development of functional analysis, linear spaces were extended to infinite-dimensional cases, resulting in the appearance of Hilbert[3] spaces and Banach[4] spaces, in which the elements are usually various functions. Linear spaces have become the most commonly used mathematical concepts in mathematics. They play an extensive and irreplaceable role in mathematics and other fields, such as natural sciences, engineering, image processing, and big data.

In this chapter, we introduce the concepts and some basic properties of linear spaces, subspaces, quotient spaces, etc.

4.2   Definitions and Properties

Roughly speaking, a linear space is a set of objects called vectors. So, the key is to know what a vector is. We find that the corresponding operations and operation rules determine it.

Definition 4.2.1. Let V be a nonempty set and F a number field. If we have the following two operations

Addition: $V \times V \longrightarrow V$, $(\alpha, \beta) \longmapsto \alpha + \beta$;

Scalar multiplication: $F \times V \longrightarrow V$, $(k, \alpha) \longmapsto k \cdot \alpha$

satisfying the following eight axioms:

(1) **Commutative law of addition:** $\alpha + \beta = \beta + \alpha$,
(2) **Associative law of addition:** $(\alpha + \beta) + \gamma = \alpha + (\beta + \gamma)$,
(3) Existence of the **zero element:** There exists an element $0 \in V$ such that for any $\alpha \in V$ we always have $0 + \alpha = \alpha$ (0 is called the zero element or the zero vector of V),
(4) Existence of the **additive inverse:** For any $\alpha \in V$, there exists $\alpha' \in V$ such that $\alpha + \alpha' = 0$ (α' is called the additive inverse of α),
(5) **Associative law of scalar multiplication:** $(k_1 k_2) \cdot \alpha = k_1 \cdot (k_2 \cdot \alpha)$,

---

[3]David Hilbert, 1862–1943, German mathematician.
[4]Stefan Banach, 1892–1945, Polish mathematician.

(6) **Unitary law:** $1 \cdot \alpha = \alpha$,

(7) **Distributive law I:** $(k_1 + k_2) \cdot \alpha = k_1 \cdot \alpha + k_2 \cdot \alpha$,

(8) **Distributive law II:** $k \cdot (\alpha + \beta) = k \cdot \alpha + k \cdot \beta$,

where $\alpha, \beta, \gamma \in V$ and $k, k_1, k_2 \in F$, then we call $(V, F, +, \cdot)$ a **linear space** or **vector space**, or, simply, we call V a linear space or vector space over F. Every element of V is called a **vector**. Every element of F is called a **scalar**.

Remark 4.2.2.

(1) A linear space comprises four ingredients. For any two linear spaces, as long as one of the four ingredients is different, then these two linear spaces are different. For example, $(\mathbb{C}, \mathbb{C}, +, \cdot)$ and $(\mathbb{C}, \mathbb{R}, +, \cdot)$ are two different linear spaces, where addition and scalar multiplication are the usual addition and multiplication of complex numbers.

(2) The linear spaces over $\mathbb{R}$ ($\mathbb{C}$) are called real (complex) linear spaces. Moreover, if there is no confusion, we call a linear space over F a linear space.

(3) We use uppercase English letters $U, V, W, \ldots$ to represent linear spaces, Greek alphabets $\alpha, \beta, \gamma, \ldots$ to represent the elements in a linear space, and lowercase English letters $k, l, \ldots$ to represent the elements in the number field. The scalar multiplication $k \cdot \alpha$ is abbreviated as $k\alpha$.

Example 4.2.3.

(1) Under the addition of polynomials and the scalar multiplication of numbers with polynomials, the polynomial ring $F[x]$ is a linear space over F. This linear space has many subsets, which are also linear spaces. For example, let $n \geqslant 1$,

$$F[x]_n = \{0\} \cup \{f(x) \in F[x] \mid \deg(f(x)) \leqslant n - 1\}.$$

Then clearly $F[x]_n$ is also a linear space and we have

$$0 \subset F[x]_1 \subset F[x]_2 \subset \cdots \subset F[x]_n \subset F[x]_{n+1} \subset \cdots \subset F[x].$$

(2) Under the usual addition of matrices and the scalar multiplication of a matrix by a number, the set $\mathrm{M}_{m \times n}(F)$ formed by all $m \times n$ matrices over the number field F is a linear space,

which has been discussed in detail in Chapter 2. This space is sometimes denoted by $F^{m \times n}$.

(3) Under the usual addition of n-dimensional vectors and the scalar multiplication of an n-dimensional vector by a number, the solution set of a homogeneous linear system with n unknowns is a linear space, which is called the **solution space** of this system.

(4) Under the addition of functions and the scalar multiplication of a function by a number, the set $C[a, b]$ consisting of all continuous real functions on the closed interval $[a, b]$ becomes a real linear space.

Example 4.2.4. Consider the set $\mathbb{R}^+ = \{a \in \mathbb{R} \mid a > 0\}$ and we define the following two operations:

$$\oplus: \ \mathbb{R}^+ \times \mathbb{R}^+ \longrightarrow \mathbb{R}^+, \ \ (a, b) \longmapsto ab$$

(the multiplication of a and b);

$$\circ: \ \mathbb{R} \times \mathbb{R}^+ \longrightarrow \mathbb{R}^+, \ \ (k, a) \longmapsto a^k \ \text{(the kth power of a)}.$$

Prove that $(\mathbb{R}^+, \mathbb{R}, \oplus, \circ)$ is a linear space.

Proof. According to the definition, we need to show that the above two operations satisfy axioms $(1) \sim (8)$. We verify these one by one. Let $a, b, c \in \mathbb{R}^+$, $k, l \in \mathbb{R}$:

(1) Since $a \oplus b = ab = ba = b \oplus a$, we have the commutative law of addition.

(2) Since $(a \oplus b) \oplus c = (ab)c = a(bc) = a \oplus (b \oplus c)$, we have the associative law of addition.

(3) Since $1 \oplus a = 1a = a$, we have the zero element for addition. Note that the zero element is 1.

(4) Since $a \oplus \frac{1}{a} = a\frac{1}{a} = 1$, we have an additive inverse. Note that the additive inverse of a is $\frac{1}{a}$.

(5) Since $l \circ (k \circ a) = l \circ a^k = a^{lk} = (lk) \circ a$, we have the associative law of scalar multiplication.

(6) Since $1 \circ a = a^1 = a$, we have the unitary law.

(7) Since $(l + k) \circ a = a^{l+k} = a^l a^k = (l \circ a) \oplus (k \circ a)$, we have the distributive law I.

(8) Since $l \circ (a \oplus b) = (ab)^l = a^l b^l = (l \circ a) \oplus (l \circ b)$, we have the distributive law II.

Next, we give some basic properties of a linear space.

Let V be a linear space and $\alpha, \beta \in V$, we call $\alpha + \beta$ the **sum** of α, β. We note that in the definition of a linear space, only the sum of two vectors is defined. How do we define the sum of more vectors? To do this, take any n vectors $\alpha_1, \alpha_2, \ldots, \alpha_n \in V$ and let's first write the following expression:

$$\alpha_1 + \alpha_2 + \cdots + \alpha_n. \tag{4.1}$$

This expression is meaningless now, but when $n = 2$, the sum $\alpha_1 + \alpha_2$ is defined. When $n = 3$, binary bracketings of the expression (4.1) yield

$$(\alpha_1 + \alpha_2) + \alpha_3 = \alpha_1 + (\alpha_2 + \alpha_3)$$

by the associative law of addition. When $n = 4$, binary bracketings of the expression (4.1) give rise to the following five different expressions: $((\alpha_1 + \alpha_2) + \alpha_3) + \alpha_4$, $(\alpha_1 + (\alpha_2 + \alpha_3)) + \alpha_4$, $(\alpha_1 + \alpha_2) + (\alpha_3 + \alpha_4)$, $\alpha_1 + ((\alpha_2 + \alpha_3) + \alpha_4)$, $\alpha_1 + (\alpha_2 + (\alpha_3 + \alpha_4))$.

Here is a natural question: When $n \geqslant 4$, are the results of the different binary bracketings of the expression (4.1) the same? Since n is finite, the number of binary bracketings of the expression (4.1) is also finite. We denote this number by N for convenience (if you feel interested, then you can show that N is equal to the Catalan[5] number $\frac{1}{n} C_{2(n-1)}^{n-1}$, such number is also called an An-Tu Ming[6] number). Suppose that the expressions corresponding to these N binary bracketings are

$$\Sigma_1(\alpha_1, \ldots, \alpha_n), \ \Sigma_2(\alpha_1, \ldots, \alpha_n), \ \ldots, \ \Sigma_N(\alpha_1, \ldots, \alpha_n).$$

Proposition 4.2.5. *Let V be a linear space. Then we have*

$$\Sigma_1(\alpha_1, \ldots, \alpha_n) = \Sigma_2(\alpha_1, \ldots, \alpha_n) = \cdots = \Sigma_N(\alpha_1, \ldots, \alpha_n)$$

for any $\alpha_1, \ldots, \alpha_n \in V$ and $n \geqslant 2$.

Proof. First, for $k \geqslant 1$ and $\alpha_1, \ldots, \alpha_k \in V$, we define the standard sum $\Sigma(\alpha_1, \ldots, \alpha_k)$ inductively as follows: When $k = 1$, $\Sigma(\alpha_1) = \alpha_1$. When $k = 2$, $\Sigma(\alpha_1, \alpha_2) = \alpha_1 + \alpha_2$. Let $k \geqslant 3$

---

[5] Eugène Charles Catalan, 1814–1894, French and Belgian mathematician.
[6] An-Tu Ming, 1692–1765, Chinese mathematician.

and assume that $\Sigma(\alpha_1,\ldots,\alpha_{k-1})$ is already defined, then define $\Sigma(\alpha_1,\ldots,\alpha_k) = \Sigma(\alpha_1,\ldots,\alpha_{k-1}) + \alpha_k$.

Second, we prove that for any positive integers l and k, one has

$$\Sigma(\alpha_1,\ldots,\alpha_l) + \Sigma(\alpha_{l+1},\ldots,\alpha_{l+k}) = \Sigma(\alpha_1,\ldots,\alpha_{l+k}). \qquad (4.2)$$

We prove this fact by induction on k. When $k = 1$, we get it from the definition of the standard sum. Assume that $k \geqslant 2$ and the expression (4.2) is verified for $k - 1$, then from the definition of the standard sum we have

$$\Sigma(\alpha_1,\ldots,\alpha_l) + \Sigma(\alpha_{l+1},\ldots,\alpha_{l+k})$$

$$= \Sigma(\alpha_1,\ldots,\alpha_l) + (\Sigma(\alpha_{l+1},\ldots,\alpha_{l+k-1}) + \alpha_{l+k})$$

$$\xlongequal{\text{Associativity}} (\Sigma(\alpha_1,\ldots,\alpha_l) + \Sigma(\alpha_{l+1},\ldots,\alpha_{l+k-1})) + \alpha_{l+k}$$

$$\xlongequal{\text{Induction}} \Sigma(\alpha_1,\ldots,\alpha_{l+k-1}) + \alpha_{l+k}$$

$$= \Sigma(\alpha_1,\ldots,\alpha_{l+k}).$$

Therefore, we get (4.2).

Finally, we prove that for $1 \leqslant i \leqslant N$, we always have

$$\Sigma_i(\alpha_1,\ldots,\alpha_n) = \Sigma(\alpha_1,\ldots,\alpha_n).$$

We prove this by induction on n. The case $n = 2$ is obvious. Assume that $n \geqslant 3$ and the conclusion is true when the number of vectors is smaller than n. Now, we consider the case $\alpha_1,\ldots,\alpha_n \in V$. First, there must exist an integer s with $1 \leqslant s < n$ such that

$$\Sigma_i(\alpha_1,\ldots,\alpha_n) = \Sigma'(\alpha_1,\ldots,\alpha_s) + \Sigma''(\alpha_{s+1},\ldots,\alpha_n),$$

where $\Sigma'(\alpha_1,\ldots,\alpha_s)$ is an expression of $\alpha_1,\ldots,\alpha_s$ obtained by some binary bracketings and $\Sigma''(\alpha_{s+1},\ldots,\alpha_n)$ is an expression of $\alpha_{s+1},\ldots,\alpha_n$ obtained by some binary bracketings. By induction,

$$\Sigma'(\alpha_1,\ldots,\alpha_s) = \Sigma(\alpha_1,\ldots,\alpha_s), \ \Sigma''(\alpha_{s+1},\ldots,\alpha_n) = \Sigma(\alpha_{s+1},\ldots,\alpha_n).$$

Then, by (4.2), we have

$$\Sigma_i(\alpha_1,\ldots,\alpha_n) = \Sigma(\alpha_1,\ldots,\alpha_s) + \Sigma(\alpha_{s+1},\ldots,\alpha_n) = \Sigma(\alpha_1,\ldots,\alpha_n).$$

Remark 4.2.6.

(1) Let V be a linear space and $\alpha_1, \alpha_2, \ldots, \alpha_n \in V$. From now on, we use $\alpha_1 + \alpha_2 + \cdots + \alpha_n$ or $\sum_{i=1}^{n} \alpha_i$ to represent the unique result in Proposition 4.2.5 regardless of the methods of binary bracketings and we call it the sum of $\alpha_1, \alpha_2, \ldots, \alpha_n$. Thus, the expression (4.1) is meaningful now. In addition, due to the commutative law of addition, for any permutation $j_1 j_1 \ldots j_n$ of $1, 2, \ldots, n$, we have $\sum_{i=1}^{n} \alpha_i = \sum_{i=1}^{n} \alpha_{j_i}$.

(2) Careful readers will find that in the process of proving Proposition 4.2.5, we only use the associative law of addition. In other words, if some operation on the set S

$$\circ: \ S \times S \to S, \ (s_1, s_2) \longmapsto s_1 \circ s_2$$

satisfies the associative law, that is, for any $s_1, s_2, s_3 \in S$, one has

$$(s_1 \circ s_2) \circ s_3 = s_1 \circ (s_2 \circ s_3),$$

then we can use $s_1 \circ s_2 \circ \ldots \circ s_n$ to denote the result of $s_1, s_2, \ldots, s_n \in S$ $(n \geqslant 2)$ under the operation $\circ$. That is the reason why associative law is so important.

Proposition 4.2.7. *Let V be a linear space, $\alpha \in V$, and $k \in F$.*

(1) *The zero element of V and the additive inverse of α are unique;*
(2) *$k\alpha = 0$ if and only if $k = 0$ or $\alpha = 0$.*

Proof. (1) Let both 0_1 and 0_2 be zero elements of V. Then

$$0_1 = 0_2 + 0_1 = 0_1 + 0_2 = 0_2.$$

Here in the first equality, we use the fact that 0_2 is a zero element, in the second, we use the commutative law of addition, and in the last, we use the fact that 0_1 is also a zero element. Therefore, the zero element of V is unique.

Similarly, let both α_1 and α_2 be additive inverses of $\alpha \in V$. Then

$$\alpha_1 = 0 + \alpha_1 = (\alpha + \alpha_2) + \alpha_1 = (\alpha_2 + \alpha) + \alpha_1$$

$$= \alpha_2 + (\alpha + \alpha_1) = \alpha_2 + 0 = 0 + \alpha_2 = \alpha_2.$$

Therefore, the additive inverse of α is unique.

(2) Sufficiency. If $k = 0$, then $0 \cdot \alpha = (0+0) \cdot \alpha = 0 \cdot \alpha + 0 \cdot \alpha$, and so $0 \cdot \alpha = 0$. If $k \neq 0$, $\alpha = 0$, then by $k \cdot 0 = k \cdot (0+0) = k \cdot 0 + k \cdot 0$, whence $k \cdot 0 = 0$.

Necessity. If $k \neq 0$, then $\alpha = 1\alpha = \left(\frac{1}{k} \times k\right)\alpha = \frac{1}{k}(k\alpha) = \frac{1}{k} \cdot 0 = 0$.

Remark 4.2.8. Let V be a linear space.

(1) By Proposition 4.2.7, the zero element is unique, and we usually denote it by 0_V, or 0 if there is no confusion about V.
(2) Let $\alpha \in V$. By Proposition 4.2.7, the additive inverse of α is unique and is denoted by $-\alpha$.
(3) Let $\alpha, \beta \in V$. We define $\alpha - \beta = \alpha + (-\beta)$. It is clear that $\alpha + \beta = \gamma$ if and only if $\alpha = \gamma - \beta$.

Corollary 4.2.9. *For any $\lambda \in F$, we have $-(\lambda\alpha) = (-\lambda)\alpha$. In particular, $-\alpha = (-1)\alpha$.*

Proof. By Remark 4.2.8 (2), $-(\lambda\alpha)$ is the additive inverse of $\lambda\alpha$. Therefore, it is enough to show that $\lambda\alpha + (-\lambda)\alpha = 0$. In fact, $\lambda\alpha + (-\lambda)\alpha = 0\alpha = 0$. Here we use Proposition 4.2.7 (2).

4.3 Dimensions, Bases, and Coordinates

Section 3.3 gives some concepts and properties of the vector space F^n. It is not difficult to find that in these concepts and all properties that do not involve components of vectors, we only use addition, scalar multiplication and the corresponding eight operation rules (see Section 2.7), which have been included in the definition of a linear space. So, these concepts and properties are also valid for any linear space and can be directly referenced. For convenience, the definition of a linear combination is given here. Based on this, readers can provide relevant concepts, such as linear expressions, linear equivalences, linear dependence, linear independence, maximal linearly independent sets, and the rank of a sequence of vectors.

Definition 4.3.1. Let V be a linear space and s a positive integer, and let $\alpha_1, \alpha_2, \ldots, \alpha_s \in V$, $k_1, k_2, \ldots, k_s \in F$. The vector $k_1\alpha_1 + k_2\alpha_2 + \cdots + k_s\alpha_s$ is called a **linear combination** of $\alpha_1, \alpha_2, \ldots, \alpha_s$.

For convenience, we list the following two important properties.

Lemma 4.3.2. *Let V be a linear space and $\alpha_1, \alpha_2, \ldots, \alpha_m, \beta \in V$. If the sequence $\alpha_1, \alpha_2, \ldots, \alpha_m$ is linearly independent, and the sequence $\alpha_1, \alpha_2, \ldots, \alpha_m, \beta$ is linearly dependent, then β can be linearly expressed by $\alpha_1, \alpha_2, \ldots, \alpha_m$ uniquely.*

Lemma 4.3.3 (Steinitz exchange lemma). *Let V be a linear space and $\alpha_1, \ldots, \alpha_r,\ \beta_1, \ldots, \beta_m \in V$. Assume that $\alpha_1, \ldots, \alpha_r$ can be linearly expressed by $\beta_1, \ldots, \beta_m$. If $\alpha_1, \ldots, \alpha_r$ are linearly independent, then $r \leqslant m$ and we can use $\alpha_1, \ldots, \alpha_r$ to replace r vectors, say $\beta_1, \ldots, \beta_r$, in $\beta_1, \ldots, \beta_m$ such that $\alpha_1, \ldots, \alpha_r, \beta_{r+1}, \ldots, \beta_m$ and $\beta_1, \ldots, \beta_r, \beta_{r+1}, \ldots, \beta_m$ are linearly equivalent.*

Definition 4.3.4. Let V be a linear space.

(1) If n is a positive integer and there exist n linearly independent vectors $\alpha_1, \alpha_2, \ldots, \alpha_n \in V$ such that all vectors in V can be linearly expressed by these n vectors, then V is called an **n-dimensional linear space**, and the sequence $\alpha_1, \alpha_2, \ldots, \alpha_n$ is called a **basis** for V.

(2) If V only has the zero vector, then V is called the **zero space** or **0-dimensional linear space**.

(3) If V is an n-dimensional space for some $n \in \mathbb{N}$, then we call V a **finite-dimensional** linear space, and n, denoted by $\dim_F V$ or just $\dim V$, is called the **dimension** of V. If V is not a finite-dimensional linear space, then we call V an **infinite-dimensional** linear space.

Remark 4.3.5. The definition of the dimension of a linear space is well defined by Lemma 4.3.3.

Example 4.3.6.

(1) The linear spaces $F[x]_n$, $\mathrm{M}_{m\times n}(F)$ in Example 4.2.3 and F^n are all finite-dimensional and

$$\dim F[x]_n = n; \quad \dim \mathrm{M}_{m\times n}(F) = mn; \quad \dim F^n = n.$$

In fact, the sequence $1, x, \ldots, x^{n-1}$ is a basis for $F[x]_n$; the sequence $E_{ij}, 1 \leqslant i \leqslant m, 1 \leqslant j \leqslant n$, is a basis for $\mathrm{M}_{m\times n}(F)$, where E_{ij} is the matrix unit defined in Exercise 25 of Chapter 2; the sequence $\varepsilon_1, \varepsilon_2, \ldots, \varepsilon_n$ (see Example 3.3.4) is a basis for F^n, and it is called the **standard basis** for F^n.

(2) We have $\dim_{\mathbb{C}}\mathbb{C} = 1$, $\dim_{\mathbb{R}}\mathbb{C} = 2$ and $\mathbb{C}$ is an infinite-dimensional linear space over $\mathbb{Q}$. In fact, any nonzero complex number is a basis for the complex linear space $\mathbb{C}$, and the sequence $1, i$ is a basis for the real linear space $\mathbb{C}$. Let n be any positive number. Then $x^n + 2$ is irreducible over $\mathbb{Q}$. Suppose that α is a root of $x^n + 2 = 0$ in $\mathbb{C}$, then clearly the sequence $1, \alpha, \alpha^2, \ldots, \alpha^{n-1}$ is linearly independent over $\mathbb{Q}$, and thus $\mathbb{C}$ is an infinite-dimensional linear space over $\mathbb{Q}$. So, the dimension of a linear space depends on which number field it is defined over.

(3) The linear spaces $F[x]$ and $C[a, b]$ are infinite-dimensional.

Remark 4.3.7. There are essential differences between finite-dimensional and infinite-dimensional linear spaces. In this book, all linear spaces are finite-dimensional unless stated otherwise. We see in Chapter 5 that dimension is the most important character of a finite-dimensional linear space, and it completely determines the structure of this linear space. By definition, a finite-dimensional nonzero linear space always has a basis. We must point out that for an infinite-dimensional linear space V, there exists a linearly independent sequence of vectors in V such that any vector in V can be linearly expressed by a finite number of vectors in this sequence. Such a sequence of vectors is usually called a **Hamel**[7] **basis.** For example, the sequence $1, x, x^2, \ldots, x^n, \ldots$ is a Hamel basis for $F[x]$. The common infinite-dimensional linear spaces include many different forms of function spaces, such as $F[x]$ and $C[a, b]$ in Example 4.3.6. These function spaces constitute the main research objects of another branch of mathematics — "Functional analysis".

Definition 4.3.8. Let the sequence $\alpha_1, \alpha_2, \ldots, \alpha_n$ be a basis for an n-dimensional linear space V and $\alpha \in V$. If

$$\alpha = k_1\alpha_1 + k_2\alpha_2 + \cdots + k_n\alpha_n, \text{ where } k_1, k_2, \ldots, k_n \in F,$$

then the column vector $(k_1, k_2, \ldots, k_n)'$ is called the **coordinate** of α with respect to the basis $\alpha_1, \alpha_2, \ldots, \alpha_n$.

Example 4.3.9. The coordinate of any vector in the 2-dimensional plane $\mathbb{R}^2$ with respect to the basis $(1, 0)$, $(0, 1)$ is just the usual Cartesian coordinate.

---

[7]Georg Karl Wilhelm Hamel, 1877–1954, German mathematician.

Next, we discuss the relationship between coordinates of the same vector with respect to different bases.

Let V be a linear space and $\alpha_1, \alpha_2, \ldots, \alpha_n \in V$, and let $k_i, a_{ij} \in F$, $1 \leqslant i \leqslant n$, $1 \leqslant j \leqslant m$ and $A = (a_{ij})_{n \times m}$. For convenience, we write $\beta = \sum_{i=1}^{n} k_i \alpha_i$ in the formal way: $\beta = (\alpha_1, \alpha_2, \ldots, \alpha_n) \begin{pmatrix} k_1 \\ k_2 \\ \vdots \\ k_n \end{pmatrix}$.

Therefore, the sequence of vectors

$$\beta_1 = \sum_{i=1}^{n} a_{i1} \alpha_i, \beta_2 = \sum_{i=1}^{n} a_{i2} \alpha_i, \ldots, \beta_m = \sum_{i=1}^{n} a_{im} \alpha_i$$

can be written formally as $(\beta_1, \beta_2, \ldots, \beta_m) = (\alpha_1, \alpha_2, \ldots, \alpha_n) A$.

Let $A, C \in \mathrm{M}_{n \times m}(F), B \in \mathrm{M}_{m \times l}(F)$. By direct verification, we can show that the above formal expression satisfies the following properties:

(1) $((\alpha_1, \alpha_2, \ldots, \alpha_n) A) B = (\alpha_1, \alpha_2, \ldots, \alpha_n)(AB)$;

(2) $(\alpha_1 + \beta_1, \alpha_2 + \beta_2, \ldots, \alpha_n + \beta_n) A$
$= (\alpha_1, \alpha_2, \ldots, \alpha_n) A + (\beta_1, \beta_2, \ldots, \beta_n) A$;

(3) $(\alpha_1, \alpha_2, \ldots, \alpha_n)(A + C) = (\alpha_1, \alpha_2, \ldots, \alpha_n) A + (\alpha_1, \alpha_2, \ldots, \alpha_n) C$.

Let $\varepsilon_1, \varepsilon_2, \ldots, \varepsilon_n$ and $\eta_1, \eta_2, \ldots, \eta_n$ be two bases for V. Then there exists $A \in \mathrm{M}_n(F)$ such that $(\eta_1, \eta_2, \ldots, \eta_n) = (\varepsilon_1, \varepsilon_2, \ldots, \varepsilon_n) A$ and we call the matrix A the **transition matrix** from the basis $\varepsilon_1, \varepsilon_2, \ldots, \varepsilon_n$ to the basis $\eta_1, \eta_2, \ldots, \eta_n$.

Moreover, let B be the transition matrix from the basis $\eta_1, \eta_2, \ldots, \eta_n$ to the basis $\varepsilon_1, \varepsilon_2, \ldots, \varepsilon_n$, that is, $(\varepsilon_1, \varepsilon_2, \ldots, \varepsilon_n) = (\eta_1, \eta_2, \ldots, \eta_n) B$. Then

$$(\eta_1, \eta_2, \ldots, \eta_n) = (\varepsilon_1, \varepsilon_2, \ldots, \varepsilon_n) A = ((\eta_1, \eta_2, \ldots, \eta_n) B) A$$

$$= (\eta_1, \eta_2, \ldots, \eta_n)(BA),$$

$$(\varepsilon_1, \varepsilon_2, \ldots, \varepsilon_n) = (\eta_1, \eta_2, \ldots, \eta_n) B = ((\varepsilon_1, \varepsilon_2, \ldots, \varepsilon_n) A) B$$

$$= (\varepsilon_1, \varepsilon_2, \ldots, \varepsilon_n)(AB).$$

By the uniqueness of coordinates, $AB = BA = I_n$. Therefore, transition matrices are invertible.

Let $\alpha \in V$ and assume that the coordinates of α with respect to the bases $\varepsilon_1, \varepsilon_2, \ldots, \varepsilon_n$ and $\eta_1, \eta_2, \ldots, \eta_n$ are $X, Y \in F^n$, respectively, that is, $\alpha = (\varepsilon_1, \varepsilon_2, \ldots, \varepsilon_n)X = (\eta_1, \eta_2, \ldots, \eta_n)Y$. Then

$$\alpha = (\eta_1, \eta_2, \ldots, \eta_n)Y = ((\varepsilon_1, \varepsilon_2, \ldots, \varepsilon_n)A)Y = (\varepsilon_1, \varepsilon_2, \ldots, \varepsilon_n)(AY).$$

Thus, by the uniqueness of coordinates, we have $X = AY$, which is the relationship between coordinates with respect to different bases.

4.4   Subspaces

Starting with this section, we study linear spaces using the idea of the set theory, that is, to study the "sub, intersection, union and complement" of linear spaces. Readers see that the "sub" and "intersection" of sets can be directly extended to the situation of linear spaces, but the "union" and "complement" cannot be done so directly. Proper treatment is required. The results of the treatment are called "(direct) sum" and "direct sum complement, quotient".

Definition 4.4.1. Let V be a linear space and W a nonempty subset of V. If W is also a linear space under the addition and scalar multiplication of V, then W is called a **subspace** of V.

Example 4.4.2.

(1) Let V be a linear space. Obviously, $\{0\}$ is a subspace of V which is called the **zero subspace** of V and is abbreviated as 0; V is also a subspace of V. V and 0 are called **trivial subspaces** of V. A subspace which is not trivial is call a **nontrivial subspace**. If a subspace $W \subsetneq V$, then W is called a **proper subspace** of V.
(2) A nontrivial subspace of $\mathbb{R}^2$ is just a line through the origin. Similarly, a nontrivial subspace of $\mathbb{R}^3$ is either a line through the origin or a plane through the origin.
(3) $\mathbb{R}$ is a proper subspace of $\mathbb{C}$ as real linear spaces.
(4) $F[x]_n$ is an n-dimensional subspace of $F[x]$.
(5) Let $A \in \mathrm{M}_{m \times n}(F)$, then the solution space of the homogeneous linear system $AX = 0$ is a subspace of F^n.
(6) Under usual addition and multiplication of real numbers, $\mathbb{R}$ is a real linear space. It is easy to see that $\mathbb{R}^+$ (Example 4.2.4) is not a subspace of $\mathbb{R}$.

Lemma 4.4.3. *Let W be a nonempty subset of the linear space V. Then W is a subspace of V if and only if W is closed under the addition and scalar multiplication of V.*

Proof. The necessity is obvious. Next, we prove the sufficiency, that is, we prove that W satisfies the eight axioms listed in the definition of a linear space. Note that the commutative law of addition, the associative law of addition, the associative law of scalar multiplication, unitary law, and distributive laws I and II hold for all elements in V and, of course, hold for elements in W. So, we only need to show that W contains the zero vector and the additive inverses. In fact, by Proposition 4.2.7 and Corollary 4.2.9, we know that for any $\alpha \in W$, we have $0 = 0\alpha \in W$, $-\alpha = (-1)\alpha \in W$.

Proposition 4.4.4. *Let W be a subspace of V, then any basis for W can be extended to a basis for V.*

Proof. This result follows from Steinitz exchange lemma.

Lemma 4.4.5. *Let $\{V_i\}_{i \in I}$ be a family of subspaces of V. Then the set $\bigcap\limits_{i \in I} V_i$ is also a subspace of V and is called the **intersection** of $\{V_i\}_{i \in I}$.*

Proof. Take any $\alpha, \beta \in \bigcap\limits_{i \in I} V_i$ and $k \in F$. Since V_i is a subspace,

$$\alpha + \beta \in V_i, \ k\alpha \in V_i, \ i \in I.$$

Thus, $\alpha + \beta \in \bigcap\limits_{i \in I} V_i$, $k\alpha \in \bigcap\limits_{i \in I} V_i$, that is, $\bigcap\limits_{i \in I} V_i$ is closed under addition and scalar multiplication. By Lemma 4.4.3, $\bigcap\limits_{i \in I} V_i$ is a subspace of V.

Example 4.4.6. Let $A \in \mathrm{M}_{m \times n}(F)$ and $B \in \mathrm{M}_{s \times n}(F)$. If V_1 and V_2 are solution spaces of $AX = 0$ and $BX = 0$, respectively, then $V_1 \cap V_2$ is the solution space of $\begin{pmatrix} A \\ B \end{pmatrix} X = 0$.

Example 4.4.7. In the 3-dimensional real space $\mathbb{R}^3$, take two different planes π_1, π_2 though the origin, and we know that π_1, π_2 are subspaces. Then their intersection $\pi_1 \cap \pi_2$ is a line through the origin and thus a 1-dimensional subspace. We recommend that the readers consider the following question: In $\mathbb{R}^4$, is the intersection of two different 2-dimensional subspaces still 1-dimensional?

Let V be a linear space and S a nonempty subset of V. Define

$$L(S) = \left\{ \sum_{i=1}^{m} k_i \alpha_i \;\middle|\; m \in \mathbb{N}^*, \alpha_i \in S, \; k_i \in F, \; i = 1, 2, \ldots, m \right\}.$$

Clearly, $L(S)$ is the set formed by all possible linear combinations of finitely many vectors in S and is closed under addition and scalar multiplication. Thus, it is a subspace of V, which is called the **subspace spanned by** S, and S is usually called a **generating set** of $L(S)$.

Proposition 4.4.8. *Let V be a linear space and $\alpha_i, \gamma_j \in V$, $1 \leqslant i \leqslant s, 1 \leqslant j \leqslant t$, and let S be a nonempty subset of V.*

(1) *$L(S)$ is the smallest subspace containing S.*
(2) *$L(\alpha_1, \alpha_2, \ldots, \alpha_s) = L(\gamma_1, \gamma_2, \ldots, \gamma_t)$ if and only if the sequences $\alpha_1, \alpha_2, \ldots, \alpha_s$ and $\gamma_1, \gamma_2, \ldots, \gamma_t$ are linearly equivalent.*
(3) *If the sequence $\alpha_1, \ldots, \alpha_s$ contains a nonzero vector, then any maximal linearly independent set of the sequence $\alpha_1, \ldots, \alpha_s$ is a basis for $L(\alpha_1, \alpha_2, \ldots, \alpha_s)$, and thus*

$$\dim L(\alpha_1, \alpha_2, \ldots, \alpha_s) = \operatorname{rank}(\alpha_1, \alpha_2, \ldots, \alpha_s).$$

Proof. (1) On one hand, by definition, $L(S)$ is a subspace containing S. On the other hand, let U be a space containing S. Since U is closed under the addition and scalar multiplication, it contains all possible linear combinations of any finitely many elements in S. Therefore, $L(S) \subseteq U$ and thus $L(S)$ is the smallest subspace containing S.

(2) Necessity. Since $L(\alpha_1, \alpha_2, \ldots, \alpha_s) = L(\gamma_1, \gamma_2, \ldots, \gamma_t)$, every α_i belongs to $L(\gamma_1, \gamma_2, \ldots, \gamma_t)$ for $1 \leqslant i \leqslant s$ and thus can be linearly expressed by $\gamma_1, \gamma_2, \ldots, \gamma_t$. Similarly, every γ_i $(1 \leqslant i \leqslant t)$ can be linearly expressed by $\alpha_1, \alpha_2, \ldots, \alpha_s$. Therefore, the sequences $\alpha_1, \alpha_2, \ldots, \alpha_s$ and $\gamma_1, \gamma_2, \ldots, \gamma_t$ are linearly equivalent.

Sufficiency. Since $\alpha_1, \alpha_2, \ldots, \alpha_s$ and $\gamma_1, \gamma_2, \ldots, \gamma_t$ are linearly equivalent, every α_i $(1 \leqslant i \leqslant s)$ belongs to $L(\gamma_1, \gamma_2, \ldots, \gamma_t)$. By (1), $L(\alpha_1, \alpha_2, \ldots, \alpha_s) \subseteq L(\gamma_1, \gamma_2, \ldots, \gamma_t)$. Similarly, we have $L(\gamma_1, \gamma_2, \ldots, \gamma_t) \subseteq L(\alpha_1, \alpha_2, \ldots, \alpha_s)$, as desired.

(3) Let $\alpha_{i_1}, \alpha_{i_2}, \ldots, \alpha_{i_r}$ be a maximal linearly independent set of the sequence $\alpha_1, \alpha_2, \ldots, \alpha_s$. By definition, the sequence $\alpha_{i_1}, \alpha_{i_2}, \ldots, \alpha_{i_r}$ is a basis for $L(\alpha_{i_1}, \alpha_{i_2}, \ldots, \alpha_{i_r})$. By (2), $L(\alpha_1, \alpha_2, \ldots, \alpha_s) = L(\alpha_{i_1}, \alpha_{i_2}, \ldots, \alpha_{i_r})$, and thus the sequence $\alpha_{i_1}, \alpha_{i_2}, \ldots, \alpha_{i_r}$ is a basis for $L(\alpha_1, \alpha_2, \ldots, \alpha_s)$. It follows that $\dim L(\alpha_1, \alpha_2, \ldots, \alpha_s) = r = \operatorname{rank}(\alpha_1, \alpha_2, \ldots, \alpha_s)$. $\qquad\square$

Example 4.4.9. Let V_1 and V_2 be two subspaces of V. Prove that $V_1 \cup V_2$ is a subspace of V if and only if $V_1 \subseteq V_2$ or $V_2 \subseteq V_1$.

Proof. The sufficiency is clear; we prove the necessity now. Assume that $V_1 \not\subseteq V_2$ and $V_2 \not\subseteq V_1$. So, there are $\alpha \in V_1$ while $\alpha \notin V_2$, and $\beta \in V_2$ while $\beta \notin V_1$. Since $V_1 \cup V_2$ is a linear space, $\alpha + \beta \in V_1 \cup V_2$, and thus $\alpha + \beta \in V_1$ or $\alpha + \beta \in V_2$. If $\alpha + \beta \in V_1$, then by $\alpha \in V_1$, we have $\beta \in V_1$ which is impossible. Similarly, it is also impossible for $\alpha + \beta \in V_2$.

This example shows that the union of two subspaces of a linear space may not be a subspace. But if we take two subspaces V_1 and V_2 of V, the subspace $L(V_1 \cup V_2)$ spanned by $V_1 \cup V_2$ is the smallest space containing V_1 and V_2. Clearly, by definition, $L(V_1 \cup V_2) = \{\alpha_1 + \alpha_2 \mid \alpha_1 \in V_1, \ \alpha_2 \in V_2\}$.

Definition 4.4.10. Let V_1 and V_2 be two subspaces of V. We call $L(V_1 \cup V_2)$ the **sum** of V_1 and V_2, denoted by $V_1 + V_2$.

Example 4.4.11.

(1) The sum of two different lines through the origin in the 2-dimensional plane $\mathbb{R}^2$ is $\mathbb{R}^2$.

(2) In the 3-dimensional real space $\mathbb{R}^3$, take a line through the origin and a plane through the origin. If this line lies in this plane, then their sum is just this plane; otherwise, their sum is the whole space $\mathbb{R}^3$. Similarly, the sum of two different planes through the origin is also $\mathbb{R}^3$.

(3) Let V be a linear space, $\alpha_i, \beta_j \in V$, $1 \leqslant i \leqslant t, 1 \leqslant j \leqslant s$. Then
$$L(\alpha_1, \alpha_2, \ldots, \alpha_t) + L(\beta_1, \beta_2, \ldots, \beta_s)$$
$$= L(\alpha_1, \alpha_2, \ldots, \alpha_t, \beta_1, \beta_2, \ldots, \beta_s).$$

By this example, we find that $\dim(V_1 + V_2)$ is less than or equal to $\dim V_1 + \dim V_2$. The exact relation between $\dim(V_1 + V_2)$ and $\dim V_1 + \dim V_2$ is given by the following.

Theorem 4.4.12 (Dimension formula). *Let V_1 and V_2 be subspaces of a linear space V. Then*

$$\dim(V_1 + V_2) + \dim(V_1 \cap V_2) = \dim V_1 + \dim V_2. \qquad (4.3)$$

Proof. Suppose that the sequence $\alpha_1, \alpha_2, \ldots, \alpha_l$ is a basis for $V_1 \cap V_2$ and extend it to a basis $\alpha_1, \alpha_2, \ldots, \alpha_l, \beta_1, \beta_2, \ldots, \beta_t$ for V_1 and a basis $\alpha_1, \alpha_2, \ldots, \alpha_l, \gamma_1, \gamma_2, \ldots, \gamma_s$ for V_2, respectively. Note that

$$V_1 = L(\alpha_1, \alpha_2, \ldots, \alpha_l, \beta_1, \beta_2, \ldots, \beta_t),$$

$$V_2 = L(\alpha_1, \alpha_2, \ldots, \alpha_l, \gamma_1, \gamma_2, \ldots, \gamma_s).$$

Thus, $V_1 + V_2 = L(\alpha_1, \alpha_2, \ldots, \alpha_l, \beta_1, \beta_2, \ldots, \beta_t, \gamma_1, \gamma_2, \ldots, \gamma_s)$, and hence any element in $V_1 + V_2$ can be linearly expressed by $\alpha_1, \alpha_2, \ldots, \alpha_l, \beta_1, \beta_2, \ldots, \beta_t, \gamma_1, \gamma_2, \ldots, \gamma_s$.

Next, we show that $\alpha_1, \alpha_2, \ldots, \alpha_l, \beta_1, \beta_2, \ldots, \beta_t, \gamma_1, \gamma_2, \ldots, \gamma_s$ are linearly independent. Let

$$a_1 \alpha_1 + \cdots + a_l \alpha_l + b_1 \beta_1 + \cdots + b_t \beta_t + c_1 \gamma_1 + \cdots + c_s \gamma_s = 0,$$

where $a_i, b_j, c_k \in F$, $1 \leqslant i \leqslant l$, $1 \leqslant j \leqslant t$, $1 \leqslant k \leqslant s$, then

$$a_1 \alpha_1 + \cdots + a_l \alpha_l + b_1 \beta_1 + \cdots + b_t \beta_t = -(c_1 \gamma_1 + \cdots + c_s \gamma_s) \in V_1 \cap V_2.$$

Therefore, there are $d_i \in F$, $1 \leqslant i \leqslant l$ such that

$$-(c_1 \gamma_1 + \cdots + c_s \gamma_s) = d_1 \alpha_1 + \cdots + d_l \alpha_l,$$

that is, $d_1 \alpha_1 + \cdots + d_l \alpha_l + c_1 \gamma_1 + \cdots + c_s \gamma_s = 0$.

Since the sequence $\alpha_1, \alpha_2, \ldots, \alpha_l, \gamma_1, \gamma_2, \ldots, \gamma_s$ is a basis for V_2, $c_1 = \cdots = c_s = 0$. Thus, $a_1 \alpha_1 + \cdots + a_l \alpha_l + b_1 \beta_1 + \cdots + b_t \beta_t = 0$. Note that the sequence $\alpha_1, \alpha_2, \ldots, \alpha_l, \beta_1, \beta_2, \ldots, \beta_t$ is a basis for V_1, and so $a_1 = \cdots = a_l = b_1 = \cdots = b_t = 0$. Therefore,

$$\alpha_1, \alpha_2, \ldots, \alpha_l, \beta_1, \beta_2, \ldots, \beta_t, \gamma_1, \gamma_2, \ldots, \gamma_s$$

are linearly independent, and thus

$$\dim(V_1 + V_2) = l + t + s = l + t + l + s - l$$

$$= \dim V_1 + \dim V_2 - \dim(V_1 \cap V_2).$$

So, the equality (4.3) holds.

If $V_1 \cap V_2 = 0$, then the sequence $\beta_1, \beta_2, \ldots, \beta_t, \gamma_1, \gamma_2, \ldots, \gamma_s$ is a basis for $V_1 + V_2$ by the proof above. So, (4.3) is also verified. $\qquad \square$

Next, we consider a particular case: The dimension of the sum of subspaces is equal to the sum of the dimensions of subspaces.

Definition 4.4.13. Let V_1 and V_2 be subspaces of a linear space V. If any vector α in $V_1 + V_2$ has a unique decomposition $\alpha = \alpha_1 + \alpha_2$ for $\alpha_1 \in V_1$, $\alpha_2 \in V_2$, then we call the sum $V_1 + V_2$ a **direct sum**, denoted by $V_1 \oplus V_2$.

Proposition 4.4.14. *Let V_1 and V_2 be subspaces of a linear space V. Then the following statements are equivalent:*

(1) *The sum $V_1 + V_2$ is a direct sum, that is, $V_1 + V_2 = V_1 \oplus V_2$.*
(2) *The decomposition of zero is unique.*
(3) $V_1 \cap V_2 = 0$.
(4) $\dim(V_1 + V_2) = \dim V_1 + \dim V_2$.

Proof. $(1) \Longrightarrow (2)$ is obvious.

$(2) \Longrightarrow (3)$. Take any $\alpha \in V_1 \cap V_2$, then $0 = \alpha + (-\alpha) \in V_1 + V_2$. By (2), we know that $\alpha = 0$ and thus $V_1 \cap V_2 = 0$.

$(3) \Longrightarrow (4)$ follows from the dimension formula.

$(4) \Longrightarrow (1)$. If $V_1 + V_2$ is not a direct sum, then there exists $\alpha \in V_1 + V_2$ such that $\alpha = \alpha_1 + \alpha_2 = \beta_1 + \beta_2$ for α_1, $\beta_1 \in V_1$, α_2, $\beta_2 \in V_2$ and $\alpha_1 \neq \beta_1$ or $\alpha_2 \neq \beta_2$. It is harmless to assume that $\alpha_1 \neq \beta_1$. Then

$$0 \neq \alpha_1 - \beta_1 = \beta_2 - \alpha_2 \in V_1 \cap V_2.$$

So $V_1 \cap V_2 \neq 0$ and thus $\dim(V_1 \cap V_2) > 0$. By the dimension formula,

$$\dim(V_1 + V_2) < \dim V_1 + \dim V_2,$$

which contradicts the condition in (4).

Remark 4.4.15.

(1) Similar to the sum and direct sum of two subspaces, we can define them for more subspaces: Let $V_1, V_2, \ldots, V_n$ be n subspaces of V, then their sum $V_1 + V_2 + \cdots + V_n$ is defined to the set

$$\{\alpha_1 + \alpha_2 + \cdots + \alpha_n \mid \alpha_i \in V_i,\ 1 \leqslant i \leqslant n\},$$

abbreviated as $\sum_{i=1}^{n} V_i$. Readers can verify that $\sum_{i=1}^{n} V_i$ is indeed a linear subspace of V. Moreover, if every vector α in $\sum_{i=1}^{n} V_i$ has a unique decomposition $\alpha = \sum\limits_{i=1}^{n} \alpha_i$ for $\alpha_i \in V_i$, $1 \leqslant i \leqslant n$, then we call the sum $V_1 + V_2 + \cdots + V_n$ a direct sum, denoted by $V_1 \oplus V_2 \oplus \cdots \oplus V_n$ or just $\bigoplus\limits_{i=1}^{n} V_i$. One can verify directly that the following statements are indeed equivalent:

(1.1) $\sum\limits_{i=1}^{n} V_i$ is a direct sum, that is, $\sum\limits_{i=1}^{n} V_i = \bigoplus\limits_{i=1}^{n} V_i$.

(1.2) The decomposition of zero is unique.

(1.3) For $1 \leqslant i \leqslant n$, $V_i \cap \left(\sum\limits_{j \neq i} V_j \right) = 0$.

(1.4) $\dim \left(\sum\limits_{i=1}^{n} V_i \right) = \sum\limits_{i=1}^{n} \dim V_i$.

(2) In the above definitions for a sum and a direct sum, we always assume that $V_1, V_2, \ldots, V_n$ lie in some linear space V, that is, such sums are formed "inside" of a big space V. In fact, the "direct sum" can be formed "outside" too: Take any two linear spaces W_1, W_2 and consider the set

$$W_1 \times W_2 = \{(w_1, w_2) \mid w_1 \in W_1,\ w_2 \in W_2\}.$$

We define two operations:

(2.1) Addition:
$$(W_1 \times W_2) \times (W_1 \times W_2) \longrightarrow W_1 \times W_2,$$
$$((w_1, w_2), (w_1', w_2')) \longmapsto (w_1 + w_1', w_2 + w_2');$$
(2.2) Scalar multiplication: $F \times (W_1 \times W_2) \longrightarrow W_1 \times W_2,$
$$(k, (w_1, w_2)) \longmapsto (kw_1, kw_2).$$

It is easy to see that these two operations satisfy the eight axioms listed in the definition of a linear space (see Exercise 25). So, $W_1 \times W_2$ is a linear space under these two operations. Take a basis $\alpha_1, \alpha_2, \ldots, \alpha_m$ for W_1 and a basis $\beta_1, \beta_2, \ldots, \beta_n$ for W_2. Then the sequence $(\alpha_1, 0), \ldots, (\alpha_m, 0), (0, \beta_1), \ldots, (0, \beta_n)$ is a basis for $W_1 \times W_2$, and thus $\dim W_1 + \dim W_2 = \dim(W_1 \times W_2)$. Compared with the equivalent conditions of direct sums, $W_1 \times W_2$ can be regarded as a kind of direct sum of W_1 and W_2. To avoid confusion, we call $W_1 \times W_2$ the **external direct sum** of W_1 and W_2 and call the previous direct sum an **internal direct sum**.

The direct sums in this book are usually internal direct sums unless stated otherwise. Similarly, one can define the external direct sum $W_1 \times W_2 \times \cdots \times W_n$ for finitely many linear spaces $W_1, W_2, \ldots, W_n$.

(3) By the definition of external direct sums, we have

$$F^n = \overbrace{F \times F \times \cdots \times F}^{n}.$$

Similarly, for a linear space W, we use W^n to denote the external direct sum of n copies of W.

Question. What is the relationship between an external direct sum and an internal direct sum?

Definition 4.4.16. Let V_1 be a subspace of a linear space V. If there exists another subspace V_2 of V such that $V = V_1 \oplus V_2$, then we call V_2 a **complement** of V_1.

Proposition 4.4.17. *Let V_1 be a subspace of an n-dimensional linear space V, then it has a complement.*

Proof. The conclusion is clear if V_1 is a trivial subspace. Without loss of generality, we may assume that the sequence $\alpha_1, \alpha_2, \ldots, \alpha_m$ is a basis for V_1. By Proposition 4.4.4, it can be extended to a basis

$$\alpha_1, \alpha_2, \ldots, \alpha_m, \alpha_{m+1}, \ldots, \alpha_n$$

for V. Let $V_2 = L(\alpha_{m+1}, \ldots, \alpha_n)$, then $V = V_1 \oplus V_2$.

Remark 4.4.18. Although a complement always exists, it is generally not unique. For example, take a line through the origin in $\mathbb{R}^2$, then any different line through the origin is a complement of this line.

4.5 Quotient Spaces

Equivalence relations have been mentioned several times before. As an application of equivalence relations, this section classifies the elements in a linear space V through a fixed subspace of V.

If $\sim$ is an equivalence relation on a set A and $a \in A$, then the set $\overline{a} = \{x \in A \mid x \sim a\}$ is called the **equivalence class** of A containing a. An element in an equivalence class is called a **representative** of this class. The set of all equivalence classes of A by $\sim$ is called the **quotient set** of A by $\sim$, denoted by $A/\sim$, i.e., $A/\sim = \{\overline{a} \mid a \in A\}$.

Let W be a subspace of a linear space V. If $v_1, v_2 \in V$, we define $v_1 \sim v_2$ if $v_1 - v_2 \in W$. Then, the relation $\sim$ is an equivalence relation, and $V/\sim = \{\overline{v} \mid v \in V\}$. Sometimes, we also call $\overline{v}$ the **congruence class** of v modulo W. Let $k \in F, v_1, v_2 \in V$. We define the following two operations:

$$\text{Addition} : \overline{v_1} + \overline{v_2} = \overline{v_1 + v_2}; \quad \text{Scalar multiplication} : k\overline{v} = \overline{kv}.$$

Lemma 4.5.1. *The two operations defined above are well defined, i.e., they do not depend on the choice of representatives.*

Proof. If $\overline{v_1} = \overline{v_1'}$, $\overline{v_2} = \overline{v_2'}$, $\overline{v} = \overline{v'}$, we need to show that $\overline{v_1 + v_2} = \overline{v_1' + v_2'}$, $\overline{kv} = \overline{kv'}$. In fact, from $\overline{v_1} = \overline{v_1'}$, $\overline{v_2} = \overline{v_2'}$, and $\overline{v} = \overline{v'}$, we know that $v_1 - v_1' \in W$, $v_2 - v_2' \in W$, and $v - v' \in W$. Since W is a subspace of V, we have $(v_1 + v_2) - (v_1' + v_2') \in W$ and $kv - kv' = k(v - v') \in W$. Therefore, $\overline{v_1 + v_2} = \overline{v_1' + v_2'}$, $\overline{kv} = \overline{kv'}$.

Note that the addition and multiplication on $V/\sim$ are determined by the addition and multiplication on V, so it is not hard to show that they satisfy the eight axioms listed in the definition of a linear space, and thus $V/\sim$ is a linear space. We denote this space by V/W.

Definition 4.5.2. Let W be a subspace of a linear space V. The linear space V/W defined above is called the **quotient space** of V modulo W or just a quotient space.

The quotient space has the following simple properties.

Proposition 4.5.3. *Let W be a subspace of a linear space V. Then the following statements hold:*

(1) *For $v \in V$, $\overline{v} = \overline{0}$ if and only if $v \in W$.*
(2) *Take a complement W_1 of W in V and assume that the sequence $\alpha_1, \alpha_2, \ldots, \alpha_t$ is a basis for W_1. Then the sequence $\overline{\alpha_1}, \overline{\alpha_2}, \ldots, \overline{\alpha_t}$ is a basis for V/W.*
(3) $\dim V/W = \dim V - \dim W$.

Proof. (1) By definition, $\overline{v} = \overline{0}$ if and only if $v = v - 0 \in W$.

(2) Let the sequence $\beta_1, \beta_2, \ldots, \beta_s$ be a basis for W. Note that $V = W \oplus W_1$, and so the sequence $\beta_1, \beta_2, \ldots, \beta_s, \alpha_1, \alpha_2, \ldots, \alpha_t$ is a basis for V. Therefore, for any $v \in V$, there are $k_i \in F$, $l_j \in F$, $1 \leqslant i \leqslant s$, $1 \leqslant j \leqslant t$ such that $v = k_1\beta_1 + k_2\beta_2 + \cdots + k_s\beta_s + l_1\alpha_1 + l_2\alpha_2 + \cdots + l_t\alpha_t$. Thus,

$$
\begin{aligned}
\overline{v} &= \overline{k_1\beta_1 + k_2\beta_2 + \cdots + k_s\beta_s + l_1\alpha_1 + l_2\alpha_2 + \cdots + l_t\alpha_t} \\
&= k_1\overline{\beta_1} + k_2\overline{\beta_2} + \cdots + k_s\overline{\beta_s} + l_1\overline{\alpha_1} + l_2\overline{\alpha_2} + \cdots + l_t\overline{\alpha_t} \\
&= k_1\overline{0} + k_2\overline{0} + \cdots + k_s\overline{0} + l_1\overline{\alpha_1} + l_2\overline{\alpha_2} + \cdots + l_t\overline{\alpha_t} \\
&= l_1\overline{\alpha_1} + l_2\overline{\alpha_2} + \cdots + l_t\overline{\alpha_t}.
\end{aligned}
$$

Here, we use (1) in the third equality. Therefore, all elements in V/W can be linearly expressed by $\overline{\alpha_1}, \overline{\alpha_2}, \ldots, \overline{\alpha_t}$. So, to get the result, it is enough to show that $\overline{\alpha_1}, \overline{\alpha_2}, \ldots, \overline{\alpha_t}$ are linearly independent. To this end, let

$$
c_1\overline{\alpha_1} + c_2\overline{\alpha_2} + \cdots + c_t\overline{\alpha_t} = 0
$$

for $c_i \in F$, $1 \leqslant i \leqslant t$. By (1), $c_1\alpha_1 + c_2\alpha_2 + \cdots + c_t\alpha_t \in W$ and thus there exist $b_j \in F$, $1 \leqslant j \leqslant s$ such that $c_1\alpha_1 + c_2\alpha_2 + \cdots + c_t\alpha_t = b_1\beta_1 + b_2\beta_2 + \cdots + b_s\beta_s$. Since the sequence $\beta_1, \beta_2, \ldots, \beta_s, \alpha_1, \alpha_2, \ldots, \alpha_t$ is a basis for V, $c_1 = c_2 = \cdots = c_t = 0$. So, $\overline{\alpha_1}, \overline{\alpha_2}, \ldots, \overline{\alpha_t}$ are linearly independent.

(3) follows from (2).

Exercises

1. Verify whether each of the following sets becomes a linear space over the real number field for the indicated operations:

   (1) Under the addition of real polynomials and the scalar multiplication of real numbers with polynomials, the set consisting of the zero polynomial and the real polynomials of degree $\geqslant n$ for $n \in \mathbb{N}$.

   (2) Under the usual addition of matrices and the scalar multiplication of a matrix by a number, the set of all upper triangular square matrices of order n.

(3) For the set $\mathbb{R}^2 = \{(a,b) \mid a,b \in \mathbb{R}\}$, define the following operations:

$$(a_1, b_1) \oplus (a_2, b_2) = (a_1 + a_2, b_1 + b_2 + a_1 a_2),$$

$$k \circ (a_1, b_1) = \left(ka_1, kb_1 + \frac{k(k-1)}{2} a_1^2 \right),$$

where $a_i, b_i, k \in \mathbb{R}$, $i = 1, 2$.

(4) For the set $\mathbb{R}^2$ in (2) and any two fixed real numbers a, b, define the following operations:

$$(a_1, b_1) \oplus (a_2, b_2) = (aa_1 + aa_2, bb_1 + bb_2),$$

$$k \circ (a_1, b_1) = (ka_1, kb_1),$$

where $a_i, b_i, k \in \mathbb{R}$, $i = 1, 2$.

(5) For the set $\mathbb{R}^2$ in (2), define the following operators:

$$(a_1, b_1) \oplus (a_2, b_2) = (a_1 + a_2, b_1 + b_2),$$

$$k \circ (a_1, b_1) = (0, 0),$$

where $a_i, b_i, k \in \mathbb{R}$, $i = 1, 2$.

(6) For the set $\mathbb{R}^2$ in (2), define the following operators:

$$(a_1, b_1) \oplus (a_2, b_2) = (a_1 + a_2, b_1 + b_2),$$

$$k \circ (a_1, b_1) = (a_1, b_1),$$

where $a_i, b_i, k \in \mathbb{R}$, $i = 1, 2$.

(7) The set $G = \{(a,b) \mid a,b \in \mathbb{R}, a^2 = b\}$ together with the usual addition and scalar multiplication for vectors in $\mathbb{R}^2$.

2.* For the eight axioms in the definition of a linear space, we keep axioms $(2) - (8)$ first and then replace the third one by

$$\alpha + 0 = \alpha$$

or replace the fourth one by

$$\alpha' + \alpha = 0.$$

Prove that now we can get axiom (1), that is, we have the commutative law of addition.

3. Let V be a linear space. Prove that $k(\alpha - \beta) = k\alpha - k\beta$ for $\alpha, \beta \in V, k \in F$.

4. In the space of real functions, prove that $\cos^2 x, \sin^2 x, \cos 2x$ are linearly dependent.

5. In the space of real functions, prove that $\cos^2 x, \sin^2 x, x$ are linearly independent.

6. Let $f_1(x), f_2(x), f_3(x)$ be three coprime polynomials in the linear space $F[x]$. If any two of $f_1(x), f_2(x), f_3(x)$ are not coprime, prove that they are linearly independent.

7. In the linear space F^4, find the coordinate of α with respect to the basis $\varepsilon_1, \varepsilon_2, \varepsilon_3, \varepsilon_4$:

(1) $\varepsilon_1 = (1, 1, 0, 0), \quad \varepsilon_2 = (0, 1, 1, 0), \quad \varepsilon_3 = (0, 0, 1, 1)$
$\varepsilon_4 = (0, 0, 0, 1), \quad \alpha = (1, 2, 4, 3)$;

(2) $\varepsilon_1 = (1, 1, 0, 1), \quad \varepsilon_2 = (2, 1, 3, 1), \quad \varepsilon_3 = (1, 1, 1, 1)$
$\varepsilon_4 = (0, 1, -1, -1), \quad \alpha = (1, 2, 0, 1)$.

8. Find a basis for each of the following linear spaces and determine the dimensions of these linear spaces:

(1) $M_n(F)$.

(2) All symmetric (skew-symmetric, upper triangular) square matrices of order n over F.

(3) Over the field of real numbers, the space formed by all real polynomials in the matrix J, where $J = \begin{pmatrix} 1 & 1 & 0 & 0 & 0 \\ 0 & 1 & 1 & 0 & 0 \\ 0 & 0 & 1 & 1 & 0 \\ 0 & 0 & 0 & 1 & 1 \\ 0 & 0 & 0 & 0 & 1 \end{pmatrix}$.

9. In the linear space F^4, find the transition matrix from the basis $\varepsilon_1, \varepsilon_2, \varepsilon_3, \varepsilon_4$ to the basis $\eta_1, \eta_2, \eta_3, \eta_4$ and determine the coordinates of ξ with respect to these two bases:

(1) $\begin{cases} \varepsilon_1 = (1, 0, 0, 0), \\ \varepsilon_2 = (0, 1, 0, 0), \\ \varepsilon_3 = (0, 0, 1, 0), \\ \varepsilon_4 = (0, 0, 0, 1), \end{cases} \begin{cases} \eta_1 = (2, 1, -1, 1), \\ \eta_2 = (0, 3, 1, 0), \\ \eta_3 = (5, 3, 2, 1), \\ \eta_4 = (6, 6, 1, 3), \end{cases} \xi = (1, 2, 3, 4)$;

(2) $\begin{cases} \varepsilon_1 = (1, 2, -1, -1), \\ \varepsilon_2 = (1, -1, 1, 1), \\ \varepsilon_3 = (0, 2, 1, 1), \\ \varepsilon_4 = (-1, -1, 0, 1), \end{cases} \begin{cases} \eta_1 = (2, 1, 0, 1), \\ \eta_2 = (0, 1, 2, 2), \\ \eta_3 = (-2, 1, 1, 2), \\ \eta_4 = (1, 1, 1, 2), \end{cases} \xi = (4, 1, 3, 3)$;

(3) $\begin{cases} \varepsilon_1 = (1,1,1,1), \\ \varepsilon_2 = (1,1,-1,-1), \\ \varepsilon_3 = (1,-1,1,-1), \\ \varepsilon_4 = (1,-1,-1,1), \end{cases} \quad \begin{cases} \eta_1 = (1,1,0,1), \\ \eta_2 = (2,1,3,1), \\ \eta_3 = (1,1,0,0), \\ \eta_4 = (0,1,-1,-1), \end{cases}$
$\xi = (1,0,0,2).$

10. Find a nonzero vector ξ such that it has the same coordinate with respect to the bases $\varepsilon_1, \varepsilon_2, \varepsilon_3, \varepsilon_4$ and $\eta_1, \eta_2, \eta_3, \eta_4$ in (1) of the exercise above.

11. Let V_1 and V_2 be subspaces of V with $V_1 \subseteq V_2$. If $\dim V_1 = \dim V_2$, prove that $V_1 = V_2$.

12. Let $\alpha_1, \alpha_2, \ldots, \alpha_n$ be a basis for a linear space V and $\beta_1, \beta_2, \ldots, \beta_m \in V$. If $(\beta_1, \beta_2, \ldots, \beta_m) = (\alpha_1, \alpha_2, \ldots, \alpha_n)A$, where $A \in M_{n \times m}(F)$, prove that $\mathrm{rank}(\beta_1, \beta_2, \ldots, \beta_m) = \mathrm{rank} A$.

13. Assume that $c_1\alpha + c_2\beta + c_3\gamma = 0$ and $c_1c_3 \neq 0$. Prove that $L(\alpha, \beta) = L(\beta, \gamma)$.

14.* Let V be an n-dimensional linear space and $m \geqslant n$. Prove that there are m vectors $\alpha_1, \alpha_2, \ldots, \alpha_m \in V$ such that any n vectors in $\alpha_1, \alpha_2, \ldots, \alpha_m$ are linearly independent.

15. Let $A \in M_n(F)$.

(1) If $C(A) = \{B \in M_n(F) | AB = BA\}$, prove that $C(A)$ is a linear subspace of $M_n(F)$.

(2) If $A = \begin{pmatrix} 1 & 1 & 0 & 0 & 0 \\ 0 & 1 & 1 & 0 & 0 \\ 0 & 0 & 1 & 1 & 0 \\ 0 & 0 & 0 & 1 & 1 \\ 0 & 0 & 0 & 0 & 1 \end{pmatrix}$, find a basis for $C(A)$ and determine the dimension of $C(A)$.

(3) If $A = \begin{pmatrix} 1 & 0 & 0 \\ 0 & 1 & 0 \\ 1 & 1 & 1 \end{pmatrix}$, find a basis for $C(A)$ and determine the dimension of $C(A)$.

16. In the space F^4, find a basis for the subspace spanned by α_i ($i = 1, 2, 3, 4$), and determine the dimension of the subspace:

(1) $\begin{cases} \alpha_1 = (2,1,3,0), \\ \alpha_2 = (1,2,0,0), \\ \alpha_3 = (-1,0,-1,1), \\ \alpha_4 = (4,-1,0,1); \end{cases} \quad$ (2) $\begin{cases} \alpha_1 = (2,1,3,2), \\ \alpha_2 = (1,-2,1,0), \\ \alpha_3 = (-1,0,0,1), \\ \alpha_4 = (0,-1,0,1). \end{cases}$

17. In the space F^4, find bases for the intersection and the sum of the subspace spanned by α_i and the subspace spanned by β_i, and determine the dimensions of the intersection and the sum:

(1) $\begin{cases} \alpha_1 = (1,2,1,0), \\ \alpha_2 = (-1,1,1,1), \end{cases}$ $\qquad$ $\begin{cases} \beta_1 = (2,1,-1,1), \\ \beta_2 = (0,3,1,0); \end{cases}$

(2) $\begin{cases} \alpha_1 = (1,1,0,0), \\ \alpha_2 = (1,0,1,1), \end{cases}$ $\qquad$ $\begin{cases} \beta_1 = (0,0,-1,1), \\ \beta_2 = (0,1,1,0); \end{cases}$

(3) $\begin{cases} \alpha_1 = (1,2,-1,-2), \\ \alpha_2 = (3,1,1,1), \\ \alpha_3 = (-1,0,-1,1), \end{cases}$ $\qquad$ $\begin{cases} \beta_1 = (2,4,-2,4), \\ \beta_2 = (-1,0,4,3). \end{cases}$

18. Let $A \in \mathrm{M}_n(F)$. Prove that there exists a polynomial $f(x) \in F[x]$ of degree $\leqslant n^2$ such that $f(A) = 0$.

19. Let $\alpha_1, \alpha_2, \ldots, \alpha_n$ be a basis for a linear space V. If $\beta \in V$ can be linearly expressed by any $n-1$ vectors in the basis $\alpha_1, \alpha_2, \ldots, \alpha_n$, prove that $\beta = 0$.

20. Consider the linear space F^n.

   (1) Prove that there exists a subspace W of F^n such that every component of any nonzero vector in W is not 0.

   (2) Let W be a nonzero subspace of F^n such that every component of any nonzero vector in W is not 0. Prove that $\dim W = 1$.

21. Let V_1 and V_2 be solution spaces of the homogeneous linear systems $\sum\limits_{i=1}^{n} x_i = 0$ and $x_1 = x_2 = \cdots = x_n$, respectively. Prove that $F^n = V_1 \oplus V_2$.

22. Prove that if $V = V_1 \oplus V_2$, $V_2 = V_{21} \oplus V_{22}$, then

$$V = V_1 \oplus V_{21} \oplus V_{22}.$$

23. Prove that any n-dimensional space is a direct sum of n subspaces of dimension 1.

24. Let V_i be subspaces of V for $i = 1, 2, \ldots, m$. Prove that $\sum\limits_{i=1}^{m} V_i = \bigoplus\limits_{i=1}^{m} V_i$ if and only if for any $2 \leqslant i \leqslant m$, we have $V_i \cap \sum\limits_{j=1}^{i-1} V_j = 0$.

25. Prove that the addition and scalar multiplication given in Remark 4.4.15 (2) satisfy the eight axioms listed in the definition of a linear space, and show that $\dim W_1 + \dim W_2 = \dim(W_1 \times W_2)$.

26. Let V_1 and V_2 be subspaces of a linear space V and $\alpha, \beta \in V$. Prove that if $\alpha + V_1 = \beta + V_2$, then $V_1 = V_2$, where $\alpha + V_1 = \{\alpha + \alpha_1 \mid \alpha_1 \in V_1\}$ and $\beta + V_2 = \{\beta + \alpha_2 \mid \alpha_2 \in V_2\}$.

27. (Comparing with Exercise 43 of Chapter 1) Let $a_1, a_2, \ldots, a_n$ be n different numbers in F and set

$$f_i(x) = (x - a_1) \cdots (x - a_{i-1})(x - a_{i+1}) \cdots (x - a_n),$$

$$i = 1, 2, \ldots, n.$$

  (1) Prove that the sequence $f_1(x), f_2(x), \ldots, f_n(x)$ is a basis for $F[x]_n$.

  (2) If $F = \mathbb{C}$ and let $a_1, a_2, \ldots, a_n$ be all nth roots of unity. Find the transition matrix from the basis $1, x, \ldots, x^{n-1}$ to the basis $f_1(x), f_2(x), \ldots, f_n(x)$.

28. Let $V_1, V_2, \ldots, V_s$ be s nontrivial subspaces of a linear space V for $s \geqslant 2$. Prove that $V \neq V_1 \cup V_2 \cup \cdots \cup V_s$.

29. (1) Let $V = M_n(F)$ and V_1 be the subspace of V consisting of all diagonal matrices. Find a basis for the quotient space V/V_1.

  (2) Let $V = M_3(F)$ and $A = \begin{pmatrix} 1 & 0 & 0 \\ 0 & 1 & 0 \\ 0 & 1 & 1 \end{pmatrix}$. Find a basis for $V/C(A)$. See Exercise 15 for the definition of $C(A)$.

30.* Assume that α is a particular solution of the nonhomogeneous linear system $AX = \beta$. Prove that the solution set of $AX = \beta$ is the congruence class of α modulo W_A, where W_A is the solution space of $AX = 0$.

31.* Let W be a nontrivial subspace of F^n. For any $\alpha \in F^n$, prove that there exist $A \in M_{m \times n}(F)$ and $\beta \in F^m$ such that $\overline{\alpha} \in F^n/W$ is the solution set of $AX = \beta$.

Chapter 5

Linear Maps

5.1　Introduction

Linear maps originated from analytic and projective geometry in the 17th and 18th centuries. The connection between linear transformations and matrices has been implied in Gauss's famous book "distributions Arithmeticae" published in 1801. In 1844, Grassmann[1] published the book *Die Lineale Ausdehnungslehre: Ein neuer Zweig der Mathematik*, which initially started the research on linear transformations of finite-dimensional linear spaces. Later, Ferdinand Gotthold Max Eisenstein and Hermite[2] began to study linear transformations as an independent mathematical object in their works. The term "linear algebra" first appeared in the classic textbook "Modern Algebra" published by Van der Walden[3] in 1930. The definition and form of the current general linear maps (linear transformations) also appeared in this book.

In this chapter, we study linear maps — the relationships between linear spaces (a linear map from a linear space to itself is called a linear transformation). If we choose a basis for each finite-dimensional linear space, then each linear map can be represented by a matrix. The matrices of the same linear map (linear transformation) with respect to different bases are exactly congruent (similar), so finding

---

[1]Hermann Günther Grassmann, 1809–1877, German mathematician.

[2]Charles Hermite, 1822–1901, French mathematician.

[3]Bartel Leendert van der Waerden, 1903–1996, Netherlands mathematician.

the simplest matrix representation of a linear map (linear transformation) is reduced to finding the simplest representative element in the congruent (similar) class of the matrices. For this reason, we first introduce the concepts of linear maps and their matrix representations. We then consider the decomposition of linear spaces and the quasi-diagonalization of square matrices, introduce eigenvalues and eigenvectors, and discuss the conditions for the diagonalization of matrices. Finally, we prove that any complex matrix is similar to a Jordan[4] matrix.

5.2   Linear Maps and Isomorphisms

From now on, unless otherwise specified, we shall speak of linear spaces over a number field F.

Definition 5.2.1. Let V and W be linear spaces.

(1) A map $\sigma : V \longrightarrow W$ is called an F **linear map**, simply a linear map, if

$$\sigma(k\alpha + l\beta) = k\sigma(\alpha) + l\sigma(\beta), \text{ for all } \alpha, \beta \in V, k, l \in F. \quad (5.1)$$

(2) Let $\sigma : V \longrightarrow W$ be a linear map.

   (2.1) σ is said to be an **injective linear map** if σ is injective as a map.
   (2.2) σ is said to be a **surjective linear map** if σ is surjective as a map.
   (2.3) σ is said to be an **isomorphism** (or **invertible linear map**) if σ is bijective as a map.

(3) V and W are said to be **isomorphic**, denoted by $V \cong W$, if there is an isomorphism $\sigma : V \longrightarrow W$.

(4) A linear map from V to itself is called a **linear transformation** (or an **endomorphism**) of V. An isomorphism from V to itself is called an **automorphism** of V.

(5) A linear map $f : V \longrightarrow F$ is called a **linear function**.

---

[4]Camille Jordan, 1838–1922, French mathematician.

Remark 5.2.2.

(1) The condition (5.1) in Definition 5.2.1 (1) is equivalent to the following conditions:

$$\sigma(\alpha + \beta) = \sigma(\alpha) + \sigma(\beta),$$
$$\sigma(k\alpha) = k\sigma(\alpha), \quad \text{for all } \alpha, \beta \in V, \quad k \in F.$$

(2) In the literature, a linear map is also called a **linear operator** or a **homomorphism**.
(3) We introduce the following common notations for convenience. The set of all F linear maps from V to W is denoted by $\mathrm{Hom}_F(V, W)$, or $\mathrm{Hom}(V, W)$ for short. In particular,

   (3.1) $\mathrm{Hom}_F(V, V)$ is usually denoted by $\mathrm{End}_F(V)$, or $\mathrm{End}(V)$ for short,
   (3.2) the set of all automorphisms of V is denoted by $\mathrm{Aut}_F(V)$, or $\mathrm{Aut}(V)$ for short,
   (3.3) $\mathrm{Hom}_F(V, F)$ is usually denoted by V^*, which is called the **dual space** of V.[5]

Example 5.2.3. Let V and W be linear spaces.

(1) The **zero map** $0 : V \longrightarrow W$, $\alpha \longmapsto 0_W$, is linear.
(2) The **identity map** $1_V : V \longrightarrow V$, $\alpha \longmapsto \alpha$, is an automorphism, which is also called the **identity transformation** of V.
(3) Let $k \in F$. Then the **scalar multiplication** $\sigma_k : V \longrightarrow V, \alpha \longmapsto k\alpha$, is a linear transformation.
(4) Let W be a subspace of V. Then the **quotient map** $\pi : V \longrightarrow V/W$, $\alpha \longmapsto \overline{\alpha}$, is a linear map.

Example 5.2.4.

(1) The **trace** $\mathrm{Tr} : M_n(F) \longrightarrow F, \quad A = (a_{ij})_{n \times n} \longmapsto \mathrm{Tr}(A) = \sum_{i=1}^{n} a_{ii}$, is a linear function.
(2) The derivation $D : F[x] \longrightarrow F[x]$, $f(x) \longmapsto f'(x)$, is a linear transformation.

---

[5]Refer to Theorem 5.3.2, $\mathrm{Hom}(V, W)$, $\mathrm{End}(V)$, and V^* are all linear spaces.

(3) Let $A \in M_{m \times n}(F)$, and let s be a positive integer. Then the matrix A induces a linear map

$$\sigma_A : M_{n \times s}(F) \longrightarrow M_{m \times s}(F), \ B \longmapsto AB.$$

In particular, the induced map $\sigma_A : F^n \longrightarrow F^m$, $X \longmapsto AX$, is linear.

(4) Let $F = \mathbb{R}$. Then the integral $I : C[a,b] \longrightarrow \mathbb{R}$, $f \longmapsto I(f) = \int_a^b f(x)dx$, is a linear function.

Example 5.2.5.

(1) Let $V = \mathbb{R}^2$, the real linear space consisting of all vectors in the plane. Let $\varphi \in \mathbb{R}$. Rotating the plane $\mathbb{R}^2$ around the origin through the angle φ in counterclockwise direction is called a **rotation** of V, denoted by R_φ, as shown in Figure 5.1. It's easy to see that R_φ is an automorphism of V. Let (x_α, y_α) be the coordinate of a vector $\alpha \in V$ in the Cartesian coordinate system and $R_\varphi(\alpha) = \beta$. From $|\alpha| = |\beta|$, we have

$$x_\beta = |\beta| \cos(\omega + \varphi) = |\alpha| \cos \omega \cos \varphi - |\alpha| \sin \omega \sin \varphi$$

$$= x_\alpha \cos \varphi - y_\alpha \sin \varphi,$$

$$y_\beta = |\beta| \sin(\omega + \varphi) = |\alpha| \cos \omega \sin \varphi + |\alpha| \sin \omega \cos \varphi$$

$$= x_\alpha \sin \varphi + y_\alpha \cos \varphi.$$

Hence $\begin{pmatrix} x_\beta \\ y_\beta \end{pmatrix} = \begin{pmatrix} \cos \varphi & -\sin \varphi \\ \sin \varphi & \cos \varphi \end{pmatrix} \begin{pmatrix} x_\alpha \\ y_\alpha \end{pmatrix}.$

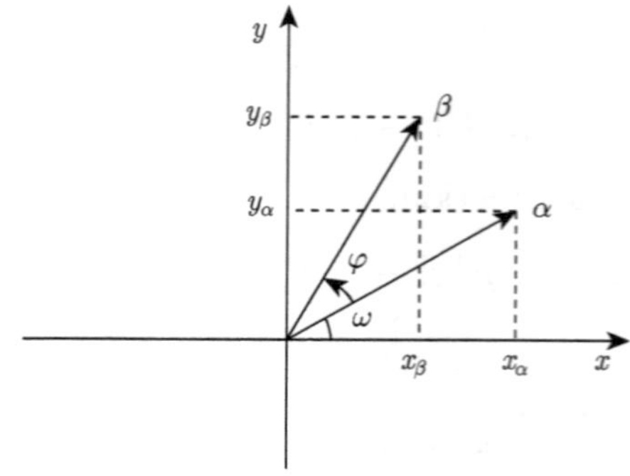

Figure 5.1.　Rotation R_φ.

(2) Let $W = \mathbb{R}^3$ be the usual geometry space. Given a unit vector $\alpha \in W$ and $\varphi \in \mathbb{R}$. Rotating the space W around the directed line where α is located through the angle φ by the right-hand rule is an automorphism of W, which is called a **rotation**, denoted by $R_{\alpha,\varphi}$. By Rodrigues'[6] rotation formula, for any $\eta \in W$,

$$R_{\alpha,\varphi}(\eta) = (\cos\varphi)\eta + (\alpha \cdot \eta)(1 - \cos\varphi)\alpha + (\sin\varphi)(\alpha \times \eta),$$
$$(5.2)$$

where $\alpha \cdot \eta$ and $\alpha \times \eta$ are respectively the **inner product** and **exterior product** of α and η. For the proof of (5.2), we refer the reader to [36] or Exercise 16 of Chapter 8. We explain the rotation in detail in Section 8.4.

(3) Let $W = \mathbb{R}^3$ be the usual geometry space and α a unit vector in W. For any $\eta \in W$, let $P_\alpha(\eta) = (\eta \cdot \alpha)\alpha$, the projection of η on the line where α is located. Then $P_\alpha : W \longrightarrow W$, $\eta \longmapsto P_\alpha(\eta)$, is a linear transformation of W.

Example 5.2.6. Let V_1 and V_2 be two subspaces of a finite-dimensional linear space V with $V = V_1 \oplus V_2$. For any $\alpha \in V$, there exist unique vectors $\alpha_i \in V_i$ $(i = 1, 2)$ such that $\alpha = \alpha_1 + \alpha_2$. Define $\pi_i(\alpha) = \alpha_i$, then $\pi_i : V \longrightarrow V_i$ is a linear map for each $i \in \{1, 2\}$.

It's easy to see that linear maps have the following properties.

Proposition 5.2.7. *Let* V, W, U *be linear spaces and* $\sigma \in \mathrm{Hom}(V, W)$, $\alpha_i \in V$, $k_i \in F$, $i = 1, 2, \ldots, n$.

(1) $\sigma(0_V) = 0_W$, *without causing confusion, simply denoted by* $\sigma(0) = 0$.

(2) $\sigma\left(\sum\limits_{i=1}^{n} k_i\alpha_i\right) = \sum\limits_{i=1}^{n} k_i\sigma(\alpha_i)$.

(3) *If* $\alpha_1, \alpha_2, \ldots, \alpha_n$ *are linearly dependent, then* $\sigma(\alpha_1)$, $\sigma(\alpha_2), \ldots,$ $\sigma(\alpha_n)$ *are linearly dependent.*

(4) *If* $\sigma(\alpha_1), \sigma(\alpha_2), \ldots, \sigma(\alpha_n)$ *are linearly independent, then* α_1, $\alpha_2, \ldots, \alpha_n$ *are linearly independent.*

(5) *If* σ *is an isomorphism, then* $\alpha_1, \alpha_2, \ldots, \alpha_n$ *are linearly dependent if and only if* $\sigma(\alpha_1)$, $\sigma(\alpha_2), \ldots,$ $\sigma(\alpha_n)$ *are linearly dependent.*

(6) *If* $\tau \in \mathrm{Hom}(W, U)$, *then* $\tau\sigma \in \mathrm{Hom}(V, U)$, *where* $\tau\sigma = \tau \circ \sigma$ *is the composition of* σ *and* τ.

---

[6]Benjamin Olinde Rodrigues, 1795–1851, French mathematician.

Theorem 5.2.8. *Let V and W be linear spaces. Assume that* $\dim V = n$ *and the sequence* $\alpha_1, \alpha_2, \ldots, \alpha_n$ *is a basis for* V.

(1) *Let* $\sigma, \tau \in \operatorname{Hom}(V, W)$. *If* $\sigma(\alpha_i) = \tau(\alpha_i)$, $i = 1, 2, \ldots, n$, *then* $\sigma = \tau$.

(2) *Let* $\beta_1, \beta_2, \ldots, \beta_n \in W$. *Then there exists a unique* $\sigma \in \operatorname{Hom}(V, W)$ *such that* $\sigma(\alpha_i) = \beta_i$, $i = 1, 2, \ldots, n$.

Proof. (1) For any $\alpha \in V$, we have $\alpha = \sum_{i=1}^{n} k_i \alpha_i$ for some $k_1, k_2, \ldots, k_n \in F$. Hence,

$$\sigma(\alpha) = \sigma\left(\sum_{i=1}^{n} k_i \alpha_i\right) = \sum_{i=1}^{n} k_i \sigma(\alpha_i) = \sum_{i=1}^{n} k_i \tau(\alpha_i)$$

$$= \tau\left(\sum_{i=1}^{n} k_i \alpha_i\right) = \tau(\alpha).$$

Therefore, $\sigma = \tau$.

(2) It is sufficient to show the existence of such σ, since the uniqueness is guaranteed by (1).

For any $\alpha \in V$, there exists a unique representation $\alpha = \sum_{i=1}^{n} k_i \alpha_i$, where $k_1, k_2, \ldots, k_n \in F$. Define $\sigma : V \longrightarrow W$ such that $\sigma(\alpha) = \sum_{i=1}^{n} k_i \beta_i$. It is clear that $\sigma(\alpha_i) = \beta_i, i = 1, 2, \ldots, n$. For any $k, l \in F$, $\gamma \in V$, then $\gamma = \sum_{i=1}^{n} l_i \alpha_i$, for some $l_1, l_2, \ldots, l_n \in F$, and $k\alpha + l\gamma = \sum_{i=1}^{n} (kk_i + ll_i)\alpha_i$. By definition, we have

$$\sigma(k\alpha + l\gamma) = \sum_{i=1}^{n} (kk_i + ll_i)\beta_i = k \sum_{i=1}^{n} k_i \beta_i + l \sum_{i=1}^{n} l_i \beta_i$$

$$= k\sigma(\alpha) + l\sigma(\gamma).$$

Hence, σ is a linear map, i.e., $\sigma \in \operatorname{Hom}(V, W)$. $\qquad\square$

Now, let us discuss the isomorphism of linear spaces.

Theorem 5.2.9. *The isomorphism relation of linear spaces is an equivalence relation.*

Proof. Let V, W, U be linear spaces.

(1) Reflexive. $V \cong V$ because the identity map $1_V : V \longrightarrow V$ is an isomorphism.

(2) Symmetric. Assume $V \cong W$, i.e., there exists an isomorphism $\sigma : V \longrightarrow W$. On the one hand, the inverse map $\sigma^{-1} : W \longrightarrow V$ is bijective. On the other hand, for any $\alpha, \beta \in W$, $k, l \in F$, we have

$$
\begin{aligned}
\sigma(\sigma^{-1}(k\alpha + l\beta)) &= (\sigma\sigma^{-1})(k\alpha + l\beta) \\
&= k\alpha + l\beta \\
&= k(\sigma\sigma^{-1})(\alpha) + l(\sigma\sigma^{-1})(\beta) \\
&= k\sigma(\sigma^{-1}(\alpha)) + l\sigma(\sigma^{-1}(\beta)) \\
&= \sigma(k\sigma^{-1}(\alpha) + l\sigma^{-1}(\beta)).
\end{aligned}
$$

Since σ is injective, $\sigma^{-1}(k\alpha + l\beta) = k\sigma^{-1}(\alpha) + l\sigma^{-1}(\beta)$, i.e., σ^{-1} is a linear map. Therefore, σ^{-1} is an isomorphism, and so $W \cong V$.

(3) Transitive. Assume that $V \cong W$, $W \cong U$, i.e., there exist isomorphisms $\sigma : V \longrightarrow W$, $\tau : W \longrightarrow U$. On the one hand, the composition $\tau\sigma$ is bijective because σ and τ are bijective. On the other hand, $\tau\sigma$ is a linear map by Proposition 5.2.7 (6). Thus, $\tau\sigma$ is an isomorphism, and so $V \cong U$. $\qquad\square$

Theorem 5.2.10. *Let V be a linear space of dimension n. Then $V \cong F^n$.*

Proof. Let $\alpha_1, \alpha_2, \ldots, \alpha_n$ be a basis for V. Define

$$
\sigma : V \longrightarrow F^n, \quad \alpha \longmapsto \sigma(\alpha) = (k_1, k_2, \ldots, k_n)',
$$

where $(k_1, k_2, \ldots, k_n)'$ is the coordinate of α with respect to the basis $\alpha_1, \alpha_2, \ldots, \alpha_n$.

On the one hand, σ is bijective. On the other hand, let $\alpha, \beta \in V$, and let $X, Y \in F^n$ be the coordinates of α and β with respect to the basis $\alpha_1, \alpha_2, \ldots, \alpha_n$, respectively. Then, for any $k, l \in F$, $kX + lY$ is the coordinate of $k\alpha + l\beta$ with respect to the basis $\alpha_1, \alpha_2, \ldots, \alpha_n$ and

$$
\sigma(k\alpha + l\beta) = kX + lY = k\sigma(\alpha) + l\sigma(\beta),
$$

i.e., σ is a linear map. Hence σ is an isomorphism, and so $V \cong F^n$.

$\qquad\square$

Theorem 5.2.11. *Let V and W be two finite-dimensional linear spaces. Then V and W are isomorphic if and only if $\dim V = \dim W$.*

Proof. Sufficiency. Assume $\dim V = \dim W = n$. By Theorem 5.2.10, V and W are isomorphic to F^n. Hence V and W are isomorphic by transitivity.

Necessity. Assume that V and W are isomorphic, i.e., there is an isomorphism $\sigma : V \longrightarrow W$. Suppose $\dim V = n$, $\dim W = m$, and let $\alpha_1, \alpha_2, \ldots, \alpha_n$ be a basis for V. By Proposition 5.2.7 (5), $\sigma(\alpha_1), \sigma(\alpha_1), \ldots, \sigma(\alpha_n)$ are linearly independent in W, hence $n \leqslant \dim W = m$. Similarly, we have $m \leqslant n$. Therefore, $n = m$, i.e., $\dim V = \dim W$. $\qquad\square$

Definition 5.2.12. Let V and W be linear spaces and $\sigma \in \mathrm{Hom}(V, W)$. The **kernel** of σ, denoted by $\mathrm{Ker}(\sigma)$, is defined by

$$\mathrm{Ker}(\sigma) = \{\alpha \in V \mid \sigma(\alpha) = 0\}.$$

$\mathrm{Ker}(\sigma)$ is also known as the **nullspace** of σ.

The **image** of σ, denoted by $\mathrm{Im}(\sigma)$, is defined by

$$\mathrm{Im}(\sigma) = \{\sigma(\alpha) \in W \mid \alpha \in V\}.$$

We often denote $\mathrm{Im}(\sigma)$ by $\sigma(V)$.

Theorem 5.2.13. *Let V and W be linear spaces and $\sigma \in \mathrm{Hom}(V, W)$. Then the following statements hold:*

(1) $\mathrm{Ker}(\sigma)$ is a subspace of V.
(2) $\mathrm{Im}(\sigma)$ is a subspace of W.
(3) $V/\mathrm{Ker}(\sigma) \cong \mathrm{Im}(\sigma)$.

Proof. (1) Let $\alpha, \beta \in \mathrm{Ker}(\sigma)$, i.e., $\sigma(\alpha) = \sigma(\beta) = 0$. For any $k, l \in F$,

$$\sigma(k\alpha + l\beta) = k\sigma(\alpha) + l\sigma(\beta) = 0.$$

Hence, $k\alpha + l\beta \in \mathrm{Ker}(\sigma)$, and so $\mathrm{Ker}(\sigma)$ is a subspace of V.

(2) Let $\beta_1, \beta_2 \in \mathrm{Im}(\sigma)$, i.e., $\beta_i = \sigma(\alpha_i), i = 1, 2$, for some $\alpha_1, \alpha_2 \in V$. For any $k, l \in F$, $k\beta_1 + l\beta_2 = k\sigma(\alpha_1) + l\sigma(\alpha_2) = \sigma(k\alpha_1 + l\alpha_2) \in \mathrm{Im}(\sigma)$. Hence, $\mathrm{Im}(\sigma)$ is a subspace of W.

(3) Define $\bar{\sigma} : V/\mathrm{Ker}(\sigma) \longrightarrow \mathrm{Im}(\sigma)$, $\bar{\alpha} \longmapsto \sigma(\alpha)$, for any $\alpha \in V$.

(a) $\overline{\sigma}$ is well defined. In fact, if $\alpha, \beta \in V$ and $\overline{\alpha} = \overline{\beta}$, then there exists $\gamma \in \text{Ker}(\sigma)$ such that $\alpha = \beta + \gamma$. Hence $\sigma(\alpha) = \sigma(\beta + \gamma) = \sigma(\beta) + \sigma(\gamma) = \sigma(\beta)$, and so $\overline{\sigma}(\overline{\alpha}) = \overline{\sigma}(\overline{\beta})$.

(b) $\overline{\sigma} \in \text{Hom}(V/\text{Ker}(\sigma), \text{Im}(\sigma))$, i.e., $\overline{\sigma}$ is a linear map. In fact, for any $\alpha, \beta \in V$, $k, l \in F$,

$$\overline{\sigma}(k\overline{\alpha} + l\overline{\beta}) = \overline{\sigma}(\overline{k\alpha + l\beta}) = \sigma(k\alpha + l\beta)$$

$$= k\sigma(\alpha) + l\sigma(\beta) = k\overline{\sigma}(\overline{\alpha}) + l\overline{\sigma}(\overline{\beta}).$$

(c) $\overline{\sigma}$ is injective. In fact, if $\alpha, \beta \in V$, and $\overline{\sigma}(\overline{\alpha}) = \overline{\sigma}(\overline{\beta})$, then $\sigma(\alpha) = \sigma(\beta)$, and so $\sigma(\alpha - \beta) = \sigma(\alpha) - \sigma(\beta) = 0$, i.e., $\alpha - \beta \in \text{Ker}(\sigma)$. Hence $\overline{\alpha} = \overline{\beta}$.

(d) $\overline{\sigma}$ is surjective. In fact, for any $\beta \in \text{Im}(\sigma)$, i.e., $\beta = \sigma(\alpha)$ for some $\alpha \in V$, and hence $\overline{\sigma}(\overline{\alpha}) = \sigma(\alpha) = \beta$.

From the above (a), (b), (c), and (d), we obtain that $\overline{\sigma} : V/\text{Ker}(\sigma) \longrightarrow \text{Im}(\sigma)$ is an isomorphism, so $V/\text{Ker}(\sigma) \cong \text{Im}(\sigma)$. $\square$

Theorem 5.2.14. *Let V and W be linear spaces and $\sigma \in \text{Hom}(V, W)$. If $\dim V < \infty$, then $\text{Ker}(\sigma)$ and $\text{Im}(\sigma)$ are finite-dimensional linear spaces and*

$$\dim\text{Ker}(\sigma) + \dim\text{Im}(\sigma) = \dim V. \tag{5.3}$$

Proof. From Theorem 5.2.13, we have $\dim\text{Ker}(\sigma) \leqslant \dim V < \infty$. From Theorem 5.2.11 and Proposition 4.5.3 (3), we obtain that

$$\dim\text{Im}(\sigma) = \dim V/\text{Ker}(\sigma) = \dim V - \dim\text{Ker}(\sigma) \leqslant \dim V < \infty.$$

Hence, $\dim\text{Ker}(\sigma) + \dim\text{Im}(\sigma) = \dim V.$ $\square$

Definition 5.2.15. Let V and W be linear spaces with $\dim V < \infty$, and let $\sigma \in \text{Hom}(V, W)$. The number $\dim\text{Ker}(\sigma)$ is called the **nullity** of σ, denoted by $N(\sigma)$, $\dim\text{Im}(\sigma)$ is called the **rank** of σ, denoted by $\text{rank}(\sigma)$.

Remark 5.2.16. Let V and W be linear spaces and $\dim V = n$.

(1) Let $\alpha_1, \alpha_2, \ldots, \alpha_n$ be a basis for V and $\sigma \in \text{Hom}(V, W)$. Then

$$\text{rank}(\sigma) = \dim L(\sigma(\alpha_1), \sigma(\alpha_2), \ldots, \sigma(\alpha_n))$$

$$= \text{rank}\,(\sigma(\alpha_1), \sigma(\alpha_2), \ldots, \sigma(\alpha_n)).$$

(2) Let $\sigma \in \text{End}(V)$. We cannot obtain that $V = \text{Ker}(\sigma) + \text{Im}(\sigma)$ from (5.3). For example, let $n \geqslant 2$ and

$$D : F[x]_n \longrightarrow F[x]_n, \ f(x) \longmapsto f'(x).$$

Then $\text{Ker}(D) = F \subseteq \text{Im}(D) = F[x]_{n-1}$ and

$$\text{Ker}(D) + \text{Im}(D) = \text{Im}(D) \subsetneq F[x]_n.$$

Exercise 8 gives some necessary and sufficient conditions for the equality $V = \text{Ker}(\sigma) + \text{Im}(\sigma)$ to hold.

Lemma 5.2.17. *Let V and W be linear spaces and $\sigma \in \text{Hom}(V, W)$.*

(1) *σ is surjective if and only if $\text{Im}(\sigma) = W$.*
(2) *σ is injective if and only if $\text{Ker}(\sigma) = \{0\}$.*

Proof. (1) is trivial by definition.
    (2) If σ is injective, then it is trivial that $\text{Ker}(\sigma) = \{0\}$. Conversely, let $\alpha, \beta \in V$ and $\sigma(\alpha) = \sigma(\beta)$. Then

$$\sigma(\alpha - \beta) = \sigma(\alpha) - \sigma(\beta) = 0, i.e., \alpha - \beta \in \text{Ker}(\sigma) = \{0\}.$$

Hence, $\alpha = \beta$, and so σ is injective. $\qquad\qquad\square$

Proposition 5.2.18. *Let V be a finite-dimensional linear space and $\sigma \in \text{End}(V)$. Then σ is injective if and only if σ is surjective if and only if σ is an automorphism.*

Proof. From $\dim V < \infty$, we know that the equality (5.3) holds. Hence

$$\begin{aligned}
\sigma \text{ is injective} &\Longleftrightarrow \text{Ker}(\sigma) = \{0\} \\
&\Longleftrightarrow \dim\text{Ker}(\sigma) = 0 \\
&\Longleftrightarrow \dim V = \dim\text{Im}(\sigma) \\
&\Longleftrightarrow V = \text{Im}(\sigma) \\
&\Longleftrightarrow \sigma \text{ is surjective.}
\end{aligned}$$

The conclusion obviously holds since σ is an automorphism of V if and only if σ is both injective and surjective. $\qquad\qquad\square$

Proposition 5.2.19. *Let W be a subspace of V and let $\pi : V \longrightarrow V/W$ be the quotient map. Then $\text{Ker}(\pi) = W$.*

Proof. By the definition of the kernel,

$$\mathrm{Ker}(\pi) = \{\alpha \in V \mid \pi(\alpha) = \overline{0}\} = \{\alpha \in V \mid \overline{\alpha} = \overline{0}\} = W. \qquad \square$$

Theorem 5.2.20. *Let* $\sigma \in \mathrm{Hom}(V, W)$, $1 \leqslant \dim\mathrm{Ker}(\sigma) = r \leqslant \dim V = n$, *and let* $\alpha_1, \ldots, \alpha_r$ *be a basis for* $\mathrm{Ker}(\sigma)$.

(1) *Extend the basis* $\alpha_1, \ldots, \alpha_r$ *to a basis* $\alpha_1, \ldots, \alpha_r, \alpha_{r+1}, \ldots, \alpha_n$ *for* V. *Then the sequence* $\sigma(\alpha_{r+1}), \ldots, \sigma(\alpha_n)$ *is a basis for* $\mathrm{Im}(\sigma)$.

(2) *Let* $\gamma_1, \ldots, \gamma_s$ *be a basis for* $\mathrm{Im}(\sigma)$, *where* $\gamma_i = \sigma(\beta_i)$ *for some* $\beta_i \in V$, $1 \leqslant i \leqslant s$. *Then the sequence* $\alpha_1, \ldots, \alpha_r, \beta_1, \ldots, \beta_s$ *is a basis for* V.

Proof. (1) Since

$$
\begin{aligned}
\mathrm{Im}(\sigma) &= \{\sigma(\alpha) \mid \alpha \in V\} \\
&= \left\{ \sigma\left(\sum_{i=1}^{n} k_i \alpha_i\right) \;\middle|\; k_i \in F, 1 \leqslant i \leqslant n \right\} \\
&= \left\{ \sum_{i=1}^{n} k_i \sigma(\alpha_i) \;\middle|\; k_i \in F, 1 \leqslant i \leqslant n \right\} \\
&= \left\{ \sum_{i=r+1}^{n} k_i \sigma(\alpha_i) \;\middle|\; k_i \in F, r+1 \leqslant i \leqslant n \right\} \\
&= L(\sigma(\alpha_{r+1}), \ldots, \sigma(\alpha_n)),
\end{aligned}
$$

it suffices to show that $\sigma(\alpha_{r+1}), \ldots, \sigma(\alpha_n)$ are linearly independent. Suppose that $k_{r+1}\sigma(\alpha_{r+1}) + \cdots + k_n\sigma(\alpha_n) = 0$, where $k_{r+1}, \ldots, k_n \in F$. Then $\sigma\left(\sum_{i=r+1}^{n} k_i \alpha_i\right) = 0$, and so $\sum_{i=r+1}^{n} k_i \alpha_i \in \mathrm{Ker}(\sigma)$. Hence there exist $k_1, \ldots, k_r \in F$ such that $\sum_{i=r+1}^{n} k_i \alpha_i = -\sum_{i=1}^{r} k_i \alpha_i$, i.e., $\sum_{i=1}^{n} k_i \alpha_i = 0$. Since $\alpha_1, \ldots, \alpha_n$ are linearly independent, we know that $k_i = 0, i = 1, 2, \ldots, n$. Therefore, $\sigma(\alpha_{r+1}), \ldots, \sigma(\alpha_n)$ are linearly independent.

(2) First, we claim that $\alpha_1, \ldots, \alpha_r, \beta_1, \ldots, \beta_s$ are linearly independent. In fact, if

$$k_1 \alpha_1 + \cdots + k_r \alpha_r + l_1 \beta_1 + \cdots + l_s \beta_s = 0, \tag{5.4}$$

where $k_1, \ldots, k_r, l_1, \ldots, l_s \in F$, then

$$\sigma(k_1\alpha_1 + \cdots + k_r\alpha_r + l_1\beta_1 + \cdots + l_s\beta_s) = \sigma(0) = 0.$$

Therefore, $k_1\sigma(\alpha_1) + \cdots + k_r\sigma(\alpha_r) + l_1\sigma(\beta_1) + \cdots + l_s\sigma(\beta_s) = 0$, and it follows that $l_1\gamma_1 + \cdots + l_s\gamma_s = 0$. Since $\gamma_1, \ldots, \gamma_s$ are linearly independent, we have $l_j = 0$, $j = 1, \ldots, s$. From (5.4), we get $k_1\alpha_1 + \cdots + k_r\alpha_r = 0$. Note that $\alpha_1, \ldots, \alpha_s$ are linearly independent, and so $k_i = 0$, $i = 1, \ldots, r$. Hence, $\alpha_1, \ldots, \alpha_r, \beta_1, \ldots, \beta_s$ are linearly independent.

Second, any vector $\alpha \in V$ can be expressed by $\alpha_1, \ldots, \alpha_r$, $\beta_1, \ldots, \beta_s$. In fact, there exist $k_1, \ldots, k_s \in F$ such that

$$\sigma(\alpha) = \sum_{i=1}^{s} k_i\gamma_i = \sum_{i=1}^{s} k_i\sigma(\beta_i) = \sigma\left(\sum_{i=1}^{s} k_i\beta_i\right).$$

Hence, $\sigma\left(\alpha - \sum_{i=1}^{s} k_i\beta_i\right) = \sigma(\alpha) - \sigma\left(\sum_{i=1}^{s} k_i\beta_i\right) = 0$, i.e., $\alpha - \sum_{i=1}^{s} k_i\beta_i \in \mathrm{Ker}(\sigma)$. It follows that $\alpha - \sum_{i=1}^{s} k_i\beta_i = \sum_{j=1}^{r} l_j\alpha_j$ for some $l_1, \ldots, l_r \in F$, and so $\alpha = \sum_{j=1}^{r} l_j\alpha_j + \sum_{i=1}^{s} k_i\beta_i$. Therefore, the sequence $\alpha_1, \ldots, \alpha_r, \beta_1, \ldots, \beta_s$ is a basis for V. $\qquad\square$

Remark 5.2.21. Theorem 5.2.20 (2) gives another proof of (5.3).

Example 5.2.22. Let V be an n-dimensional linear space and let $\sigma_1, \sigma_2 \in \mathrm{End}(V)$. Show that $\sigma_2(V) \subseteq \sigma_1(V)$ if and only if there exists $\sigma \in \mathrm{End}(V)$ such that $\sigma_2 = \sigma_1\sigma$.

Proof. The sufficiency is trivial. Next, we prove the necessity. Suppose $\sigma_2(V) \subseteq \sigma_1(V)$. Let $\dim\mathrm{Im}(\sigma_2) = r$ and let $\sigma_2(\alpha_1), \ldots, \sigma_2(\alpha_r)$ be a basis for $\mathrm{Im}(\sigma_2)$, where $\alpha_1, \ldots, \alpha_r \in V$. By formula (5.3), $\dim\mathrm{Ker}(\sigma_2) = n - r$. Let $\alpha_{r+1}, \ldots, \alpha_n$ be a basis for $\mathrm{Ker}(\sigma_2)$. Then the sequence $\alpha_1, \ldots, \alpha_r, \alpha_{r+1}, \ldots, \alpha_n$ is basis for V by Theorem 5.2.20 (2).

Since $\sigma_2(V) \subseteq \sigma_1(V)$, there exists $\beta_i \in V$ such that $\sigma_1(\beta_i) = \sigma_2(\alpha_i)$, $i = 1, 2, \ldots, r$. By Theorem 5.2.8 (2), there exists a unique $\sigma \in \mathrm{End}(V)$ such that

$$\sigma(\alpha_i) = \beta_i, i = 1, 2, \ldots, r; \quad \sigma(\alpha_j) = 0, \quad j = r + 1, \ldots, n.$$

Hence, $(\sigma_1\sigma)(\alpha_i) = \sigma_2(\alpha_i), i = 1, 2, \ldots, n$. From Theorem 5.2.8 (1), we obtain that $\sigma_2 = \sigma_1\sigma$. $\square$

5.3 Matrix Representations of Linear Maps

In the previous section, we discussed the properties of linear maps. In this section, we study the matrix representations of linear maps.

Lemma 5.3.1. *Let V and W be two linear spaces. For any $\sigma, \tau \in \mathrm{Hom}(V, W)$, $k \in F$, define*

Addition $\qquad\qquad\qquad \sigma + \tau : V \longrightarrow W, \alpha \longmapsto \sigma(\alpha) + \tau(\alpha);$

Scalar multiplication $\quad k\sigma : V \longrightarrow W, \alpha \longmapsto k\sigma(\alpha).$

Then $\sigma + \tau$, $k\sigma \in \mathrm{Hom}(V, W)$.

Proof. Let $\alpha, \beta \in V$, $a, b, k \in F$. Then

$$
\begin{aligned}
(\sigma + \tau)(a\alpha + b\beta) &= \sigma(a\alpha + b\beta) + \tau(a\alpha + b\beta) \\
&= a\sigma(\alpha) + b\sigma(\beta) + a\tau(\alpha) + b\tau(\beta) \\
&= a(\sigma(\alpha) + \tau(\alpha)) + b(\sigma(\beta) + \tau(\beta)) \\
&= a(\sigma + \tau)(\alpha) + b(\sigma + \tau)(\beta),
\end{aligned}
$$

and

$$
\begin{aligned}
(k\sigma)(a\alpha + b\beta) &= k\sigma(a\alpha + b\beta) \\
&= k(a\sigma(\alpha) + b\sigma(\beta)) \\
&= ka\sigma(\alpha) + kb\sigma(\beta) \\
&= ak\sigma(\alpha) + bk\sigma(\beta) \\
&= a(k\sigma)(\alpha) + b(k\sigma)(\beta).
\end{aligned}
$$

Hence, $\sigma + \tau \in \mathrm{Hom}(V, W)$ and $k\sigma \in \mathrm{Hom}(V, W)$. $\square$

It's easy to prove that the addition and scalar multiplication defined in the above lemma meet the eight axioms listed in the definition of a linear space. Hence we obtain the following.

Theorem 5.3.2. *Let V and W be two linear spaces. Then* $\mathrm{Hom}(V, W)$ *is a linear space. In particular,* $\mathrm{End}(V)$ *and* V^* *are linear spaces.*

Example 5.3.3. Let V be an n-dimensional linear space and $\sigma \in \mathrm{End}(V)$. Let $f(x) = a_m x^m + a_{m-1} x^{m-1} + \cdots + a_1 x + a_0 \in F[x]$. Define

$$f(\sigma) = a_m \sigma^m + a_{m-1} \sigma^{m-1} + \cdots + a_1 \sigma + a_0 1_V.$$

Then $f(\sigma) \in \mathrm{End}(V)$ and we say that $f(\sigma)$ is a **polynomial in** σ. Let $g(x) \in F[x]$ and $h(x) = f(x)g(x)$. Then $f(\sigma)g(\sigma) = h(\sigma) = g(\sigma)f(\sigma)$. Hence, polynomials in σ are mutually commuting.

Remark 5.3.4. For convenience, we introduce the following symbol. Let

$$\sigma(\gamma_1, \gamma_2, \ldots, \gamma_r) = (\sigma(\gamma_1), \sigma(\gamma_2), \ldots, \sigma(\gamma_r)) \in W^r$$

for $\sigma \in \mathrm{Hom}(V, W)$ and $\gamma_1, \gamma_2, \ldots, \gamma_r \in V$.

Let $A = (a_{ij}) \in \mathrm{M}_{n \times r}(F)$, $k \in F$, $\alpha_1, \alpha_2, \ldots, \alpha_n \in V$. Then

$$\sigma\left((\alpha_1, \alpha_2, \ldots, \alpha_n)A\right) = \sigma(\alpha_1, \alpha_2, \ldots, \alpha_n)A, \tag{5.5}$$

$$(k\sigma)\left((\alpha_1, \alpha_2, \ldots, \alpha_n)A\right) = \sigma(\alpha_1, \alpha_2, \ldots, \alpha_n)(kA). \tag{5.6}$$

In fact, set $(\gamma_1, \gamma_2, \ldots, \gamma_r) = (\alpha_1, \alpha_2, \ldots, \alpha_n)A$, then $\gamma_j = \sum_{t=1}^{n} a_{tj}\alpha_t$, $1 \leqslant j \leqslant r$. Hence

$$\sigma(\gamma_j) = \sum_{t=1}^{n} a_{tj}\sigma(\alpha_t)$$

$$= (\sigma(\alpha_1), \sigma(\alpha_2), \ldots, \sigma(\alpha_n)) \begin{pmatrix} a_{1j} \\ \vdots \\ a_{nj} \end{pmatrix}$$

$$= \sigma(\alpha_1, \alpha_2, \ldots, \alpha_n) \begin{pmatrix} a_{1j} \\ \vdots \\ a_{nj} \end{pmatrix}.$$

Therefore,

$$\sigma\left((\alpha_1, \alpha_2, \ldots, \alpha_n)A\right) = \sigma(\gamma_1, \gamma_2, \ldots, \gamma_r)$$
$$= (\sigma(\gamma_1), \sigma(\gamma_2), \ldots, \sigma(\gamma_r))$$
$$= \sigma(\alpha_1, \alpha_2, \ldots, \alpha_n)A,$$

i.e., (5.5) holds. Similarly, (5.6) also holds.

Theorem 5.3.5. *Let* $\alpha_1, \alpha_2, \ldots, \alpha_n$ *be a basis for a linear space* V, *and let* $\beta_1, \beta_2, \ldots, \beta_m$ *be a basis for a linear space* W. *Then the following statements hold:*

(1) *The map* $\psi : \mathrm{Hom}(V, W) \longrightarrow \mathrm{M}_{m \times n}(F)$, $\sigma \longmapsto A$, *is an isomorphism, where* A *is determined by*

$$\sigma(\alpha_1, \alpha_2, \ldots, \alpha_n) = (\beta_1, \beta_2, \ldots, \beta_m)A. \tag{5.7}$$

Hence, $\mathrm{Hom}(V, W) \cong \mathrm{M}_{m \times n}(F)$ *and* $\dim\mathrm{Hom}(V, W) = mn$.

(2) *The map* $\psi : \mathrm{End}(V) \longrightarrow \mathrm{M}_n(F)$, $\sigma \longmapsto A$, *is an isomorphism, where* A *is determined by*

$$\sigma(\alpha_1, \alpha_2, \ldots, \alpha_n) = (\alpha_1, \alpha_2, \ldots, \alpha_n)A. \tag{5.8}$$

Hence $\mathrm{End}(V) \cong \mathrm{M}_n(F)$ *and* $\dim\mathrm{End}(V) = n^2$.

(3) $V^* \cong \mathrm{M}_{1 \times n}(F)$, $\dim V^* = \dim V$, $V^* \cong V$.

Proof. It's sufficient to prove (1) since that (2) and (3) are special cases of (1).

(i) ψ is injective. Let $\sigma, \tau \in \mathrm{Hom}(V, W)$ and $\psi(\sigma) = \psi(\tau)$. Then for any $j = 1, 2, \ldots, n$, $\sigma(\alpha_j)$ and $\tau(\alpha_j)$ have the same coordinates with respect to the basis $\beta_1, \beta_2, \ldots, \beta_m$, and so $\sigma(\alpha_j) = \tau(\alpha_j)$. Hence $\sigma = \tau$ by Theorem 5.2.8 (1).

(ii) ψ is surjective. In fact, for any $A \in \mathrm{M}_{m \times n}(F)$, put

$$(\gamma_1, \gamma_2, \ldots, \gamma_n) = (\beta_1, \beta_2, \ldots, \beta_m)A.$$

From Theorem 5.2.8 (2), there exists $\sigma \in \mathrm{Hom}(V, W)$ such that $\sigma(\alpha_i) = \gamma_i$, $i = 1, 2, \ldots, n$. Hence $\psi(\sigma) = A$.

(iii) ψ is a linear map. Let $\sigma, \tau \in \mathrm{Hom}(V, W)$ and $\psi(\sigma) = A$, $\psi(\tau) = B$, i.e.,

$$\sigma(\alpha_1, \alpha_2, \ldots, \alpha_n) = (\beta_1, \beta_2, \ldots, \beta_m)A,$$
$$\tau(\alpha_1, \alpha_2, \ldots, \alpha_n) = (\beta_1, \beta_2, \ldots, \beta_m)B.$$

For any $k, l \in F$, we have

$$(k\sigma + l\tau)(\alpha_1, \alpha_2, \ldots, \alpha_n)$$
$$= ((k\sigma + l\tau)(\alpha_1), (k\sigma + l\tau)(\alpha_2), \ldots, (k\sigma + l\tau)(\alpha_n))$$
$$= (k\sigma(\alpha_1) + l\tau(\alpha_1), k\sigma(\alpha_2) + l\tau(\alpha_2), \ldots, k\sigma(\alpha_n) + l\tau(\alpha_n))$$
$$= (k\sigma(\alpha_1), k\sigma(\alpha_2), \ldots, k\sigma(\alpha_n)) + (l\tau(\alpha_1), l\tau(\alpha_2), \ldots, l\tau(\alpha_n))$$
$$= (k\sigma)(\alpha_1, \alpha_2, \ldots, \alpha_n) + (l\tau)(\alpha_1, \alpha_2, \ldots, \alpha_n)$$
$$= (\beta_1, \beta_2, \ldots, \beta_m)(kA) + (\beta_1, \beta_2, \ldots, \beta_m)(lB)$$
$$= (\beta_1, \beta_2, \ldots, \beta_m)(kA + lB).$$

Hence, $\psi(k\sigma + l\tau) = kA + lB = k\psi(\sigma) + l\psi(\tau)$.

By (i), (ii), and (iii), ψ is an isomorphism. $\qquad\qquad\square$

Definition 5.3.6. The matrix A in (5.7) is called **the matrix of the linear map** σ with respect to the bases $\alpha_1, \alpha_2, \ldots, \alpha_n$ and $\beta_1, \beta_2, \ldots, \beta_m$. The matrix A in (5.8) is called **the matrix of the linear transformation** σ with respect to the basis $\alpha_1, \alpha_2, \ldots, \alpha_n$.

Example 5.3.7.

(1) Let $V = \mathbb{R}^2$. Given $\varphi \in \mathbb{R}$, then the rotation R_φ is an automorphism of V (see Example 5.2.5 (1)). Let $\epsilon_1, \epsilon_2 \in V$ be two unit vectors perpendicular to each other, as shown in Figure 5.2. It's clear that the sequence ϵ_1, ϵ_2 is a basis for V and

$$R_\varphi(\epsilon_1) = (\cos\varphi)\epsilon_1 + (\sin\varphi)\epsilon_2, \quad R_\varphi(\epsilon_2) = -(\sin\varphi)\epsilon_1 + (\cos\varphi)\epsilon_2.$$

Hence, $\begin{pmatrix} \cos\varphi & -\sin\varphi \\ \sin\varphi & \cos\varphi \end{pmatrix}$ is the matrix of R_φ with respect to the basis ϵ_1, ϵ_2.

(2) Let $W = \mathbb{R}^3$, and let $\alpha, \beta, \gamma \in W$ be orthogonal unit vectors such that $\beta \times \gamma = \alpha$. Then the sequence α, β, γ is a basis for W.

Given $\varphi \in \mathbb{R}$, let $R_{\alpha,\varphi}$ be the rotation in Example 5.2.5 (2). By Rodrigues' formula (5.2), we have

$$R_{\alpha,\varphi}(\alpha) = \alpha,$$
$$R_{\alpha,\varphi}(\beta) = (\cos\varphi)\beta + (\sin\varphi)\gamma,$$
$$R_{\alpha,\varphi}(\gamma) = -(\sin\varphi)\beta + (\cos\varphi)\gamma.$$

Hence, $\begin{pmatrix} 1 & 0 & 0 \\ 0 & \cos\varphi & -\sin\varphi \\ 0 & \sin\varphi & \cos\varphi \end{pmatrix}$ is the matrix of $R_{\alpha,\varphi}$ with respect to the basis α, β, γ.

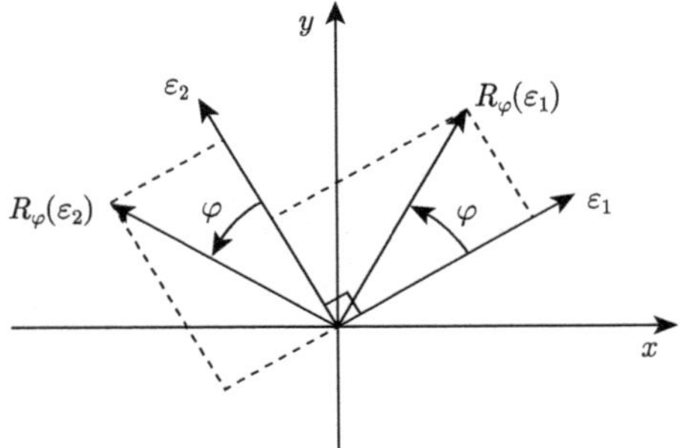

Figure 5.2.   Perpendicular vectors under R_φ.

Example 5.3.8. Let $A \in \mathrm{M}_{m\times n}(F)$. Then A induces a linear map $\sigma_A : F^n \longrightarrow F^m,\ \alpha \longmapsto A\alpha$. Conversely, if $\sigma \in \mathrm{Hom}(F^n, F^m)$, then there exists $A \in \mathrm{M}_{m\times n}(F)$ such that $\sigma = \sigma_A$. In fact, let $\varepsilon_1, \varepsilon_2, \ldots, \varepsilon_n$ and $\eta_1, \eta_2, \ldots, \eta_m$ be the standard bases for F^n and F^m, respectively. Then $I_n = (\varepsilon_1, \varepsilon_2, \ldots, \varepsilon_n)$, $I_m = (\eta_1, \eta_2, \ldots, \eta_m)$. Let A be the matrix of σ with respect to the bases $\varepsilon_1, \varepsilon_2, \ldots, \varepsilon_n$ and $\eta_1, \eta_2, \ldots, \eta_m$. Thus, for any $\alpha \in F^n$,

$$\begin{aligned}
\sigma(\alpha) &= \sigma\left((\varepsilon_1, \varepsilon_2, \ldots, \varepsilon_n)\alpha\right) \\
&= \sigma(\varepsilon_1, \varepsilon_2, \ldots, \varepsilon_n)\alpha \\
&= (\eta_1, \eta_2, \ldots, \eta_m)A\alpha \\
&= A\alpha \\
&= \sigma_A(\alpha).
\end{aligned}$$

Hence, $\sigma = \sigma_A$.

Corollary 5.3.9. *Let* $\psi : \mathrm{End}(V) \longrightarrow \mathrm{M}_n(F)$ *be the isomorphism determined by the basis* $\alpha_1, \alpha_2, \ldots, \alpha_n$ *for* V *in Theorem 5.3.5 (2). Then we have*

(1) $\psi(1_V) = I_n$;

(2) $\psi(\sigma\tau) = \psi(\sigma)\psi(\tau)$ *for* $\sigma, \tau \in \mathrm{End}(V)$;

(3) ψ *induces a bijection:* $\mathrm{Aut}(V) \longrightarrow \mathrm{GL}_n(F)$, *also denoted by* ψ, *and* $\psi(\sigma^{-1}) = \psi(\sigma)^{-1}$ *for* $\sigma \in \mathrm{Aut}(V)$.

Proof.   (1) is obvious by the definition of ψ.

   (2) Suppose $\psi(\sigma) = A = (a_{ij})_{n\times n}$, $\psi(\tau) = B = (b_{ij})_{n\times n}$, i.e.,

$$\sigma(\alpha_1, \alpha_2, \ldots, \alpha_n) = (\alpha_1, \alpha_2, \ldots, \alpha_n)A,$$
$$\tau(\alpha_1, \alpha_2, \ldots, \alpha_n) = (\alpha_1, \alpha_2, \ldots, \alpha_n)B.$$

Then

$$\begin{aligned}
(\sigma\tau)(\alpha_1, \alpha_2, \ldots, \alpha_n) &= \sigma(\tau(\alpha_1, \alpha_2, \ldots, \alpha_n)) \\
&= \sigma((\alpha_1, \alpha_2, \ldots, \alpha_n)B) \\
&= \sigma(\alpha_1, \alpha_2, \ldots, \alpha_n)B \\
&= ((\alpha_1, \alpha_2, \ldots, \alpha_n)A)\,B \\
&= (\alpha_1, \alpha_2, \ldots, \alpha_n)(AB).
\end{aligned}$$

Hence, $\psi(\sigma\tau) = AB = \psi(\sigma)\psi(\tau)$.

   (3) Note that the sequence $\alpha_1, \alpha_2, \ldots, \alpha_n$ is a basis for V and $\sigma(\alpha_1, \alpha_2, \ldots, \alpha_n) = (\alpha_1, \alpha_2, \ldots, \alpha_n)A$. Then

$$\mathrm{Im}(\sigma) = L(\sigma(\alpha_1), \sigma(\alpha_2), \ldots, \sigma(\alpha_n)),$$
$$\mathrm{rank}(\sigma(\alpha_1), \sigma(\alpha_2), \ldots, \sigma(\alpha_n)) = \mathrm{rank}A \text{ (Exercise 12 of Chapter 4).}$$

Hence, we have

$$\begin{aligned}
A \in \mathrm{GL}_n(F) &\Longleftrightarrow \mathrm{rank}A = n \\
&\Longleftrightarrow \mathrm{rank}\,(\sigma(\alpha_1), \sigma(\alpha_2), \ldots, \sigma(\alpha_n)) = n \\
&\Longleftrightarrow \dim\mathrm{Im}(\sigma) = n \\
&\Longleftrightarrow \mathrm{Im}(\sigma) = V \\
&\Longleftrightarrow \sigma \text{ is a surjection} \\
&\Longleftrightarrow \sigma \text{ is an automorphism.}
\end{aligned}$$

   Let $\sigma \in \mathrm{Aut}(V)$. By (1) and (2), $\psi(\sigma)\psi(\sigma^{-1}) = \psi(\sigma\sigma^{-1}) = \psi(1_V) = I_n$. So $\psi(\sigma^{-1}) = \psi(\sigma)^{-1}$. $\square$

Theorem 5.3.10 (Coordinate transformation). *Let* $\alpha_1, \alpha_2, \ldots,$ α_n *be a basis for a linear space V, and let $\beta_1, \beta_2, \ldots, \beta_m$ be a basis for a linear space W. For $\alpha \in V$, $\beta \in W$, denote by $X_\alpha \in F^n$ the coordinate of α with respect to $\alpha_1, \alpha_2, \ldots, \alpha_n$ and $Y_\beta \in F^m$ the coordinate of β with respect to $\beta_1, \beta_2, \ldots, \beta_m$, i.e.,*

$$\alpha = (\alpha_1, \alpha_2, \ldots, \alpha_n) X_\alpha, \quad \beta = (\beta_1, \beta_2, \ldots, \beta_m) Y_\beta.$$

(1) *Let A be the matrix of $\sigma \in \mathrm{Hom}(V, W)$ with respect to the bases $\alpha_1, \alpha_2, \ldots, \alpha_n$ and $\beta_1, \beta_2, \ldots, \beta_m$. Then $Y_{\sigma(\alpha)} = AX_\alpha$.*
(2) *Let B be the matrix of $\tau \in \mathrm{End}(V)$ with respect to the basis $\alpha_1, \alpha_2, \ldots, \alpha_n$. Then $X_{\tau(\alpha)} = BX_\alpha$.*

Proof. It's sufficient to prove (1) since (2) is a special case of (1). From $\alpha = (\alpha_1, \alpha_2, \ldots, \alpha_n) X_\alpha$, we obtain that

$$\begin{aligned}
\sigma(\alpha) &= \sigma\left((\alpha_1, \alpha_2, \ldots, \alpha_n) X_\alpha\right) \\
&= \sigma(\alpha_1, \alpha_2, \ldots, \alpha_n) X_\alpha \\
&= \left((\beta_1, \beta_2, \ldots, \beta_m) A\right) X_\alpha \\
&= (\beta_1, \beta_2, \ldots, \beta_m) (AX_\alpha).
\end{aligned}$$

Note that $\sigma(\alpha) = (\beta_1, \beta_2, \ldots, \beta_m) Y_{\sigma(\alpha)}$. The uniqueness of coordinates yields $Y_{\sigma(\alpha)} = AX_\alpha$. $\qquad\square$

Theorem 5.3.11. *Let $\alpha_1, \alpha_2, \ldots, \alpha_n$ and $\beta_1, \beta_2, \ldots, \beta_n$ be two bases for an n-dimensional linear space V, and let $\gamma_1, \gamma_2, \ldots, \gamma_m$ and $\delta_1, \delta_2, \ldots, \delta_m$ be two bases for an m-dimensional linear space W. Assume that*

$$(\beta_1, \beta_2, \ldots, \beta_n) = (\alpha_1, \alpha_2, \ldots, \alpha_n) P, \quad P \in \mathrm{GL}_n(F),$$
$$(\delta_1, \delta_2, \ldots, \delta_m) = (\gamma_1, \gamma_2, \ldots, \gamma_m) Q, \quad Q \in \mathrm{GL}_m(F).$$

(1) *If $\sigma \in \mathrm{Hom}(V, W)$ and*

$$\sigma(\alpha_1, \alpha_2, \ldots, \alpha_n) = (\gamma_1, \gamma_2, \ldots, \gamma_m) A, \quad A \in \mathrm{M}_{m \times n}(F),$$
$$\sigma(\beta_1, \beta_2, \ldots, \beta_n) = (\delta_1, \delta_2, \ldots, \delta_m) B, \quad B \in \mathrm{M}_{m \times n}(F),$$

then $B = Q^{-1}AP$.

(2) *If $\sigma \in \mathrm{End}(V)$ and*

$$\sigma(\alpha_1, \alpha_2, \ldots, \alpha_n) = (\alpha_1, \alpha_2, \ldots, \alpha_n)A, \quad A \in \mathrm{M}_n(F),$$
$$\sigma(\beta_1, \beta_2, \ldots, \beta_n) = (\beta_1, \beta_2, \ldots, \beta_n)B, \quad B \in \mathrm{M}_n(F),$$

then $B = P^{-1}AP$.

Proof. Since (2) is a special case of (1), it's sufficient to prove (1). Note that

$$\begin{aligned}
\sigma(\beta_1, \beta_2, \ldots, \beta_n) &= \sigma\left((\alpha_1, \alpha_2, \ldots, \alpha_n)P\right) \\
&= \sigma(\alpha_1, \alpha_2, \ldots, \alpha_n)P \\
&= \left((\gamma_1, \gamma_2, \ldots, \gamma_m)A\right)P \\
&= (\delta_1, \delta_2, \ldots, \delta_m)\left(Q^{-1}AP\right).
\end{aligned}$$

From $\sigma(\beta_1, \beta_2, \ldots, \beta_n) = (\delta_1, \delta_2, \ldots, \delta_m)B$ and the uniqueness of coordinates, we obtain $B = Q^{-1}AP$. $\qquad\square$

Definition 5.3.12. Let $A, B \in \mathrm{M}_n(F)$. We say that A is **similar** to B over F if there exists $P \in \mathrm{GL}_n(F)$ such that $B = P^{-1}AP$.

Proposition 5.3.13. *Similarity is an equivalence relation on the set* $\mathrm{M}_n(F)$.

We leave the proof of this fact to the reader.

From Theorem 5.3.11, we know that matrices of a linear map between two finite-dimensional linear spaces with respect to different bases are equivalent, and matrices of a linear transformation of a finite-dimensional linear space with respect to different bases are similar. The following theorem tells us that the inverse of Theorem 5.3.11 also holds.

Theorem 5.3.14.

(1) *Let $A, B \in \mathrm{M}_{m \times n}(F)$. Assume that A and B are equivalent, then there exist linear spaces V and W with $\dim V = n, \dim W = m$, two bases $\alpha_1, \alpha_2, \ldots, \alpha_n$ and $\beta_1, \beta_2, \ldots, \beta_n$ for V, two bases $\gamma_1, \gamma_2, \ldots, \gamma_m$ and $\delta_1, \delta_2, \ldots, \delta_m$ for W, and a linear map $\sigma \in \mathrm{Hom}(V, W)$ such that*

$$\sigma(\alpha_1, \alpha_2, \ldots, \alpha_n) = (\gamma_1, \gamma_2, \ldots, \gamma_m)A,$$
$$\sigma(\beta_1, \beta_2, \ldots, \beta_n) = (\delta_1, \delta_2, \ldots, \delta_m)B.$$

(2) *Let $A, B \in M_n(F)$. Assume that A and B are similar, then there exists an n-dimensional linear space V and its two bases $\alpha_1, \alpha_2, \ldots, \alpha_n$ and $\beta_1, \beta_2, \ldots, \beta_n$ such that*

$$\sigma(\alpha_1, \alpha_2, \ldots, \alpha_n) = (\alpha_1, \alpha_2, \ldots, \alpha_n)A,$$
$$\sigma(\beta_1, \beta_2, \ldots, \beta_n) = (\beta_1, \beta_2, \ldots, \beta_n)B.$$

Proof. Since (2) is a special case of (1), it's sufficient to prove (1).

Let V be an n-dimensional linear space with a basis $\alpha_1, \alpha_2, \ldots, \alpha_n$ and let W be an m-dimensional linear space with a basis $\gamma_1, \gamma_2, \ldots, \gamma_m$. By Theorem 5.2.8, there exists $\sigma \in \mathrm{Hom}(V, W)$ such that

$$\sigma(\alpha_1, \alpha_2, \ldots, \alpha_n) = (\gamma_1, \gamma_2, \ldots, \gamma_m)A.$$

By assumption, $B = QAP$ for some $P \in \mathrm{GL}_n(F)$ and $Q \in \mathrm{GL}_m(F)$. Put

$$(\beta_1, \beta_2, \ldots, \beta_n) = (\alpha_1, \alpha_2, \ldots, \alpha_n)P,$$
$$(\delta_1, \delta_2, \ldots, \delta_m) = (\gamma_1, \gamma_2, \ldots, \gamma_m)Q^{-1}.$$

It's obvious that the sequence $\beta_1, \beta_2, \ldots, \beta_n$ is a basis for V and the sequence $\delta_1, \delta_2, \ldots, \delta_m$ is a basis for W. By Theorem 5.3.11, we obtain

$$\sigma(\beta_1, \beta_2, \ldots, \beta_n) = (\delta_1, \delta_2, \ldots, \delta_m)QAP = (\delta_1, \delta_2, \ldots, \delta_m)B. \quad \square$$

Example 5.3.15. Let $A, B \in M_n(F)$ and $f(x) \in F[x]$. If A and B are similar, then $f(A)$ and $f(B)$ are similar.

Proof. By assumption, $B = P^{-1}AP$ for some $P \in \mathrm{GL}_n(F)$. It's clear that $B^k = (P^{-1}AP)^k = P^{-1}A^k P$ for any $k \in \mathbb{N}$. Let $f(x) = \sum_{i=0}^{m} a_i x^i$. Then we have

$$f(B) = a_0 I_n + a_1 B + a_2 B^2 + \cdots + a_m B^m$$
$$= a_0 I_n + a_1 P^{-1}AP + a_2 P^{-1}A^2 P + \cdots + a_m P^{-1}A^m P$$
$$= P^{-1}\left(a_0 I_n + a_1 A + a_2 A^2 + \cdots + a_m A^m\right) P$$
$$= P^{-1}f(A)P. \quad \square$$

Question. Under the similarity relation, $M_n(F)$ is partitioned into equivalence classes, called **similarity classes**. How do you find a "simple" representative element in each class?

5.4   Invariant Subspaces

Diagonal matrices and block diagonal matrices are relatively simple matrices. In this section, we use linear transformations to study the relationship between the decomposition of an n-dimensional linear space and the similarity of a square matrix of order n to a block diagonal matrix.

Definition 5.4.1. Let W be a subspace of a linear space V and $\sigma \in \mathrm{End}(V)$.

(1) We say that W is a σ-**invariant subspace** or σ-**subspace** if $\sigma(\alpha) \in W$ for any $\alpha \in W$.
(2) If W is a σ-subspace, then we can restrict σ on W to arrive at a linear transformation of W, denoted by $\sigma|_W$, which is called the **restriction** of σ on W.

Example 5.4.2. Let $\sigma \in \mathrm{End}(V)$. Then $\mathrm{Ker}(\sigma)$ and $\mathrm{Im}(\sigma)$ are both σ-subspaces.

Example 5.4.3. Let $\sigma, \tau \in \mathrm{End}(V)$ and $\sigma\tau = \tau\sigma$. Then $\mathrm{Ker}(\sigma)$ and $\mathrm{Im}(\sigma)$ are τ-subspaces. Similarly, $\mathrm{Ker}(\tau)$ and $\mathrm{Im}(\tau)$ are σ-subspaces.

Proof.   Let $\alpha \in \mathrm{Ker}(\sigma)$, i.e., $\sigma(\alpha) = 0$. Then $\sigma(\tau(\alpha)) = \tau(\sigma(\alpha)) = \tau(0) = 0$. Hence, $\tau(\alpha) \in \mathrm{Ker}(\sigma)$ and $\mathrm{Ker}(\sigma)$ is a τ-subspace.
    Let $\beta \in \mathrm{Im}(\sigma)$, i.e., $\beta = \sigma(\alpha)$ for some $\alpha \in V$. Then

$$\tau(\beta) = \tau(\sigma(\alpha)) = \sigma(\tau(\alpha)) \in \mathrm{Im}(\sigma).$$

Therefore, $\mathrm{Im}(\sigma)$ is a τ-subspace. $\square$

Proposition 5.4.4. *Let V be a linear space and $\sigma \in \mathrm{End}(V)$.*

(1) *If $W_1, W_2, \ldots, W_m \subseteq V$ are σ-subspaces, then $\sum\limits_{i=1}^{m} W_i$ is a σ-subspace.*
(2) *Let W_i be a σ-subspace of V for every $i \in I$, where I is some index set. Then $\bigcap\limits_{i \in I} W_i$ is a σ-subspace.*

(3) *Let W be a σ-subspace of V. Define $\overline{\sigma} : V/W \longrightarrow V/W$, $\overline{\alpha} \longmapsto \overline{\sigma(\alpha)}$. Then $\overline{\sigma}$ is a linear transformation of V/W, i.e., $\overline{\sigma} \in \mathrm{End}(V/W)$.*

Proof. (1) and (2) are trivial by definition. It's sufficient to show (3).

First, we claim that $\overline{\sigma}$ is well defined. In fact, let $\alpha, \beta \in V$ and $\overline{\alpha} = \overline{\beta}$, i.e., $\alpha = \beta + \gamma$ for some $\gamma \in W$. Then $\sigma(\alpha) - \sigma(\beta) = \sigma(\alpha - \beta) = \sigma(\gamma) \in W$ since W is a σ-subspace. Hence $\overline{\sigma(\alpha)} = \overline{\sigma(\beta)}$, i.e., $\overline{\sigma}(\overline{\alpha}) = \overline{\sigma}(\overline{\beta})$.

Second, $\overline{\sigma}$ is linear. In fact, for any $\alpha, \beta \in V$ and $k, l \in F$, then

$$\begin{aligned}
\overline{\sigma}(k\overline{\alpha} + l\overline{\beta}) &= \overline{\sigma}(\overline{k\alpha + l\beta}) \\
&= \overline{\sigma(k\alpha + l\beta)} = \overline{k\sigma(\alpha) + l\sigma(\beta)} = k\overline{\sigma(\alpha)} + l\overline{\sigma(\beta)} \\
&= k\overline{\sigma}(\overline{\alpha}) + l\overline{\sigma}(\overline{\beta}).
\end{aligned}$$

Hence, $\overline{\sigma} \in \mathrm{End}(V/W)$. $\qquad\square$

Proposition 5.4.5. *Let V be an n-dimensional linear space and $\sigma \in \mathrm{End}(V)$. Assume that W is a σ-subspace with $\dim W = r$. Let $\alpha_1, \ldots, \alpha_r$ be a basis for W and let $\alpha_1, \ldots, \alpha_r, \alpha_{r+1}, \ldots, \alpha_n$ be a basis for V. Then*

(1) *the sequence $\overline{\alpha_{r+1}}, \ldots, \overline{\alpha_n}$ is a basis for V/W,*
(2) *$\sigma(\alpha_1, \ldots, \alpha_r, \alpha_{r+1}, \ldots, \alpha_n)$*

$$= (\alpha_1, \ldots, \alpha_r, \alpha_{r+1}, \ldots, \alpha_n) \begin{pmatrix} A_{11} & A_{12} \\ 0 & A_{22} \end{pmatrix},$$

where A_{11} is the matrix of $\sigma|_W$ with respect to the basis $\alpha_1, \ldots, \alpha_r$ for W and A_{22} is the matrix of $\overline{\sigma}$ with respect to the basis $\overline{\alpha_{r+1}}, \ldots, \overline{\alpha_n}$ for V/W.

Proof. (1) follows from Proposition 4.5.3.

(2) On the one hand, for $i = 1, 2, \ldots, r$, $\sigma(\alpha_i) \in W$, we have

$$\sigma(\alpha_i) = (\alpha_1, \ldots, \alpha_r, \alpha_{r+1}, \ldots, \alpha_n) X_i,$$

where $X_i = (a_{1i}, \ldots, a_{ri}, 0, \ldots, 0)' \in F^n$.

On the other hand, for $j = r+1, \ldots, n$, there exist $a_{1j}, \ldots, a_{rj}, a_{r+1,j}, \ldots, a_{nj} \in F$ such that $\sigma(\alpha_j) = (\alpha_1, \ldots, \alpha_r, \alpha_{r+1}, \ldots, \alpha_n) X_j$, where $X_j = (a_{1j}, \ldots, a_{rj}, a_{r+1,j}, \ldots, a_{nj})' \in F^n$. Therefore,

$$\sigma(\alpha_1, \ldots, \alpha_r, \alpha_{r+1}, \ldots, \alpha_n)$$

$$= (\alpha_1, \ldots, \alpha_r, \alpha_{r+1}, \ldots, \alpha_n) \begin{pmatrix} a_{11} \cdots a_{1r} & a_{1,r+1} & \cdots & a_{1n} \\ \vdots & \vdots & \vdots & & \vdots \\ a_{r1} \cdots a_{rr} & a_{r,r+1} & \cdots & a_{rn} \\ 0 \cdots 0 & a_{r+1,r+1} & \cdots & a_{r+1,n} \\ \vdots & \vdots & \vdots & & \vdots \\ 0 \cdots 0 & a_{n,r+1} & \cdots & a_{nn} \end{pmatrix}$$

$$= (\alpha_1, \ldots, \alpha_r, \alpha_{r+1}, \ldots, \alpha_n) \begin{pmatrix} A_{11} & A_{12} \\ 0 & A_{22} \end{pmatrix},$$

where $A_{11} = (a_{ij})_{r \times r}$. Consequently,

$$\sigma|_W (\alpha_1, \ldots, \alpha_r) = \sigma(\alpha_1, \ldots, \alpha_r) = (\alpha_1, \ldots, \alpha_r) A_{11},$$

i.e., A_{11} is the matrix of $\sigma|_W$ with respect to the basis $\alpha_1, \ldots, \alpha_r$ for W. It's easy to see that for $r+1 \leqslant j \leqslant n$,

$$\overline{\sigma(\alpha_j)} = \overline{a_{1j}\alpha_1 + \cdots + a_{rj}\alpha_r + a_{r+1,j}\alpha_{r+1} + \cdots + a_{nj}\alpha_n}$$

$$= a_{1j}\overline{\alpha_1} + \cdots + a_{rj}\overline{\alpha_r} + a_{r+1,j}\overline{\alpha_{r+1}} + \cdots + a_{nj}\overline{\alpha_n}$$

$$= a_{r+1,j}\overline{\alpha_{r+1}} + \cdots + a_{nj}\overline{\alpha_n}$$

$$= (\overline{\alpha_{r+1}}, \ldots, \overline{\alpha_n}) \begin{pmatrix} a_{r+1,j} \\ \vdots \\ a_{nj} \end{pmatrix},$$

hence we have

$$\overline{\sigma}(\overline{\alpha_{r+1}}, \ldots, \overline{\alpha_n}) = (\overline{\alpha_{r+1}}, \ldots, \overline{\alpha_n}) \begin{pmatrix} a_{r+1,r+1} & \cdots & a_{r+1,n} \\ \vdots & & \vdots \\ a_{n,r+1} & \cdots & a_{nn} \end{pmatrix}$$

$$= (\overline{\alpha_{r+1}}, \ldots, \overline{\alpha_n}) A_{22},$$

i.e., A_{22} is the matrix of $\overline{\sigma}$ with respect to the basis $\overline{\alpha_{r+1}}, \ldots, \overline{\alpha_n}$ for V/W. $\qquad\square$

The converse of the above proposition is also true, that is, we have the following result, whose proof is left to the reader.

Proposition 5.4.6. *Let V be an n-dimensional linear space with a basis $\alpha_1, \alpha_2, \ldots, \alpha_n$. Assume that $\sigma \in \mathrm{End}(V)$ and*
$$\sigma(\alpha_1, \alpha_2, \ldots, \alpha_n) = (\alpha_1, \alpha_2, \ldots, \alpha_n) \begin{pmatrix} A_{11} & A_{12} \\ 0 & A_{22} \end{pmatrix}, \quad \text{where } A_{11} \in$$
$\mathrm{M}_r(F)$, $A_{22} \in \mathrm{M}_{n-r}(F)$. *Then*

(1) $W = L(\alpha_1, \ldots, \alpha_r)$ *is a σ-subspace,*
(2) *the sequence $\overline{\alpha_{r+1}}, \ldots, \overline{\alpha_n}$ is a basis for V/W and*
$$\overline{\sigma}(\overline{\alpha_{r+1}}, \ldots, \overline{\alpha_n}) = (\overline{\alpha_{r+1}}, \ldots, \overline{\alpha_n}) A_{22}.$$

Theorem 5.4.7. *Let V be an n-dimensional linear space and $\sigma \in \mathrm{End}(V)$. If A is the matrix of σ with respect to a basis $\alpha_1, \alpha_2, \ldots, \alpha_n$ for V, then A is similar to a block diagonal matrix $\mathrm{diag}(A_1, A_2, \ldots, A_s)$ with $A_i \in \mathrm{M}_{n_i}(F)$ ($1 \leqslant i \leqslant s$) if and only if there exist nonzero σ-subspaces $W_1, W_2, \ldots, W_s$ of V such that $V = \bigoplus_{i=1}^{s} W_i$, where $n_i = \dim W_i$, $i = 1, 2, \ldots, s$.*

Proof. Suppose that A is similar to a block diagonal matrix $B = \mathrm{diag}(A_1, A_2, \ldots, A_s)$ with $A_i \in \mathrm{M}_{n_i}(F), i = 1, 2, \ldots, s$, i.e., there exists $P \in \mathrm{GL}_n(F)$ such that $P^{-1}AP = B$. Set
$$(\beta_1, \beta_2, \ldots, \beta_n) = (\alpha_1, \alpha_2, \ldots, \alpha_n)P.$$
Then the sequence $\beta_1, \beta_2, \ldots, \beta_n$ is a basis for V and
$$\sigma(\beta_1, \beta_2, \ldots, \beta_n) = (\beta_1, \beta_2, \ldots, \beta_n)B. \tag{5.9}$$
Put
$$t_i = \begin{cases} 0, & \text{if } i = 1, \\ \sum_{j=1}^{i-1} n_j, & \text{if } i = 2, 3, \ldots, s. \end{cases}$$

By (5.9), we have $\sigma(\beta_{t_i+1}, \ldots, \beta_{t_i+n_i}) = (\beta_{t_i+1}, \ldots, \beta_{t_i+n_i})A_i$, and
$$W_i = L(\beta_{t_i+1}, \beta_{t_i+2}, \ldots, \beta_{t_i+n_i})$$
is a nonzero σ-subspace, $\dim W_i = n_i$, $i = 1, 2, \ldots, s$, and $V = \bigoplus_{i=1}^{s} W_i$.

Conversely, let $V = \bigoplus_{i=1}^{s} W_i$, where W_i is a nonzero n_i-dimensional σ-subspace, $i = 1, 2, \ldots, s$. Let $A_i \in \mathrm{M}_{n_i}(F)$ be the matrix of $\sigma|_{W_i}$ with respect to a basis $\beta_{i1}, \ldots, \beta_{in_i}$ for W_i, $i = 1, 2, \ldots, s$. Then the sequence $\beta_{11}, \ldots, \beta_{1n_1}, \beta_{21}, \ldots, \beta_{2n_2}, \ldots, \beta_{s1}, \ldots, \beta_{sn_s}$ is a basis for V and

$$\sigma(\beta_{11}, \ldots, \beta_{1n_1}, \beta_{21}, \ldots, \beta_{2n_2}, \ldots, \beta_{s1}, \ldots, \beta_{sn_s})$$
$$= (\beta_{11}, \ldots, \beta_{1n_1}, \beta_{21}, \ldots, \beta_{2n_2}, \ldots, \beta_{s1}, \ldots, \beta_{sn_s})B,$$

where $B = \mathrm{diag}(A_1, A_2, \ldots, A_s)$. Let P be the transition matrix from the basis $\alpha_1, \ldots, \alpha_n$ to the basis $\beta_{11}, \ldots, \beta_{1n_1}, \beta_{21}, \ldots, \beta_{2n_2}, \ldots, \beta_{s1}, \ldots, \beta_{sn_s}$, i.e.,

$$(\beta_{11}, \ldots, \beta_{1n_1}, \beta_{21}, \ldots, \beta_{2n_2}, \ldots, \beta_{s1}, \ldots, \beta_{sn_s}) = (\alpha_1, \ldots, \alpha_n)P.$$

Then $P \in \mathrm{GL}_n(F)$ and $P^{-1}AP = \mathrm{diag}(A_1, A_2, \ldots, A_s)$, i.e., A is similar to a block diagonal matrix. $\qquad\square$

Corollary 5.4.8. *Let V be an n-dimensional linear space over a number field F and $\sigma \in \mathrm{End}(V)$, and let A be the matrix of σ with respect to a basis $\alpha_1, \alpha_2, \ldots, \alpha_n$ for V. Then A is similar to a diagonal matrix over F if and only if $V = \bigoplus_{i=1}^{n} W_i$, where W_i is a σ-subspace of dimension 1, $i = 1, 2, \ldots, n$.*

Definition 5.4.9. Let $A \in \mathrm{M}_n(F)$. We say that A is **diagonalizable** if there exists $P \in \mathrm{GL}_n(F)$ such that $P^{-1}AP$ is a diagonal matrix.

In view of Corollary 5.4.8, a matrix A is diagonalizable if and only if V is a direct sum of n σ-subspaces of dimension 1. Let W be a σ-subspace of dimension 1. Then $W = \{k\alpha \mid k \in F\}$ for any nonzero $\alpha \in W$. Hence $\sigma(\alpha) = \lambda\alpha$ for some $\lambda \in F$, where $\alpha \in W$. This is an important phenomenon, which we shall study in detail in the following section.

5.5   Eigenvalues and Eigenvectors

Definition 5.5.1. Let V be a linear space over a number field F and $\sigma \in \mathrm{End}(V)$. If there exist $\lambda \in F$ and $0 \neq \alpha \in V$ such that

$\sigma(\alpha) = \lambda\alpha$, then λ is called an **eigenvalue** of σ and α is called an **eigenvector** of σ associated with the eigenvalue λ.

Remark 5.5.2. Let V be a linear space and $\sigma \in \text{End}(V)$.

(1) 0 is not an eigenvector although $\sigma(0) = \lambda 0$ for any $\lambda \in F$.
(2) The eigenvalues and eigenvectors of σ occur simultaneously: If there is an eigenvalue, then there is an eigenvector. An eigenvector must be associated with an eigenvalue.

Let $\alpha_1, \alpha_2, \ldots, \alpha_n$ be a basis for a linear space V and $\sigma \in \text{End}(V)$. Suppose that $A \in M_n(F)$ is the matrix of σ with respect to the basis $\alpha_1, \alpha_2, \ldots, \alpha_n$, i.e., $\sigma(\alpha_1, \alpha_2, \ldots, \alpha_n) = (\alpha_1, \alpha_2, \ldots, \alpha_n)A$. For $\alpha \in V$, denote by $X_\alpha \in F^n$ the coordinate of α with respect to the basis $\alpha_1, \alpha_2, \ldots, \alpha_n$. By Theorem 5.3.10, we know that $X_{\sigma(\alpha)} = AX_\alpha$ is the coordinate of $\sigma(\alpha)$ with respect to the basis $\alpha_1, \alpha_2, \ldots, \alpha_n$. Hence

$$\sigma(\alpha) = \lambda\alpha \iff X_{\sigma(\alpha)} = X_{\lambda\alpha} \iff AX_\alpha = \lambda X_\alpha. \qquad (5.10)$$

Note that $\alpha \neq 0$ if and only if $X_\alpha \neq 0$. It's natural to give the following definition.

Definition 5.5.3. Let $A \in M_n(F)$. If there exist $\lambda \in F$ and $0 \neq X \in F^n$ such that $AX = \lambda X$, then λ is called an **eigenvalue** of A and X is called an **eigenvector** of A associated with the eigenvalue λ.

In view of (5.10), we have the following.

Lemma 5.5.4. *Let V be an n-dimensional linear space and $\sigma \in \text{End}(V)$. Assume that A is the matrix of σ with respect to a basis $\alpha_1, \alpha_2, \ldots, \alpha_n$ for V and $\lambda \in F$. Then*

(1) *λ is an eigenvalue of σ if and only if λ is an eigenvalue of A,*
(2) *$\alpha \in V$ is an eigenvector of σ associated with the eigenvalue λ if and only if the coordinate $X_\alpha \in F^n$ of α with respect to the basis $\alpha_1, \alpha_2, \ldots, \alpha_n$ for V is an eigenvector of A associated with the eigenvalue λ.*

Let V be an n-dimensional linear space. Given $\sigma \in \text{End}(V)$ or $A \in M_n(F)$, there are two natural questions:

1. (Qualitative) Is there an eigenvalue of σ (respectively, A)?

2. (Quantitative) If σ (respectively, A) has eigenvalues, how many are there? How many eigenvectors are there for each eigenvalue?

Due to Lemma 5.5.4, for the above two questions, we only need to consider $A \in M_n(F)$.

Theorem 5.5.5. *Let $A \in M_n(F)$ and $\lambda_0 \in F$. Then*

(1) *λ_0 is an eigenvalue of A if and only if λ_0 is a root of the equation $|\lambda I_n - A| = 0$,*
(2) *$X_0 \in F^n$ is an eigenvector of A associated with the eigenvalue λ_0 if and only if X_0 is a nonzero solution of the homogeneous linear system $(\lambda_0 I_n - A)X = 0$.*

Proof. It's clear that

λ_0 is an eigenvalue of A

$\Longleftrightarrow$ there exists a nonzero vector $X_0 \in F^n$ such that $AX_0 = \lambda_0 X_0$

$\Longleftrightarrow$ there exists a nonzero vector $X_0 \in F^n$ such that

$(\lambda_0 I_n - A)X_0 = 0$

$\Longleftrightarrow$ the homogeneous linear system $(\lambda_0 I_n - A)X = 0$

has a nonzero solution $X_0 \in F^n$

$\Longleftrightarrow |\lambda_0 I_n - A| = 0$

$\Longleftrightarrow \lambda_0$ is a root of $|\lambda I_n - A| = 0$ in F.

Hence, (1) and (2) hold. $\qquad\square$

Example 5.5.6. Let $A = \begin{pmatrix} -1 & -4 & -6 \\ -5 & -13 & -20 \\ 3 & 10 & 15 \end{pmatrix} \in M_3(\mathbb{R})$. Find the real eigenvalues and the associated eigenvectors of A.

Solution. It's clear that the equation $|\lambda I_3 - A| = (\lambda - 1)(\lambda^2 + 1) = 0$ has a unique real solution $\lambda = 1$. Solving the homogeneous linear system $(I_3 - A)X = 0$, we obtain the solution $X = k(2, 5, -4)'$, where $k \in \mathbb{R}$. From Theorem 5.5.5, the matrix A has a unique real eigenvalue 1, and for any real number $k \neq 0$, the vector $k(2, 5, -4)'$ is an eigenvector of A associated with the eigenvalue 1. $\qquad\square$

Lemma 5.5.7. *Let $A \in M_n(F)$. Then $f_A(\lambda) = |\lambda I_n - A|$ is a monic polynomial of degree n over F. Furthermore, let*

$$f_A(\lambda) = \lambda^n - p_1 \lambda^{n-1} + \cdots + (-1)^{n-1} p_{n-1} \lambda + (-1)^n p_n \quad (5.11)$$

and let $\lambda_1, \lambda_2, \ldots, \lambda_n$ be the n complex roots of $f_A(\lambda)$. Then

$$p_k = \sum_{1 \leqslant i_1 < i_2 < \cdots < i_k \leqslant n} \lambda_{i_1} \lambda_{i_2} \cdots \lambda_{i_k} = s_k, \tag{5.12}$$

where s_k is the sum of all principal minors of order k of A, $k = 1, 2, \ldots, n$. In particular,

$$\mathrm{Tr}(A) = \sum_{i=1}^{n} \lambda_i, \ |A| = \prod_{i=1}^{n} \lambda_i, \ \mathrm{Tr}(A^*) = \sum_{i=1}^{n} \lambda_i^*, \tag{5.13}$$

where A^* is the adjoint matrix of A, and

$$\lambda_i^* = \prod_{\substack{j=1 \\ j \neq i}}^{n} \lambda_j = \begin{cases} \lambda_2 \lambda_3 \cdots \lambda_n, & i = 1, \\ \lambda_1 \cdots \lambda_{i-1} \lambda_{i+1} \cdots \lambda_n, & i = 2, \ldots, n-1, \\ \lambda_1 \lambda_2 \cdots \lambda_{n-1}, & i = n. \end{cases} \tag{5.14}$$

Proof. Comparing the coefficients of the equation

$$\lambda^n - p_1 \lambda^{n-1} + \cdots + (-1)^{n-1} p_{n-1} \lambda + (-1)^n p_n = \prod_{i=1}^{n} (\lambda - \lambda_i)$$

$$= \lambda^n - \left(\sum_{i=1}^{n} \lambda_i \right) \lambda^{n-1} + \cdots + (-1)^{n-1} \left(\sum_{i=1}^{n} \lambda_i^* \right) \lambda + (-1)^n \prod_{i=1}^{n} \lambda_i,$$

we obtain the first equality of (5.12).

Let $I_n = (\varepsilon_1, \varepsilon_2, \ldots, \varepsilon_n)$ and $A = (\alpha_1, \alpha_2, \ldots, \alpha_n)$. Then

$$f_A(\lambda) = |\lambda \varepsilon_1 - \alpha_1, \lambda \varepsilon_2 - \alpha_2, \ldots, \lambda \varepsilon_n - \alpha_n|$$

$$= |\lambda \varepsilon_1, \lambda \varepsilon_2, \ldots, \lambda \varepsilon_n| + g(\lambda) + |-\alpha_1, -\alpha_2, \ldots, -\alpha_n|$$

$$= \lambda^n + \sum_{t=1}^{n-1} (-1)^{n-t} p_{n-t} \lambda^t + (-1)^n |A|$$

$$= \lambda^n - p_1 \lambda^{n-1} + \cdots + (-1)^{n-1} p_{n-1} \lambda + (-1)^n p_n,$$

where $p_n = |A|$, $g(\lambda)$ is the sum

$$\sum_{\substack{1 \leqslant t \leqslant n-1 \\ 1 \leqslant i_1 < \cdots < i_t \leqslant n}} |-\alpha_1, \ldots, -\alpha_{i_1-1}, \lambda \varepsilon_{i_1}, -\alpha_{i_1+1}, \ldots, -\alpha_{i_t-1}, \lambda \varepsilon_{i_t}, -\alpha_{i_t+1}, \ldots, -\alpha_n|,$$

and for $1 \leqslant t \leqslant n-1$, $p_{n-t}\lambda^t$ is the sum

$$\sum_{1 \leqslant i_1 < \cdots < i_t \leqslant n} |\alpha_1, \ldots, \alpha_{i_1-1}, \lambda\varepsilon_{i_1}, \alpha_{i_1+1}, \ldots, \alpha_{i_t-1}, \lambda\varepsilon_{i_t}, \alpha_{i_t+1}, \ldots, \alpha_n|$$

which is exactly $s_{n-t}\lambda^t$ by applying the Laplace expansion to the $i_1, \ldots, i_t$ columns of each determinant in the sum. Hence, $f_A(\lambda)$ is a monic polynomial of degree n over F, and $p_k = s_k$ for any $1 \leqslant k \leqslant n$.
$\square$

Definition 5.5.8. Let $A \in M_n(F)$.

(1) The polynomial $f_A(\lambda) = |\lambda I_n - A|$ is called the **characteristic polynomial** of A.

(2) The equation $f_A(\lambda) = 0$ is called the **characteristic equation** of A.

(3) Let $\lambda \in F$ be an eigenvalue of A. The subspace

$$V_\lambda = \mathrm{Ker}(\lambda I_n - A) = \{\alpha \in F^n \mid (\lambda I_n - A)\alpha = 0\}$$

is called the **eigenspace** of A associated with the eigenvalue λ.

Corollary 5.5.9.

(1) *Let $A \in M_n(F)$. Then $A \in GL_n(F)$ if and only if $f_A(0) \neq 0$, i.e., 0 is not an eigenvalue of A.*

(2) *Let $A \in GL_n(F)$ and $0 \neq \lambda \in F$. Then λ is an eigenvalue of A if and only if λ^{-1} is an eigenvalue of A^{-1} if and only if $\lambda^* = \frac{|A|}{\lambda}$ is an eigenvalue of A^*. Furthermore, α is an eigenvector of A associated with λ if and only if α is an eigenvector of A^{-1} associated with λ^{-1} if and only if α is an eigenvector of A^* associated with λ^*. Hence A, A^{-1}, and A^* have the same eigenvectors.*

Proof. (1) follows immediately from (5.13).

(2) comes from the fact $A^* = |A|A^{-1}$ and the fact that $A\alpha = \lambda\alpha$ if and only if $A^{-1}\alpha = \lambda^{-1}\alpha$.
$\square$

Lemma 5.5.10. *Let $A, B \in M_n(F)$. If A and B are similar, then*

$$f_A(\lambda) = f_B(\lambda), \ \mathrm{Tr}(A) = \mathrm{Tr}(B), \ |A| = |B|.$$

Proof. Let $B = P^{-1}AP, \ P \in \mathrm{GL}_n(F)$. Then

$$f_B(\lambda) = |\lambda I_n - B|$$

$$= |\lambda I_n - P^{-1}AP|$$

$$= |P^{-1}(\lambda I_n - A)P|$$

$$= |P^{-1}||\lambda I_n - A||P|$$

$$= f_A(\lambda).$$

By Lemma 5.5.7, we have $\mathrm{Tr}(A) = \mathrm{Tr}(B), \ |A| = |B|.$ $\qquad\square$

Remark 5.5.11.

(1) The converse of Lemma 5.5.10 is not true, for example,

$$A = \begin{pmatrix} 0 & 0 \\ 0 & 0 \end{pmatrix}, \quad B = \begin{pmatrix} 0 & 1 \\ 0 & 0 \end{pmatrix}.$$

It's obvious that $f_A(\lambda) = f_B(\lambda) = \lambda^2$, but A and B are not similar.

(2) Let n be a positive integer. Define a map

$$f : \mathrm{M}_n(F) \longrightarrow F[\lambda], \ A \longmapsto f_A(\lambda).$$

Then f is not injective as shown by the above example in (1), and

$$\mathrm{Im}(f) = \{\lambda^n + a_{n-1}\lambda^{n-1} + \cdots + a_1\lambda + a_0 \mid a_0, a_1, \ldots, a_{n-1} \in F\}.$$

In fact, each monic polynomial of degree n is a characteristic polynomial of some $A \in \mathrm{M}_n(F)$ by Exercise 14 (1) of Chapter 2.

(3) Let $A \in \mathrm{M}_{n \times m}(F)$ and $B \in \mathrm{M}_{m \times n}(F)$. From Example 2.10.21, we obtain that $\lambda^m f_{AB}(\lambda) = \lambda^n f_{BA}(\lambda)$. Hence, AB and BA have the same nonzero eigenvalues.

(4) By Remark 2.10.22 (3), if $A, B \in \mathrm{M}_n(F)$, then AB and BA have the same characteristic polynomials and eigenvalues.

Definition 5.5.12. Let $\alpha_1, \alpha_2, \ldots, \alpha_n$ be a basis for an n-dimensional linear space V and $\sigma \in \mathrm{End}(V)$. Suppose that $A \in \mathrm{M}_n(F)$ is the matrix of σ with respect to the basis $\alpha_1, \alpha_2, \ldots, \alpha_n$, i.e.,

$$\sigma(\alpha_1, \alpha_2, \ldots, \alpha_n) = (\alpha_1, \alpha_2, \ldots, \alpha_n)A.$$

(1) The characteristic polynomial $f_A(\lambda)$ of A is called the **characteristic polynomial** of σ, denoted by $f_\sigma(\lambda)$.
(2) The equation $f_\sigma(\lambda) = 0$ is called the **characteristic equation** of σ.
(3) $\mathrm{Tr}(A)$ is called the **trace** of σ, denoted by $\mathrm{Tr}(\sigma)$.
(4) $|A|$ is called the **determinant** of σ, denoted by $|\sigma|$ or $\det(\sigma)$.
(5) Let $\lambda_0 \in F$ be an eigenvalue of σ. The subspace

$$\mathrm{Ker}(\lambda_0 1_V - \sigma) = \{\alpha \in V \mid \sigma(\alpha) = \lambda_0 \alpha\}$$

is called the **eigenspace** of σ associated with λ_0.

Remark 5.5.13. From Theorem 5.3.11 and Lemma 5.5.10, we know that the characteristic polynomial $f_\sigma(\lambda)$, the trace $\mathrm{Tr}(\sigma)$, and the determinant $\det(\sigma)$ of $\sigma \in \mathrm{End}(V)$ are independent of the choice of bases for V.

By Lemma 5.5.4 and Theorem 5.5.5, we get the method of calculating the eigenvalues and the associated eigenvectors of a linear transformation as follows. Let V be an n-dimensional linear space and $\sigma \in \mathrm{End}(V)$. Choose a basis $\alpha_1, \alpha_2, \ldots, \alpha_n$ for V and let $A \in \mathrm{M}_n(F)$ such that

$$\sigma(\alpha_1, \alpha_2, \ldots, \alpha_n) = (\alpha_1, \alpha_2, \ldots, \alpha_n)A.$$

(1) Calculate the eigenvalues of σ by solving the characteristic equation $f_A(\lambda) = 0$.
(2) Let λ_0 be an eigenvalue of σ. We can obtain a basis $X_1, X_2, \ldots, X_t$ for the eigenspace V_{λ_0} of A associated with λ_0 by solving the homogeneous linear system $(\lambda_0 I_n - A)X = 0$, where $t = n - \mathrm{rank}(\lambda_0 I_n - A)$. Let

$$\beta_i = (\alpha_1, \alpha_2, \ldots, \alpha_n)X_i, \ i = 1, 2, \ldots, t.$$

Then the sequence $\beta_1, \beta_2, \ldots, \beta_t$ is a basis for the eigenspace W_{λ_0} of σ associated with the eigenvalue λ_0.

Example 5.5.14. Let $a \in F$. Define

$$\sigma : F[x]_3 \longrightarrow F[x]_3, \ f(x) \longmapsto xf'(x) + f(x + a).$$

It's easy to see that σ is a linear transformation of $F[x]_3$. Find the eigenvalues and bases for the associated eigenspaces of σ.

Solution. We know that the sequence $1, x, x^2$ is a basis for $F[x]_3$, and

$$\sigma(1) = 1, \quad \sigma(x) = 2x + a, \quad \sigma(x^2) = 3x^2 + 2ax + a^2.$$

Hence, $\sigma(1, x, x^2) = (1, x, x^2)A$, where $A = \begin{pmatrix} 1 & a & a^2 \\ 0 & 2 & 2a \\ 0 & 0 & 3 \end{pmatrix}$. By solving

the characteristic equation

$$f_\sigma(\lambda) = f_A(\lambda) = |\lambda I_3 - A| = (\lambda - 1)(\lambda - 2)(\lambda - 3) = 0,$$

we get the eigenvalues $\lambda_1 = 1$, $\lambda_2 = 2$, $\lambda_3 = 3$.

Solving the homogeneous linear system $(I_3 - A)X = 0$, we obtain a basis $X_1 = (1, 0, 0)'$ for the eigenspace V_1 of A associated with the eigenvalue 1. Hence $f_1(x) = (1, x, x^2)X_1 = 1$ is a basis for the eigenspace W_1 of σ associated with the eigenvalue 1.

Similarly, we obtain that $f_2(x) = a + x$ is a basis for the eigenspace W_2 of σ associated with the eigenvalue 2 and $f_3(x) = \frac{3a^2}{2} + 2ax + x^2$ is a basis for the eigenspace W_3 of σ associated with the eigenvalue 3.

$\square$

Theorem 5.5.15. *Let $\alpha_1, \alpha_2, \ldots, \alpha_n$ be a basis for an n-dimensional linear space V and $\sigma \in \mathrm{End}(V)$. Assume that A is the matrix of σ with respect to the basis $\alpha_1, \alpha_2, \ldots, \alpha_n$. Then the following statements are equivalent:*

(1) *There exists a σ-subspace W of V with $\dim W = 1$.*
(2) *The characteristic equation $f_\sigma(\lambda) = 0$ has a root in F.*
(3) *σ has an eigenvalue.*
(4) *There exists $P \in \mathrm{GL}_n(F)$ such that $P^{-1}AP = \begin{pmatrix} \lambda_0 & \beta \\ 0 & A_1 \end{pmatrix}$, where*

$\beta \in \mathrm{M}_{1 \times (n-1)}(F)$, $A_1 \in \mathrm{M}_{n-1}(F)$, $\lambda_0 \in F$.

Proof. It's trivial that $(2) \iff (3) \iff (4)$. Hence it is sufficient to show that (1) and (3) are equivalent.

$(1) \implies (3)$. Let W be a σ-subspace with $\dim W = 1$ and $0 \neq \alpha_0 \in W$. Then

$$W = L(\alpha_0) = \{k\alpha_0 \mid k \in F\}.$$

Since $\sigma(\alpha_0) \in W$, there exists $\lambda_0 \in F$ such that $\sigma(\alpha_0) = \lambda_0 \alpha_0$. Hence λ_0 is an eigenvalue of σ.

$(3) \implies (1)$. Let $\lambda_0 \in F$ be an eigenvalue of σ. Then there exists $\alpha_0 \in V \setminus \{0\}$ such that $\sigma(\alpha_0) = \lambda_0 \alpha_0$. Put $W = L(\alpha_0) = \{k\alpha_0 \mid k \in F\}$. Then $\dim W = 1$. Let $\alpha \in W$. Then there is $k \in F$ such that $\alpha = k\alpha_0$, and hence

$$\sigma(\alpha) = \sigma(k\alpha_0) = k\sigma(\alpha_0) = k(\lambda_0\alpha_0) = (k\lambda_0)\alpha_0 \in W.$$

Therefore, W is a σ-subspace of V with $\dim W = 1$. $\square$

Example 5.5.16. Let $A \in M_n(F)$ and $g(x) \in F[x]$. Assume that λ is an eigenvalue of A and $\alpha \in F^n$ is an eigenvector of A associated with λ. Then $g(\lambda)$ is an eigenvalue of $g(A)$ and α is an eigenvector of $g(A)$ associated with $g(\lambda)$.

Proof. From $A\alpha = \lambda\alpha$, we obtain that for any positive integer i,

$$A^i\alpha = \lambda A^{i-1}\alpha = \cdots = \lambda^i \alpha.$$

Let $g(x) = a_0 + a_1 x + a_2 x^2 + \cdots + a_m x^m$. Then

$$\begin{aligned}
g(A)\alpha &= \left(a_0 I_n + a_1 A + a_2 A^2 + \cdots + a_m A^m\right)\alpha \\
&= a_0\alpha + a_1 A\alpha + a_2 A^2\alpha + \cdots + a_m A^m\alpha \\
&= a_0\alpha + a_1 \lambda\alpha + a_2 \lambda^2\alpha + \cdots + a_m \lambda^m\alpha \\
&= \left(a_0 + a_1\lambda + \cdots + a_m\lambda^m\right)\alpha \\
&= g(\lambda)\alpha.
\end{aligned}$$

So, $g(\lambda)$ is an eigenvalue of $g(A)$ and α is an eigenvector of $g(A)$ associated with $g(\lambda)$. $\square$

Theorem 5.5.17. *Let $A \in M_n(F)$ and $f_A(\lambda) = |\lambda I_n - A|$ be the characteristic polynomial of A.*

(1) Assume that $f_A(\lambda) = g(\lambda) \prod_{i=1}^{t} (\lambda - \lambda_i)$, where $\lambda_1, \lambda_2, \ldots, \lambda_t \in F$, $g(\lambda) \in F[\lambda]$. Then there exists $P \in \mathrm{GL}_n(F)$ such that $P^{-1}AP = \begin{pmatrix} A_{11} & A_{12} \\ 0 & A_{22} \end{pmatrix}$, where $A_{11} = \begin{pmatrix} \lambda_1 & \star & \star \\ & \ddots & \star \\ & & \lambda_t \end{pmatrix}$ is an upper triangular matrix, $A_{12} \in M_{t\times(n-t)}(F), A_{22} \in M_{n-t}(F)$.

(2) *Let $F = \mathbb{C}$. Then $f_A(\lambda) = \prod\limits_{i=1}^{n}(\lambda - \lambda_i)$, and there exists $P \in$*

$$\mathrm{GL}_n(\mathbb{C}) \text{ such that } P^{-1}AP = \begin{pmatrix} \lambda_1 & \star & \star \\ & \ddots & \star \\ & & \lambda_n \end{pmatrix} \text{ is an upper triangular}$$

matrix.

Proof. (1) We proceed by induction on t. When $t = 1$, (1) is true by Theorem 5.5.15. Assume that $t \geqslant 2$ and the conclusion holds for $t-1$. Consider the case that the characteristic polynomial $f_A(\lambda)$ has t roots $\lambda_1, \lambda_2, \ldots, \lambda_t \in F$, i.e., $f_A(\lambda) = g(\lambda) \prod\limits_{i=1}^{t}(\lambda - \lambda_i)$. By Theorem 5.5.15, there exists $P_1 \in \mathrm{GL}_n(F)$ such that $B = P_1^{-1}AP_1 = \begin{pmatrix} \lambda_1 & \star \\ 0 & C \end{pmatrix}$, where $C \in \mathrm{M}_{n-1}(F)$. Hence we have $g(\lambda) \prod\limits_{i=1}^{t}(\lambda - \lambda_i) = f_A(\lambda) = f_B(\lambda) = (\lambda - \lambda_1)f_C(\lambda)$. We obtain that $f_C(\lambda) = g(\lambda) \prod\limits_{i=2}^{t}(\lambda - \lambda_i)$. By induction, there exists $P_2 \in \mathrm{GL}_{n-1}(F)$ such that $P_2^{-1}CP_2 = \begin{pmatrix} C_{11} & C_{12} \\ 0 & C_{22} \end{pmatrix}$, where

$$C_{11} = \begin{pmatrix} \lambda_2 & \star & \star \\ & \ddots & \star \\ & & \lambda_t \end{pmatrix} \text{ is an upper triangular matrix of order } t - 1.$$

Set $P = P_1 \begin{pmatrix} 1 & 0 \\ 0 & P_2 \end{pmatrix}$. Then $P \in \mathrm{GL}_n(F)$ and

$$P^{-1}AP = \begin{pmatrix} 1 & 0 \\ 0 & P_2^{-1} \end{pmatrix} \begin{pmatrix} \lambda_1 & \star \\ 0 & C \end{pmatrix} \begin{pmatrix} 1 & 0 \\ 0 & P_2 \end{pmatrix} = \begin{pmatrix} \lambda_1 & \star \\ 0 & P_2^{-1}CP_2 \end{pmatrix}$$

$$= \begin{pmatrix} A_{11} & A_{12} \\ 0 & A_{22} \end{pmatrix},$$

where $A_{11} = \begin{pmatrix} \lambda_1 & \star & \star \\ & \ddots & \star \\ & & \lambda_t \end{pmatrix}$ is an upper triangular matrix, $A_{12} \in$

$\mathrm{M}_{t \times (n-t)}(F)$, $A_{22} = C_{22} \in \mathrm{M}_{n-t}(F)$.

(2) Since the characteristic polynomial $f_A(\lambda)$ of A has n roots in $\mathbb{C}$, the conclusion is obvious from (1). $\qquad\square$

Corollary 5.5.18. *Let $A \in M_n(F)$ and $f_A(\lambda) = \prod_{i=1}^{n}(\lambda - \lambda_i)$, where $\lambda_1, \lambda_2, \ldots, \lambda_n \in F$.*

(1) *For any $g(\lambda) \in F[\lambda]$, we have*

$$f_{g(A)}(\lambda) = |\lambda I_n - g(A)| = \prod_{i=1}^{n}(\lambda - g(\lambda_i)),$$

i.e., $g(\lambda_1), g(\lambda_2), \ldots, g(\lambda_n)$ are all eigenvalues of $g(A)$.

(2) $f_{A^*}(\lambda) = |\lambda I_n - A^*| = \prod_{i=1}^{n}(\lambda - \lambda_i^*)$, *where $\lambda_1^*, \lambda_2^*, \ldots, \lambda_n^*$ are defined by* (5.14).

Proof. By Theorem 5.5.17, there exists $P \in \mathrm{GL}_n(F)$ such that

$$P^{-1}AP = \begin{pmatrix} \lambda_1 & \star & \star \\ & \ddots & \star \\ & & \lambda_n \end{pmatrix} \quad \text{is an upper triangular matrix.} \quad (5.15)$$

(1) Let $g(\lambda) \in F[\lambda]$. Then $P^{-1}g(A)P = g(P^{-1}AP) = \begin{pmatrix} g(\lambda_1) & \star & \star \\ & \ddots & \star \\ & & g(\lambda_n) \end{pmatrix}$ is an upper triangular matrix. Hence $f_{g(A)}(\lambda) = \prod_{i=1}^{n}(\lambda - g(\lambda_i))$, i.e., $g(\lambda_1), g(\lambda_2), \ldots, g(\lambda_n) \in F$ are exactly n eigenvalues of $g(A)$.

(2) From $(P^*)^{-1} = (P^{-1})^*$, Exercise 36 of Chapter 2, and (5.15), we obtain that

$$P^*A^*(P^*)^{-1} = (P^{-1}AP)^*$$

$$= \begin{pmatrix} \lambda_1^* & \star & \star \\ & \ddots & \star \\ & & \lambda_n^* \end{pmatrix} \quad \text{is an upper triangular matrix.}$$

Hence, $f_{A^*}(\lambda) = f_{P^*A^*(P^*)^{-1}}(\lambda) = \prod_{i=1}^{n}(\lambda - \lambda_i^*)$, i.e., $\lambda_1^*, \lambda_2^*, \ldots, \lambda_n^* \in F$ are exactly n eigenvalues of A^*. $\qquad\square$

Example 5.5.19. Apply Corollary 5.5.18 to prove Example 2.6.14, *i.e*, prove that

$$D = \begin{vmatrix} a_0 & a_1 & a_2 & \cdots\cdots\cdots & a_{n-3} & a_{n-2} & a_{n-1} \\ a_{n-1} & a_0 & a_1 & \ddots & & a_{n-3} & a_{n-2} \\ a_{n-2} & a_{n-1} & a_0 & \ddots & & & a_{n-3} \\ \vdots & & & & & & \vdots \\ \vdots & & & & & & \vdots \\ a_3 & & & & a_0 & a_1 & a_2 \\ a_2 & a_3 & & & a_{n-1} & a_0 & a_1 \\ a_1 & a_2 & a_3 & \cdots\cdots\cdots & a_{n-2} & a_{n-1} & a_0 \end{vmatrix} = \prod_{i=0}^{n-1} f(\xi^i),$$

where $f(x) = a_0 + a_1 x + \cdots + a_{n-1} x^{n-1}$ and ξ is a primitive nth root of unit.

Proof. Put

$$B = \begin{pmatrix} 0 & 1 & 0 & \cdots & \cdots & \cdots & 0 \\ 0 & 0 & 1 & \ddots & & & \vdots \\ 0 & 0 & 0 & \ddots & \ddots & & \vdots \\ \vdots & \vdots & \ddots & \ddots & \ddots & \ddots & \vdots \\ \vdots & \vdots & & \ddots & 0 & 1 & 0 \\ 0 & 0 & \cdots & \cdots & 0 & 0 & 1 \\ 1 & 0 & \cdots & \cdots & 0 & 0 & 0 \end{pmatrix}$$

and $A = f(B) = a_0 I_n + a_1 B + a_2 B^2 + \cdots + a_{n-1} B^{n-1}$. Then $D = |A|$ and the characteristic polynomial of B is given by

$$|\lambda I_n - B| = \begin{vmatrix} \lambda & -1 & 0 & \cdots\cdots\cdots & 0 \\ 0 & \lambda & -1 & \ddots & \vdots \\ 0 & 0 & \lambda & \ddots \ddots & \vdots \\ \vdots & \vdots & \ddots & \ddots \ddots & \vdots \\ \vdots & \vdots & & \ddots & \lambda & -1 & 0 \\ 0 & 0 & \cdots\cdots & 0 & \lambda & -1 \\ -1 & 0 & \cdots\cdots & 0 & 0 & \lambda \end{vmatrix} = \lambda^n - 1.$$

Hence, the n eigenvalues of B are $\xi^i, i = 0, 1, \ldots, n-1$, where ξ is a primitive nth root of unit. From Corollary 5.5.18, $A = f(B)$ has n eigenvalues $f(\xi^i), i = 0, 1, \ldots, n-1$. Therefore, we obtain that

$$D = |A| = \prod_{i=0}^{n-1} f(\xi^i).$$

$\square$

5.6 Minimal Polynomials

Let $A \in \mathrm{M}_n(F)$. There exists a nonzero polynomial $f(\lambda) \in F[\lambda]$ such that $f(A) = 0$ by Exercise 18 of Chapter 4.

Definition 5.6.1. Let $A \in \mathrm{M}_n(F)$.

(1) If $f(\lambda) \in F[\lambda]$ satisfies $f(A) = 0$, then $f(\lambda)$ is called an **annihilating polynomial** of A.
(2) The **minimal polynomial** of A is the monic polynomial of least degree in the set of nonzero polynomials annihilating A, denoted by $m_A(\lambda)$.

Theorem 5.6.2 (Cayley–Hamilton). *Let $A \in \mathrm{M}_n(F)$. Then the characteristic polynomial*

$$f_A(\lambda) = \lambda^n + a_1 \lambda^{n-1} + \cdots + a_{n-1}\lambda + a_n \in F[\lambda] \qquad (5.16)$$

is an annihilating polynomial of A, i.e.,

$$f_A(A) = A^n + a_1 A^{n-1} + \cdots + a_{n-1}A + a_n I_n = 0. \qquad (5.17)$$

Proof. By the definition of an adjoint matrix, we may assume that (refer to Lemma 6.4.1)

$$(\lambda I_n - A)^* = \lambda^{n-1} B_0 + \lambda^{n-2} B_1 + \cdots + \lambda B_{n-2} + B_{n-1},$$

where $B_i \in \mathrm{M}_n(F)$, $i = 0, 1, \ldots, n-1$. From $(\lambda I_n - A)^*(\lambda I_n - A) = f_A(\lambda) I_n$, we get that

$$\lambda^n B_0 + \sum_{i=1}^{n-1} \lambda^i (B_{n-i} - B_{n-i-1}A) - B_{n-1}A = \lambda^n I_n + \sum_{i=0}^{n-1} a_{n-i}\lambda^i I_n.$$

Hence,

$$B_0 = I_n,$$
$$B_1 - B_0 A = a_1 I_n,$$
$$B_2 - B_1 A = a_2 I_n,$$
$$\vdots \qquad \vdots$$
$$B_{n-1} - B_{n-2} A = a_{n-1} I_n,$$
$$-B_{n-1} A = a_n I_n.$$

Multiplying the $n + 1$ equalities above on the right by $A^n, A^{n-1}, \ldots, A, I_n$ from the top to the bottom respectively and adding them, we get the quality $f_A(A) = 0$. $\qquad\square$

Remark 5.6.3. Ferdinand Georg Frobenius gave the first complete proof of the Cayley–Hamilton theorem in 1878.

Assume that the characteristic polynomial $f_A(\lambda)$ of A is given by (5.16). If A is invertible, then $f_A(0) = a_n \neq 0$ and

$$I_n = -\frac{1}{a_n} \left(A^{n-1} + a_1 A^{n-2} + \cdots + a_{n-1} I_n \right) A$$

by (5.17). Hence $A^{-1} = -\frac{1}{a_n} \left(A^{n-1} + a_1 A^{n-2} + \cdots + a_{n-1} I_n \right) = g(A)$, where

$$g(\lambda) = -\frac{1}{a_n} \left(\lambda^{n-1} + a_1 \lambda^{n-2} + \cdots + a_{n-1} \right) \in F[\lambda].$$

Therefore, we have the following result.

Corollary 5.6.4. *Let $A \in \mathrm{GL}_n(F)$ and let $\lambda_0 \in F$ be an eigenvalue of A. Then there exist $g(\lambda), h(\lambda) \in F[\lambda]$ such that $A^{-1} = g(A)$, $A^* = h(A)$, and $\lambda_0^{-1} = g(\lambda_0)$, $\lambda_0^* = \frac{|A|}{\lambda_0} = h(\lambda_0)$.*

The Cayley–Hamilton theorem implies that the minimal polynomial of $A \in \mathrm{M}_n(F)$ has degree at most n. In the following theorem, we prove that the minimum polynomial of A is not only a factor of its characteristic polynomial, but also they have the same irreducible factors.

Theorem 5.6.5. *Let $A \in \mathrm{M}_n(F)$ and*

$$f_A(\lambda) = p_1(\lambda)^{e_1} p_2(\lambda)^{e_2} \cdots p_s(\lambda)^{e_s},$$

where $p_1(\lambda), p_2(\lambda), \ldots, p_s(\lambda) \in F[\lambda]$ are different monic irreducible polynomials, $e_1, e_2, \ldots, e_s \in \mathbb{N}^*$. Then the following statements hold:

(1) *The minimal polynomial $m_A(\lambda)$ exists.*

(2) *Let $g(\lambda) \in F[\lambda]$. Then $g(A) = 0$ if and only if $m_A(\lambda)|g(\lambda)$.*

(3) *$m_A(\lambda)$ is unique and $m_A(\lambda) = p_1(\lambda)^{r_1} p_2(\lambda)^{r_2} \cdots p_s(\lambda)^{r_s}$, where $1 \leqslant r_i \leqslant e_i$, $i = 1, 2, \ldots, s$.*

(4) *Let $B \in \mathrm{M}_n(F)$. If A and B are similar, then $m_A(\lambda) = m_B(\lambda)$.*

(5) *Let $C \in \mathrm{M}_t(F)$ and $D = \begin{pmatrix} A & 0 \\ 0 & C \end{pmatrix}$. Then $m_D(\lambda) = [m_A(\lambda), m_C(\lambda)]$.*

Proof.   (1) The Cayley–Hamilton theorem implies the existence of the minimal polynomial of A.

(2) Assume $g(A) = 0$. By the Euclidean algorithm, $g(\lambda) = q(\lambda)m_A(\lambda) + r(\lambda)$, where $r(\lambda) = 0$ or $\deg(r(\lambda)) < \deg(m_A(\lambda))$. Then $0 = g(A) = q(A)m_A(A) + r(A) = r(A)$. From the definition of $m_A(\lambda)$, we get $r(\lambda) = 0$. Hence, $m_A(\lambda)|g(\lambda)$.

Assume $m_A(\lambda)|g(\lambda)$. Then $g(\lambda) = q(\lambda)m_A(\lambda)$ for some $q(\lambda) \in F[\lambda]$. Hence, $g(A) = q(A)m_A(A) = 0$.

(3) Assume that $m_1(\lambda), m_2(\lambda) \in F[\lambda]$ are the minimal polynomials of A. By (1), $m_1(\lambda)|m_2(\lambda)$, $m_2(\lambda)|m_1(\lambda)$. This implies that $m_1(\lambda) = m_2(\lambda)$ since $m_1(\lambda)$ and $m_2(\lambda)$ are monic.

By the Cayley–Hamilton theorem and (2), we get $m_A(\lambda)|f_A(\lambda)$. Hence

$$m_A(\lambda) = p_1(\lambda)^{r_1} p_2(\lambda)^{r_2} \cdots p_s(\lambda)^{r_s}, \quad 0 \leqslant r_i \leqslant e_i, \ i = 1, \ldots, s.$$

For any $i \in \{1, 2, \ldots, s\}$, it's sufficient to show $r_i \geqslant 1$. Let $c \in \mathbb{C}$ be a root of $p_i(\lambda) = 0$. Then $f_A(c) = 0$. Note that A can be viewed as a complex matrix, then c is an eigenvalue of A. Let α be an eigenvector of A associated with c, i.e., $\alpha \neq 0$, $A\alpha = c\alpha$. By Example 5.5.16, we know that α is an eigenvector of $m_A(A)$ associated with $m_A(c)$, i.e., $m_A(A)\alpha = m_A(c)\alpha$. From $m_A(A) = 0$ and $\alpha \neq 0$, we get $m_A(c) = 0$. Since $p_i(\lambda) \in F[\lambda]$ is irreducible and $m_A(\lambda) \in F[\lambda]$, we obtain that $p_i(\lambda)|m_A(\lambda)$. Hence $1 \leqslant r_i \leqslant e_i$.

(4) Let $B = P^{-1}AP$ for some $P \in \mathrm{GL}_n(F)$. Then

$$m_A(B) = m_A(P^{-1}AP) = P^{-1}m_A(A)P = 0.$$

By (1), $m_B(\lambda)|m_A(\lambda)$. Similarly, $m_A(\lambda)|m_B(\lambda)$. Hence $m_A(\lambda) = m_B(\lambda)$ since $m_A(\lambda)$ and $m_B(\lambda)$ are monic.

(5) Put $m(\lambda) = [m_A(\lambda), m_C(\lambda)]$. On the one hand,

$$m(D) = \begin{pmatrix} m(A) & 0 \\ 0 & m(C) \end{pmatrix} = \begin{pmatrix} 0 & 0 \\ 0 & 0 \end{pmatrix} = 0.$$

Hence, $m_D(\lambda)|m(\lambda)$ by (2). On the other hand,

$$0 = m_D(D) = \begin{pmatrix} m_D(A) & 0 \\ 0 & m_D(C) \end{pmatrix}.$$

This implies that $m_D(A) = 0$ and $m_D(C) = 0$. By (2), we get $m_A(\lambda)|m_D(\lambda)$, $m_C(\lambda)|m_D(\lambda)$, and so $[m_A(\lambda), m_C(\lambda)]|m_D(\lambda)$, i.e., $m(\lambda)|m_D(\lambda)$. Hence we obtain that $m_D(\lambda) = m(\lambda) = [m_A(\lambda), m_C(\lambda)]$ since $m_D(\lambda)$ and $m(\lambda)$ are monic. $\square$

Let $A, B \in \mathrm{M}_n(F)$. From Remark 5.5.11, we know that AB and BA have the same characteristic polynomials, but AB and BA don't necessarily have to have the same minimal polynomials. For example, $A = \begin{pmatrix} 0 & 0 \\ 0 & 1 \end{pmatrix}$, $B = \begin{pmatrix} 0 & 1 \\ 0 & 0 \end{pmatrix}$, then

$$AB = \begin{pmatrix} 0 & 0 \\ 0 & 0 \end{pmatrix}, \quad BA = \begin{pmatrix} 0 & 1 \\ 0 & 0 \end{pmatrix}, \quad f_{AB}(\lambda) = \lambda^2 = f_{BA}(\lambda),$$

and $m_{AB}(\lambda) = \lambda \neq \lambda^2 = m_{BA}(\lambda)$.

Example 5.6.6. Let $A = aI_n$, $a \in F$. Then $f_A(\lambda) = (\lambda - a)^n$, $m_A(\lambda) = \lambda - a$.

Example 5.6.7. Let $A = J_n(\lambda_0) = \begin{pmatrix} \lambda_0 & & & \\ 1 & \ddots & & \\ & \ddots & \ddots & \\ & & 1 & \lambda_0 \end{pmatrix} \in \mathrm{M}_n(F)$.

Prove $m_A(\lambda) = f_A(\lambda) = (\lambda - \lambda_0)^n$.

Proof. It's obvious that $f_A(\lambda) = (\lambda - \lambda_0)^n$. Hence

$$m_A(\lambda) = (\lambda - \lambda_0)^r, \ 1 \leqslant r \leqslant n.$$

Put

$$N = \begin{pmatrix} 0 & & & \\ 1 & \ddots & & \\ & \ddots & \ddots & \\ & & 1 & 0 \end{pmatrix}_{n \times n}.$$

Then $A - \lambda_0 I_n = N$ and $N^n = 0$. But for any $1 \leqslant k \leqslant n-1$, $N^k \neq 0$. Hence $m_A(\lambda) = (\lambda - \lambda_0)^n$. $\qquad\square$

Example 5.6.8. Let $f(\lambda) = c_0 + c_1\lambda + \cdots + c_{n-1}\lambda^{n-1} + \lambda^n \in F[\lambda]$. Put

$$C = C(f) = \begin{pmatrix} 0 & & & & -c_0 \\ 1 & \ddots & & & -c_1 \\ & \ddots & \ddots & & \vdots \\ & & \ddots & 0 & -c_{n-2} \\ & & & 1 & -c_{n-1} \end{pmatrix}_{n \times n}.$$

Then $m_C(\lambda) = f_C(\lambda) = f(\lambda)$. The matrix $C(f)$ is called the **companion matrix** of the monic polynomial $f(\lambda)$.

Proof. It's easy to calculate that $f_C(\lambda) = |\lambda I_n - C| = f(\lambda)$ (see Exercise 14 (1) of Chapter 2).

Let $I_n = (\varepsilon_1, \varepsilon_2, \ldots, \varepsilon_n)$. Then

$$C\varepsilon_1 = \varepsilon_2, C^2\varepsilon_1 = C\varepsilon_2 = \varepsilon_3, \ldots, C^{n-1}\varepsilon_1 = C\varepsilon_{n-1} = \varepsilon_n.$$

If $g(\lambda) = a_0 + a_1\lambda + \cdots + a_{n-1}\lambda^{n-1} \in F[\lambda]$ satisfies $g(C) = 0$, then

$$0 = g(C)\varepsilon_1 = a_0\varepsilon_1 + a_1 C\varepsilon_1 + \cdots + a_{n-1}C^{n-1}\varepsilon_1$$

$$= a_0\varepsilon_1 + a_1\varepsilon_2 + \cdots + a_{n-1}\varepsilon_n.$$

Since $\varepsilon_1, \varepsilon_2, \ldots, \varepsilon_n$ are linearly independent, $a_i = 0$, $i = 0, 1, \ldots, n - 1$, whence $g(\lambda) = 0$. Thus $\deg(m_C(\lambda)) \geqslant n$, and so $m_C(\lambda) = f_C(\lambda) = f(\lambda)$. $\qquad\square$

Remark 5.6.9. Let V be an n-dimensional linear space and $\sigma \in \mathrm{End}(V)$. We can define an **annihilating polynomial** and the **minimal polynomial** of σ by replacing A with σ in Definition 5.6.1. Usually, we denote by $m_\sigma(\lambda)$ the minimal polynomial of σ.

Theorem 5.6.10. *Let $\alpha_1, \alpha_2, \ldots, \alpha_n$ be a basis for an n-dimensional linear space V and $\sigma \in \mathrm{End}(V)$. Assume that A is the matrix of σ with respect to the basis $\alpha_1, \alpha_2, \ldots, \alpha_n$.*

(1) *$g(\lambda) \in F[\lambda]$ is an annihilating polynomial of σ if and only if $g(\lambda)$ is an annihilating polynomial of A.*

(2) *(Cayley–Hamilton Theorem) $f_\sigma(\sigma) = 0$.*

(3) *$m(\lambda) \in F[\lambda]$ is the minimal polynomial of σ if and only if $m(\lambda)$ is the minimal polynomial of A, i.e., $m_\sigma(\lambda) = m_A(\lambda)$.*

Proof. (1) Let $g(\lambda) \in F[\lambda]$. By Theorem 5.3.5 (2) and Corollary 5.3.9 (1), we have

$$g(\sigma)(\alpha_1, \alpha_2, \ldots, \alpha_n) = (\alpha_1, \alpha_2, \ldots, \alpha_n)g(A).$$

Hence, $g(\sigma) = 0 \iff g(A) = 0$.

(2) and (3) come form (1). $\qquad\qquad\qquad\qquad\qquad\qquad\qquad\qquad\square$

5.7　Diagonalization of Matrices

In order to study the invariants of a square matrix under similarity, such as the characteristic polynomial, trace, determinant, etc., we hope to find a relatively simple representative element with good properties in a similarity class. Obviously, the diagonal matrix is simple and has good properties, but not every square matrix is similar to a diagonal matrix. Therefore, in this section, we study the conditions under which a square matrix is similar to a diagonal matrix.

Definition 5.7.1. Let $A \in \mathrm{M}_n(F)$ with an eigenvalue $\lambda_0 \in F$ and let V_{λ_0} be the eigenspace of A associated with λ_0. The characteristic polynomial of A has the form $f_A(\lambda) = (\lambda - \lambda_0)^e g(\lambda)$, where $g(\lambda) \in F[\lambda]$ and $g(\lambda_0) \neq 0$. The number e is called the **algebraic multiplicity** of λ_0, and $\dim V_{\lambda_0}$ is called the **geometric multiplicity** of λ_0.

Similarly, we can define the algebraic multiplicity and geometric multiplicity of an eigenvalue of a linear transformation.

Lemma 5.7.2. *Let λ_0 be an eigenvalue of $A \in \mathrm{M}_n(F)$ (or a linear transformation σ of an n-dimensional linear space V). Then the geometric multiplicity of λ_0 is less than or equal to the algebraic multiplicity of λ_0.*

Proof. Let e be the algebraic multiplicity of λ_0 and $r = \dim V_{\lambda_0}$, where $V_{\lambda_0} \subseteq F^n$ is the eigenspace of A associated with λ_0. Then

$$f_A(\lambda) = (\lambda - \lambda_0)^e g(\lambda),$$

where $g(\lambda) \in F[\lambda]$ and $g(\lambda_0) \neq 0$. Extend a basis $\alpha_1, \ldots, \alpha_r$ for V_{λ_0} to a basis $\alpha_1, \ldots, \alpha_r, \alpha_{r+1}, \ldots, \alpha_n$ for F^n. Put

$$P = (\alpha_1, \ldots, \alpha_r, \alpha_{r+1}, \ldots, \alpha_n).$$

Then $P \in \mathrm{GL}_n(F)$ and

$$AP = A(\alpha_1, \ldots, \alpha_r, \alpha_{r+1}, \ldots, \alpha_n)$$

$$= (\alpha_1, \ldots, \alpha_r, \alpha_{r+1}, \ldots, \alpha_n) \begin{pmatrix} \lambda_0 I_r & \star \\ 0 & A_1 \end{pmatrix}$$

$$= P \begin{pmatrix} \lambda_0 I_r & \star \\ 0 & A_1 \end{pmatrix}.$$

Hence, $P^{-1} A P = \begin{pmatrix} \lambda_0 I_r & \star \\ 0 & A_1 \end{pmatrix}$ and we obtain that

$$(\lambda - \lambda_0)^e g(\lambda) = f_A(\lambda) = f_{P^{-1}AP}(\lambda) = (\lambda - \lambda_0)^r f_{A_1}(\lambda).$$

Since $g(\lambda_0) \neq 0$, we get $r \leqslant e$. $\qquad\qquad\square$

Lemma 5.7.3. *Let $A \in \mathrm{M}_n(F)$ and let $\lambda_1, \lambda_2, \ldots, \lambda_s \in F$ be s different eigenvalues of A. Denote by V_{λ_i} the eigenspace of A associated with λ_i, $1 \leqslant i \leqslant s$. Then $\sum_{i=1}^{s} V_{\lambda_i} = \bigoplus_{i=1}^{s} V_{\lambda_i}$, i.e., the eigenvectors of A associated with different eigenvalues are linearly independent.*

Proof. We do an induction on s. The case $s = 1$ is trivial. Assume $s \geqslant 2$ and the result holds for $s - 1$. Consider the case that A has s different eigenvalues $\lambda_1, \lambda_2, \ldots, \lambda_s$ with the associated eigenspaces $V_{\lambda_1}, \ldots, V_{\lambda_s}$. For $i \in \{1, 2, \ldots, s\}$, choose a basis $\alpha_{i1}, \ldots, \alpha_{ir_i}$ for V_{λ_i}, where $r_i = \dim V_{\lambda_i}$. It's sufficient to show that $\alpha_{11}, \ldots, \alpha_{1r_1}, \alpha_{21}, \ldots, \alpha_{2r_2}, \ldots, \alpha_{s1}, \ldots, \alpha_{sr_s}$ are linearly independent.

Suppose

$$\sum_{i=1}^{s} \left(k_{i1}\alpha_{i1} + k_{i2}\alpha_{i2} + \cdots + k_{ir_i}\alpha_{ir_i}\right) = 0, \qquad (5.18)$$

where $k_{i1}, k_{i2}, \ldots, k_{ir_i} \in F$, $i = 1, 2, \ldots, s$. Multiplying both sides of (5.18) by A and λ_s on the left, respectively, we have

$$\sum_{i=1}^{s} \lambda_i \left(k_{i1}\alpha_{i1} + k_{i2}\alpha_{i2} + \cdots + k_{ir_i}\alpha_{ir_i}\right) = 0 \qquad (5.19)$$

and

$$\lambda_s \sum_{i=1}^{s} \left(k_{i1}\alpha_{i1} + k_{i2}\alpha_{i2} + \cdots + k_{ir_i}\alpha_{ir_i}\right) = 0. \qquad (5.20)$$

Subtracting (5.19) from (5.20) yields

$$\sum_{i=1}^{s-1} (\lambda_s - \lambda_i)\left(k_{i1}\alpha_{i1} + k_{i2}\alpha_{i2} + \cdots + k_{ir_i}\alpha_{ir_i}\right) = 0. \qquad (5.21)$$

By induction, $\alpha_{11}, \ldots, \alpha_{1r_1}, \ldots, \alpha_{s-1,1}, \ldots, \alpha_{s-1,r_{s-1}}$ are linearly independent, and hence

$$(\lambda_s - \lambda_i)k_{i1} = (\lambda_s - \lambda_i)k_{i2} = \cdots = (\lambda_s - \lambda_i)k_{ir_i} = 0, \quad i = 1, 2, \ldots, s-1.$$

Since $\lambda_s - \lambda_i \neq 0, i = 1, 2, \ldots, s - 1$, we get

$$k_{i1} = k_{i2} = \cdots = k_{ir_i} = 0, \quad i = 1, 2, \ldots, s - 1.$$

Thus, by (5.18), we have $k_{s1}\alpha_{s1} + k_{s2}\alpha_{s2} + \cdots + k_{sr_s}\alpha_{sr_s} = 0$. The linear independence of $\alpha_{s1}, \alpha_{s2}, \ldots, \alpha_{sr_s}$ implies that $k_{s1} = k_{s2} = \cdots = k_{sr_s} = 0$. Hence $\alpha_{11}, \ldots, \alpha_{1r_1}, \alpha_{21}, \ldots, \alpha_{2r_2}, \ldots, \alpha_{s1}, \ldots, \alpha_{sr_s}$ are linearly independent and $V_{\lambda_1} + V_{\lambda_2} + \cdots + V_{\lambda_s} = V_{\lambda_1} \oplus V_{\lambda_2} \oplus \cdots \oplus V_{\lambda_s}$. $\square$

Let $A \in M_n(F)$. We say that A is **diagonalizable** if $P^{-1}AP$ is a diagonal matrix for some $P \in GL_n(F)$. Let σ be a linear transformation of a finite-dimensional linear space V. We say that σ is **diagonalizable** if there is a basis for V such that the matrix A of σ with respect to this basis is diagonal, i.e., there is a basis for V such that each vector in this basis is an eigenvector of σ. Hence it's sufficient to discuss the conditions for the diagonalization of matrices.

Theorem 5.7.4. *Let $A \in \mathrm{M}_n(F)$ and $f_A(\lambda) = \prod\limits_{i=1}^{s}(\lambda - \lambda_i)^{e_i}$, where $\lambda_1, \lambda_2, \ldots, \lambda_s \in F$ are different, $e_1, e_2, \ldots, e_s \in \mathbb{N}^*$. Let $V_{\lambda_i} \subseteq F^n$ be the eigenspace of A associated with λ_i, $i = 1, 2, \ldots, s$. Then the following statements are equivalent:*

(1) *A is diagonalizable.*

(2) *There are n linearly independent eigenvectors of A.*

(3) *$\sum\limits_{i=1}^{s} \dim V_{\lambda_i} = n$.*

(4) *For any $i = 1, 2, \ldots, s$, $\dim V_{\lambda_i} = e_i$.*

(5) *$\sum\limits_{i=1}^{s} (n - \mathrm{rank}(\lambda_i I_n - A)) = n$.*

(6) *$\sum\limits_{i=1}^{s} \mathrm{rank}(\lambda_i I_n - A) = (s - 1)n$.*

Proof.   It's easy to see that $(1) \Longleftrightarrow (2)$. In fact,

A is diagonalizable

$\Longleftrightarrow$ there is a matrix $P \in \mathrm{GL}_n(F)$ such that
$P^{-1}AP = \mathrm{diag}(\lambda_1, \lambda_2, \ldots, \lambda_n)$

$\Longleftrightarrow$ there is a matrix $P \in \mathrm{GL}_n(F)$ such that
$AP = P\mathrm{diag}(\lambda_1, \lambda_2, \ldots, \lambda_n)$

$\Longleftrightarrow$ there are n linearly independent vectors $\alpha_1, \alpha_2, \ldots, \alpha_n \in F^n$
such that
$$A(\alpha_1, \alpha_2, \ldots, \alpha_n) = (\alpha_1, \alpha_2, \ldots, \alpha_n)\mathrm{diag}(\lambda_1, \lambda_2, \ldots, \lambda_n)$$
$$= (\lambda_1\alpha_1, \lambda_2\alpha_2, \ldots, \lambda_n\alpha_n)$$

$\Longleftrightarrow$ there are n linearly independent vectors $\alpha_1, \alpha_2, \ldots, \alpha_n \in F^n$
such that $A\alpha_i = \lambda_i\alpha_i, i = 1, 2, \ldots, n$.

$\Longleftrightarrow$ there are n linearly independent eigenvectors of A.

It's obvious that

$$(3) \Longleftrightarrow (5) \Longleftrightarrow (6) \text{ and } (2) \xrightarrow{\text{Lemma } 5.7.3} (3) \xleftrightarrow{\text{Lemma } 5.7.2} (4).$$

To sum up, we have proved that $(1) \sim (6)$ are equivalent to each other. $\qquad\square$

Corollary 5.7.5. *If $A \in \mathrm{M}_n(F)$ has n different eigenvalues, then A is diagonalizable.*

Proof. By hypothesis, the algebraic multiplicity and geometric multiplicity of each eigenvalue are 1. Hence A is diagonalizable. $\square$

Corollary 5.7.6. *Let $A \in \mathrm{M}_n(F)$ and $g(\lambda) = \prod\limits_{i=1}^{t} (\lambda - x_i)^{r_i} \in F[\lambda]$, where $x_1, x_2, \ldots, x_t \in F$ are different, $r_1, r_2, \ldots, r_t \in \mathbb{N}^*$. If $g(A) = 0$, then*

$$\sum_{i=1}^{t} (n - \mathrm{rank}(x_i I_n - A)) = \sum_{i=1}^{t} \dim\mathrm{Ker}(x_i I_n - A) \leqslant n,$$

and the following statements are equivalent:

(1) *A is diagonalizable.*

(2) *$\sum\limits_{i=1}^{t} \dim\mathrm{Ker}(x_i I_n - A) = n$.*

(3) *$\sum\limits_{i=1}^{t} (n - \mathrm{rank}(x_i I_n - A)) = n$.*

(4) *$\sum\limits_{i=1}^{t} \mathrm{rank}(x_i I_n - A) = (t - 1)n$.*

Proof. On the one hand, by Theorem 5.6.5, we know that $m_A(\lambda)$ and $f_A(\lambda)$ have the same irreducible factors and $m_A(\lambda)|g(\lambda)$. Hence there exist $\lambda_1, \lambda_2, \ldots, \lambda_s \in \{x_1, x_2, \ldots, x_t\}$, $e_1, e_2, \ldots, e_s \in \mathbb{N}^*$ such that

$$f_A(\lambda) = (\lambda - \lambda_1)^{e_1}(\lambda - \lambda_2)^{e_2} \cdots (\lambda - \lambda_s)^{e_s}.$$

On the other hand, if x_i is not an eigenvalue of A, i.e., $x_i \neq \lambda_j$, $j = 1, 2, \ldots, s$, then $|x_i I_n - A| \neq 0$, i.e., $\mathrm{rank}(x_i I_n - A) = n$, and hence $\dim\mathrm{Ker}(x_i I_n - A) = 0$. By Lemma 5.7.2, we get that

$$\sum_{i=1}^{t} \dim\mathrm{Ker}(x_i I_n - A) = \sum_{j=1}^{s} \dim\mathrm{Ker}(\lambda_j I_n - A) \leqslant \sum_{j=1}^{s} e_j = n.$$

Note that

$$\sum_{j=1}^{s} \dim\mathrm{Ker}(\lambda_j I_n - A) = n \iff \sum_{i=1}^{t} \dim\mathrm{Ker}(x_i I_n - A) = n,$$

$$\sum_{j=1}^{s} (n - \mathrm{rank}(\lambda_j I_n - A)) = n \iff \sum_{i=1}^{t} (n - \mathrm{rank}(x_i I_n - A)) = n,$$

$$\sum_{j=1}^{s} \operatorname{rank}(\lambda_j I_n - A) = (s-1)n \iff \sum_{i=1}^{t} \operatorname{rank}(x_i I_n - A) = (t-1)n.$$

Therefore, from Theorem 5.7.4, we obtain that $(1) \sim (4)$ are equivalent to each other. $\square$

Corollary 5.7.7. *Let $A = \operatorname{diag}(A_1, A_2, \ldots, A_t)$, where $A_i \in \mathrm{M}_{n_i}(F)$, $i = 1, 2, \ldots, t$. Then A is diagonalizable if and only if $A_1, A_2, \ldots, A_t$ are all diagonalizable.*

Proof. Sufficiency. Assume that $A_1, A_2, \ldots, A_t$ are all diagonalizable. Then for each $i = 1, 2, \ldots, t$, there exists $P_i \in \mathrm{GL}_{n_i}(F)$ such that $P_i^{-1} A_i P_i = \Lambda_i$ is a diagonal matrix. Put $P = \operatorname{diag}(P_1, P_2, \ldots, P_t)$. Then P is invertible and $P^{-1}AP = \operatorname{diag}(\Lambda_1, \Lambda_2, \ldots, \Lambda_t)$. Hence A is diagonalizable.

Necessity. Assume that A is diagonalizable. Let

$$f_A(\lambda) = (\lambda - \lambda_1)^{e_1}(\lambda - \lambda_2)^{e_2} \cdots (\lambda - \lambda_s)^{e_s},$$

where $\lambda_1, \lambda_2, \ldots, \lambda_s \in F$ are different, $e_1, e_2, \ldots, e_s \in \mathbb{N}^*$, $n = \sum_{i=1}^{s} e_i$. By the Cayley–Hamilton theorem, $f_A(A) = 0$, hence, for any $j = 1, 2, \ldots, t$, $f_A(A_j) = 0$. By Corollary 5.7.6,

$$\sum_{i=1}^{s}(n_j - \operatorname{rank}(\lambda_i I_{n_j} - A_j)) \leqslant n_j, \quad j = 1, 2, \ldots, t.$$

By assumption and Theorem 5.7.4,

$$n = \sum_{i=1}^{s}(n - \operatorname{rank}(\lambda_i I_n - A))$$

$$= \sum_{i=1}^{s}\sum_{j=1}^{t}(n_j - \operatorname{rank}(\lambda_i I_{n_j} - A_j))$$

$$= \sum_{j=1}^{t}\sum_{i=1}^{s}(n_j - \operatorname{rank}(\lambda_i I_{n_j} - A_j))$$

$$\leqslant \sum_{j=1}^{t} n_j = n.$$

Hence, for any $j = 1, 2, \ldots, t$, $\sum_{i=1}^{s}(n_j - \operatorname{rank}(\lambda_i I_{n_j} - A_j)) = n_j$, and thus A_j is diagonalizable by Corollary 5.7.6. $\square$

Example 5.7.8 (Comparing with Exercise 57 of Chapter 2).
Let $A^2 = A \in \mathrm{M}_n(F)$ and $r = \mathrm{rank}A$. Prove that A is diagonalizable

and there exists a matrix $P \in \mathrm{GL}_n(F)$ such that $P^{-1}AP = \begin{pmatrix} I_r & 0 \\ 0 & 0 \end{pmatrix}$,

hence $f_A(\lambda) = \lambda^{n-r}(\lambda - 1)^r$.

Proof. Let $g(\lambda) = \lambda^2 - \lambda = \lambda(\lambda - 1)$. Then $g(A) = 0$. By Theorem
5.6.5, we know that the minimal polynomial $m_A(\lambda)$ of A is λ, or
$\lambda - 1$, or $\lambda(\lambda - 1)$. Hence, the characteristic polynomial of A is given
by $f_A(\lambda) = \lambda^{n-t}(\lambda - 1)^t$, where $0 \leqslant t \leqslant n$. If $t = 0$ or $t = n$, then
$A = 0$ or $A = I_n$, and so the conclusion is true. Suppose $0 < t < n$.
By Corollary 5.7.6, we know that A is diagonalizable if and only if

$$\mathrm{rank}(0I_n - A) + \mathrm{rank}(I_n - A) = \mathrm{rank}A + \mathrm{rank}(I_n - A) = n.$$

From $A^2 = A$, i.e., $A(I_n - A) = 0$, we get $\mathrm{rank}A + \mathrm{rank}(I_n - A) \leqslant n$.
It's easy to see that

$$n = \mathrm{rank}(I_n) = \mathrm{rank}(A + I_n - A) \leqslant \mathrm{rank}A + \mathrm{rank}(I_n - A),$$

Hence, $\mathrm{rank}A + \mathrm{rank}(I_n - A) = n$. Therefore, A is diagonalizable and

there exists a matrix $P \in \mathrm{GL}_n(F)$ such that $P^{-1}AP = \begin{pmatrix} I_t & 0 \\ 0 & 0 \end{pmatrix}$. It

follows that $t = \mathrm{rank}(P^{-1}AP) = \mathrm{rank}A = r$. $\qquad\square$

Replacing A by a transformation σ of a linear space in Theorem
5.7.4, we have the following theorem.

Theorem 5.7.9. *Let V be an n-dimensional linear space and $\sigma \in$
$\mathrm{End}(V)$. Assume that the characteristic polynomial $f_\sigma(\lambda)$ of σ has
a factorization $f_\sigma(\lambda) = \prod_{i=1}^{s}(\lambda - \lambda_i)^{e_i}$, where $\lambda_1, \lambda_2, \ldots, \lambda_s \in F$ are
different, $e_1, e_2, \ldots, e_s \in \mathbb{N}^*$. Let $V_{\lambda_i} = \mathrm{Ker}(\lambda_i 1_V - \sigma) \subseteq V$ be the
eigenspace of σ associated with λ_i, $i = 1, 2, \ldots, s$. Then the following
statements are equivalent:*

(1) *σ is diagonalizable.*
(2) *There are n linearly independent eigenvectors of σ.*
(3) $\sum_{i=1}^{s} \dim V_{\lambda_i} = n.$
(4) *For any $i \in \{1, 2, \ldots, s\}$, $\dim V_{\lambda_i} = e_i$.*

(5) $\displaystyle\sum_{i=1}^{s}\left(n - \operatorname{rank}(\lambda_i 1_V - \sigma)\right) = n.$

(6) $\displaystyle\sum_{i=1}^{s}\operatorname{rank}(\lambda_i 1_V - \sigma) = (s-1)n.$

For $A \in \mathrm{M}_n(F)$, Theorem 5.7.4 not only provides a criterion for determining whether A can be diagonalized, but also its proof provides a method of finding an invertible matrix P such that $P^{-1}AP$ is diagonal if A can be diagonalized. The process is as follows:

(1) Calculate the characteristic polynomial $f_A(\lambda)$ of A and factor $f_A(\lambda)$ over F.

(2) If $f_A(\lambda)$ has an irreducible factor over F of degree greater than one, then A cannot be diagonalized.

(3) Assume $f_A(\lambda) = \displaystyle\prod_{i=1}^{s}(\lambda - \lambda_i)^{e_i}$, where $\lambda_1, \lambda_2, \ldots, \lambda_s \in F$ are different and $e_1, e_2, \ldots, e_s \in \mathbb{N}^*$. For every eigenvalue λ_i, solving the homogeneous linear system $(\lambda_i I_n - A)\,X = 0$, we obtain a basis $\alpha_{i1}, \alpha_{i2}, \ldots, \alpha_{in_i}$ for the eigenspace V_{λ_i} of A associated with λ_i, $i = 1, 2, \ldots, s$.

(a) If there exists $i_0 \in \{1, 2, \ldots, s\}$ such that $n_{i_0} \neq e_{i_0}$, then A cannot be diagonalized.

(b) If $n_i = e_i$ for any $i \in \{1, 2, \ldots, s\}$, then A is diagonalizable. In this case, $P = (\alpha_{11}, \ldots, \alpha_{1e_1}, \alpha_{21}, \ldots, \alpha_{2e_2}, \ldots, \alpha_{s1}, \ldots, \alpha_{se_s}) \in$ $\mathrm{GL}_n(F)$ and

$$P^{-1}AP = \begin{pmatrix} \lambda_1 I_{e_1} & & & \\ & \lambda_2 I_{e_2} & & \\ & & \ddots & \\ & & & \lambda_s I_{e_s} \end{pmatrix}.$$

Example 5.7.10. Determine whether the following matrix $A = A_i$ $(i = 1, 2, 3)$ can be diagonalized over $\mathbb{R}$. If it can be diagonalized, find the invertible real matrix P so that $P^{-1}AP$ is a diagonal matrix:

$$A_1 = \begin{pmatrix} 0 & 0 & 1 \\ 1 & 0 & -1 \\ 0 & 1 & 1 \end{pmatrix}, \quad A_2 = \begin{pmatrix} 2 & 0 & 1 \\ 0 & 1 & 1 \\ 0 & 0 & 1 \end{pmatrix}, \quad A_3 = \begin{pmatrix} 28 & -4 & -8 \\ 6 & 10 & -4 \\ 9 & -13 & 10 \end{pmatrix}.$$

Solution. (1) The characteristic polynomial $f_{A_1}(\lambda) = |\lambda I_3 - A_1| = (\lambda - 1)\left(\lambda^2 + 1\right)$ has an irreducible factor $\lambda^2 + 1$ over $\mathbb{R}$, hence A_1 cannot be diagonalized over $\mathbb{R}$.

(2) The characteristic polynomial of A_2 is $f_{A_2}(\lambda) = |\lambda I_3 - A_2| = (\lambda - 2)(\lambda - 1)^2$, the algebraic multiplicity and geometric multiplicity of the eigenvalue 1 are 2 and $1 (= 3 - \text{rank}(I_3 - A_2))$, respectively. Hence A_2 cannot be diagonalized over $\mathbb{R}$.

(3) The characteristic polynomial

$$f_{A_3}(\lambda) = |\lambda I_3 - A_3| = (\lambda - 8)(\lambda - 16)(\lambda - 24)$$

has three different roots $8, 16$, and 24. Hence A_3 can be diagonalized over $\mathbb{R}$.

Solving $(8I_3 - A_3)X = 0$ yields a basis $\alpha_1 = (1, 1, 2)'$ for the eigenspace V_8 of A_3 associated with 8.

Solving $(16I_3 - A_3)X = 0$ yields a basis $\alpha_2 = (2, 0, 3)'$ for the eigenspace V_{16} of A_3 associated with 16.

Solving $(24I_3 - A_3)X = 0$ yields a basis $\alpha_3 = (3, 1, 1)'$ for the eigenspace V_{24} of A_3 associated with 24.

Let $P = (\alpha_1, \alpha_2, \alpha_3)$. Then $P \in \text{GL}_3(\mathbb{R})$ and $P^{-1}A_3P = \text{diag}(8, 16, 24)$. $\qquad\square$

Remark 5.7.11.

(1) The matrix A_1 in Example 5.7.10 is diagonalizable over $\mathbb{C}$ since A_1 has three different complex roots: $1, i, -i$.

(2) Diagonalization of matrices brings convenience in calculating powers of a matrix: If $P^{-1}AP = \text{diag}(\lambda_1, \lambda_2, \ldots, \lambda_n)$, then for any positive integer m, we have

$$A^m = P\text{diag}(\lambda_1^m, \lambda_2^m, \ldots, \lambda_n^m)P^{-1}.$$

(3) If A is diagonalizable, then the invertible matrix P such that $P^{-1}AP$ is a diagonal matrix is not unique.

5.8 The First Decomposition Theorem

In Section 5.4, we use the theory of invariant subspaces of linear transformations to discuss sufficient and necessary conditions for a square matrix to be similar to a block diagonal matrix. In this section, we use the standard factorization of a characteristic polynomial to decompose the linear space into a direct sum of some invariant subspaces. Consequently, all square matrices are similar to block diagonal matrices.

Let V be an n-dimensional linear space and $\sigma \in \text{End}(V)$. By Example 5.3.3, for any

$$f(\lambda) = a_m\lambda^m + a_{m-1}\lambda^{m-1} + \cdots + a_1\lambda + a_0 \in F[\lambda],$$

$f(\sigma) = a_m\sigma^m + a_{m-1}\sigma^{m-1} + \cdots + a_1\sigma + a_0 1_V \in \text{End}(V)$. The following lemma gives some properties of the kernel $\text{Ker}(f(\sigma))$ and the image $\text{Im}(f(\sigma))$.

Lemma 5.8.1. *Let V be a linear space, $\sigma \in \text{End}(V)$, and $f(\lambda), g(\lambda) \in F[\lambda]$. Then the following statements are true:*

(1) $\text{Ker}(f(\sigma))$ *and* $\text{Im}(f(\sigma))$ *are both σ-subspaces.*
(2) *If $f(\sigma) = 0$, then $\text{Ker}(f(\sigma)) = V$. If $f(\sigma)$ is invertible, then $\text{Ker}(f(\sigma)) = 0$.*
(3) *If $g(\lambda)|f(\lambda)$, then $\text{Ker}(g(\sigma)) \subseteq \text{Ker}(f(\sigma))$.*
(4) $\text{Ker}(g(\sigma)) \cap \text{Ker}(f(\sigma)) = \text{Ker}(d(\sigma))$, *where* $d(\lambda) = (f(\lambda), g(\lambda))$.
(5) $\text{Ker}(g(\sigma)) + \text{Ker}(f(\sigma)) = \text{Ker}(M(\sigma))$, *where* $M(\lambda) = [f(\lambda), g(\lambda)]$.
(6) *If* $(f(\lambda), g(\lambda)) = 1$, *then*

$$\text{Ker}(f(\sigma)g(\sigma)) = \text{Ker}(f(\sigma)) + \text{Ker}(g(\sigma)) = \text{Ker}(f(\sigma)) \oplus \text{Ker}(g(\sigma)).$$

(7) *If* $f(\lambda) = \prod_{i=1}^{s} f_i(\lambda)$, *where* $f_1(\lambda), f_2(\lambda), \ldots, f_s(\lambda) \in F[\lambda]$ *are pairwise coprime, then*

$$\text{Ker}(f(\sigma)) = \bigoplus_{i=1}^{s} \text{Ker}(f_i(\sigma)).$$

Furthermore, if $f(\sigma) = 0$, then

$$V = \bigoplus_{i=1}^{s} \text{Ker}(f_i(\sigma)) = \bigoplus_{i=1}^{s} \text{Im}(g_i(\sigma)),$$

where $g_i(\lambda) = \frac{f(\lambda)}{f_i(\lambda)}$, $i = 1, 2, \ldots, s$.

Proof. (1) From $f(\sigma)\sigma = \sigma f(\sigma)$ and Example 5.4.3, it's easy to see that $\text{Ker}(f(\sigma))$ and $\text{Im}(f(\sigma))$ are both σ-subspaces.

(2) is obviously true.

(3) By assumption, we have $f(\lambda) = h(\lambda)g(\lambda)$ for some $h(\lambda) \in F[\lambda]$. Let $\alpha \in \mathrm{Ker}(g(\sigma))$, i.e., $g(\sigma)(\alpha) = 0$. Then

$$f(\sigma)(\alpha) = (h(\sigma)g(\sigma))(\alpha) = h(\sigma)(g(\sigma)(\alpha)) = h(\sigma)(0) = 0.$$

Hence, $\alpha \in \mathrm{Ker}(f(\sigma))$ and so $\mathrm{Ker}(g(\sigma)) \subseteq \mathrm{Ker}(f(\sigma))$.

(4) Let $d(\lambda) = (f(\lambda), g(\lambda))$. Then

$$\mathrm{Ker}(d(\sigma)) \subseteq \mathrm{Ker}(f(\sigma)) \cap \mathrm{Ker}(g(\sigma))$$

by (3), and there exist $u(\lambda), v(\lambda) \in F[\lambda]$ such that

$$u(\lambda)f(\lambda) + v(\lambda)g(\lambda) = d(\lambda)$$

by Theorem 1.6.11. Thus, for any $\alpha \in \mathrm{Ker}(f(\sigma)) \cap \mathrm{Ker}(g(\sigma))$, we have

$$\begin{aligned}
d(\sigma)(\alpha) &= (u(\sigma)f(\sigma) + v(\sigma)g(\sigma))(\alpha) \\
&= u(\sigma)(f(\sigma)(\alpha)) + v(\sigma)(g(\sigma)(\alpha)) \\
&= u(\sigma)(0) + v(\sigma)(0) = 0,
\end{aligned}$$

and so $\alpha \in \mathrm{Ker}(d(\sigma))$. It follows that $\mathrm{Ker}(f(\sigma)) \cap \mathrm{Ker}(g(\sigma)) \subseteq \mathrm{Ker}(d(\sigma))$, and hence $\mathrm{Ker}(g(\sigma)) \cap \mathrm{Ker}(f(\sigma)) = \mathrm{Ker}(d(\sigma))$.

(5) Let $M(\lambda) = [f(\lambda), g(\lambda)]$. Then $\mathrm{Ker}(g(\sigma))$ and $\mathrm{Ker}(f(\sigma))$ are both subspaces of $\mathrm{Ker}(M(\sigma))$ by (3), thus $\mathrm{Ker}(g(\sigma)) + \mathrm{Ker}(f(\sigma)) \subseteq \mathrm{Ker}(M(\sigma))$.

Let $M(\lambda) = f(\lambda)f_1(\lambda) = g(\lambda)g_1(\lambda)$ for some $f_1(\lambda), g_1(\lambda) \in F[\lambda]$. Then $(f_1(\lambda), g_1(\lambda)) = 1$. From Theorem 1.6.11, there exist $u(\lambda), v(\lambda) \in F[\lambda]$ such that $f_1(\lambda)u(\lambda) + g_1(\lambda)v(\lambda) = 1$. Let $\alpha \in \mathrm{Ker}(M(\sigma))$. Then

$$\alpha = (f_1(\sigma)u(\sigma) + g_1(\sigma)v(\sigma))(\alpha) = (f_1(\sigma)u(\sigma))(\alpha) + (g_1(\sigma)v(\sigma))(\alpha).$$

Since

$$\begin{aligned}
f(\sigma)((f_1(\sigma)u(\sigma))(\alpha)) &= M(\sigma)u(\sigma)(\alpha) = u(\sigma)(M(\sigma)(\alpha)) \\
&= u(\sigma)(0) = 0,
\end{aligned}$$

we have $(f_1(\sigma)u(\sigma))(\alpha) \in \mathrm{Ker}(f(\sigma))$. Similarly, $(g_1(\sigma)v(\sigma))(\alpha) \in \mathrm{Ker}(g(\sigma))$. So, $\alpha = (f_1(\sigma)u(\sigma))(\alpha) + (g_1(\sigma)v(\sigma))(\alpha) \in \mathrm{Ker}(f(\sigma)) +$

$\mathrm{Ker}(g(\sigma))$. Therefore, $\mathrm{Ker}(M(\sigma) \subseteq \mathrm{Ker}(f(\sigma)) + \mathrm{Ker}(g(\sigma))$, and so
$$\mathrm{Ker}(M(\sigma)) = \mathrm{Ker}(f(\sigma)) + \mathrm{Ker}(g(\sigma)).$$

(6) If $(f(\lambda), g(\lambda)) = 1$, then $\mathrm{Ker}(f(\sigma)) \cap \mathrm{Ker}(g(\sigma)) = 0$ by (4). Hence
$$\mathrm{Ker}(f(\sigma)) + \mathrm{Ker}(g(\sigma)) = \mathrm{Ker}(f(\sigma)) \oplus \mathrm{Ker}(g(\sigma)).$$
Since $M(\lambda) = [f(\lambda), g(\lambda)]$ is a nonzero constant multiple of $f(\lambda)g(\lambda)$, by (5), we have
$$\mathrm{Ker}(f(\sigma)g(\sigma)) = \mathrm{Ker}(M(\sigma)) = \mathrm{Ker}(f(\sigma)) + \mathrm{Ker}(g(\sigma))$$
$$= \mathrm{Ker}(f(\sigma)) \oplus \mathrm{Ker}(g(\sigma)).$$

(7) By assumption, $(f_i(\lambda), f_j(\lambda)) = 1$ for any $1 \leqslant i \neq j \leqslant s$. Due to (6), we get
$$\mathrm{Ker}(f(\sigma)) = \bigoplus_{i=1}^{s} \mathrm{Ker}(f_i(\sigma)).$$

Furthermore, if $f(\sigma) = 0$, then $V = \mathrm{Ker}(f(\sigma)) = \bigoplus_{i=1}^{s} \mathrm{Ker}(f_i(\sigma))$.

Let $g_i(\lambda) = \frac{f(\lambda)}{f_i(\lambda)}$. Then $(f_i(\lambda), g_i(\lambda)) = 1$. Next we prove that
$$\mathrm{Ker}(f_i(\sigma)) = \mathrm{Im}(g_i(\sigma)), \quad i = 1, 2, \ldots, s.$$

Let $\beta \in \mathrm{Im}(g_i(\sigma))$. Then there exists $\alpha \in V$ such that $\beta = g_i(\sigma)(\alpha)$. So
$$f_i(\sigma)(\beta) = f_i(\sigma)(g_i(\sigma)(\alpha)) = (f_i(\sigma)g_i(\sigma))(\alpha) = f(\sigma)(\alpha) = 0.$$
Therefore, $\beta \in \mathrm{Ker}(f_i(\sigma))$, and hence
$$\mathrm{Im}(g_i(\sigma)) \subseteq \mathrm{Ker}(f_i(\sigma)). \tag{5.22}$$
Since $(f_i(\lambda), g_i(\lambda)) = 1$, by (6), we get
$$V = \mathrm{Ker}(f(\sigma)) = \mathrm{Ker}(f_i(\sigma)) \oplus \mathrm{Ker}(g_i(\sigma)),$$
and so $\dim V = \dim\mathrm{Ker}(f_i(\sigma)) + \dim\mathrm{Ker}(g_i(\sigma))$. Since
$$\dim V = \dim\mathrm{Im}(g_i(\sigma)) + \dim\mathrm{Ker}(g_i(\sigma)),$$
it follows that
$$\dim\mathrm{Ker}(f_i(\sigma)) = \dim\mathrm{Im}(g_i(\sigma)). \tag{5.23}$$
From (5.22) and (5.23), we get $\mathrm{Ker}(f_i(\sigma)) = \mathrm{Im}(g_i(\sigma))$. Hence
$$V = \bigoplus_{i=1}^{s} \mathrm{Ker}(f_i(\sigma)) = \bigoplus_{i=1}^{s} \mathrm{Im}(g_i(\sigma)). \qquad \square$$

Theorem 5.8.2 (First decomposition theorem). *Let V be an n-dimensional linear space and $\sigma \in \mathrm{End}(V)$. Assume that the characteristic polynomial and minimal polynomial of σ have the following factorizations:*

$$f_\sigma(\lambda) = p_1(\lambda)^{e_1} p_2(\lambda)^{e_2} \cdots p_s(\lambda)^{e_s},$$

$$m_\sigma(\lambda) = p_1(\lambda)^{r_1} p_2(\lambda)^{r_2} \cdots p_s(\lambda)^{r_s},$$

where $p_1(\lambda), p_2(\lambda), \ldots, p_s(\lambda) \in F[\lambda]$ are different monic irreducible polynomials, $1 \leqslant r_i \leqslant e_i$, $i = 1, 2, \ldots, s$. Then the following statements hold:

(1) *For any $i \in \{1, 2, \ldots, s\}$, $l \in \mathbb{N}$, we have*

$$\mathrm{Ker}\,(p_i(\sigma)^{r_i})$$

$$= \mathrm{Ker}\left(p_i(\sigma)^{r_i+l}\right)$$

$$= \left\{\alpha \in V \mid \text{there exists } k \in \mathbb{N}^* \text{ such that } p_i(\sigma)^k(\alpha) = 0\right\}.$$

In particular, $\mathrm{Ker}\,(p_i(\sigma)^{r_i}) = \mathrm{Ker}\,(p_i(\sigma)^{e_i})$.

(2) *$V = \bigoplus\limits_{i=1}^{s} W_i$, where $W_i = \mathrm{Ker}\,(p_i(\sigma)^{r_i})$, $i = 1, 2, \ldots, s$.*

(3) *For any $i \in \{1, 2, \ldots, s\}$, let $\sigma_i = \sigma|_{W_i}$, then $f_{\sigma_i}(\lambda) = p_i(\lambda)^{e_i}$ and $m_{\sigma_i}(\lambda) = p_i(\lambda)^{r_i}$ are the characteristic polynomial and minimal polynomial of σ_i, respectively. Hence $\dim W_i = \deg(f_{\sigma_i}(\lambda)) = \deg(p_i(\lambda)^{e_i})$.*

(4) *For any $i \in \{1, 2, \ldots, s\}$, let*

$$\pi_i : V = \bigoplus_{j=1}^{s} W_j \longrightarrow W_i \subset V, \quad \alpha = \sum_{j=1}^{s} \alpha_j \longmapsto \alpha_i.$$

Then there exists $h_i(x) \in F[x]$ such that $\pi_i = h_i(\sigma)$.

(5) *Let $\tau \in \mathrm{End}(V)$ and $\sigma\tau = \tau\sigma$. Then W_i is a τ-subspace, $i = 1, 2, \ldots, s$.*

Proof. (1) Let $1 \leqslant i \leqslant s$ and $1 \leqslant k \leqslant r_i$. Then $p_i(\lambda)^k | p_i(\lambda)^{r_i}$, and so $\mathrm{Ker}\,(p_i(\lambda)^k) \subseteq \mathrm{Ker}\,(p_i(\lambda)^{r_i})$ by Lemma 5.8.1 (3).

For any $k \geqslant r_i$, $p_i(\lambda)^{r_i} = (m_\sigma(\lambda), p_i(\lambda)^k)$, it follows that

$$\mathrm{Ker}\,(p_i(\sigma)^{r_i}) = \mathrm{Ker}\,(m_\sigma(\sigma)) \cap \mathrm{Ker}\,\left(p_i(\sigma)^k\right) = V \cap \mathrm{Ker}\,\left(p_i(\sigma)^k\right)$$

$$= \mathrm{Ker}\,\left(p_i(\sigma)^k\right),$$

and so

$$\mathrm{Ker}\,(p_i(\sigma)) \subseteq \mathrm{Ker}\,(p_i(\sigma)^2) \subseteq \cdots \subseteq \mathrm{Ker}\,(p_i(\sigma)^{r_i})$$

$$= \mathrm{Ker}\,(p_i(\sigma)^{r_i+1}) = \cdots.$$

Hence

$$\mathrm{Ker}\,(p_i(\sigma)^{e_i}) = \mathrm{Ker}\,(p_i(\sigma)^{r_i})$$

$$= \bigcup_{k=1}^{r_i} \mathrm{Ker}\,\left(p_i(\sigma)^k\right)$$

$$= \bigcup_{k=1}^{\infty} \mathrm{Ker}\,\left(p_i(\sigma)^k\right)$$

$$= \left\{\alpha \in V \mid \text{there exists } k \in \mathbb{N}^* \text{ such that } p_i(\sigma)^k(\alpha) = 0\right\}.$$

(2) follows from Lemma 5.8.1 (7).

(3) Let $1 \leqslant i \leqslant s$. We first prove that $m_{\sigma_i}(\lambda) = p_i(\lambda)^{r_i}$. Obviously, $p_i(\sigma_i)^{r_i} = p_i(\sigma)^{r_i}|_{W_i} = 0$. Let $g(\lambda) \in F[\lambda]$ and $g(\sigma_i) = 0$. Put $g_i(\lambda) = \dfrac{m_\sigma(\lambda)}{p_i(\lambda)^{r_i}}$ and $h(\lambda) = g(\lambda)g_i(\lambda)$. For any $\alpha \in V = \bigoplus_{j=1}^{s} W_j$, then

$$\alpha = \sum_{j=1}^{s} \alpha_j, \text{ where } \alpha_j \in W_j = \mathrm{Ker}\,(p_j(\sigma)^{r_j}), \text{ i.e., } p_j(\sigma)^{r_j}(\alpha_j) = 0,$$

$1 \leqslant j \leqslant s$.

If $j \neq i$, then $p_j(\lambda)^{r_j}|g_i(\lambda)$ and so $g_i(\sigma)(\alpha_j) = 0$. Therefore,

$$h(\sigma)(\alpha) = \sum_{j=1}^{s} h(\sigma)(\alpha_j)$$

$$= (g(\sigma)g_i(\sigma))(\alpha_i) + \sum_{j \neq i}(g(\sigma)g_i(\sigma))(\alpha_j)$$

$$= g_i(\sigma)(g(\sigma)(\alpha_i)) + \sum_{j \neq i}g(\sigma)(g_i(\sigma)(\alpha_j))$$

$$= g_i(\sigma)(g(\sigma_i)(\alpha_i)) + \sum_{j \neq i}g(\sigma)(0)$$

$$= 0.$$

From the arbitrariness of α, we know that $h(\sigma) = 0$. Hence $m_\sigma(\lambda)|h(\lambda)$, i.e., $p_i(\lambda)^{r_i}g_i(\lambda)|g(\lambda)g_i(\lambda)$, consequently $p_i(\lambda)^{r_i}|g(\lambda)$. Therefore, $m_{\sigma_i}(\lambda) = p_i(\lambda)^{r_i}$.

Let $n_i = \dim W_i$ and let $\alpha_{i1}, \ldots, \alpha_{in_i}$ be a basis for W_i. Note that the characteristic polynomial of σ_i has the form $f_{\sigma_i}(\lambda) = p_i(\lambda)^{k_i}$ for some $k_i \geqslant r_i$ by Theorem 5.6.5. Hence, $n_i = \deg(p_i(\lambda)^{k_i})$. Since $V = \bigoplus_{i=1}^{s} W_i$, we know that the sequence $\alpha_{11}, \ldots, \alpha_{1n_1}, \alpha_{21}, \ldots, \alpha_{2n_2}, \ldots, \alpha_{s1}, \ldots, \alpha_{sn_s}$ is a basis for V and

$$\sigma(\alpha_{11}, \ldots, \alpha_{1n_1}, \alpha_{21}, \ldots, \alpha_{2n_2}, \ldots, \alpha_{s1}, \ldots, \alpha_{sn_s})$$

$$= (\alpha_{11}, \ldots, \alpha_{1n_1}, \alpha_{21}, \ldots, \alpha_{2n_2}, \ldots, \alpha_{s1}, \ldots, \alpha_{sn_s})$$

$$\times \operatorname{diag}(A_1, A_2, \ldots, A_s),$$

where $\sigma(\alpha_{i1}, \ldots, \alpha_{in_i}) = (\alpha_{i1}, \ldots, \alpha_{in_i})A_i$, $A_i \in \mathrm{M}_{n_i}(F)$, $i = 1, 2, \ldots, s$. It's easy to see that $f_{A_i}(\lambda) = f_{\sigma_i}(\lambda) = p_i(\lambda)^{k_i}$ $(1 \leqslant i \leqslant s)$ and

$$\prod_{i=1}^{s} p_i(\lambda)^{e_i} = f_\sigma(\lambda) = \prod_{i=1}^{s} f_{A_i}(\lambda) = \prod_{i=1}^{s} p_i(\lambda)^{k_i}.$$

Hence, $e_i = k_i$, i.e., $f_{\sigma_i}(\lambda) = p_i(\lambda)^{e_i}$, and so

$$\dim W_i = \deg\left(f_{\sigma_i}(\lambda)\right) = \deg\left(p_i(\lambda)^{e_i}\right), \ 1 \leqslant i \leqslant s.$$

(4) By assumption and the definition of $g_i(\lambda)$ given in (3), we know that $(g_1(\lambda), g_2(\lambda), \ldots, g_s(\lambda)) = 1$. By Theorem 1.6.11, there exist $u_1(\lambda), u_2(\lambda), \ldots, u_s(\lambda) \in F[\lambda]$ such that $\sum_{i=1}^{s} g_i(\lambda)u_i(\lambda) = 1$. Hence $\sum_{i=1}^{s} g_i(\sigma)u_i(\sigma) = 1_V$. Let $\alpha \in V$. Then $\alpha = \sum_{i=1}^{s} g_i(\sigma)$ $u_i(\sigma)(\alpha) = \sum_{i=1}^{s} \alpha_i$, where $\alpha_i = (g_i(\sigma)u_i(\sigma))(\alpha) = g_i(\sigma)(u_i(\sigma)(\alpha)) \in \mathrm{Im}(g_i(\sigma)) = \mathrm{Ker}(p_i(\sigma)^{r_i}) = W_i$, $1 \leqslant i \leqslant s$. By definition, $\pi_i(\alpha) = \alpha_i = (g_i(\sigma)u_i(\sigma))(\alpha)$. Therefore, $\pi_i = g_i(\sigma)u_i(\sigma) = h_i(\sigma)$, where $h_i(\lambda) = g_i(\lambda)u_i(\lambda) \in F[\lambda], 1 \leqslant i \leqslant s$.

(5) Let $\tau \in \mathrm{End}(V)$ and $\sigma\tau = \tau\sigma$. It's obvious that $\tau p_i(\sigma)^{r_i} = p_i(\sigma)^{r_i}\tau$. By Example 5.4.3, we know that $W_i = \mathrm{Ker}\left(p_i(\sigma)^{r_i}\right)$ is a τ-subspace. $\qquad\square$

Corollary 5.8.3. *Let $A \in \mathrm{M}_n(F)$. Assume that the characteristic polynomial $f_A(\lambda)$ and minimal polynomial $m_A(\lambda)$ of A have the following factorizations:*

$$f_A(\lambda) = \prod_{i=1}^{s} p_i(\lambda)^{e_i}, \ m_A(\lambda) = \prod_{i=1}^{s} p_i(\lambda)^{r_i}, \ 1 \leqslant r_i \leqslant e_i, i = 1, 2, \ldots, s,$$

where $p_1(\lambda), p_2(\lambda), \ldots, p_s(\lambda) \in F[\lambda]$ are different monic irreducible polynomials. Then the following statements hold:

(1) *There exists $P \in \mathrm{GL}_n(F)$ such that*

$$A = P\mathrm{diag}(A_1, A_2, \ldots, A_s)P^{-1},$$

     where $A_i \in \mathrm{M}_{n_i}(F)$, $n_i = \deg\left(p_i(\lambda)^{e_i}\right)$, and $p_i(\lambda)^{e_i}$ and $p_i(\lambda)^{r_i}$ are respectively the characteristic polynomial and minimal polynomial of A_i, $i = 1, 2, \ldots, s$.

(2) *For any $i \in \{1, 2, \ldots, s\}$, there exists $h_i(\lambda) \in F[\lambda]$ such that*

$$P\mathrm{diag}(0, \ldots, 0, I_{n_i}, 0, \ldots, 0)P^{-1} = h_i(A).$$

(3) *If $B \in \mathrm{M}_n(F)$ satisfies $AB = BA$, then $B = P\mathrm{diag}(B_1, \ldots, B_s)P^{-1}$, where $B_i \in \mathrm{M}_{n_i}(F)$, $i = 1, 2, \ldots, s$.*

Proof. Let $\alpha_1, \alpha_2, \ldots, \alpha_n$ be a basis for an n-dimensional linear space V. From Theorem 5.3.5, there exist $\sigma, \tau \in \mathrm{End}(V)$ such that

$$\sigma(\alpha_1, \alpha_2, \ldots, \alpha_n) = (\alpha_1, \alpha_2, \ldots, \alpha_n)A,$$

$$\tau(\alpha_1, \alpha_2, \ldots, \alpha_n) = (\alpha_1, \alpha_2, \ldots, \alpha_n)B.$$

Hence, $f_\sigma(\lambda) = f_A(\lambda)$, $m_\sigma(\lambda) = m_A(\lambda)$, $\sigma\tau = \tau\sigma$ if and only if $AB = BA$.

By Theorem 5.8.2, we know that $V = \displaystyle\bigoplus_{i=1}^{s} W_i$, where $W_i = \mathrm{Ker}\left(p_i(\sigma)^{e_i}\right)$ is an n_i-dimensional σ-subspace, where $n_i = \deg\left(p_i(\lambda)^{e_i}\right)$, $1 \leqslant i \leqslant s$.

Let $\alpha_{i1}, \ldots, \alpha_{in_i}$ be a basis for W_i, $1 \leqslant i \leqslant s$.

(i) We have $\sigma(\alpha_{i1}, \ldots, \alpha_{in_i}) = (\alpha_{i1}, \ldots, \alpha_{in_i})A_i$, where $A_i \in \mathrm{M}_{n_i}(F)$, since $\sigma(W_i) \subseteq W_i$.

(ii) It's obvious that the sequence

$$\alpha_{11}, \ldots, \alpha_{1n_1}, \alpha_{21}, \ldots, \alpha_{2n_2}, \ldots, \alpha_{s1}, \ldots, \alpha_{sn_s}$$

is a basis for V and

$$\sigma(\alpha_{11}, \ldots, \alpha_{1n_1}, \alpha_{21}, \ldots, \alpha_{2n_2}, \ldots, \alpha_{s1}, \ldots, \alpha_{sn_s})$$
$$= (\alpha_{11}, \ldots, \alpha_{1n_1}, \alpha_{21}, \ldots, \alpha_{2n_2}, \ldots, \alpha_{s1}, \ldots, \alpha_{sn_s})$$
$$\times \operatorname{diag}(A_1, A_2, \ldots, A_s).$$

Let

$$(\alpha_{11}, \ldots, \alpha_{1n_1}, \alpha_{21}, \ldots, \alpha_{2n_2}, \ldots, \alpha_{s1}, \ldots, \alpha_{sn_s}) = (\alpha_1, \ldots, \alpha_n)P.$$
Then $P \in \operatorname{GL}_n(F)$ and

$$P^{-1}AP = \operatorname{diag}(A_1, A_2, \ldots, A_s), \text{ i.e., } A = P\operatorname{diag}(A_1, A_2, \ldots, A_s)P^{-1}.$$

From $AB = BA$, we know $\sigma\tau = \tau\sigma$, hence W_i is a τ-subspace, i.e., $\tau(W_i) \subseteq W_i$. Let

$$\tau(\alpha_{i1}, \ldots, \alpha_{in_i}) = (\alpha_{i1}, \ldots, \alpha_{in_i})B_i,$$
$$\text{where } B_i \in \operatorname{M}_{n_i}(F), \ i = 1, 2, \ldots, s.$$

Then

$$\tau(\alpha_{11}, \ldots, \alpha_{1n_1}, \alpha_{21}, \ldots, \alpha_{2n_2}, \ldots, \alpha_{s1}, \ldots, \alpha_{sn_s})$$
$$= (\alpha_{11}, \ldots, \alpha_{1n_1}, \alpha_{21}, \ldots, \alpha_{2n_2}, \ldots, \alpha_{s1}, \ldots, \alpha_{sn_s})$$
$$\times \operatorname{diag}(B_1, B_2, \ldots, B_s).$$

Therefore, $B = P\operatorname{diag}(B_1, B_2, \ldots, B_s)P^{-1}$.

For any $i \in \{1, 2, \ldots, s\}$, let

$$\pi_i : \ V = \bigoplus_{j=1}^{s} W_j \longrightarrow W_i \subset V, \ \alpha = \sum_{j=1}^{s} \alpha_j \longmapsto \alpha_i,$$

and $\pi_i(\alpha_1, \alpha_2, \ldots, \alpha_n) = (\alpha_1, \alpha_2, \ldots, \alpha_n)E_i$. Then for any $j = 1, 2, \ldots, s$, W_j is a π_i-invariant subspace, and $\pi_i|_{W_j} = \begin{cases} 1_{W_i}, & \text{if } j = i, \\ 0, & \text{if } j \neq i. \end{cases}$
Thus

$$\pi_i(\alpha_{i1}, \ldots, \alpha_{in_i}) = (\alpha_{i1}, \ldots, \alpha_{in_i})I_{n_i},$$
$$\pi_i(\alpha_{j1}, \ldots, \alpha_{jn_j}) = (\alpha_{i1}, \ldots, \alpha_{in_j})0, \ j \neq i,$$

and so

$$\pi_i(\alpha_{11}, \ldots, \alpha_{1n_1}, \ldots, \alpha_{i1}, \ldots, \alpha_{in_i}, \ldots, \alpha_{s1}, \ldots, \alpha_{sn_s})$$
$$= (\alpha_{11}, \ldots, \alpha_{1n_1}, \ldots, \alpha_{i1}, \ldots, \alpha_{in_i}, \ldots, \alpha_{s1}, \ldots, \alpha_{sn_s})D_i,$$

where $D_i = \mathrm{diag}(0, \ldots, 0, I_{n_i}, 0, \ldots, 0)$. Therefore, $PD_iP^{-1} = E_i$. From Theorem 5.8.2 (4), we know that there exists $h_i(\lambda) \in F[\lambda]$ such that $\pi_i = h_i(\sigma)$. Hence $E_i = h_i(A)$. $\qquad\square$

Corollary 5.8.4. *Let V be an n-dimensional linear space and $\sigma \in$ End(V). Assume that the characteristic polynomial and minimal polynomial of σ have the following factorizations:*

$$f_\sigma(\lambda) = \prod_{i=1}^{s}(\lambda - \lambda_i)^{e_i}, \quad m_\sigma(\lambda) = \prod_{i=1}^{s}(\lambda - \lambda_i)^{r_i},$$

where $\lambda_1, \lambda_2, \ldots, \lambda_s \in F$ are different, $1 \leqslant r_i \leqslant e_i$, $i = 1, 2, \ldots, s$. Then the following statements hold:

(1) *$V = \bigoplus_{i=1}^{s} W_i$, where $W_i = \mathrm{Ker}(\sigma - \lambda_i 1_V)^{r_i}$ with $\dim W_i = e_i, i = 1, 2, \ldots, s$.*

(2) *$f_{\sigma_i}(\lambda) = (\lambda - \lambda_i)^{e_i}$, $m_{\sigma_i}(\lambda) = (\lambda - \lambda_i)^{r_i}$, where $\sigma_i = \sigma|_{W_i}$, $i = 1, 2, \ldots, s$.*

(3) *Regard the projection $\pi_i : V \longrightarrow W_i$ as a linear transformation of V through $W_i \subseteq V$. Then π_i is a polynomial in σ, i.e., there exists $h_i(\lambda) \in F[\lambda]$ such that $\pi_i = h_i(\sigma)$, $i = 1, 2, \ldots, s$.*

(4) *If $\tau \in$ End(V) and $\sigma\tau = \tau\sigma$, then W_i is a τ-subspace, $i = 1, 2, \ldots, s$.*

(5) *For any $i \in \{1, 2, \ldots, s\}$, the eigenspace V_i of σ associated with the eigenvalue λ_i is a subspace of W_i. Furthermore, $V_i = W_i$ if and only if $\dim V_i = e_i$. Hence σ can be diagonalized if and only if $V_i = W_i$ for any $i \in \{1, 2, \ldots, s\}$.*

Definition 5.8.5. Let V be an n-dimensional linear space and $\sigma \in$ End(V). Assume that λ is an eigenvalue of σ with the algebraic multiplicity e. The subspace $W = \mathrm{Ker}(\sigma - \lambda 1_V)^e$ of V is called the **root subspace** of σ associated with the eigenvalue λ.

Corollary 5.8.6. *Let $A \in \mathrm{M}_n(F)$. Then A is diagonalizable if and only if $m_A(\lambda) = \prod_{i=1}^{s}(\lambda - \lambda_i)$, where $\lambda_1, \lambda_2, \ldots, \lambda_s \in F$ are different. In particular, when $F = \mathbb{C}$, A is diagonalizable if and only if the minimal polynomial $m_A(\lambda)$ of A has no multiple roots.*

Proof. Necessity. Assume that A is diagonalizable, i.e., there exists $P \in \mathrm{GL}_n(F)$ such that

$$P^{-1}AP = \Lambda = \mathrm{diag}(\lambda_1 I_{n_1}, \lambda_2 I_{n_2}, \ldots, \lambda_s I_{n_s}),$$

where $\lambda_1, \lambda_2, \ldots, \lambda_s \in F$ are different. The minimal polynomial of $\lambda_i I_{n_i}$ is $m_{\lambda_i I_{n_i}}(\lambda) = \lambda - \lambda_i$, $i = 1, 2, \ldots, s$. Hence, we have

$$m_A(\lambda) = m_\Lambda(\lambda) = \left[m_{\lambda_1 I_{n_1}}(\lambda), m_{\lambda_2 I_{n_2}}(\lambda), \ldots, m_{\lambda_s I_{n_s}}(\lambda) \right]$$

$$= \prod_{i=1}^{s} (\lambda - \lambda_i).$$

Sufficiency. Assume $m_A(\lambda) = \prod\limits_{i=1}^{s} (\lambda - \lambda_i)$, where $\lambda_1, \lambda_2, \ldots, \lambda_s \in F$ are different. By Corollary 5.8.3, there exists $P \in \mathrm{GL}_n(F)$ such that

$$P^{-1}AP = \mathrm{diag}(A_1, A_2, \ldots, A_s), \quad \text{and} \quad m_{A_i}(\lambda) = \lambda - \lambda_i, 1 \leqslant i \leqslant s.$$

Hence, $0 = m_{A_i}(A_i) = A_i - \lambda_i I_{n_i}$, i.e., $A_i = \lambda_i I_{n_i}$, $1 \leqslant i \leqslant s$. Therefore, $P^{-1}AP = \mathrm{diag}(\lambda_1 I_{n_1}, \lambda_2 I_{n_2}, \ldots, \lambda_s I_{n_s})$, i.e., A is diagonalizable. $\qquad\square$

Example 5.8.7. Let $A = \begin{pmatrix} 0 & -1 & 2 & 0 \\ 1 & 0 & -2 & 0 \\ 0 & 0 & 1 & 0 \\ 1 & 1 & -2 & 1 \end{pmatrix}$. Then A induces a linear transformation $A : \mathbb{R}^4 \longrightarrow \mathbb{R}^4$, $\alpha \longmapsto A\alpha$. Find the first decomposition $V = \bigoplus\limits_{i=1}^{s} W_i$ of $\mathbb{R}^4$ with respect to A. Regard every projection $\pi_i : \mathbb{R}^4 \longrightarrow W_i$ as a linear transformation of $\mathbb{R}^4$, and find a polynomial $h_i(\lambda) \in \mathbb{R}[\lambda]$ such that $\pi_i = h_i(A)$, $i = 1, 2, \ldots, s$.

Solution. It's easy to calculate that the characteristic polynomial of A is $f_A(\lambda) = (\lambda - 1)^2 (\lambda^2 + 1)$. Hence A has a unique real eigenvalue 1 with multiplicity two. Assume that V_1 is the eigenspace of A associated with the eigenvalue 1 and W_1 is the root subspace of A associated with eigenvalue 1. From $V_1 \subseteq W_1$ and $\dim_{\mathbb{R}} V_1 = 4 - \mathrm{rank}(I_4 - A) = 4 - 2 = 2 = \dim_{\mathbb{R}} W_1$, we obtain that

$$W_1 = V_1 = \{\alpha \in \mathbb{R}^4 \mid A\alpha = \alpha\}.$$

Therefore, $\mathbb{R}^4 = W_1 \oplus W_2$, where

$$W_2 = \mathrm{Ker}(A^2 + I_4) = \{\alpha \in \mathbb{R}^4 \mid (A^2 + I_4)\alpha = 0\}.$$

Solving the homogeneous linear systems $(A - I_4)X = 0$ and $(A^2 + I_4)X = 0$ respectively, we get a basis $\alpha_1 = (2, 0, 1, 0)', \alpha_2 = (0, 0, 0, 1)'$ for W_1 and a basis $\alpha_3 = (-1, 0, 0, 1)', \alpha_4 = (0, 1, 0, 0)'$ for W_2. Thus

$$A(\alpha_1, \alpha_2, \alpha_3, \alpha_4) = (\alpha_1, \alpha_2, \alpha_3, \alpha_4) \begin{pmatrix} 1 & 0 & 0 & 0 \\ 0 & 1 & 0 & 0 \\ 0 & 0 & 0 & 1 \\ 0 & 0 & -1 & 0 \end{pmatrix}$$

$$= (\alpha_1, \alpha_2, \alpha_3, \alpha_4) \begin{pmatrix} A_1 & 0 \\ 0 & A_2 \end{pmatrix},$$

where $A_1 = \begin{pmatrix} 1 & 0 \\ 0 & 1 \end{pmatrix}$, $A_2 = \begin{pmatrix} 0 & 1 \\ -1 & 0 \end{pmatrix}$. It's obvious that $m_{A_1}(\lambda) = \lambda - 1$, $m_{A_2}(\lambda) = \lambda^2 + 1$, and so $m_A(\lambda) = [m_{A_1}(\lambda), m_{A_2}(\lambda)] = (\lambda - 1)(\lambda^2 + 1)$. Put $g_1(\lambda) = \lambda^2 + 1$, $g_2(\lambda) = \lambda - 1$, $u_1(\lambda) = \frac{1}{2}, u_2(\lambda) = -\frac{1}{2}(\lambda + 1)$. Then $u_1(\lambda)g_1(\lambda) + u_2(\lambda)g_2(\lambda) = 1$.

Letting $h_1(\lambda) = u_1(\lambda)g_1(\lambda)$, $h_2(\lambda) = u_2(\lambda)g_2(\lambda)$, we have

$$\pi_1 = \frac{1}{2}(A^2 + I_4) : V \longrightarrow W_1, \quad \pi_2 = -\frac{1}{2}(A + I_4)(A - I_4) : V \longrightarrow W_2.$$

$\square$

5.9* The Second Decomposition Theorem

In this section, we decompose those σ-invariant subspaces that appear in the first decomposition theorem into a direct sum of some smaller σ-subspaces. Then, applying it to the complex linear spaces, we finally prove that every complex square matrix is similar to a Jordan matrix.

Definition 5.9.1. Let V be an n-dimensional linear space, $0 \neq \alpha \in V$ and $\sigma \in \mathrm{End}(V)$.

(1) If $f(\lambda) \in F[\lambda]$ satisfies $f(\sigma)(\alpha) = 0$, then we say that $f(\lambda)$ is a σ-**annihilating polynomial** or σ-**annihilator** of α.

(2) Among all σ-annihilators of α, the nonzero monic polynomial with the lowest degree is called the σ-**minimal polynomial** of α, denoted by $m_{\sigma,\alpha}(\lambda)$.

Remark 5.9.2. The Cayley–Hamilton theorem implies the existence of the σ-minimal polynomial of any nonzero vector α.

Proposition 5.9.3. *Let* $\dim V = n$ *and* $\sigma \in \mathrm{End}(V)$. *Assume that* $0 \neq \alpha \in V$ *and* $m_{\sigma,\alpha}(\lambda)$ *is the* σ*-minimal polynomial of* α *with* $\deg(m_{\sigma,\alpha}(\lambda)) = k$. *Then the following statements hold:*

(1) *Let* $f(\lambda) \in F[\lambda]$. *Then* $f(\sigma)(\alpha) = 0$ *if and only if* $m_{\sigma,\alpha}(\lambda) | f(\lambda)$. *Hence the* σ*-minimal polynomial of* α *is unique.*

(2) $m_{\sigma,\alpha}(\lambda) = a_k + a_{k-1}\lambda + \cdots + a_1\lambda^{k-1} + \lambda^k$ *if and only if* α, $\sigma(\alpha)$, $\ldots$, $\sigma^{k-1}(\alpha)$ *are linear independent and*

$$a_k\alpha + a_{k-1}\sigma(\alpha) + \cdots + a_1\sigma^{k-1}(\alpha) + \sigma^k(\alpha) = 0.$$

(3) *Let* $W_\sigma(\alpha) = L(\alpha, \sigma(\alpha), \ldots, \sigma^{k-1}(\alpha))$. *Then*

$$W_\sigma(\alpha) = \{g(\sigma)(\alpha) \in V \mid g(\lambda) \in F[\lambda]\}$$

and $W_\sigma(\alpha)$ *is the minimal* σ*-subspace of* V *containing* α.

(4) *The sequence* $\alpha, \sigma(\alpha), \ldots, \sigma^{k-1}(\alpha)$ *is a basis for* $W_\sigma(\alpha)$, *hence* $\dim W_\sigma(\alpha) = k$ *and the matrix of* $\sigma|_{W_\sigma(\alpha)}$ *with respect to the basis* $\alpha, \sigma(\alpha), \ldots, \sigma^{k-1}(\alpha)$ *is the companion matrix of* $m_{\sigma,\alpha}(\lambda)$, *i.e.,*

$$\sigma(\alpha, \sigma(\alpha), \ldots, \sigma^{k-1}(\alpha))$$

$$= (\alpha, \sigma(\alpha), \ldots, \sigma^{k-1}(\alpha)) \begin{pmatrix} 0 & \cdots & \cdots & \cdots & 0 & -a_k \\ 1 & \ddots & & & \vdots & \vdots \\ 0 & 1 & \ddots & & \vdots & \vdots \\ \vdots & \ddots & \ddots & \ddots & \vdots & \vdots \\ \vdots & & \ddots & \ddots & 0 & 0 & -a_3 \\ \vdots & & & \ddots & 1 & 0 & -a_2 \\ 0 & \cdots & \cdots & & 0 & 1 & -a_1 \end{pmatrix}.$$

(5) *The minimal polynomial and characteristic polynomial of* $\sigma|_{W_\sigma(\alpha)}$ *are both equal to* $m_{\sigma,\alpha}(\lambda)$, *i.e.,*

$$f_{\sigma|_{W_\sigma(\alpha)}}(\lambda) = m_{\sigma|_{W_\sigma(\alpha)}}(\lambda) = m_{\sigma,\alpha}(\lambda).$$

Proof. For convenience, let $m(\lambda) = m_{\sigma,\alpha}(\lambda)$.

(1) The sufficiency is obvious. It's sufficient to prove the necessity. Let

$$f(\lambda) = q(\lambda)m(\lambda) + r(\lambda), \text{ where } r(\lambda) = 0 \text{ or } \deg(r(\lambda)) < \deg(m(\lambda)).$$

Then $0 = f(\sigma)(\alpha) = q(\sigma)m(\sigma)(\alpha) + r(\sigma)(\alpha) = r(\sigma)(\alpha)$. By the definition of $m(\lambda)$, we know that $r(\lambda) = 0$.

(2) follows from the definition of $m(\lambda) = m_{\sigma,\alpha}(\lambda)$.

(3) Put $U_\sigma(\alpha) = \{g(\sigma)(\alpha) \in V \mid g(\lambda) \in F[\lambda]\}$. It's clear that $W_\sigma(\alpha) \subseteq U_\sigma(\alpha)$. Conversely, for any $\beta \in U_\sigma(\alpha)$, there exists $g(\lambda) \in F[\lambda]$ such that $\beta = g(\sigma)(\alpha)$. Let $g(\lambda) = q(\lambda)m(\lambda) + r(\lambda)$, where $q(\lambda), r(\lambda) \in F[\lambda]$, $r(\lambda) = 0$ or $\deg(r(\lambda)) \leqslant \deg m(\lambda) - 1 = k - 1$. Then we have

$$\beta = g(\sigma)(\alpha) = q(\sigma)(m(\sigma)(\alpha)) + r(\sigma)(\alpha) = r(\sigma)(\alpha) \in W_\sigma(\alpha).$$

Hence, $W_\sigma(\alpha) = U_\sigma(\alpha) = \{g(\sigma)(\alpha) \in V \mid g(\lambda) \in F[\lambda]\}$.

Let W be any σ-subspace of V containing α. Then for any $g(\lambda) \in F[\lambda]$, $g(\sigma)(\alpha) \in W$, and so $W_\sigma(\alpha) \subseteq W$. Hence $W_\sigma(\alpha)$ is the minimal σ-subspace of V containing α.

(4) follows immediately from (2) and (3).

(5) comes immediately from (4). $\square$

Definition 5.9.4. Let V be an n-dimensional linear space, $\alpha \in V$, and $\sigma \in \mathrm{End}(V)$.

(1) The σ-subspace $W_\sigma(\alpha)$ is said to be the **cyclic σ-subspace** generated by α.

(2) If $V = W_\sigma(\alpha)$, i.e., the sequence $\alpha, \sigma(\alpha), \ldots, \sigma^{n-1}(\alpha)$ is a basis for V, then we say that α is a **cyclic vector** of σ.

Lemma 5.9.5. *Let* $\dim V = m$ *and let* $\tau \in \mathrm{End}(V)$ *be a nilpotent linear transformation, i.e., there exists a positive integer N such that* $\tau^N = 0$. *Then there exist* $\alpha_1, \alpha_2, \ldots, \alpha_t \in V$ *and positive integers* $k_1, k_2, \ldots, k_t$ *such that the sequence*

$$\begin{array}{cccc} \alpha_1, & \alpha_2, & \cdots, & \alpha_t, \\ \tau(\alpha_1), & \tau(\alpha_2), & \cdots, & \tau(\alpha_t), \\ \vdots & \vdots & & \vdots \\ \tau^{k_1-1}(\alpha_1), & \tau^{k_2-1}(\alpha_2), & \cdots, & \tau^{k_t-1}(\alpha_t) \end{array}$$

is a basis for V, and $\tau^{k_1}(\alpha_1) = 0, \tau^{k_2}(\alpha_2) = 0, \ldots, \tau^{k_t}(\alpha_t) = 0$. *Hence*

$$V = C_1 \oplus C_2 \oplus \cdots \oplus C_t,$$

where $C_j = W_\tau(\alpha_j)$ *is the cyclic τ-subspace generated by* α_j, $j = 1, 2, \ldots, t$.

Proof. We proceed by induction on $m = \dim V$.

When $m = 1$, any nonzero vector $\alpha_1 \in V$ is a basis for V, i.e., $V = F\alpha_1 = \{k\alpha_1 \mid k \in F\}$, hence there exists $\lambda_1 \in F$ such that $\tau(\alpha_1) = \lambda_1 \alpha_1$. From τ being nilpotent, we know there exists $N \geqslant 1$ such that $\tau^N = 0$. Therefore,

$$\lambda_1^N \alpha_1 = \tau^N(\alpha_1) = 0.$$

From $\alpha_1 \neq 0$, we get $\lambda_1 = 0$, and so $\tau(\alpha_1) = 0$, i.e., $k_1 = 1$. Hence Lemma 5.9.5 holds for $m = 1$.

Let $m \geqslant 2$ and assume that Lemma 5.9.5 holds for nilpotent linear transformations of a linear space with dimension $\leqslant m - 1$.

Let V be an m-dimensional linear space and $\tau \in \mathrm{End}(V)$ a nilpotent transformation, i.e., there exists $N \geqslant 1$ such that $\tau^N = 0$. We claim that $\mathrm{Im}(\tau) \subsetneqq V$. Otherwise, $\mathrm{Im}(\tau) = V$, i.e., $\tau(V) = V$, thus

$$V = \tau(V) = \tau^2(V) = \cdots = \tau^{N-1}(V) = \tau^N(V) = \{0\},$$

which contradicts $\dim V = m \geqslant 2$.

Thus, $\dim \mathrm{Im}(\tau) \leqslant m-1$ and τ is a nilpotent linear transformation of $\mathrm{Im}(\tau)$. By induction, there exist $\beta_1, \beta_2, \ldots, \beta_s \in \mathrm{Im}(\tau)$ and positive integers $k_1, k_2, \ldots, k_s$ such that the sequence

$$
\begin{array}{cccc}
\beta_1, & \beta_2, & \ldots, & \beta_s, \\
\tau(\beta_1), & \tau(\beta_2), & \ldots, & \tau(\beta_s), \\
\vdots & \vdots & & \vdots \\
\tau^{k_1-1}(\beta_1), & \tau^{k_2-1}(\beta_2), & \ldots, & \tau^{k_s-1}(\beta_s)
\end{array}
\tag{5.24}
$$

is a basis for $\mathrm{Im}(\tau)$ and $\tau^{k_1}(\beta_1) = 0$, $\tau^{k_2}(\beta_2) = 0$, $\ldots, \tau^{k_s}(\beta_s) = 0$. Note that there exist $\alpha_1, \alpha_2, \ldots, \alpha_s \in V$ such that $\tau(\alpha_1) = \beta_1, \tau(\alpha_2) = \beta_2, \ldots, \tau(\alpha_s) = \beta_s$, and hence the sequence

$$
\begin{array}{cccc}
\alpha_1, & \alpha_2, & \ldots, & \alpha_s, \\
\tau(\alpha_1), & \tau(\alpha_2), & \ldots, & \tau(\alpha_s), \\
\vdots & \vdots & & \vdots \\
\tau^{k_1-1}(\alpha_1), & \tau^{k_2-1}(\alpha_2), & \ldots, & \tau^{k_s-1}(\alpha_s)
\end{array}
$$

is a sequence of inverse images of the vectors in the basis (5.24) under τ.

Since the sequence

$$\tau^{k_1}(\alpha_1) = \tau^{k_1-1}(\beta_1), \tau^{k_2}(\alpha_2) = \tau^{k_2-1}(\beta_1), \ldots, \tau^{k_t}(\alpha_t)$$
$$= \tau^{k_s-1}(\beta_s) \in \mathrm{Ker}(\tau)$$

is linearly independent, it can be extended to a basis

$$\tau^{k_1}(\alpha_1), \tau^{k_2}(\alpha_2), \ldots, \tau^{k_s}(\alpha_s), \alpha_{s+1}, \ldots, \alpha_t$$

for $\mathrm{Ker}(\tau)$. By Theorem 5.2.20, we know that the sequence

$$
\begin{array}{ccccc}
\alpha_1, & \alpha_2, & \ldots, & \alpha_s, & \alpha_{s+1}, \ldots, \alpha_t, \\
\tau(\alpha_1), & \tau(\alpha_2), & \ldots, & \tau(\alpha_s), & \\
\vdots & \vdots & & \vdots & \\
\tau^{k_1-1}(\alpha_1), & \tau^{k_2-1}(\alpha_2), & \ldots, & \tau^{k_s-1}(\alpha_s), & \\
\tau^{k_1}(\alpha_1), & \tau^{k_2}(\alpha_2), & \ldots, & \tau^{k_s}(\alpha_s), &
\end{array}
$$

is a basis for V, and

$$\tau^{k_1+1}(\alpha_1) = 0, \quad \tau^{k_2+1}(\alpha_2) = 0, \ldots, \tau^{k_s+1}(\alpha_s) = 0,$$
$$\tau(\alpha_{s+1}) = 0, \ldots, \tau(\alpha_t) = 0.$$

Let $C_i = W_\tau(\alpha_i)$, $i = 1, 2, \ldots, t$. Then $V = C_1 \oplus C_2 \oplus \cdots \oplus C_t$. $\square$

Theorem 5.9.6 (Second decomposition theorem). *Let V be an n-dimensional linear space and $\sigma \in \mathrm{End}(V)$. Assume that $f_\sigma(\lambda) = \prod_{i=1}^{s}(\lambda - \lambda_i)^{e_i}$, where $\lambda_1, \lambda_2, \ldots, \lambda_s \in F$ are different, $e_1, e_2, \ldots, e_s \in \mathbb{N}^*$.*
  Let $W_i = \mathrm{Ker}(\sigma - \lambda_i 1_V)^{e_i}$, $i = 1, 2, \ldots, s$.

(1) *For any $i \in \{1, 2, \ldots, s\}$, there exist $\alpha_{i1}, \alpha_{i2}, \ldots, \alpha_{it_i} \in V$ such that $W_i = \bigoplus_{j=1}^{t_i} C_{ij}$, where C_{ij} is the cyclic σ-subspace $W_\sigma(\alpha_{ij})$ generated by α_{ij}, $j = 1, 2, \ldots, t_i$.*

(2) *$V = \bigoplus_{i=1}^{s} \bigoplus_{j=1}^{t_i} C_{ij}$.*

Proof. By the first decomposition theorem, (1) implies (2), so it's sufficient to prove (1).

For $1 \leqslant i \leqslant s$, set $\tau_i = \sigma - \lambda_i 1_V$. It's obvious that $W_i = \mathrm{Ker}\,(\tau_i^{e_i})$ is a τ_i-subspace and $\tau_i|_{W_i}$ is a nilpotent transformation of W_i, in fact, $\tau_i^{e_i} = 0$. Hence, by Lemma 5.9.5, there exist $\alpha_{i1}, \alpha_{i2}, \ldots, \alpha_{it_i} \in W_i \subseteq V$ such that

$$W_i = C_{i1} \oplus C_{i2} \oplus \cdots \oplus C_{it_i},$$

where $C_{ij} = W_{\tau_i}(\alpha_{ij})$ is the cyclic τ_i-subspace generated by α_{ij}, $j = 1, 2, \ldots, t_i$. For $1 \leqslant i \leqslant s$; $1 \leqslant j \leqslant t_i$, from Proposition 5.9.3 (3) we know that

$$C_{ij} = W_{\tau_i}(\alpha_{ij}) = \{g(\tau_i)(\alpha_{ij}) \mid g(\lambda) \in F[\lambda]\}$$
$$= \{h(\sigma)(\alpha_{ij}) \mid h(\lambda) \in F[\lambda]\} = W_\sigma(\alpha_{ij}),$$

i.e., $C_{ij} = W_\sigma(\alpha_{ij})$ is the cyclic σ-subspace generated by α_{ij}. $\qquad\square$

Definition 5.9.7.

(1) The matrix $J_n(\lambda) = \begin{pmatrix} \lambda & & & \\ 1 & \lambda & & \\ & \ddots & \ddots & \\ & & 1 & \lambda \end{pmatrix} \in \mathrm{M}_n(F)$ is called the **Jordan block** of order n with eigenvalue λ.

(2) The block diagonal matrix $J = \mathrm{diag}(J_1, J_2, \ldots, J_s)$ is called a **Jordan matrix**, where J_i is a Jordan block of order n_i, $i = 1, 2, \ldots, s$.

Remark 5.9.8. The transpose $J_n(\lambda)' = \begin{pmatrix} \lambda & 1 & & & \\ & \lambda & 1 & & \\ & & \ddots & \ddots & \\ & & & \lambda & 1 \\ & & & & \lambda \end{pmatrix}_{n \times n}$ of

$J_n(\lambda)$ is also called the Jordan block of order n with eigenvalue λ. In fact, $J_n(\lambda)$ and $J_n(\lambda)'$ are similar:

$$J_n(\lambda)' = S^{-1} J_n(\lambda) S, \quad \text{where } S = \begin{pmatrix} & & 1 \\ & \iddots & \\ 1 & & \end{pmatrix}_{n \times n}.$$

Theorem 5.9.9 (Jordan canonical form).

(1) *Let $A \in \mathrm{M}_n(F)$. Assume that the characteristic polynomial $f_A(\lambda)$ of A has n roots in F (possibly multiple roots). Then there exists $P \in \mathrm{GL}_n(F)$ such that $P^{-1}AP = J = \mathrm{diag}(J_1, J_2, \ldots, J_t)$ is a Jordan matrix, where J_i is a Jordan block of order n_i, $i = 1, 2, \ldots, t$, and J is determined uniquely by A regardless of the order of Jordan blocks $J_1, \ldots, J_t$. The Jordan matrix J is called the **Jordan canonical form** of A.*

(2) *Let $A \in \mathrm{M}_n(\mathbb{C})$. Then A is similar to a Jordan matrix, i.e., there exists $P \in \mathrm{GL}_n(\mathbb{C})$ such that $P^{-1}AP = J$ is a Jordan matrix, and J is determined uniquely by A regardless of the order of Jordan blocks.*

Proof. It's obvious that (2) is immediately from (1), hence it's sufficient to prove (1).

We first prove the existence. Let $\alpha_1, \alpha_2, \ldots, \alpha_n$ be a basis for an n-dimensional linear space V. By Theorem 5.3.5 (2), there exists $\sigma \in \mathrm{End}(V)$ such that $\sigma(\alpha_1, \alpha_2, \ldots, \alpha_n) = (\alpha_1, \alpha_2, \ldots, \alpha_n)A$. Hence

$$f_\sigma(\lambda) = f_A(\lambda) = (\lambda - \lambda_1)^{e_1}(\lambda - \lambda_2)^{e_2} \cdots (\lambda - \lambda_s)^{e_s},$$

where $\lambda_1, \lambda_2, \ldots, \lambda_s \in F$ are different, $e_1, e_2, \ldots, e_s \in \mathbb{N}^*$, and $\sum\limits_{i=1}^{s} e_i = n$. By the second decomposition theorem, we have

$$V = \bigoplus_{j=1}^{s} \bigoplus_{i=1}^{t_j} W_{\tau_j}(\alpha_{ji}),$$

where $\tau_j = \sigma - \lambda_j 1_V$, $m_{\tau_j, \alpha_{ji}}(\lambda) = \lambda^{k_{ji}}$ and the sequence

$$\alpha_{ji}, \tau_j(\alpha_{ji}), \ldots, \tau_j^{k_{ji}-1}(\alpha_{ji})$$

is a basis for the τ_j-cyclic subspace $W_{\tau_j}(\alpha_{ji})$, furthermore,

$$\tau_j(\alpha_{ji}, \tau_j(\alpha_{ji}), \ldots, \tau_j^{k_{ji}-1}(\alpha_{ji}))$$

$$= (\alpha_{ji}, \tau_j(\alpha_{ji}), \ldots, \tau_j^{k_{ji}-1}(\alpha_{ji})) \begin{pmatrix} 0 & & & \\ 1 & \ddots & & \\ & \ddots & \ddots & \\ & & 1 & 0 \end{pmatrix}_{k_{ji} \times k_{ji}}.$$

Hence from $\sigma = \tau_j + \lambda_j 1_V$ we obtain that

$$\sigma(\alpha_{ji}, \tau_j(\alpha_{ji}), \ldots, \tau_j^{k_{ji}-1}(\alpha_{ji}))$$

$$= (\alpha_{ji}, \tau_j(\alpha_{ji}), \ldots, \tau_j^{k_{ji}-1}(\alpha_{ji})) \begin{pmatrix} \lambda_j & & & \\ 1 & \ddots & & \\ & \ddots & \ddots & \\ & & 1 & \lambda_j \end{pmatrix}_{k_{ji} \times k_{ji}} .$$

Put $J_{k_{ji}}(\lambda_j) = \begin{pmatrix} \lambda_j & & & & \\ 1 & \ddots & & & \\ & \ddots & \ddots & & \\ & & \ddots & \ddots & \\ & & & 1 & \lambda_j \end{pmatrix}_{k_{ji} \times k_{ji}}$, $j = 1, 2, \ldots, s;\ i = 1, 2, \ldots, t_j$. Then the Jordan matrix

$$J = \mathrm{diag}\left(J_{k_{11}}(\lambda_1), \ldots, J_{k_{1t_1}}(\lambda_1), \ldots, J_{k_{s1}}(\lambda_s), \ldots, J_{k_{st_s}}(\lambda_s) \right)$$

is the matrix of σ with respect to the basis

$$\alpha_{11}, \tau_1(\alpha_{11}), \ldots, \tau_1^{k_{11}-1}(\alpha_{11}), \ldots, \alpha_{st_s}, \tau_s(\alpha_{st_s}), \ldots, \tau_s^{k_{st_s}-1}(\alpha_{st_s})$$

for V. Let

$$\left(\alpha_{11}, \tau_1(\alpha_{11}), \ldots, \tau_1^{k_{11}-1}(\alpha_{11}), \ldots, \alpha_{st_s}, \tau_s(\alpha_{st_s}), \ldots, \tau_s^{k_{st_s}-1}(\alpha_{st_s}) \right)$$

$$= (\alpha_1, \alpha_2, \ldots, \alpha_n)P.$$

Thus, $P \in \mathrm{GL}_n(F)$ and $P^{-1}AP = J$ is a Jordan matrix.

For the uniqueness, please refer to Exercise 66 of this chapter or Theorem 6.5.3 of Chapter 6. $\qquad\square$

Corollary 5.9.10. *Let $A \in \mathrm{M}_3(\mathbb{C})$. Then the Jordan canonical form of A is determined uniquely by the dimensions of eigenspaces of A, i.e., the Jordan block associated with the eigenvalue λ of A is uniquely determined by $\dim V_\lambda$.*

Proof. Consider the characteristic polynomial $f_A(\lambda)$ of A as follows:

(1) If $f_A(\lambda) = (\lambda - \lambda_1)^3$, then

$$
A \text{ is similar to}
\begin{cases}
\begin{pmatrix} \lambda_1 & & \\ & \lambda_1 & \\ & & \lambda_1 \end{pmatrix} & \text{if and only if } \dim V_{\lambda_1} = 3; \\[2em]
\begin{pmatrix} \lambda_1 & & \\ & \lambda_1 & \\ & 1 & \lambda_1 \end{pmatrix} & \text{if and only if } \dim V_{\lambda_1} = 2; \\[2em]
\begin{pmatrix} \lambda_1 & & \\ 1 & \lambda_1 & \\ & 1 & \lambda_1 \end{pmatrix} & \text{if and only if } \dim V_{\lambda_1} = 1.
\end{cases}
$$

(2) If $f_A(\lambda) = (\lambda - \lambda_1)(\lambda - \lambda_2)^2$, where $\lambda_1 \neq \lambda_2$, then $\dim V_{\lambda_1} = 1$ and

$$
A \text{ is similar to}
\begin{cases}
\begin{pmatrix} \lambda_1 & & \\ & \lambda_2 & \\ & & \lambda_2 \end{pmatrix} & \text{if and only if } \dim V_{\lambda_2} = 2; \\[2em]
\begin{pmatrix} \lambda_1 & & \\ & \lambda_2 & \\ & 1 & \lambda_2 \end{pmatrix} & \text{if and only if } \dim V_{\lambda_2} = 1.
\end{cases}
$$

(3) If $f_A(\lambda) = (\lambda - \lambda_1)(\lambda - \lambda_2)(\lambda - \lambda_3)$, where $\lambda_1, \lambda_2, \lambda_3$ are different, then $\dim V_{\lambda_i} = 1$, $i = 1, 2, 3$, and A is similar to $\begin{pmatrix} \lambda_1 & & \\ & \lambda_2 & \\ & & \lambda_3 \end{pmatrix}$. $\qquad\square$

Corollary 5.9.11. *Let $A \in M_n(\mathbb{C})$. Then A is diagonalizable if and only if $\operatorname{rank}(A - \lambda I_n) = \operatorname{rank}(A - \lambda I_n)^2$ for every eigenvalue λ of A.*

Proof. Let $J_m(\rho)$ be the Jordan block of order m with eigenvalue ρ. On the one hand, it's easy to see that

$$
\operatorname{rank}(J_m(\rho) - \rho I_m) = \operatorname{rank}(J_m(\rho) - \rho I_m)^2
$$
$$
\Longleftrightarrow m = 1 \tag{5.25}
$$
$$
\Longleftrightarrow J_m(\rho) \text{ is diagonalizable.}
$$

On the other hand, we have

$$\mathrm{rank}(J_m(\rho) - \lambda I_m) = \mathrm{rank}(J_m(\rho) - \lambda I_m)^2, \quad \lambda \neq \rho. \qquad (5.26)$$

Let $J = \mathrm{diag}(J_1, J_2, \ldots, J_s)$ be the Jordan canonical form of the matrix A, where $J_i = J_{n_i}(\lambda_i)$ is the Jordan block of order n_i with eigenvalue λ_i. Then

$$A \text{ is diagonalizable}$$

$$\Longleftrightarrow J \text{ is diagonalizable}$$

$$\xleftrightarrow{\text{Corollary } 5.7.7} J_i \text{ is diagonalizable}, \ i = 1, 2, \ldots, s,$$

$$\xleftrightarrow{(5.25)} \mathrm{rank}(J_i - \lambda_i I_{n_i}) = \mathrm{rank}(J_i - \lambda_i I_{n_i})^2, \ i = 1, 2, \ldots, s,$$

$$\xleftrightarrow{(5.26)} \sum_{j=1}^{s} \mathrm{rank}(J_j - \lambda_i I_{n_j}) = \sum_{j=1}^{s} \mathrm{rank}(J_j - \lambda_i I_{n_j})^2,$$
$$i = 1, 2, \ldots, s,$$

$$\Longleftrightarrow \mathrm{rank}(J - \lambda_i I_n) = \mathrm{rank}(J - \lambda_i I_n)^2, \ i = 1, 2, \ldots, s,$$

$$\Longleftrightarrow \mathrm{rank}(A - \lambda I_n) = \mathrm{rank}(A - \lambda I_n)^2$$
$$\text{for any eigenvalue } \lambda \text{ of } A. \qquad \qquad \square$$

Exercises

1. Decide which of the maps defined in the following are linear and which are not.

   (1) Translation map: Let α be a fixed vector in V. Define

   $$\sigma_\alpha : V \longrightarrow V, \ \xi \longmapsto \xi + \alpha.$$

   (2) $\sigma : F^3 \longrightarrow F^3, (a_1, a_2, a_3) \longmapsto (a_1 - a_2, a_2 - a_3, a_3 - a_1)$.
   (3) $\sigma : F^3 \longrightarrow F^3, (a_1, a_2, a_3) \longmapsto (a_1^2, a_2^2, a_3^2)$.
   (4) $\sigma : F[x] \longrightarrow F[x], f(x) \longmapsto f(x+1)$.
   (5) $\sigma : F[x] \longrightarrow F[x], f(x) \longmapsto f(x_0)$, where $x_0 \in F$ is a fixed number.

    (6) The complex field is a complex linear space, $\sigma : \mathbb{C} \longrightarrow \mathbb{C}$, $\alpha \longmapsto \overline{\alpha}$, where $\alpha \in \mathbb{C}$, $\overline{\alpha}$ is the complex conjugate of α.

    (7) Let $V = M_n(F)$. For a fixed matrix $A \in M_n(F)$, define

$$\sigma : V \longrightarrow V, \; X \longmapsto AX - XA.$$

    (8) Let $V = M_n(F)$. For fixed matrices $A, B \in M_n(F)$, define

$$\sigma : V \longrightarrow V, \; X \longmapsto AXB.$$

2. Let V be a linear space and let $\sigma_1, \sigma_2, \ldots, \sigma_s \in \mathrm{End}(V)$ be s different transformations. Prove that there exists $\alpha \in V$ such that $\sigma_1(\alpha), \sigma_2(\alpha), \ldots, \sigma_s(\alpha)$ are also different.

3. Let $f : M_n(F) \longrightarrow F$ be a linear function such that $f(AB) = f(BA)$ for any $A, B \in M_n(F)$. Prove that there exists $\lambda \in F$ such that $f = \lambda \mathrm{Tr}$, i.e., $f(A) = \lambda \mathrm{Tr}(A)$ for any $A \in M_n(F)$.

4.* Let W be a subspace of the real linear space $M_n(\mathbb{R})$.

    (1) Put $p = \max\{\mathrm{rank} A \mid A \in W\}$. Prove that $\dim W \leqslant np$.

    (2) Assume $\dim W = n^2 - n + 1$. Prove that $W \cap \mathrm{GL}_n(\mathbb{R}) \neq \emptyset$.

5. Let $V = \left\{ \begin{pmatrix} a & b \\ -b & a \end{pmatrix} \; \middle| \; a, b \in \mathbb{R} \right\}$. Define

$$f : \mathbb{C} \longrightarrow V, \; a + bi \longmapsto \begin{pmatrix} a & b \\ -b & a \end{pmatrix}.$$

    Prove the following:

    (1) V is a subspace of the real linear space $M_2(\mathbb{R})$.

    (2) f is an isomorphism and $f(\alpha\beta) = f(\alpha)f(\beta)$ for any $\alpha, \beta \in \mathbb{C}$.

6. Let V be an n-dimensional linear space and $\sigma \in \mathrm{End}(V)$. Assume that there exist subspaces V_1, V_2 of V such that $\mathrm{Ker}(\sigma) = V_1 \cap V_2$. Prove that there exist $\sigma_1, \sigma_2 \in \mathrm{End}(V)$ such that $V_i \subseteq \mathrm{Ker}(\sigma_i)$, $i = 1, 2$, and $\sigma_1 + \sigma_2 = \sigma$.

7. Let $\sigma : \mathbb{R}^3 \longrightarrow \mathbb{R}^3$ be defined by $(a, b, c)' \longmapsto (b, c, 0)'$. Find a basis for $\mathrm{Im}(\sigma)$, $\mathrm{Ker}(\sigma)$ and $\mathrm{Im}(\sigma) \cap \mathrm{Ker}(\sigma)$, respectively.

8. Let V be an n-dimensional linear space and $\sigma \in \mathrm{End}(V)$. Prove that the following statements are equivalent:

    (1) $V = \mathrm{Ker}(\sigma) + \mathrm{Im}(\sigma)$.

    (2) $V = \mathrm{Ker}(\sigma) \oplus \mathrm{Im}(\sigma)$.

    (3) $\mathrm{Im}(\sigma) = \mathrm{Im}(\sigma^2)$.

    (4) $\mathrm{rank}(\sigma) = \mathrm{rank}(\sigma^2)$.

    (5) $\mathrm{Ker}(\sigma) = \mathrm{Ker}(\sigma^2)$.

    (6) $N(\sigma) = N(\sigma^2)$.

9. Let V be an n-dimensional linear space and $\sigma \in \text{End}(V)$.

   (1) Let W be a subspace of V. Prove that

   $$\sigma(W) = \{\sigma(\alpha) \mid \alpha \in W\}$$

   is also a subspace of V and

   $$\dim W = \dim \sigma(W) + \dim \left(\text{Ker}(\sigma) \bigcap W\right).$$

   (2) Let W_1, W_2 be two subspaces of V and $V = W_1 \oplus W_2$. Prove that σ is an automorphism of V if and only if $V = \sigma(W_1) \oplus \sigma(W_2)$.

10. Let V be an n-dimensional linear space and $\sigma, \tau \in \text{End}(V)$. Prove that

    $$\dim\text{Ker}(\sigma\tau) \leqslant \dim\text{Ker}(\sigma) + \dim\text{Ker}(\tau).$$

11. Let V_i be a finite-dimensional linear space, $i = 0, 1, \ldots, n + 1$. Assume that $V_0 = V_{n+1} = \{0\}$ and for any $i \in \{0, 1, \ldots, n\}$, there exists $\sigma_i \in \text{Hom}(V_i, V_{i+1})$ such that $\text{Ker}(\sigma_{j+1}) = \text{Im}(\sigma_j)$ for $j = 0, 1, \ldots, n - 1$. Prove that

    $$\sum_{i=1}^{n} (-1)^i \dim V_i = 0.$$

12. Let V be an n-dimensional linear space and $\sigma, \tau \in \text{End}(V)$. If $\sigma\tau = \tau\sigma$ and there exists $\alpha \in V$ such that the sequence $\alpha, \sigma(\alpha), \sigma^2(\alpha), \ldots, \sigma^{n-1}(\alpha)$ is a basis for V, prove that there exists $g(x) \in F[x]$ such that $\tau = g(\sigma)$.

13. Let U, V be linear spaces with $\dim V = n$ and $\dim U = m$. For $\alpha \in V$, define $K(\alpha) = \{\sigma \in \text{Hom}(V, U) \mid \sigma(\alpha) = 0\}$. Prove that $K(\alpha)$ is a subspace of $\text{Hom}(V, U)$ and find $\dim K(\alpha)$.

14. Let V be an n-dimensional linear space and $\sigma \in \text{End}(V)$. Define

    $$K(\sigma) = \{\tau \in \text{End}(V) \mid \sigma\tau = 0\}.$$

   (1) Prove that $K(\sigma)$ is a subspace of $\text{End}(V)$.
   (2) Find $\sigma_1, \sigma_2, \sigma_3 \in \text{End}(V)$ such that

   $$\dim K(\sigma_1) = 0, \ \dim K(\sigma_2) = n, \ \dim K(\sigma_3) = n^2.$$

15. Let V be an n-dimensional linear space and $\sigma, \tau \in \text{End}(V)$. Assume that $\sigma^2 = \sigma$ and $\tau^2 = \tau$. Prove the following:

   (1) $\text{Im}(\sigma) = \text{Im}(\tau)$ if and only if $\sigma\tau = \tau, \ \tau\sigma = \sigma$.

(2) $\mathrm{Ker}(\sigma) = \mathrm{Ker}(\tau)$ if and only if $\sigma\tau = \sigma$, $\tau\sigma = \tau$.

(3) If $(\sigma + \tau)^2 = \sigma + \tau$, then $\sigma\tau = \tau\sigma = 0$.

(4) If $\sigma\tau = \tau\sigma$, then $(\sigma + \tau - \sigma\tau)^2 = \sigma + \tau - \sigma\tau$.

16. Let V be an n-dimensional linear space and let $\sigma_1, \sigma_2, \ldots, \sigma_t \in \mathrm{End}(V)$ such that

$$\sigma_i^2 = \sigma_i, \ i = 1, 2, \ldots, t; \ \sigma_i\sigma_j = 0, \ 1 \leqslant i \neq j \leqslant t.$$

Prove the following:

(1) $\mathrm{Im}(\sigma_1 + \sigma_2 + \cdots + \sigma_t) = \mathrm{Im}(\sigma_1) \oplus \mathrm{Im}(\sigma_2) \oplus \cdots \oplus \mathrm{Im}(\sigma_t)$.

(2) $\mathrm{Ker}(\sigma_1 + \sigma_2 + \cdots + \sigma_t) = \bigcap_{i=1}^{t} \mathrm{Ker}(\sigma_i)$.

(3) $V = \mathrm{Im}(\sigma_1) \oplus \mathrm{Im}(\sigma_2) \oplus \cdots \oplus \mathrm{Im}(\sigma_t) \oplus \left(\bigcap_{i=1}^{t} \mathrm{Ker}(\sigma_i) \right)$.

17. Let $\alpha_1, \alpha_2, \ldots, \alpha_n$ be a basis for an n-dimensional linear space V. For any $i = 1, 2, \ldots, n$, define $\alpha_i^* \in V^*$ as follows: $\alpha_i^*(\alpha_j) = \delta_{ij}$, $j = 1, 2, \ldots, n$, where δ_{ij} is the Kronecker symbol. Prove that the sequence $\alpha_1^*, \alpha_2^*, \ldots, \alpha_n^*$ is a basis for V^*, which is called the **dual basis** of the basis $\alpha_1, \alpha_2, \ldots, \alpha_n$, and for any $f \in V^*$,

$$f = \sum_{i=1}^{n} f(\alpha_i)\alpha_i^*, \ i.e., \ f(\alpha) = \sum_{i=1}^{n} f(\alpha_i)\alpha_i^*(\alpha), \ \alpha \in V.$$

18. Let V be an n-dimensional linear space and let $V^{**} = (V^*)^*$ be the dual space of V^*. For $\alpha \in V$, define $\alpha^{**} : V^* \longrightarrow F$, $f \longmapsto f(\alpha)$. Prove the following:

(1) $\alpha^{**} \in V^{**}$, and so $\delta_V : V \longrightarrow V^{**}$, $\alpha \longmapsto \alpha^{**}$, is a map.

(2) δ_V is an isomorphism, and hence V and V^{**} are isomorphic.

19. Let σ, τ be two linear transformations of an n-dimensional linear space V. Prove that $\sigma\tau - \tau\sigma \neq 1_V$. Does the conclusion hold if V is an infinite-dimensional linear space?

20. Let $V = F^4$. Define $\sigma \in \mathrm{End}(V)$ as follows:

$$\sigma(\alpha) = \begin{pmatrix} a_1 - 2a_2 + a_3 + a_4 \\ -a_1 + 2a_2 + a_3 + 2a_4 \\ 2a_1 + a_2 - a_3 - 2a_4 \\ a_1 + 3a_2 + 2a_3 + 3a_4 \end{pmatrix} \text{ for } \alpha = \begin{pmatrix} a_1 \\ a_2 \\ a_3 \\ a_4 \end{pmatrix} \in V.$$

(1) Find a basis for V and the matrix of σ with respect to this basis.
(2) Find a basis for $\mathrm{Ker}(\sigma)$ and $\dim\mathrm{Ker}(\sigma)$.
(3) Find a basis for $\mathrm{Im}(\sigma)$ and $\dim\mathrm{Im}(\sigma)$.

21. Let V be an n-dimensional linear space and $\sigma \in \mathrm{End}(V)$. Assume that $\mathrm{Im}(\sigma) = \mathrm{Ker}(\sigma)$. Prove that n is even and there exists a basis $\varepsilon_1, \varepsilon_2, \ldots, \varepsilon_n$ for V such that the matrix of σ with respect to this basis is $\begin{pmatrix} 0 & I_{\frac{n}{2}} \\ 0 & 0 \end{pmatrix}$.

22. Let V be an n-dimensional linear space and $\sigma \in \mathrm{End}(V)$. Assume that there exists $\alpha \in V$ such that $\sigma^{n-1}(\alpha) \neq 0$ and $\sigma^n(\alpha) = 0$. Prove that the sequence $\alpha, \sigma(\alpha), \sigma^2(\alpha), \ldots, \sigma^{n-1}(\alpha)$ is a basis for V, and find the matrix of σ with respect to this basis.

23. Let V be an n-dimensional linear space and $\sigma, \tau \in \mathrm{End}(V)$. Assume that $\sigma^2 = \sigma$. Prove that $\mathrm{Ker}(\sigma)$ and $\mathrm{Im}(\sigma)$ are both τ-subspaces if and only if $\sigma\tau = \tau\sigma$.

24. Let V be an n-dimensional linear space and $\sigma \in \mathrm{End}(V)$. Put

$$V_1 = \{\alpha \in V \mid \text{there exists } m \in \mathbb{N}^* \text{ such that } \sigma^m(\alpha) = 0\},$$

$$V_2 = \bigcap_{i=1}^{\infty} \sigma^i(V).$$

Prove the following:

(1) V_1 and V_2 are both σ-subspaces and $V = V_1 \oplus V_2$.
(2) $\sigma|_{V_1}$ is a nilpotent transformation of V_1 and $\sigma|_{V_2}$ is an automorphism of V_2.

25. Let V be a 2-dimensional real linear space and $\sigma \in \mathrm{End}_{\mathbb{R}}(V)$. Assume that the matrix of σ with respect to a basis $\varepsilon_1, \varepsilon_2$ for V is

$$A = \begin{pmatrix} \cos\theta & \sin\theta \\ -\sin\theta & \cos\theta \end{pmatrix}, \quad \theta \neq k\pi, \ k \in \mathbb{Z}.$$

Prove that σ does not have nontrivial invariant subspaces.

26. Let V be an n-dimensional complex linear subspace and $\sigma, \tau \in \mathrm{End}_{\mathbb{C}}(V)$. Assume $\sigma + \tau + \sigma\tau = 0$. Prove the following:

(1) $\sigma\tau = \tau\sigma$.
(2) σ and τ have a common eigenvector.

(3) There exists a basis for V such that the matrices of σ and τ with respect to this basis are both upper triangular matrices.

27. Let V be an n-dimensional linear space and $\sigma, \tau \in \mathrm{End}(V)$. Assume that $\sigma\tau = \tau\sigma$ and σ has n different eigenvalues. Prove that there exists $g(x) \in F[x]$ such that $\tau = g(\sigma)$.

28. Let $A \in \mathrm{M}_n(F)$ and let $\lambda_1, \lambda_2, \ldots, \lambda_n$ be the n complex roots of $f_A(\lambda)$. If there exist some $\lambda_i \in F$ $(1 \leqslant i \leqslant n)$ and $0 \neq \alpha \in F^n$ such that $A\alpha = \lambda_i\alpha$, prove that $\lambda_i^* \in F$ and $A^*\alpha = \lambda_i^*\alpha$, where λ_i^* is defined by (5.14).

29. Let $A \in \mathrm{M}_m(\mathbb{C})$ and $B \in \mathrm{M}_n(\mathbb{C})$. Prove that the equation $AX = XB$ only has the zero solution if and only if A and B have no common eigenvalues.

30. Let $A, B \in \mathrm{M}_n(\mathbb{C})$. Define a map

$$\varphi : \mathrm{M}_n(\mathbb{C}) \longrightarrow \mathrm{M}_n(\mathbb{C}), \ X \longmapsto AX - XB.$$

Prove the following:

(1) $\varphi \in \mathrm{End}_{\mathbb{C}}(\mathrm{M}_n(\mathbb{C}))$, i.e., φ is linear.
(2) $\varphi \in \mathrm{Aut}_{\mathbb{C}}(\mathrm{M}_n(\mathbb{C}))$ if and only if A and B have no common eigenvalues.

31. Let $A, B \in \mathrm{M}_n(\mathbb{C})$. Assume that λ is a nonzero eigenvalue of AB and BA. Let W_λ and V_λ be the eigenspaces of AB and BA associated with the eigenvalue λ, respectively. Prove that $\dim W_\lambda = \dim V_\lambda$.

32. Let $A, B \in \mathrm{M}_n(\mathbb{C})$. Prove that $f_A(B)$ is invertible if and only if A and B have no common eigenvalues.

33. Let $A \in \mathrm{M}_n(\mathbb{R})$. Assume that

$$f_A(\lambda) = \prod_{i=1}^{t}(\lambda - \lambda_i) \prod_{j=1}^{r}(\lambda^2 + a_j\lambda + b_j),$$

where $t, r \in \mathbb{N}$, $t + 2r = n$, $\lambda_i, a_j, b_j \in \mathbb{R}$, $a_j^2 - 4b_j < 0$, $1 \leqslant i \leqslant t$, $1 \leqslant j \leqslant r$. We know that A induces a linear transformation of $\mathbb{R}^n$, also denoted by $A : \mathbb{R}^n \longrightarrow \mathbb{R}^n, X \longmapsto AX$. Let $0 \neq \alpha = X + iY \in \mathbb{C}^n$ and $\lambda = a + bi \in \mathbb{C}$ such that $A\alpha = \lambda\alpha$, where $a, b \in \mathbb{R}$, $X, Y \in \mathbb{R}^n$ and $b \neq 0$. Prove the following:

(1) X, Y are $\mathbb{R}$-linear independent vectors.

(2) $W = L(X, Y)$ is an A-subspace of $\mathbb{R}^n$ and there exists $P \in \mathrm{GL}_n(\mathbb{R})$ such that

$$P^{-1}AP = \begin{pmatrix} A_{11} & A_{12} \\ 0 & A_{22} \end{pmatrix},$$

where $A_{11} = \begin{pmatrix} a & b \\ -b & a \end{pmatrix}$, $A_{12} \in \mathrm{M}_{2 \times (n-2)}(\mathbb{R})$, $A_{22} \in \mathrm{M}_{n-2}(\mathbb{R})$.

(3) There exists $P \in \mathrm{GL}_n(\mathbb{R})$ such that

$$P^{-1}AP = \begin{pmatrix} \lambda_1 & \star & \star & \star & \star & \star \\ & \ddots & \star & \star & \star & \star \\ & & \lambda_t & \star & \star & \star \\ & & & A_1 & \star & \star \\ & & & & \ddots & \star \\ & & & & & A_r \end{pmatrix}$$

is a block upper triangular matrix, where $A_j = \begin{pmatrix} c_j & d_j \\ -d_j & c_j \end{pmatrix} \in \mathrm{M}_2(\mathbb{R})$, and $f_{A_j}(\lambda) = \lambda^2 + a_j \lambda + b_j$ is the characteristic polynomial of A_j, $1 \leqslant j \leqslant r$.

34. Let V be an n-dimensional linear space and $\sigma \in \mathrm{End}(V)$.

(1) Assume that V is a real linear space. Prove that there exist σ-subspaces $V_1, V_2, \ldots, V_m = V$ such that

$$V_1 \subseteq V_2 \subseteq \cdots \subseteq V_m = V,$$
$$\dim V_{i+1}/V_i = 1 \text{ or } 2, \quad i = 1, 2, \ldots, m - 1.$$

(2) Assume that V is a complex linear space. Prove that there exist σ-subspaces $V_1, V_2, \ldots, V_n = V$ such that

$$V_1 \subseteq V_2 \subseteq \cdots \subseteq V_n = V, \dim V_{i+1}/V_i = 1, \quad i = 1, 2, \ldots, n - 1.$$

35. Let $A \in \mathrm{M}_2(\mathbb{C})$. Prove that A is similar to a matrix of the form $\begin{pmatrix} a & 0 \\ 0 & b \end{pmatrix}$ or $\begin{pmatrix} a & 1 \\ 0 & a \end{pmatrix}$.

36.[*] Let $A \in \mathrm{M}_n(F)$ and $\mathrm{Tr}(A) = 0$. Prove the following:

    (1) There exists $B = (b_{ij})_{n \times n} \in \mathrm{M}_n(F)$ such that A is similar to B and $b_{ii} = 0, i = 1, 2, \ldots, n$.

    (2) There exist $X, Y \in \mathrm{M}_n(F)$ such that $A = XY - YX$.

37. Let $A \in \mathrm{M}_n(F)$. Prove that A is a nilpotent matrix if and only if $f_A(\lambda) = \lambda^n$.

38. Let $A, B \in \mathrm{M}_n(\mathbb{C})$ and $C = AB - BA$. If $AC = CA$, prove that C is a nilpotent matrix.

39. Let V be an n-dimensional linear space and $\sigma \in \mathrm{End}(V)$. Assume that the matrix of σ with respect to a basis for V is the companion matrix of a monic polynomial $f(\lambda) \in F[\lambda]$. Prove that there exists $\alpha \in V$ such that $V = \{g(\sigma)(\alpha) \in V \mid g(\lambda) \in F[\lambda]\}$ and $f_\sigma(\lambda) = m_\sigma(\lambda) = f(\lambda)$.

40. Let σ be a linear transformation of an n-dimensional linear space V. Prove that σ has exactly 2^n invariant subspaces if σ has n different eigenvalues.

41. Let $A, B \in \mathrm{M}_n(\mathbb{C})$ and assume that all eigenvalues of A are different. Prove that every eigenvector of A is an eigenvector of B if and only if $AB = BA$.

42. Let $A = \begin{pmatrix} 2 & 0 & 0 \\ a & 2 & 0 \\ b & c & -1 \end{pmatrix} \in \mathrm{M}_3(\mathbb{C})$. Prove that A is diagonalizable if and only if $a = 0$.

43. Let $A \in V = \mathrm{M}_n(F)$. Define a transformation

$$T : V \longrightarrow V, \ B \longmapsto T(B) = AB.$$

    (1) Prove that T and A have the same minimal polynomials.

    (2) Prove that T is diagonalizable if and only if A is diagonalizable.

    (3) Find the trace and determinant of T.

44. Let $A \in V = \mathrm{M}_n(F)$. Define a transformation

$$T : V \longrightarrow V, \ B \longmapsto T(B) = AB - BA.$$

Prove the following:

    (1) If A is a nilpotent matrix, then T is a nilpotent linear transformation.

    (2) If A is diagonalizable, then T is diagonalizable.

45. Let V be an n-dimensional linear space and $\sigma \in \text{End}(V)$. Assume that σ is diagonalizable and $\lambda_1, \lambda_2, \ldots, \lambda_t$ are all different eigenvalues of σ. Prove that there exist $\sigma_1, \sigma_2, \ldots, \sigma_t \in \text{End}(V)$ such that

   (1) $\sigma = \lambda_1\sigma_1 + \lambda_2\sigma_2 + \cdots + \lambda_t\sigma_t$,
   (2) $\sigma_1 + \sigma_2 + \cdots + \sigma_t = 1_V$,
   (3) $\sigma_i\sigma_j = 0$, $1 \leqslant i \neq j \leqslant t$,
   (4) $\sigma_i(V) = V_{\lambda_i}$, $i = 1, 2, \ldots, t$.

46. Let $\alpha, \beta \in \mathbb{R}^n$ be two column vectors and $A = \alpha\beta'$. Find the characteristic polynomial of A and discuss whether A is diagonalizable.

47. Let $A \in \text{M}_n(F)$ and $A^2 = I_n$. Prove that A is diagonalizable and there exists $P \in \text{GL}_n(F)$ such that

$$P^{-1}AP = \begin{pmatrix} I_r & 0 \\ 0 & -I_{n-r} \end{pmatrix}, \quad 0 \leqslant r \leqslant n.$$

48. Let $A = \begin{pmatrix} 4 & 6 & 0 \\ -3 & -5 & 0 \\ -3 & -6 & 1 \end{pmatrix}$. Find A^{10}.

49. Let $A, B \in \text{M}_n(F)$ and $AB = BA$. If A and B are both diagonalizable, prove that there exists $P \in \text{GL}_n(F)$ such that $P^{-1}AP$ and $P^{-1}BP$ are both diagonal matrices.

50*. Let $\{A_j\}_{j \in J}$ be a family of complex square matrices of order n and $A_iA_j = A_jA_i$ for any $i, j \in J$. If every A_j is diagonalizable for any $j \in J$, prove that there exists $P \in \text{GL}_n(\mathbb{C})$ such that $P^{-1}A_jP$ is a diagonal matrix for any $j \in J$.

51. Let k be a positive integer and

$$C(k) = \begin{pmatrix} 0 & \cdots & \cdots & \cdots & \cdots & 0 & 1 \\ 1 & 0 & \cdots & \cdots & \cdots & 0 & 0 \\ 0 & 1 & \ddots & & & \vdots & \vdots \\ \vdots & \ddots & \ddots & \ddots & & \vdots & \vdots \\ \vdots & & \ddots & \ddots & \ddots & \vdots & \vdots \\ \vdots & & & \ddots & 1 & 0 & 0 \\ 0 & \cdots & \cdots & \cdots & 0 & 1 & 0 \end{pmatrix}_{k \times k}.$$

Prove that $C(k)$ is diagonalizable over the field $\mathbb{C}$ of complex numbers.

52. Let A be a **permutation matrix** of order n over the field $\mathbb{C}$ of complex numbers, i.e., A is the product of finitely many elementary matrices of the form $E(i,j)$. Prove that A is diagonalizable.

53. Prove that there exists $P \in \mathrm{GL}_n(\mathbb{C})$ such that

$$
P^{-1}
\begin{pmatrix}
a_0 & a_1 & a_2 & \cdots & & a_{n-2} & a_{n-1} \\
a_{n-1} & a_0 & a_1 & \cdots & & & a_{n-2} \\
a_{n-2} & a_{n-1} & a_0 & \cdots & & & \ddots \\
\vdots & & & & & & \\
& & & & a_0 & a_1 & a_2 \\
a_2 & & \cdots & & a_{n-1} & a_0 & a_1 \\
a_1 & a_2 & \cdots & & a_{n-2} & a_{n-1} & a_0
\end{pmatrix}
P
$$

is a diagonal matrix for any $a_0, a_1, \ldots, a_{n-1} \in \mathbb{C}$.

54. Decide whether the following matrices $A = A_i \quad (i = 1, 2, 3)$ are diagonalizable over $\mathbb{R}$. If A is diagonalizable, find an invertible real matrix P such that $P^{-1}AP$ is a diagonal matrix:

$$
A_1 = \begin{pmatrix} 13 & -4 & -3 \\ 61 & -24 & -24 \\ -33 & 13 & 13 \end{pmatrix}, \quad
A_2 = \begin{pmatrix} -5 & -2 & -6 \\ -39 & -10 & -35 \\ 21 & 6 & 20 \end{pmatrix},
$$

$$
A_3 = \begin{pmatrix} 1 & -2 & -4 \\ -12 & -10 & -26 \\ 6 & 6 & 15 \end{pmatrix}.
$$

55. Let V be an n-dimensional linear space and $\sigma \in \mathrm{End}(V)$. Assume that the minimal polynomial of σ has the form

$$
m_\sigma(\lambda) = p_1(\lambda)^{r_1} p_2(\lambda)^{r_2} \cdots p_s(\lambda)^{r_s},
$$

where $p_1(\lambda), p_2(\lambda), \ldots, p_s(\lambda) \in F[\lambda]$ are different monic irreducible polynomials, $r_i \geqslant 1$, $i = 1, 2, \ldots, s$. Set $W_i = \mathrm{Ker}\,(p_i(\sigma)^{r_i})$, $i = 1, 2, \ldots, s$. Prove that

$$
W = (W \cap W_1) \oplus (W \cap W_2) \oplus \cdots \oplus (W \cap W_s)
$$

for any σ-subspace W of V.

56. Let $\sigma \in \mathrm{End}_{\mathbb{R}}(\mathbb{R}^3)$. Assume that the matrix of σ with respect to the standard basis $\{\varepsilon_1, \varepsilon_2, \varepsilon_3\}$ for $\mathbb{R}^3$ is $A = \begin{pmatrix} 6 & -3 & -2 \\ 4 & -1 & -2 \\ 10 & -5 & -3 \end{pmatrix}$. Find the first decomposition of $\mathbb{R}^3$ determined by σ.

57. Let V be a 3-dimensional complex linear space. Assume the matrix of $\sigma \in \mathrm{End}_{\mathbb{C}}(V)$ with respect to the basis $\alpha_1, \alpha_2, \alpha_3$ for V is $\begin{pmatrix} 4 & -5 & 2 \\ 5 & -7 & 3 \\ 6 & -9 & 4 \end{pmatrix}$. Find the eigenvalues of σ and the associated root subspaces.

58. Let $A \in \mathrm{M}_n(\mathbb{C})$ and $A^m = I_n$, where m is a positive integer. Prove that A is diagonalizable.

59. Let V be an n-dimensional linear space over a number field F and $\sigma \in \mathrm{End}(V)$. Assume that the minimal polynomial $m_\sigma(\lambda)$ of σ is irreducible over F. Let W be a σ-subspace of V. Prove the following:

    (1) $\deg\,(m_\sigma(\lambda)) \mid n$.
    (2) For any $\alpha_1, \alpha_2 \in V$,
    $$W_\sigma(\alpha_1) = W_\sigma(\alpha_2) \text{ or } W_\sigma(\alpha_1) \cap W_\sigma(\alpha_2) = 0.$$

    (3) There exists $\alpha_1, \alpha_2, \ldots, \alpha_r \in W$ such that $W = \bigoplus_{i=1}^{r} W_\sigma(\alpha_i)$.

    (4) There exists a σ-subspace U of V such that $V = W \oplus U$.

60. Let V be an n-dimensional complex linear space and $\sigma \in \mathrm{End}_{\mathbb{C}}(V)$. Assume that the matrix of σ with respect to a basis $\alpha_1, \alpha_2, \ldots, \alpha_n$ for V is the Jordan block of order n with eigenvalue λ. Set $V_i = L(\alpha_{n-i+1}, \alpha_{n-i+2}, \ldots, \alpha_n)$, $1 \leqslant i \leqslant n$. Prove the following:

    (1) $V_i = \mathrm{Ker}(\sigma - \lambda\,1_V)^i$, hence V_i is a σ-subspace, $i = 1, 2, \ldots, n$.
    (2) The σ-subspace of V containing α_1 is $V = V_n$.
    (3) Every σ-subspace of V contains α_n.
    (4) Let W be a σ-subspace of V. If $\dim W = i \geqslant 1$, then $W = V_i$. Therefore, V has no nonzero σ-subspaces other than $V_1, V_2, \ldots, V_n$.
    (5) V cannot be decomposed into a direct sum of two nontrivial σ-subspaces.

61. (Jordan decomposition theorem) Let $A \in M_n(F)$ and
$$f_A(\lambda) = (\lambda - \lambda_1)^{e_1}(\lambda - \lambda_2)^{e_2} \cdots (\lambda - \lambda_s)^{e_s},$$
where $\lambda_1, \lambda_2, \ldots, \lambda_s \in F$ are different, $e_1, e_2, \ldots, e_s \in \mathbb{N}^*$. Prove that $A = D + N$, where $D, N \in M_n(F)$, D is diagonalizable, N is nilpotent, and D and N are both polynomials in A and determined uniquely by A.

62. Let $A, B \in M_3(\mathbb{C})$. Assume that A and B both have only one eigenvalue λ. Prove that A is similar to B if and only if $\dim V_\lambda(A) = \dim V_\lambda(B)$, where $V_\lambda(A)$ and $V_\lambda(B)$ are the eigenspaces of A and B associated with λ, respectively.

63. Find the Jordan canonical forms of the following matrices:
$$\begin{pmatrix} 4 & 5 & -2 \\ -2 & -2 & 1 \\ -1 & -1 & 1 \end{pmatrix}; \quad \begin{pmatrix} 1 & 2 & 0 \\ 0 & 2 & 0 \\ -2 & -2 & 1 \end{pmatrix}; \quad \begin{pmatrix} -4 & 2 & 10 \\ 4 & 3 & 7 \\ -3 & 1 & 7 \end{pmatrix}.$$

64.* Let V be a finite-dimensional linear space over a number field F and $\sigma \in \mathrm{End}(V)$. If for any σ-subspace W of V, there exists a σ-subspace U of V such that $V = W \oplus U$, then we say that σ is a **semi-simple linear transformation**. Prove the following:

(1) Let $m_\sigma(\lambda) = p(\lambda)^e$ be the minimal polynomial of σ, where $p(\lambda) \in F[\lambda]$ is monic and irreducible. Then σ is a semi-simple linear transformation if and only if $e = 1$, i.e., $m_\sigma(\lambda)$ is irreducible.

(2) σ is a semi-simple linear transformation if and only if the minimal polynomial $m_\sigma(\lambda)$ of σ has the factorization
$$m_\sigma(\lambda) = p_1(\lambda)p_2(\lambda) \cdots p_s(\lambda),$$
where $p_1(\lambda), p_2(\lambda), \ldots, p_s(\lambda) \in F[\lambda]$ are different monic irreducible polynomials.

(3) If F is the field $\mathbb{C}$ of complex numbers, then σ is a semi-simple linear transformation if and only if σ is diagonalizable.

(4) σ is a semi-simple linear transformation if and only if the matrix of σ with respect to a basis for V is diagonalizable as a complex matrix.

65.* Let $N \in M_n(F)$ be a nilpotent matrix and let m be the **nilpotency index** of N (i.e., the minimal positive integer m such that $N^m = 0$). Let $J = \mathrm{diag}(J_1, J_2, \ldots, J_s)$ be the Jordan canonical form of N. Prove the following:

(1) The number of Jordan blocks is $s = n - \mathrm{rank}N = \dim\mathrm{Ker}(N) = \dim V_0$, where V_0 is the eigenspace of N associated with the eigenvalue 0.

(2) $m = \max_{1 \leqslant i \leqslant s}\{n_i \mid J_i \text{ is a Jordan block of order } n_i\}$.

(3) The number of Jordan blocks of order k in J is

$$\mathrm{rank}\big(N^{k-1}\big) - 2\mathrm{rank}\big(N^k\big) + \mathrm{rank}\big(N^{k+1}\big),$$

where $k \in \{1, 2, \ldots, m\}$.

(4) J is determined uniquely by N regardless of the order of Jordan blocks $J_1, J_2, \ldots, J_s$.

66.* Let $A \in \mathrm{M}_n(F)$ and $f_A(\lambda) = \prod_{i=1}^{s}(\lambda - \lambda_i)^{e_i}$, where $\lambda_1, \lambda_2, \ldots, \lambda_s \in F$ are different, $e_1, e_2, \ldots, e_s \in \mathbb{N}^*$. Prove the following:

(1) The number of Jordan blocks in the Jordan canonical form J of A is $t = \sum_{i=1}^{s} \dim V_{\lambda_i}$, where V_{λ_i} is the eigenspace of A associated with the eigenvalue λ_i, $i = 1, 2, \ldots, s$.

(2) The number r_k of Jordan blocks of order k $(1 \leqslant k \leqslant n)$ in the Jordan canonical form J of A is

$$\sum_{j=1}^{s} \Big(\mathrm{rank}(A - \lambda_j I_n)^{k-1} - 2\mathrm{rank}(A - \lambda_j I_n)^k$$

$$+ \mathrm{rank}(A - \lambda_j I_n)^{k+1}\Big).$$

(3) The Jordan canonical form $J = \mathrm{diag}(J_1, J_2, \ldots, J_t)$ of A is determined uniquely by A regardless of the order of Jordan blocks $J_1, J_2, \ldots, J_t$.

Chapter 6

Λ-Matrices

6.1 Introduction

Smith[1] gave the Smith normal form of a matrix over integers in 1861. Ferdinand Georg Frobenius gave the concepts of invariant factors and elementary divisors about matrices in 1878. With the establishment of abstract algebraic theory, it is found that these concepts and theories exist for matrices over principal ideal domains (the ring of integers and the monadic polynomial ring over a field are both principal ideal domains); see Chapter 3 in [20].

In this chapter, we mainly focus on matrices over the monadic polynomial ring over a number field F, which are called λ-matrices. We first introduce the Smith normal form of a λ-matrix. Then we study the invariants of λ-matrices under equivalence and prove that two number matrices are similar if and only if their characteristic matrices are equivalent. On this basis, we give a concise proof of Theorem 5.9.9 on the Jordan canonical form of a matrix. Finally, we introduce integer matrices and their Smith normal forms.

6.2 The Canonical Forms of λ-Matrices Under Equivalence

Let F be a number field and let $F[\lambda]$ be the ring of polynomials over F in the indeterminate λ.

---

[1]Henry John Stephen Smith, 1826–1883, British mathematician.

Definition 6.2.1. Let m, n be positive integers and $a_{ij}(\lambda) \in F[\lambda]$, $i = 1, \ldots, m$, $j = 1, \ldots, n$. Then the matrix

$$A(\lambda) = \begin{pmatrix} a_{11}(\lambda) & a_{12}(\lambda) & \cdots & a_{1n}(\lambda) \\ a_{21}(\lambda) & a_{22}(\lambda) & \cdots & a_{2n}(\lambda) \\ \vdots & \vdots & & \vdots \\ a_{m1}(\lambda) & a_{m2}(\lambda) & \cdots & a_{mn}(\lambda) \end{pmatrix}$$

is called an $m \times n$ λ**-matrix** over F. The set of all $m \times n$ λ-matrices is denoted by $\mathrm{M}_{m \times n}(F[\lambda])$. In particular, an $n \times n$ λ-matrix is called a square λ-matrix of order n. $\mathrm{M}_{n \times n}(F[\lambda])$ is often denoted by $\mathrm{M}_n(F[\lambda])$. For distinction, hereafter we call matrices over F **number matrices**.

The addition, the scalar multiplication, the multiplication of matrices, and the determinant of a square matrix over a number field F are all defined by the addition, subtraction, and multiplication in F; therefore, the addition of λ-matrices, the scalar multiplication of a λ-matrix by an element in $F[\lambda]$, the multiplication of λ-matrices, as well as the determinant of a square λ-matrix can be defined by the addition, subtraction, and multiplication of polynomials in $F[\lambda]$ in a similar way as the operations of number matrices over a number field F.

The rank of a number matrix is defined as the largest order of its nonzero minors, so we can define the rank of a λ-matrix analogously. The **rank** of a λ-matrix $A(\lambda)$ is the largest order of the nonzero minors of $A(\lambda)$, denoted by $\mathrm{rank}(A(\lambda))$. Obviously, a square λ-matrix $A(\lambda)$ of order n has rank n if and only if $|A(\lambda)| \neq 0$.

Properties of number matrices over F only concerning the addition, subtraction, and multiplication in F are still valid for λ-matrices. For example, the adjoint matrix of a square λ-matrix is still a square λ-matrix, and the Laplace expansion and Cauchy–Binet formula remain true for λ-matrices.

Invertible λ-matrices can be defined similarly to invertible number matrices. Let $A(\lambda) \in \mathrm{M}_n(F[\lambda])$. If there exists $B(\lambda) \in \mathrm{M}_n(F[\lambda])$ such that $A(\lambda)B(\lambda) = B(\lambda)A(\lambda) = I_n$, where I_n is the identity matrix of order n, then we call $A(\lambda)$ **an invertible λ-matrix** and $B(\lambda)$ the **inverse** of $A(\lambda)$. It's easy to prove that $B(\lambda)$ is uniquely determined by $A(\lambda)$, denoted by $A(\lambda)^{-1}$. The set of all invertible λ-matrices of order n is denoted by $\mathrm{GL}_n(F[\lambda])$.

Over a number field F, a number matrix A of order n is invertible if and only if $|A| \neq 0$, i.e., $|A| \in F \setminus \{0\} = \mathrm{GL}_1(F)$ (every nonzero element in a number field F is invertible in F), meanwhile $A^{-1} = \frac{1}{|A|}A^*$, where A^* is the adjoint matrix of A. Let $f(\lambda) \in F[\lambda]$ be invertible in $F[\lambda]$, i.e., there exists $g(\lambda) \in F[\lambda]$ such that $f(\lambda)g(\lambda) = 1$. Since the degree of the product of polynomials equals the sum of the degrees of factors, $f(\lambda)$ is a nonzero constant in F, i.e., $f(\lambda) \in F^* = F \setminus \{0\} = \mathrm{GL}_1(F)$. Therefore, $\mathrm{GL}_1(F[\lambda]) = F^*$.

Theorem 6.2.2. *Let $A(\lambda) \in \mathrm{M}_n(F[\lambda])$. Then $A(\lambda) \in \mathrm{GL}_n(F[\lambda])$ if and only if $|A(\lambda)| \in F^*$, meanwhile $A(\lambda)^{-1} = \frac{1}{|A(\lambda)|}A(\lambda)^*$.*

Proof. Necessity. Let $A(\lambda)$ be invertible. Then, by definition, there exists $B(\lambda) \in \mathrm{M}_n(F[\lambda])$ such that $A(\lambda)B(\lambda) = I_n$. Taking determinants on both sides, we have $|A(\lambda)||B(\lambda)| = 1$. Therefore, $|A(\lambda)|$ is invertible in $F[\lambda]$, so $|A(\lambda)| \in F^*$.

Sufficiency. Let $A(\lambda) \in \mathrm{M}_n(F[\lambda])$ and $|A(\lambda)| \in F^*$. Since the adjoint matrix $A(\lambda)^*$ of $A(\lambda)$ is also a λ-matrix, we have $B(\lambda) = \frac{1}{|A(\lambda)|}A(\lambda)^* \in \mathrm{M}_n(F[\lambda])$. On the other hand, $A(\lambda)B(\lambda) = B(\lambda)A(\lambda) = I_n$, therefore $A(\lambda)$ is invertible and $A(\lambda)^{-1} = \frac{1}{|A(\lambda)|}A(\lambda)^*$.
$\square$

Next, we study the canonical form of a λ-matrix under equivalence by using the idea and method of studying the canonical form of a number matrix under equivalence. For this purpose, we first introduce elementary operations of λ-matrices.

Definition 6.2.3. The following three operations are called the **elementary row (column) operations** of λ-matrices:

(1) Multiply a row (column) of a λ-matrix by a nonzero constant k in F.
(2) Add $f(\lambda)$ times a row (column) of a λ-matrix to another row (column), where $f(\lambda) \in F[\lambda]$.
(3) Interchange two rows (columns) of a λ-matrix.

The elementary row operations and column operations of λ-matrices are collectively called the **elementary operations** of λ-matrices.

Remark 6.2.4. Since we require all the elementary operations to be invertible and $\mathrm{GL}_1(F[\lambda]) = F \setminus \{0\}$, in the first type of elementary operations, the row (column) is multiplied by an invertible polynomial, i.e., a nonzero constant.

Like number matrices, we can introduce three types of **elementary λ-matrices**, i.e., the matrices obtained from the identity matrix by performing one single elementary operation:

(1) Multiplying the ith row (column) of the identity matrix by a nonzero constant k yields the elementary λ-matrices of the form

$$
E(i(k)) = \begin{pmatrix} 1 & & & & & & \\ & \ddots & & & & & \\ & & 1 & & & & \\ & & & k & & & \\ & & & & 1 & & \\ & & & & & \ddots & \\ & & & & & & 1 \end{pmatrix} \quad \text{the } i\text{th row.}
$$

(2) Adding $f(\lambda)$ times the jth row to the ith row (or $f(\lambda)$ times the ith column to the jth column) of the identity matrix yields the elementary λ-matrices of the form

$$
E(i, j(f(\lambda))) = \begin{pmatrix} 1 & & & & & & \\ & \ddots & & & & & \\ & & 1 & f(\lambda) & & & \\ & & & \ddots & & & \\ & & & 1 & & & \\ & & & & \ddots & & \\ & & & & & 1 \end{pmatrix} \begin{matrix} \\ \\ \text{the } i\text{th row} \\ \\ \text{the } j\text{th row} \\ \\ \end{matrix} , \; i < j,
$$

or

$$
E(i, j(f(\lambda))) = \begin{pmatrix} 1 & & & & & & \\ & \ddots & & & & & \\ & & 1 & & & & \\ & & & \ddots & & & \\ & & f(\lambda) & 1 & & & \\ & & & & \ddots & & \\ & & & & & 1 \end{pmatrix} \begin{matrix} \\ \\ \text{the } j\text{th row} \\ \\ \text{the } i\text{th row} \\ \\ \end{matrix} , \; i > j.
$$

(3) Interchanging the ith and jth rows (or columns) of the identity
matrix $(i \neq j)$ yields the elementary λ-matrices of the form

$$
E(i,j) =
\begin{pmatrix}
1 & & & & & & & & & \\
 & \ddots & & & & & & & & \\
 & & 1 & & & & & & & \\
 & & & 0 & & 1 & & & & \\
 & & & 1 & & & & & & \\
 & & & & & \ddots & & & & \\
 & & & & & & 1 & & & \\
 & & & 1 & & & & 0 & & \\
 & & & & & & & & 1 & \\
 & & & & & & & & & \ddots \\
 & & & & & & & & & & 1
\end{pmatrix}.
$$

Elementary λ-matrices are all invertible, and

$$
E(i(k))^{-1} = E(i(k^{-1})), \; E(i,j(f(\lambda)))^{-1} = E(i,j(-f(\lambda))),
$$
$$
E(i,j)^{-1} = E(i,j),
$$

so the inverse of an elementary λ-matrix is the same type of elementary λ-matrix as itself. Like elementary operations of number
matrices, the elementary operations of λ-matrices also do not change
the rank of a λ-matrix, and applying an elementary row (column)
operation to a λ-matrix $A(\lambda)$ is equivalent to multiplying $A(\lambda)$ on
the left (right) by the corresponding elementary λ-matrix.

Definition 6.2.5. Let $A(\lambda), B(\lambda) \in M_{m \times n}(F[\lambda])$. If $B(\lambda)$ can be
obtained by applying a finite sequence of elementary operations of
λ-matrices to $A(\lambda)$, then we say that $A(\lambda)$ and $B(\lambda)$ are **equivalent**,
denoted by $A(\lambda) \sim B(\lambda)$.

According to the relationship between the elementary operations
of λ-matrices and elementary λ-matrices, two $m \times n$ λ-matrices
$A(\lambda)$ and $B(\lambda)$ are equivalent if and only if there exist elementary
λ-matrices $P_1(\lambda), \ldots, P_s(\lambda)$ of order m and elementary λ-matrices
$Q_1(\lambda), \ldots, Q_t(\lambda)$ of order n such that

$$
B(\lambda) = P_s(\lambda) \cdots P_1(\lambda) A(\lambda) Q_1(\lambda) \cdots Q_t(\lambda). \tag{6.1}
$$

Theorem 6.2.6. *The equivalence of λ-matrices is an equivalence relation.*

Proof. This result can be directly verified by definition. $\square$

Lemma 6.2.7. *Let $A(\lambda) \in \mathrm{M}_{m \times n}(F[\lambda])$ and $A(\lambda) \neq 0$. Then $A(\lambda)$ is equivalent to $B(\lambda) = (b_{ij}(\lambda))$, where $b_{11}(\lambda) \neq 0$, and $b_{11}(\lambda)|b_{ij}(\lambda)$ for any $1 \leqslant i \leqslant m, 1 \leqslant j \leqslant n$.*

Proof. Since $A(\lambda) = (a_{ij}(\lambda)) \neq 0$, there exists $a_{ij}(\lambda) \neq 0$, and so

$$E(1, i)A(\lambda)E(1, j) = \begin{pmatrix} a_{ij}(\lambda) & \star \\ \star & \star \end{pmatrix}.$$

Without loss of generality, we may assume that $a_{11}(\lambda) \neq 0$. We shall prove Lemma 6.2.7 by induction on the degree of $\deg(a_{11}(\lambda))$. If $\deg(a_{11}(\lambda)) = 0$, i.e., $a_{11}(\lambda) \in F^*$, then it's obvious that $a_{11}(\lambda)|a_{ij}(\lambda)$, $1 \leqslant i \leqslant m, 1 \leqslant j \leqslant n$. Suppose $t \geqslant 1$ and that Lemma 6.2.7 holds for any λ-matrix with $\deg(a_{11}(\lambda)) \leqslant t - 1$. Now consider the λ-matrix with $\deg(a_{11}(\lambda)) = t$.

If $a_{11}(\lambda)|a_{ij}(\lambda)$ for any $1 \leqslant i \leqslant m, 1 \leqslant j \leqslant n$, then Lemma 6.2.7 holds. Therefore, we may assume that there exist $i_0 \in \{1, \ldots, m\}, j_0 \in \{1, \ldots, n\}$ such that $a_{11}(\lambda) \nmid a_{i_0 j_0}(\lambda)$.

(1) Assume $i_0 = 1$, i.e., $a_{11}(\lambda) \nmid a_{1j_0}(\lambda)$. According to the division algorithm, there exist $q(\lambda), r(\lambda) \in F[\lambda]$ such that $a_{1j_0}(\lambda) = q(\lambda)a_{11}(\lambda) + r(\lambda)$, where $r(\lambda) \neq 0$, and $0 \leqslant \deg(r(\lambda)) < t$. Therefore,

$$A(\lambda)E(1, j_0(-q(\lambda)))E(1, j_0) = \begin{pmatrix} r(\lambda) & \star \\ \star & \star \end{pmatrix},$$

i.e., $A(\lambda)$ is equivalent to $\begin{pmatrix} r(\lambda) & \star \\ \star & \star \end{pmatrix}$. Since $r(\lambda) \neq 0, \deg(r(\lambda)) \leqslant t - 1$,

by induction, Lemma 6.2.7 holds for the λ-matrix $\begin{pmatrix} r(\lambda) & \star \\ \star & \star \end{pmatrix}$, and so it holds for $A(\lambda)$.

(2) Assume $j_0 = 1$, i.e., $a_{11}(\lambda) \nmid a_{i_0 1}(\lambda)$. The proof is similar to that of (1), only changing the columns in (1) into rows.

(3) Assume that $a_{11}(\lambda)|a_{1k}(\lambda)$, $a_{11}(\lambda)|a_{l1}(\lambda)$, $k = 1, 2, \ldots, n$, $l = 1, 2, \ldots, m$, and $a_{11}(\lambda) \nmid a_{i_0 j_0}(\lambda)$, $i_0 \neq 1$, $j_0 \neq 1$. Let

$a_{i_0 1}(\lambda) = q_{i_0 1}(\lambda)a_{11}(\lambda)$. Then

$$A(\lambda) = \begin{pmatrix} a_{11}(\lambda) & a_{12}(\lambda) & \cdots & a_{1j_0}(\lambda) & \cdots & a_{1n}(\lambda) \\ \vdots & \vdots & & \vdots & & \vdots \\ a_{i_0 1}(\lambda) & a_{i_0 2}(\lambda) & \cdots & a_{i_0 j_0}(\lambda) & \cdots & a_{i_0 n}(\lambda) \\ \vdots & \vdots & & \vdots & & \vdots \\ a_{m1}(\lambda) & a_{m2}(\lambda) & \cdots & a_{mj_0}(\lambda) & \cdots & a_{mn}(\lambda) \end{pmatrix}$$

$$\xrightarrow{\ (-q_{i_0 1}(\lambda))\times r_1+r_{i_0}\ } \begin{pmatrix} a_{11}(\lambda) & a_{12}(\lambda) & \cdots & a_{1j_0}(\lambda) & \cdots & a_{1n}(\lambda) \\ \vdots & \vdots & & \vdots & & \vdots \\ 0 & b_{i_0 2}(\lambda) & \cdots & b_{i_0 j_0}(\lambda) & \cdots & b_{i_0 n}(\lambda) \\ \vdots & \vdots & & \vdots & & \vdots \\ a_{m1}(\lambda) & a_{m2}(\lambda) & \cdots & a_{mj_0}(\lambda) & \cdots & a_{mn}(\lambda) \end{pmatrix}$$

$$\xrightarrow{\ 1\times r_{i_0}+r_1\ } \begin{pmatrix} a_{11}(\lambda) & c_{12}(\lambda) & \cdots & c_{1j_0}(\lambda) & \cdots & c_{1n}(\lambda) \\ \vdots & \vdots & & \vdots & & \vdots \\ 0 & b_{i_0 2}(\lambda) & \cdots & b_{i_0 j_0}(\lambda) & \cdots & b_{i_0 n}(\lambda) \\ \vdots & \vdots & & \vdots & & \vdots \\ a_{m1}(\lambda) & a_{m2}(\lambda) & \cdots & a_{mj_0}(\lambda) & \cdots & a_{mn}(\lambda) \end{pmatrix},$$

where

$$b_{i_0 k}(\lambda) = a_{i_0 k}(\lambda) - q_{i_0 1}(\lambda)a_{1k}(\lambda),$$

$$c_{1k}(\lambda) = a_{1k}(\lambda) + b_{i_0 k}(\lambda) = a_{i_0 k}(\lambda) + (1 - q_{i_0 1}(\lambda))a_{1k}(\lambda),$$

$$k = 2,\ldots,n.$$

Since $a_{11}(\lambda)|a_{1j_0}(\lambda)$, $a_{11}(\lambda) \nmid a_{i_0 j_0}(\lambda)$, we have $a_{11}(\lambda) \nmid c_{1j_0}(\lambda)$, reducing into the case (1). Therefore, Lemma 6.2.7 holds. $\qquad\square$

Theorem 6.2.8. *Let* $A(\lambda) \in \mathrm{M}_{m\times n}(F[\lambda])$ *and* $r = \mathrm{rank}(A(\lambda))$. *If* $r \geqslant 1$, *then* $A(\lambda)$ *is equivalent to the matrix*

$$\Lambda(\lambda) = \begin{pmatrix} d_1(\lambda) & & & & \\ & d_2(\lambda) & & & \\ & & \ddots & & \\ & & & d_r(\lambda) & \\ & & & & 0_{(m-r)\times(n-r)} \end{pmatrix},$$

where $d_1(\lambda), d_2(\lambda), \ldots, d_r(\lambda)$ *are all monic polynomials over* F, *and* $d_i(\lambda)|d_{i+1}(\lambda)$, $i = 1, 2, \ldots, r - 1$. *We call* $\Lambda(\lambda)$ *the* **Smith normal form** *of the matrix* $A(\lambda)$. *If* $r = 0$, *we call* $\Lambda(\lambda) = 0$ *the Smith normal form of* $A(\lambda) = 0$.

Proof. Let $A(\lambda) \neq 0$. Then $A(\lambda)$ is equivalent to the λ-matrix $B(\lambda) = (b_{ij}(\lambda))$ by Lemma 6.2.7, where

$$b_{11}(\lambda) \neq 0 \text{ and } b_{11}(\lambda) \text{ divides every element of } B(\lambda). \qquad (6.2)$$

According to the transitivity of equivalence, we only need to prove that $B(\lambda)$ is equivalent to the Smith normal form $\Lambda(\lambda)$. We prove this result by induction on the number m of rows of $B(\lambda)$.

Assume that $m = 1$ and that $B(\lambda) = (b_{11}(\lambda), b_{12}(\lambda), \ldots, b_{1n}(\lambda))$ satisfies (6.2) for any positive integer n. Let $b_{1k}(\lambda) = f_k(\lambda)b_{11}(\lambda)$, $f_k(\lambda) \in F[\lambda]$, $k = 2, \ldots, n$. Adding the first column multiplied by $-f_k(\lambda)$ to the kth $(k = 2, \ldots, n)$ column, we have

$$B(\lambda)E(1, 2(-f_2(\lambda))) \cdots E(1, n(-f_n(\lambda))) = (b_{11}(\lambda), 0, \ldots, 0).$$

Let b be the leading coefficient of $b_{11}(\lambda)$ and $d_1(\lambda) = b^{-1}b_{11}(\lambda)$. Then $d_1(\lambda) \in F[\lambda]$ is a monic polynomial and

$$E(1(b^{-1}))B(\lambda)E(1, 2(-f_2(\lambda))) \cdots E(1, n(-f_n(\lambda)))$$
$$= (d_1(\lambda), 0, \ldots, 0).$$

So, $B(\lambda)$ is equivalent to the Smith normal form $\Lambda(\lambda)$. Therefore, when $m = 1$, the result is true for any positive integer n.

Assume that $m \geqslant 2$, and the statement holds for $m - 1$, i.e., for any positive integer n, all $(m-1) \times n$ λ-matrices $B(\lambda)$ satisfying (6.2) are equivalent to the normal forms $\Lambda(\lambda)$. Now consider the $m \times n$ λ-matrix $B(\lambda)$ satisfying (6.2).

If $n = 1$, $B(\lambda)$ is equivalent to the Smith normal form $\Lambda(\lambda)$ by an argument similar to that in the proof of $m = 1$.

If $n \geqslant 2$, from (6.2) we may assume $b_{ij}(\lambda) = b_{11}(\lambda)f_{ij}(\lambda)$, where $f_{ij}(\lambda) \in F[\lambda]$, $1 \leqslant i \leqslant m, 1 \leqslant j \leqslant n$. Note that $b_{11}(\lambda) \neq 0$. Applying the second type of elementary row and column operations to $B(\lambda)$

respectively, we get

$$P(\lambda)B(\lambda)Q(\lambda) = \begin{pmatrix} b_{11}(\lambda) & 0 & \cdots & 0 \\ 0 & c_{22}(\lambda) & \cdots & c_{2n}(\lambda) \\ \vdots & \vdots & & \vdots \\ 0 & c_{m2}(\lambda) & \cdots & c_{mn}(\lambda) \end{pmatrix},$$

where

$$P(\lambda) = E(n, 1(-f_{n1}(\lambda))) \cdots E(2, 1(-f_{21}(\lambda))),$$
$$Q(\lambda) = E(1, 2(-f_{12}(\lambda))) \cdots E(1, n(-f_{1n}(\lambda))),$$

and $c_{kl}(\lambda) = b_{kl}(\lambda) - f_{k1}(\lambda)b_{1l}(\lambda)$, therefore $b_{11}(\lambda)|c_{kl}(\lambda)$, $k = 2, \ldots, m$, $l = 2, \ldots, n$.

Let $A_1(\lambda) = \begin{pmatrix} c_{22}(\lambda) & \cdots & c_{2n}(\lambda) \\ \vdots & & \vdots \\ c_{m2}(\lambda) & \cdots & c_{mn}(\lambda) \end{pmatrix}$. If $A_1(\lambda) = 0$, the statement is true. Otherwise, by Lemma 6.2.7, $A_1(\lambda)$ is equivalent to an $(m-1) \times (n-1)$ λ-matrix $B_1(\lambda)$ satisfying (6.2), meanwhile $b_{11}(\lambda)$ divides every element of $B_1(\lambda)$. By induction, $B_1(\lambda)$ is equivalent to the Smith normal form

$$\Lambda_1(\lambda) = \begin{pmatrix} d_2(\lambda) & & & \\ & \ddots & & \\ & & d_r(\lambda) & \\ & & & 0_{(m-r)\times(n-r)} \end{pmatrix},$$

where $d_2(\lambda), \ldots, d_r(\lambda) \in F[\lambda]$ are monic polynomials, and $d_i(\lambda)|d_{i+1}(\lambda)$, $i = 2, \ldots, r-1$. Let $b \in F^*$ be the leading coefficient of $b_{11}(\lambda)$ and $d_1(\lambda) = b^{-1}b_{11}(\lambda)$. Then $d_1(\lambda)|d_k(\lambda)$, $k = 2, \ldots, r$, and $B(\lambda)$ is equivalent to the λ-matrix

$$\begin{pmatrix} d_1(\lambda) & & & & \\ & d_2(\lambda) & & & \\ & & \ddots & & \\ & & & d_r(\lambda) & \\ & & & & 0_{(m-r)\times(n-r)} \end{pmatrix} = \Lambda(\lambda).$$

To sum up, $A(\lambda)$ is equivalent to the Smith normal form $\Lambda(\lambda)$.

$\square$

Corollary 6.2.9. *Let $A(\lambda) \in M_{m \times n}(F[\lambda])$. Then there exist elementary λ-matrices $P_1(\lambda), \ldots, P_s(\lambda)$ of order m and elementary λ-matrices $Q_1(\lambda), \ldots, Q_t(\lambda)$ of order n such that $P_s(\lambda) \cdots P_1(\lambda) A(\lambda) Q_1(\lambda) \cdots Q_t(\lambda)$ is the Smith normal form of $A(\lambda)$.*

Corollary 6.2.10. *Let $A(\lambda) \in M_n(F[\lambda])$. Then $A(\lambda)$ is invertible if and only if $A(\lambda)$ is a product of a finite number of elementary λ-matrices.*

Proof. The sufficiency is obvious because elementary λ-matrices are all invertible, and a product of a finite number of elementary λ-matrices is still an invertible λ-matrix. For the necessity, let $A(\lambda)$ be an invertible λ-matrix of order n, then there exist elementary λ-matrices $P_1(\lambda), \ldots, P_s(\lambda)$, $Q_1(\lambda), \ldots, Q_t(\lambda)$ of order n such that

$$P_s(\lambda) \cdots P_1(\lambda) A(\lambda) Q_1(\lambda) \cdots Q_t(\lambda) = \operatorname{diag}(d_1(\lambda), d_2(\lambda), \ldots, d_n(\lambda)),$$

where $d_1(\lambda), d_2(\lambda), \ldots, d_n(\lambda)$ are monic polynomials and $d_i(\lambda) | d_{i+1}(\lambda)$, $i = 1, 2, \ldots, n-1$. Since $|P_1(\lambda)|, \ldots, |P_s(\lambda)|$, $|A(\lambda)|$, $|Q_1(\lambda)|, \ldots, |Q_t(\lambda)| \in F^*$, we have $d_i(\lambda) \in F^*$, thus $d_i(\lambda) = 1$, $i = 1, 2, \ldots, n$. Therefore,

$$A(\lambda) = P_1(\lambda)^{-1} P_2(\lambda)^{-1} \cdots P_s(\lambda)^{-1} Q_t(\lambda)^{-1} \cdots Q_2(\lambda)^{-1} Q_1(\lambda)^{-1}.$$

Since the inverses of elementary λ-matrices are also elementary λ-matrices, we obtain that $A(\lambda)$ is a product of a finite number of elementary λ-matrices. $\qquad\square$

From (6.1) and Corollary 6.2.10, we have the following.

Corollary 6.2.11. *Let $A(\lambda), B(\lambda) \in M_{m \times n}(F[\lambda])$. Then $A(\lambda)$ is equivalent to $B(\lambda)$ if and only if there exist $P(\lambda) \in GL_m(F[\lambda])$ and $Q(\lambda) \in GL_n(F[\lambda])$ such that $B(\lambda) = P(\lambda) A(\lambda) Q(\lambda)$.*

6.3   The Invariants of λ-Matrices Under Equivalence

In this section, we study the invariants of λ-matrices under equivalence and prove the uniqueness of the Smith normal form of a λ-matrix.

Definition 6.3.1. Let $A(\lambda) \in M_{m \times n}(F[\lambda])$ and $\mathrm{rank}(A(\lambda)) = r \geqslant 1$. For every positive integer $k \leqslant r$, the monic greatest common divisor of all minors of order k of $A(\lambda)$ is called the kth **determinantal divisor** of $A(\lambda)$, denoted by $D_k(A(\lambda))$, without causing confusion, abbreviated as $D_k(\lambda)$.

Obviously, the determinantal divisors of a λ-matrix $A(\lambda)$ with rank $r\,(\geqslant 1)$ are all monic polynomials. Note that every minor of order $k + 1$ is a combination of minors of order k, thus $D_k(\lambda)$ divides every minor of order $k + 1$, therefore $D_k(\lambda) | D_{k+1}(\lambda)$, $k = 1, 2, \ldots, r - 1$.

Theorem 6.3.2. *Elementary operations of λ-matrices do not change their determinantal divisors.*

Proof. Let $A(\lambda), B(\lambda) \in M_{m \times n}(F[\lambda])$ be equivalent. Then $A(\lambda)$ and $B(\lambda)$ have the same rank, denoted by $r \geqslant 1$. We only need to prove that $A(\lambda)$ and $B(\lambda)$ have the same determinantal divisors when $A(\lambda)$ is transformed into $B(\lambda)$ by one single elementary operation.

Let $1 \leqslant k \leqslant r$. Then every minor of order k of $B(\lambda)$ is a combination of some minors of order k of $A(\lambda)$, and so it can be divided by $D_k(A(\lambda))$, whence $D_k(A(\lambda)) | D_k(B(\lambda))$. Since the equivalence is symmetric, we have $D_k(B(\lambda)) | D_k(A(\lambda))$ in the same way. From $D_k(A(\lambda))$ and $D_k(B(\lambda))$ being monic polynomials, we get $D_k(A(\lambda)) = D_k(B(\lambda))$. $\qquad\square$

Theorem 6.3.3. *The Smith normal form of a λ-matrix is unique.*

Proof. Let $A(\lambda) \in M_{m \times n}(F[\lambda])$ and $\mathrm{rank}\,(A(\lambda)) = r \geqslant 1$. From Theorem 6.2.8, there exist $P(\lambda) \in GL_m(F[\lambda])$ and $Q(\lambda) \in GL_n(F[\lambda])$ such that

$$
P(\lambda)A(\lambda)Q(\lambda) = \Lambda(\lambda) = \begin{pmatrix} d_1(\lambda) & & & & \\ & d_2(\lambda) & & & \\ & & \ddots & & \\ & & & d_r(\lambda) & \\ & & & & 0_{(m-r) \times (n-r)} \end{pmatrix},
$$

where $d_1(\lambda), d_2(\lambda), \ldots, d_r(\lambda)$ are all monic polynomials, and $d_i(\lambda)|d_{i+1}(\lambda)$, $i = 1, 2, \ldots, r - 1$. By Theorem 6.3.2, $D_k(A(\lambda)) = D_k(\Lambda(\lambda))$, $k = 1, 2, \ldots, r$. It's obvious that $D_k(\Lambda(\lambda)) = d_1(\lambda)d_2(\lambda) \cdots d_k(\lambda)$, $k = 1, 2, \ldots, r$. Therefore,

$$d_1(\lambda) = D_1(A(\lambda)), \quad d_k(\lambda) = \frac{D_k(A(\lambda))}{D_{k-1}(A(\lambda))}, \quad k = 2, \ldots, r, \quad (6.3)$$

i.e., $d_1(\lambda), d_2(\lambda), \ldots, d_r(\lambda)$ are determined uniquely by the determinantal divisors, so the Smith normal form $\Lambda(\lambda)$ is unique. $\qquad\square$

Definition 6.3.4. Let $A(\lambda) \in \mathrm{M}_{m \times n}(F[\lambda])$ and $\mathrm{rank}(A(\lambda)) = r \geqslant 1$. Then the r monic polynomials $d_1(\lambda), d_2(\lambda), \ldots, d_r(\lambda)$ in the Smith normal form $\Lambda(\lambda)$ of $A(\lambda)$ are called the **invariant factors** of $A(\lambda)$.

Theorem 6.3.5. *Let* $A(\lambda), B(\lambda) \in \mathrm{M}_{m \times n}(F[\lambda])$ *and* $A(\lambda) \neq 0$, $B(\lambda) \neq 0$. *Then the following statements are equivalent:*

(1) $A(\lambda)$ *is equivalent to* $B(\lambda)$.
(2) $A(\lambda)$ *and* $B(\lambda)$ *have the same invariant factors.*
(3) $A(\lambda)$ *and* $B(\lambda)$ *have the same determinantal divisors.*

Proof. By (6.3), the invariant factors and determinantal divisors are uniquely determined by each other, so (2) $\Longleftrightarrow$ (3) holds. On the one hand, (1) $\Longrightarrow$ (3) follows from Theorem 6.3.2. On the other hand, if $A(\lambda)$ and $B(\lambda)$ have the same invariant factors, then $A(\lambda)$ and $B(\lambda)$ are equivalent to the same Smith normal form, therefore $A(\lambda)$ is equivalent to $B(\lambda)$, so (2) $\Longrightarrow$ (1) follows. $\qquad\square$

So far, we have given the canonical form theory of λ-matrices over a number field F and two ways of computing the invariant factors of λ-matrices: (1) We first calculate the determinantal divisors and then obtain the invariant factors. (2) Using elementary operations of λ-matrices, we simplify a λ-matrix to its Smith normal form and obtain the invariant factors. We can also transform a λ-matrix into a diagonal λ-matrix by elementary operations and then calculate the determinantal divisors and invariant factors.

Example 6.3.6. Find the invariant factors of $A(\lambda) \in \mathrm{GL}_n(F[\lambda])$.

Solution. From $A(\lambda) \in \mathrm{GL}_n(F[\lambda])$, we know that $\mathrm{rank}(A(\lambda)) = n$ and $|A(\lambda)| \in F^*$. Hence $D_n(\lambda) = 1$, and so $D_1(\lambda) = D_2(\lambda) = \cdots = D_n(\lambda) = 1$. Consequently, the n invariant factors are all 1. $\qquad\square$

Example 6.3.7. Let $\lambda_0 \in F$. Find the invariant factors of the

λ-matrix $A(\lambda) = \begin{pmatrix} \lambda - \lambda_0 & & & \\ -1 & & \ddots & \\ & \ddots & \ddots & \\ & & -1 & \lambda - \lambda_0 \end{pmatrix}$ of order n.

Solution. Obviously, $\mathrm{rank}(A(\lambda)) = n$, and $D_n(\lambda) = (\lambda - \lambda_0)^n$. Note that $A(\lambda)$ has a minor of order $n - 1$:

$$A(\lambda) \begin{pmatrix} 2 \; 3 \; \cdots & n \\ 1 \; 2 \; \cdots & n-1 \end{pmatrix} = \begin{vmatrix} -1 & \lambda - \lambda_0 & & & \\ & -1 & \lambda - \lambda_0 & & \\ & & \ddots & \ddots & \\ & & & -1 & \lambda - \lambda_0 \\ & & & & -1 \end{vmatrix} = (-1)^{n-1}.$$

Therefore, $D_{n-1}(\lambda) = 1$, and so the n determinantal divisors are

$$D_i(\lambda) = 1, \; i = 1, 2, \ldots, n-1, \; D_n(\lambda) = (\lambda - \lambda_0)^n.$$

Hence, the n invariant factors are

$$d_i(\lambda) = 1, \; i = 1, 2, \ldots, n-1, \; d_n(\lambda) = (\lambda - \lambda_0)^n. \qquad \square$$

Example 6.3.8. Find the invariant factors of the λ-matrix

$$A(\lambda) = \begin{pmatrix} \lambda + 1 & 2 & -6 \\ 1 & \lambda & -3 \\ 1 & 1 & \lambda - 4 \end{pmatrix}.$$

Solution. Applying elementary operations to $A(\lambda)$, we have

$$A(\lambda) = \begin{pmatrix} \lambda + 1 & 2 & -6 \\ 1 & \lambda & -3 \\ 1 & 1 & \lambda - 4 \end{pmatrix}$$

$$\xrightarrow[\;(-1)\times r_3 + r_2\;]{\;(-\lambda-1)\times r_3 + r_1\;} \begin{pmatrix} 0 & -\lambda + 1 & -\lambda^2 + 3\lambda - 2 \\ 0 & \lambda - 1 & -\lambda + 1 \\ 1 & 1 & \lambda - 4 \end{pmatrix}$$

$$\xrightarrow[\substack{(-1)\times c_1 + c_2 \\ (4-\lambda)\times c_1 + c_3}]{\;(r_1, r_3)\;} \begin{pmatrix} 1 & 0 & 0 \\ 0 & \lambda - 1 & -\lambda + 1 \\ 0 & -\lambda + 1 & -\lambda^2 + 3\lambda - 2 \end{pmatrix}$$

$$\xrightarrow{\ 1\times r_2+r_3\ } \begin{pmatrix} 1 & 0 & 0 \\ 0 & \lambda-1 & -\lambda+1 \\ 0 & 0 & -\lambda^2+2\lambda-1 \end{pmatrix}$$

$$\xrightarrow[\ (-1)\times r_3\ ]{\ 1\times c_2+c_3\ } \begin{pmatrix} 1 & 0 & 0 \\ 0 & \lambda-1 & 0 \\ 0 & 0 & (\lambda-1)^2 \end{pmatrix}.$$

Therefore, $\mathrm{rank}A(\lambda) = 3$ and the invariant factors of $A(\lambda)$ are $d_1(\lambda) = 1$, $d_2(\lambda) = \lambda - 1$, $d_3(\lambda) = (\lambda - 1)^2$. $\qquad\square$

The first chapter tells us that monic polynomials over a number field F can be factored uniquely into a product of powers of different monic irreducible factors. Next, we use the powers of monic irreducible factors to give other invariants of λ-matrices under equivalence.

Definition 6.3.9. Let $A(\lambda) \in \mathrm{M}_{m\times n}(F[\lambda])$ and $\mathrm{rank}(A(\lambda)) = r \geqslant 1$. Factorizing the invariant factors not equal to 1 of $A(\lambda)$ into a product of powers of different monic irreducible factors, all these powers of monic irreducible factors (counted by the number of occurrences) are called the **elementary divisors** of $A(\lambda)$.

Example 6.3.10. Let $d_1(\lambda) = \lambda(\lambda - 1)$, $d_2(\lambda) = \lambda(\lambda - 1)(\lambda + 1)$, $d_3(\lambda) = \lambda(\lambda - 1)^2(\lambda + 1)^2$ be the invariant factors of $A(\lambda)$. Then the elementary divisors of $A(\lambda)$ are λ, λ, λ, $\lambda - 1$, $\lambda - 1$, $(\lambda - 1)^2$, $\lambda + 1$, $(\lambda + 1)^2$.

Theorem 6.3.11. *Let $A(\lambda), B(\lambda) \in \mathrm{M}_{m\times n}(F[\lambda])$ be nonzero λ-matrices. Then $A(\lambda)$ is equivalent to $B(\lambda)$ if and only if they have the same rank and elementary divisors.*

Proof. Necessity. Suppose that $A(\lambda)$ is equivalent to $B(\lambda)$. Obviously, $\mathrm{rank}(A(\lambda)) = \mathrm{rank}(B(\lambda))$, and they have the same invariant factors. Hence, $A(\lambda)$ and $B(\lambda)$ have the same elementary divisors by definition.

Sufficiency. Let $\operatorname{rank}(A(\lambda)) = \operatorname{rank}(B(\lambda)) = r \geqslant 1$ and $A(\lambda)$ and $B(\lambda)$ have the same elementary divisors:

$$p_1(\lambda)^{m_{11}}, p_1(\lambda)^{m_{12}}, \ldots, p_1(\lambda)^{m_{1k_1}},$$
$$p_2(\lambda)^{m_{21}}, p_2(\lambda)^{m_{22}}, \ldots, p_2(\lambda)^{m_{2k_2}},$$
$$\vdots \qquad \vdots \qquad \qquad \vdots$$
$$p_s(\lambda)^{m_{s1}}, p_s(\lambda)^{m_{s2}}, \ldots, p_s(\lambda)^{m_{sk_s}},$$

where $p_1(\lambda), p_2(\lambda), \ldots, p_s(\lambda) \in F[\lambda]$ are different monic irreducible polynomials, $m_{j1} \geqslant m_{j2} \geqslant \cdots \geqslant m_{jk_j} > 0, j = 1, 2, \ldots, s$. Since every elementary divisor must be a factor of some invariant factor, for $1 \leqslant j \leqslant s$, we have $k_j \leqslant r$. If $k_j < r$, set $m_{j,k_j+1} = \cdots = m_{jr} = 0$. Let $d_1(\lambda), d_2(\lambda), \ldots, d_r(\lambda)$ be the invariant factors of $A(\lambda)$. Since $d_i(\lambda)|d_{i+1}(\lambda), i = 1, 2, \ldots, r-1$, $d_r(\lambda)$ must be the product of $p_1(\lambda)^{m_{11}}, p_2(\lambda)^{m_{21}}, \ldots, p_s(\lambda)^{m_{s1}}$, i.e.,

$$d_r(\lambda) = p_1(\lambda)^{m_{11}} p_2(\lambda)^{m_{21}} \cdots p_s(\lambda)^{m_{s1}}.$$

Similarly,

$$d_{r-1}(\lambda) = p_1(\lambda)^{m_{12}} p_2(\lambda)^{m_{22}} \cdots p_s(\lambda)^{m_{s2}},$$
$$d_{r-2}(\lambda) = p_1(\lambda)^{m_{13}} p_2(\lambda)^{m_{23}} \cdots p_s(\lambda)^{m_{s3}},$$
$$\vdots \qquad \qquad \vdots$$
$$d_1(\lambda) \quad = p_1(\lambda)^{m_{1r}} p_2(\lambda)^{m_{2r}} \cdots p_s(\lambda)^{m_{sr}}.$$

Thus, the invariant factors of the λ-matrix $A(\lambda)$ are uniquely determined by the elementary divisors and the rank of $A(\lambda)$. Hence $A(\lambda)$ and $B(\lambda)$ have the same invariant factors. By Theorem 6.3.5, $A(\lambda)$ is equivalent to $B(\lambda)$. $\qquad\square$

Remark 6.3.12. The requirement on rank is essential in Theorem 6.3.11. For example, let

$$A(\lambda) = \begin{pmatrix} (\lambda-1)^2 & 0 & 0 & 0 \\ 0 & 0 & 0 & 0 \\ 0 & 0 & 0 & 0 \end{pmatrix}, \quad B(\lambda) = \begin{pmatrix} 1 & 0 & 0 & 0 \\ 0 & (\lambda-1)^2 & 0 & 0 \\ 0 & 0 & 0 & 0 \end{pmatrix}.$$

Then $A(\lambda)$ and $B(\lambda)$ have the same elementary divisor $(\lambda - 1)^2$. But $\text{rank}(A(\lambda)) = 1$, $\text{rank}(B(\lambda)) = 2$, so $A(\lambda)$ and $B(\lambda)$ are not equivalent.

Example 6.3.13. Let $\lambda_0 \in \mathbb{C}$. Find the elementary divisors of the

$$\lambda\text{-matrix } A(\lambda) = \begin{pmatrix} \lambda - \lambda_0 & & & \\ -1 & & \ddots & \\ & & \ddots & \ddots \\ & & & -1 & \lambda - \lambda_0 \end{pmatrix} \text{ of order } n.$$

Solution. From Example 6.3.7, the λ-matrix $A(\lambda)$ has invariant factors $1, \ldots, 1, (\lambda - \lambda_0)^n$. Therefore, $(\lambda - \lambda_0)^n$ is the elementary divisor of $A(\lambda)$. $\qquad\square$

To find the elementary divisors of a λ-matrix, according to the definition, we need to calculate its invariant factors and then factorize them. Next, we give one method to find the elementary divisors of a λ-matrix without calculating the invariant factors.

Theorem 6.3.14. *Suppose that $A(\lambda) \in M_{m \times n}(F[\lambda])$ is equivalent to*

$$B(\lambda) = \begin{pmatrix} f_1(\lambda) & & & & \\ & f_2(\lambda) & & & \\ & & \ddots & & \\ & & & f_r(\lambda) & \\ & & & & 0_{(m-r) \times (n-r)} \end{pmatrix}, \qquad (6.4)$$

where $f_i(\lambda) \in F[\lambda]$ are monic polynomials, $1 \leqslant i \leqslant r$. If we factorize the nonconstant ones of $f_1(\lambda), f_2(\lambda), \ldots, f_r(\lambda)$ into products of powers of different monic irreducible factors, then these powers (counted by the number of occurrences) are exactly the elementary divisors of $A(\lambda)$.

Proof. Let $f_i(\lambda) = \prod_{j=1}^{s} p_j(\lambda)^{f_{ij}}$, where $p_1(\lambda), p_2(\lambda), \ldots, p_s(\lambda) \in F[\lambda]$ are different monic irreducible polynomials, $f_{ij} \in \mathbb{N}$, $1 \leqslant i \leqslant r, 1 \leqslant j \leqslant s$. By assumption, $\text{rank}(A(\lambda)) = \text{rank}(B(\lambda)) = r$. For $k = 1, 2, \ldots, r$, from Theorem 6.3.5, we have $D_k(A(\lambda)) = D_k(B(\lambda))$, denoted by $D_k(\lambda)$. So, $D_k(\lambda)$ is the monic greatest common divisor of the C_r^k polynomials $f_{i_1}(\lambda)f_{i_2}(\lambda)\cdots f_{i_k}(\lambda)$ $(1 \leqslant i_1 < i_2 < \cdots < i_k \leqslant r)$. Hence $D_k(\lambda) = \prod_{j=1}^{s} p_j(\lambda)^{\alpha_{kj}}$, where

$$\alpha_{kj} = \min_{1 \leqslant i_1 < i_2 < \cdots < i_k \leqslant r} (f_{i_1 j} + f_{i_2 j} + \cdots + f_{i_k j}), \quad j = 1, 2, \ldots, s.$$

For a fixed j, rearranging $f_{1j}, f_{2j}, \ldots, f_{rj}$ from small to large, denoted by $e_{1j}, e_{2j}, \ldots, e_{rj}$, we have

$$\alpha_{kj} = \min_{1 \leqslant i_1 < i_2 < \cdots < i_k \leqslant r} (f_{i_1 j} + f_{i_2 j} + \cdots + f_{i_k j})$$

$$= e_{1j} + e_{2j} + \cdots + e_{kj}, \quad 1 \leqslant k \leqslant r,$$

and so $D_k(\lambda) = \prod_{j=1}^{s} p_j(\lambda)^{e_{1j}+e_{2j}+\cdots+e_{kj}}, 1 \leqslant k \leqslant r$. Thus the kth invariant factor $d_k(\lambda) = \frac{D_k(\lambda)}{D_{k-1}(\lambda)} = \prod_{j=1}^{s} p_j(\lambda)^{e_{kj}}, 1 \leqslant k \leqslant r$, where $D_0(\lambda) = 1$. Therefore, the elementary divisors (counted by the number of occurrences) are $p_j(\lambda)^{e_{kj}}, e_{kj} > 0, 1 \leqslant j \leqslant s, 1 \leqslant k \leqslant r$, i.e., $p_i(\lambda)^{f_{li}}, f_{li} > 0, 1 \leqslant i \leqslant s, 1 \leqslant l \leqslant r$. $\qquad\square$

Theorem 6.3.15. *Let* $A_1(\lambda) \in \mathrm{M}_{m \times n}(F[\lambda]), A_2(\lambda) \in \mathrm{M}_{p \times q}(F[\lambda])$ *and*

$$A(\lambda) = \begin{pmatrix} A_1(\lambda) & 0 \\ 0 & A_2(\lambda) \end{pmatrix} \in \mathrm{M}_{(m+p) \times (n+q)}(F[\lambda]).$$

Then the elementary divisors of $A(\lambda)$ *are formed by merging all elementary divisors of* $A_1(\lambda)$ *and* $A_2(\lambda)$ *(counted by the number of occurrences).*

Proof. From Theorem 6.2.8 and Corollary 6.2.11, there exist $P_1(\lambda) \in \mathrm{GL}_m(F[\lambda])$, $Q_1(\lambda) \in \mathrm{GL}_n(F[\lambda])$, $P_2(\lambda) \in \mathrm{GL}_p(F[\lambda])$, $Q_2(\lambda) \in \mathrm{GL}_q(F[\lambda])$ such that

$$P_1(\lambda)A_1(\lambda)Q_1(\lambda) = \Lambda_1(\lambda) = \begin{pmatrix} \mathrm{diag}(d_1(\lambda), \ldots, d_s(\lambda)) & 0_{s \times (n-s)} \\ 0_{(m-s) \times s} & 0_{(m-s) \times (n-s)} \end{pmatrix},$$

$$P_2(\lambda)A_2(\lambda)Q_2(\lambda) = \Lambda_2(\lambda) = \begin{pmatrix} \mathrm{diag}(d_{s+1}(\lambda), \ldots, d_{s+r}(\lambda)) & 0_{r \times (q-r)} \\ 0_{(p-r) \times r} & 0_{(p-r) \times (q-r)} \end{pmatrix},$$

where $\operatorname{rank}(A_1(\lambda)) = s$, $\operatorname{rank}(A_2(\lambda)) = r$, and $\Lambda_1(\lambda)$ and $\Lambda_2(\lambda)$ are the Smith normal forms of $A_1(\lambda)$ and $A_2(\lambda)$, respectively. Set

$$P(\lambda) = \begin{pmatrix} P_1(\lambda) & 0 \\ 0 & P_2(\lambda) \end{pmatrix}, \quad Q(\lambda) = \begin{pmatrix} Q_1(\lambda) & 0 \\ 0 & Q_2(\lambda) \end{pmatrix}.$$

Then $P(\lambda) \in \operatorname{GL}_{m+p}(F[\lambda])$, $Q(\lambda) \in \operatorname{GL}_{n+q}(F[\lambda])$, and

$$P(\lambda)A(\lambda)Q(\lambda) = \begin{pmatrix} d_1(\lambda) \\ & \ddots \\ & & d_s(\lambda) \\ & & & 0_1 \\ & & & & d_{s+1}(\lambda) \\ & & & & & \ddots \\ & & & & & & d_{s+r}(\lambda) \\ & & & & & & & 0_2 \end{pmatrix},$$

where $0_1, 0_2$ are $(m-s) \times (n-s)$ and $(p-r) \times (q-r)$ zero matrices respectively. It's clear that $A(\lambda)$ is equivalent to the matrix

$$\begin{pmatrix} d_1(\lambda) \\ & \ddots \\ & & d_s(\lambda) \\ & & & d_{s+1}(\lambda) \\ & & & & \ddots \\ & & & & & d_{s+r}(\lambda) \\ & & & & & & 0_{(m+p-r-s) \times (n+q-r-s)} \end{pmatrix}.$$

Therefore, from Theorem 6.3.14, we know that the elementary divisors of $A(\lambda)$ are formed by merging all elementary divisors of $A_1(\lambda)$ and $A_2(\lambda)$ (counted by the number of occurrences). $\qquad \square$

6.4   The Similarity and Rational Canonical Forms of Number Matrices

In this section, we use the theory of λ-matrices to solve the problem of the similarity of matrices over a number field F. Let $A \in \operatorname{M}_n(F)$.

The λ-matrix $\lambda I_n - A$ is called the **characteristic matrix** of A. We prove that two number matrices of order n are similar if and only if their characteristic matrices are equivalent.

Lemma 6.4.1. *Let $A(\lambda) \in M_{n \times m}(F[\lambda])$ and $A(\lambda) \neq 0$. Then there exists a unique natural number d and $A_0, A_1, \ldots, A_d \in M_{n \times m}(F)$ such that*

$$A(\lambda) = A_d \lambda^d + A_{d-1} \lambda^{d-1} + \cdots + A_1 \lambda + A_0 \qquad (6.5)$$

*(or $A(\lambda) = \lambda^d A_d + \lambda^{d-1} A_{d-1} + \cdots + \lambda A_1 + A_0$), where $A_d \neq 0$. The number d is called the **degree** of $A(\lambda)$, denoted by $\deg(A(\lambda)) = d$. For any $B = (b_{ij})_{n \times m}$ and indeterminate λ, $B\lambda^k = \lambda^k B$ represents the λ-matrix $(b_{ij}\lambda^k) \in M_{n \times m}(F[\lambda])$, $k \in \mathbb{N}$.*

Proof. We first prove the existence. Let $A(\lambda) = (a_{ij}(\lambda))_{n \times m}$, where $a_{ij}(\lambda) \in F[\lambda]$, $1 \leqslant i \leqslant n, 1 \leqslant j \leqslant m$. Set

$$d = \max\{\deg(a_{ij}(\lambda)) \mid 1 \leqslant i \leqslant n, 1 \leqslant j \leqslant m, a_{ij}(\lambda) \neq 0\}.$$

Then, for $1 \leqslant i \leqslant n, 1 \leqslant j \leqslant m$, we have

$$a_{ij}(\lambda) = a_{ij}^{(d)}\lambda^d + a_{ij}^{(d-1)}\lambda^{d-1} + \cdots + a_{ij}^{(1)}\lambda + a_{ij}^{(0)}, a_{ij}^{(k)} \in F, k = 0, 1, \ldots, d.$$

Let $A_k = (a_{ij}^{(k)}) \in M_{n \times m}(F)$, $k = 0, 1, 2, \ldots, d$. Then

$$A(\lambda) = A_d \lambda^d + A_{d-1} \lambda^{d-1} + \cdots + A_1 \lambda + A_0,$$

and we know that $A_d \neq 0$ by the choice of d.

Next, we prove the uniqueness. Assume that $A(\lambda)$ has two representations:

$$A(\lambda) = A_d \lambda^d + A_{d-1} \lambda^{d-1} + \cdots + A_1 \lambda + A_0,$$
$$A(\lambda) = B_l \lambda^l + B_{l-1} \lambda^{l-1} + \cdots + B_1 \lambda + B_0,$$

where $A_d \neq 0, B_l \neq 0$, $A_i, B_j \in M_{n \times m}(F)$, $0 \leqslant i \leqslant d, 0 \leqslant j \leqslant l$. Since two matrices are equal if and only if their corresponding elements are equal, and two polynomials are equal if and only if their corresponding coefficients are equal, we have $d = l$, and $A_i = B_i$, $0 \leqslant i \leqslant d$, i.e., $A(\lambda)$ has the only representation (6.5). $\qquad \square$

Lemma 6.4.2. *Let $A(\lambda) \in \mathrm{M}_{n \times m}(F[\lambda]), B(\lambda) \in \mathrm{M}_{m \times p}(F[\lambda])$ and*

$$A(\lambda) = A_d \lambda^d + A_{d-1} \lambda^{d-1} + \cdots + A_1 \lambda + A_0,$$
$$B(\lambda) = B_l \lambda^l + B_{l-1} \lambda^{l-1} + \cdots + B_1 \lambda + B_0,$$

where $A_0, A_1, \ldots, A_d \in \mathrm{M}_{n \times m}(F)$, $A_d \neq 0$, $B_0, B_1, \ldots, B_l \in \mathrm{M}_{m \times p}(F)$, $B_l \neq 0$. If $A(\lambda)B(\lambda) \neq 0$, then

$$\deg\left(A(\lambda)B(\lambda)\right) \leqslant \deg\left(A(\lambda)\right) + \deg\left(B(\lambda)\right),$$

and the equality holds if and only if $A_d B_l \neq 0$.

Proof. Note that

$$A(\lambda)B(\lambda) = A_d B_l \lambda^{d+l} + (A_d B_{l-1} + A_{d-1} B_l)\lambda^{d+l-1} + \cdots + A_0 B_0.$$

It's obvious that if $A(\lambda)B(\lambda) \neq 0$, then

$$\deg\left(A(\lambda)B(\lambda)\right) \leqslant \deg\left(A(\lambda)\right) + \deg\left(B(\lambda)\right),$$

and

$$\deg\left(A(\lambda)B(\lambda)\right) = \deg\left(A(\lambda)\right) + \deg\left(B(\lambda)\right) \text{ if and only if } A_d B_l \neq 0.$$

$\square$

Remark 6.4.3.
(1) Let $A(\lambda)$ and $B(\lambda)$ be two square λ-matrices of order n. Then $\deg(A(\lambda)B(\lambda))$ and $\deg(B(\lambda)A(\lambda))$ may not be equal.
(2) Let $A(\lambda) \in \mathrm{M}_n(F[\lambda])$, $N \in \mathrm{M}_n(F)$.
   If $A(\lambda) = 0$, define $A(N)_L = A(N)_R = 0$.
   If $A(\lambda) \neq 0$, let $\deg(A(\lambda)) = d$, from Lemma 6.4.1,

$$A(\lambda) = A_d \lambda^d + A_{d-1} \lambda^{d-1} + \cdots + A_1 \lambda + A_0$$
$$= \lambda^d A_d + \lambda^{d-1} A_{d-1} + \cdots + \lambda A_1 + A_0,$$

where $A_0, A_1, \ldots, A_d \in \mathrm{M}_n(F)$. Define

$$A(N)_R = A_d N^d + A_{d-1} N^{d-1} + \cdots + A_1 N + A_0 \in \mathrm{M}_n(F),$$

$$A(N)_L = N^d A_d + N^{d-1} A_{d-1} + \cdots + N A_1 + A_0 \in \mathrm{M}_n(F).$$

$A(N)_R$ and $A(N)_L$ are called the **right value** and **left value** of the λ-matrix $A(\lambda)$ at N, respectively. Let $B(\lambda) \in \mathrm{M}_n(F[\lambda])$.

In general,

$$A(N)_R \neq A(N)_L,$$

$$(A(\lambda)B(\lambda))(N)_X \neq A(N)_Y B(N)_Z, \quad X, Y, Z \in \{R, L\}.$$

Let $f(\lambda) \in F[\lambda]$ and $c \in F$. From the division algorithm for polynomials, there exist unique $q(\lambda) \in F[\lambda]$ and $r \in F$ such that $f(\lambda) = (\lambda - c)q(\lambda) + r$ with $r = f(c)$. Similarly, we have the following result for λ-matrices.

Lemma 6.4.4. *Let $A \in M_n(F)$ and $U(\lambda) \in M_n(F[\lambda])$. Then there exist unique $Q(\lambda), R(\lambda) \in M_n(F[\lambda])$ and $W, V \in M_n(F)$ such that*

$$(\lambda I_n - A)Q(\lambda) + W = U(\lambda) = R(\lambda)(\lambda I_n - A) + V, \qquad (6.6)$$

and $W = U(A)_L$, $V = U(A)_R$.

Proof. Since the two equalities in (6.6) are symmetric, we only need to prove the first one.

We first prove the existence. If $U(\lambda) = 0$, set $Q(\lambda) = 0$, $W = 0$, and we are done. If $U(\lambda) \neq 0$, let $m = \deg(U(\lambda))$. From Lemma 6.4.1,

$$U(\lambda) = U_m \lambda^m + U_{m-1} \lambda^{m-1} + \cdots + U_1 \lambda + U_0,$$

where $U_m, U_{m-1}, \ldots, U_1, U_0 \in M_n(F)$.

If $m = 0$, then $Q(\lambda) = 0$, $W = U_m$ will do.

If $m \geqslant 1$, let $Q(\lambda) = Q_{m-1} \lambda^{m-1} + Q_{m-2} \lambda^{m-2} + \cdots + Q_1 \lambda + Q_0$, where $Q_{m-1}, Q_{m-2}, \ldots, Q_1, Q_0 \in M_n(F)$ are matrices to be determined. Then

$$(\lambda I_n - A)Q(\lambda) = Q_{m-1} \lambda^m + (Q_{m-2} - AQ_{m-1})\lambda^{m-1} + \cdots$$

$$+ (Q_0 - AQ_1)\lambda - AQ_0.$$

To make the first equality in (6.6) hold, we only need to choose

$$\begin{aligned}
Q_{m-1} &= U_m, \\
Q_{m-2} &= U_{m-1} + AQ_{m-1}, \\
&\ \ \vdots \\
Q_k &= U_{k+1} + AQ_{k+1}, \\
&\ \ \vdots \\
Q_0 &= U_1 + AQ_1, \\
W &= U_0 + AQ_0,
\end{aligned}$$

and in this case,

$$U(A)_L = A^m U_m + A^{m-1} U_{m-1} + \cdots + A U_1 + U_0$$
$$= A^m Q_{m-1} + A^{m-1}(Q_{m-2} - A Q_{m-1}) + \cdots$$
$$+ A(Q_0 - A Q_1) - A Q_0 + W$$
$$= W.$$

Next, we prove the uniqueness. Suppose that there are $Q(\lambda)$, $Q_1(\lambda) \in M_n(F[\lambda])$ and $W, W_1 \in M_n(F)$ such that

$$U(\lambda) = (\lambda I_n - A)Q(\lambda) + W = (\lambda I_n - A)Q_1(\lambda) + W_1.$$

Then $(\lambda I_n - A)(Q(\lambda) - Q_1(\lambda)) = W_1 - W$. If $Q(\lambda) - Q_1(\lambda) \neq 0$, then from Lemma 6.4.2, we have

$$\deg\left((\lambda I_n - A)(Q(\lambda) - Q_1(\lambda))\right)$$
$$= \deg(\lambda I_n - A) + \deg(Q(\lambda) - Q_1(\lambda)) \geqslant 1.$$

This contradicts that $W_1 - W$ is a number matrix. Therefore, $Q(\lambda) - Q_1(\lambda) = 0$, i.e., $Q(\lambda) = Q_1(\lambda)$, and so $W = W_1$. $\square$

Theorem 6.4.5. *Let $A, B \in M_n(F)$. Then A and B are similar over F if and only if the characteristic matrices $\lambda I_n - A$ and $\lambda I_n - B$ are equivalent.*

Proof. If A and B are similar over F, i.e., there exists $P \in GL_n(F)$ such that $B = P^{-1}AP$, then $P, P^{-1} \in GL_n(F[\lambda])$ and $\lambda I_n - B = \lambda I_n - P^{-1}AP = P^{-1}(\lambda I_n - A)P$. Hence $\lambda I_n - A$ and $\lambda I_n - B$ are equivalent.

Conversely, suppose that $\lambda I_n - A$ and $\lambda I_n - B$ are equivalent, by Corollary 6.2.11, there exist $P(\lambda), Q(\lambda) \in GL_n(F[\lambda])$ such that

$$P(\lambda)(\lambda I_n - A) = (\lambda I_n - B)Q(\lambda). \tag{6.7}$$

By Lemma 6.4.4, there exist $P_1(\lambda), Q_1(\lambda) \in M_n(F[\lambda])$ and $P, Q \in M_n(F)$ such that

$$P(\lambda) = (\lambda I_n - B)P_1(\lambda) + P, \quad Q(\lambda) = Q_1(\lambda)(\lambda I_n - A) + Q, \tag{6.8}$$

and $P = P(B)_L, Q = Q(A)_R$. Substituting (6.8) into (6.7), we have

$$\lambda(P - Q) + BQ - PA = (\lambda I_n - B)(Q_1(\lambda) - P_1(\lambda))(\lambda I_n - A).$$
$$(6.9)$$

If $Q_1(\lambda) \neq P_1(\lambda)$, from Lemma 6.4.2, we have

$$\deg\left((\lambda I_n - B)(Q_1(\lambda) - P_1(\lambda))(\lambda I_n - A)\right)$$
$$= \deg(\lambda I_n - B) + \deg(Q_1(\lambda) - P_1(\lambda)) + \deg(\lambda I_n - A) \geqslant 2,$$

contradicting that the left side of (6.9) is either the zero matrix or with a degree of not more than 1. Hence $Q_1(\lambda) = P_1(\lambda)$. From (6.9), we get

$$P = Q, \ BQ = PA. \tag{6.10}$$

Therefore, it's sufficient to prove that P is invertible. Since $P(\lambda)$ is invertible, from Lemma 6.4.4, we have

$$P(\lambda)^{-1} = (\lambda I_n - A)C(\lambda) + \widetilde{P}, \tag{6.11}$$

where $C(\lambda) \in M_n(F[\lambda])$, $\widetilde{P} \in M_n(F)$ is the left value of $P(\lambda)^{-1}$ at A. So

$$
\begin{aligned}
I_n &== P(\lambda)P(\lambda)^{-1} \\
&\overset{(6.11)}{====} P(\lambda)(\lambda I_n - A)C(\lambda) + P(\lambda)\widetilde{P} \\
&\overset{\substack{(6.7) \\ (6.8)}}{====} (\lambda I_n - B)Q(\lambda)C(\lambda) + (\lambda I_n - B)P_1(\lambda)\widetilde{P} + P\widetilde{P} \\
&==== (\lambda I_n - B)\left(Q(\lambda)C(\lambda) + P_1(\lambda)\widetilde{P}\right) + P\widetilde{P}.
\end{aligned}
$$

Comparing the degrees of both sides of the equality above, we get that $Q(\lambda)C(\lambda) + P_1(\lambda)\widetilde{P} = 0$, and hence $I_n = P\widetilde{P}$, i.e., P is invertible. Thus $B = PAP^{-1}$ by (6.10), and so A is similar to B. $\quad\square$

By the argument in the proof above, we have the following.

Corollary 6.4.6. *Let* $A, B \in M_n(F)$, $P(\lambda), Q(\lambda) \in GL_n(F[\lambda])$, *and*

$$P(\lambda)(\lambda I_n - A)Q(\lambda) = \lambda I_n - B.$$

Put $M(\lambda) = P(\lambda)^{-1}, N(\lambda) = Q(\lambda)^{-1}$. *Then*

(1) $P(B)_L M(A)_L = I_n$, $Q(B)_R N(A)_R = I_n$,

(2) $P(B)_L = N(A)_R$,

(3) $B = P^{-1}AP$, where $P = Q(B)_R = (P(B)_L)^{-1} = M(A)_L$.

Example 6.4.7. Let $A = \begin{pmatrix} -1 & -4 & -6 \\ -5 & -13 & -20 \\ 3 & 10 & 15 \end{pmatrix}$ and $B = \begin{pmatrix} 1 & 0 & 0 \\ 0 & 0 & 1 \\ 0 & -1 & 0 \end{pmatrix}$.

Show that A is similar to B over $\mathbb{Q}$, and find $P \in \mathrm{GL}_3(\mathbb{Q})$ such that $B = P^{-1}AP$.

Proof. Applying elementary operations of λ-matrices to $\lambda I_3 - A$, we have

$$\lambda I_3 - A = \begin{pmatrix} \lambda + 1 & 4 & 6 \\ 5 & \lambda + 13 & 20 \\ -3 & -10 & \lambda - 15 \end{pmatrix}$$

$$\xrightarrow[\;(c_2,c_3)\;]{\;(r_1,r_2)\;} \begin{pmatrix} 5 & 20 & \lambda + 13 \\ \lambda + 1 & 6 & 4 \\ -3 & \lambda - 15 & -10 \end{pmatrix}$$

$$\xrightarrow[\substack{(-4)\times c_1 + c_2 \\ \left(-\frac{1}{5}(\lambda+13)\right)\times c_1 + c_3}]{\substack{\frac{3}{5}\times r_1 + r_3 \\ \left(-\frac{1}{5}(\lambda+1)\right)\times r_1 + r_2}} \begin{pmatrix} 5 & 0 & 0 \\ 0 & -4\lambda + 2 & -\frac{1}{5}(\lambda^2 + 14\lambda - 7) \\ 0 & \lambda - 3 & \frac{1}{5}(3\lambda - 11) \end{pmatrix}$$

$$\xrightarrow[\left(-\frac{1}{50}(\lambda^2+2\lambda+37)\right)\times c_2 + c_3]{\substack{\frac{1}{5}\times r_1 \\ 4\times r_3 + r_2}} \begin{pmatrix} 1 & 0 & 0 \\ 0 & -10 & 0 \\ 0 & \lambda - 3 & -\frac{1}{50}(\lambda - 1)(\lambda^2 + 1) \end{pmatrix}$$

$$\xrightarrow{\frac{1}{10}(\lambda - 3)\times r_2 + r_3} \begin{pmatrix} 1 & 0 & 0 \\ 0 & -10 & 0 \\ 0 & 0 & -\frac{1}{50}(\lambda - 1)(\lambda^2 + 1) \end{pmatrix}$$

$$\xrightarrow[(-50)\times r_3]{\left(-\frac{1}{10}\right)\times r_2} \begin{pmatrix} 1 & 0 & 0 \\ 0 & 1 & 0 \\ 0 & 0 & (\lambda - 1)(\lambda^2 + 1) \end{pmatrix} = \Lambda(\lambda).$$

Writing the process above as a matrix equation, we get

$$P_1(\lambda)(\lambda I_3 - A)Q_1(\lambda) = \Lambda(\lambda),$$

where

$$P_1(\lambda) = E(3(-50)) \cdot E\left(2\left(-\frac{1}{10}\right)\right) \cdot E\left(3, 2\left(\frac{\lambda - 3}{10}\right)\right)$$

$$\cdot E(2, 3(4)) E\left(1\left(\frac{1}{5}\right)\right)$$

$$\cdot E\left(2, 1\left(-\frac{1}{5}(\lambda + 1)\right)\right) \cdot E\left(3, 1\left(\frac{3}{5}\right)\right) \cdot E(1, 2),$$

$$Q_1(\lambda) = E(2, 3) \cdot E\left(1, 2(-4)\right) \cdot E\left(1, 3\left(-\frac{1}{5}(\lambda + 13)\right)\right)$$

$$\cdot E\left(2, 3\left(-\frac{1}{50}(\lambda^2 + 2\lambda + 37)\right)\right)$$

$$= \begin{pmatrix} 1 & -4 & \frac{1}{25}(2\lambda^2 - \lambda + 9) \\ 0 & 0 & 1 \\ 0 & 1 & -\frac{1}{50}(\lambda^2 + 2\lambda + 37) \end{pmatrix}.$$

Similarly, applying elementary operations of λ-matrices to $\lambda I_3 - B$ yields

$$\lambda I_3 - B = \begin{pmatrix} \lambda - 1 & 0 & 0 \\ 0 & \lambda & -1 \\ 0 & 1 & \lambda \end{pmatrix}$$

$$\xrightarrow[\;(r_1, r_3)\;]{(c_1, c_2)} \begin{pmatrix} 1 & 0 & \lambda \\ \lambda & 0 & -1 \\ 0 & \lambda - 1 & 0 \end{pmatrix}$$

$$\xrightarrow[(-\lambda) \times c_1 + c_3]{\substack{(-\lambda) \times r_1 + r_2 \\ (\lambda + 1) \times r_3 + r_2}} \begin{pmatrix} 1 & 0 & 0 \\ 0 & \lambda^2 - 1 & -1 - \lambda^2 \\ 0 & \lambda - 1 & 0 \end{pmatrix}$$

$$\xrightarrow[\frac{\lambda - 1}{2} \times r_2 + r_3]{1 \times c_3 + c_2} \begin{pmatrix} 1 & 0 & 0 \\ 0 & -2 & -1 - \lambda^2 \\ 0 & 0 & -\frac{1}{2}(\lambda - 1)(\lambda^2 + 1) \end{pmatrix}$$

$$\xrightarrow[\left(-\frac{\lambda^2 + 1}{2}\right) \times c_2 + c_3]{\substack{\left(-\frac{1}{2}\right) \times r_2 \\ (-2) \times r_3}} \begin{pmatrix} 1 & 0 & 0 \\ 0 & 1 & 0 \\ 0 & 0 & (\lambda - 1)(\lambda^2 + 1) \end{pmatrix} = \Lambda(\lambda).$$

Writing the process above as a matrix equation, we obtain

$$P_2(\lambda)(\lambda I_3 - B)Q_2(\lambda) = \Lambda(\lambda),$$

where

$$P_2(\lambda) = E\left(3\left(-2\right)\right) \cdot E\left(3\left(-\frac{1}{2}\right)\right) \cdot E\left(3, 2\left(\frac{\lambda-1}{2}\right)\right)$$

$$\cdot\, E\left(2, 3\left(\lambda+1\right)\right) \cdot E\left(2, 1\left(-\lambda\right)\right) \cdot E(1, 3),$$

$$Q_2(\lambda) = E(1, 2)E\left(1, 3(-\lambda)\right)E(3, 2(1))E\left(2, 3\left(-\frac{\lambda^2+1}{2}\right)\right)$$

$$= \begin{pmatrix} 0 & 1 & -\frac{\lambda^2+1}{2} \\ 1 & -\lambda & \frac{\lambda^3-\lambda}{2} \\ 0 & 1 & \frac{1-\lambda^2}{2} \end{pmatrix}.$$

Let $P(\lambda) = P_2(\lambda)^{-1}P_1(\lambda)$ and $Q(\lambda) = Q_1(\lambda)Q_2(\lambda)^{-1}$. Then

$$P(\lambda), Q(\lambda) \in \mathrm{GL}_3(\mathbb{Q}[\lambda]) \quad \text{and} \quad \lambda I_3 - B = P(\lambda)(\lambda I_3 - A)Q(\lambda).$$

Hence, $\lambda I_3 - A$ is equivalent to $\lambda I_3 - B$ in $\mathrm{M}_3(\mathbb{Q}[\lambda])$, and so A is similar to B over $\mathbb{Q}$.

From $Q_2(\lambda)^{-1} = \begin{pmatrix} 0 & 1 & \lambda \\ \frac{1-\lambda^2}{2} & 0 & \frac{\lambda^2+1}{2} \\ -1 & 0 & 1 \end{pmatrix}$, we have

$$Q(\lambda) = \begin{pmatrix} 1 & -4 & \frac{1}{25}(2\lambda^2 - \lambda + 9) \\ 0 & 0 & 1 \\ 0 & 1 & -\frac{1}{50}(\lambda^2 + 2\lambda + 37) \end{pmatrix} \begin{pmatrix} 0 & 1 & \lambda \\ \frac{1-\lambda^2}{2} & 0 & \frac{\lambda^2+1}{2} \\ -1 & 0 & 1 \end{pmatrix}$$

$$= \begin{pmatrix} \frac{1}{25}(48\lambda^2 + \lambda - 59) & 1 & \frac{1}{25}(-48\lambda^2 + 24\lambda - 41) \\ -1 & 0 & 1 \\ \frac{1}{25}(-12\lambda^2 + \lambda + 31) & 0 & \frac{1}{25}(12\lambda^2 - \lambda - 6) \end{pmatrix}$$

$$= \begin{pmatrix} \frac{48}{25} & 0 & -\frac{48}{25} \\ 0 & 0 & 0 \\ -\frac{12}{25} & 0 & \frac{12}{25} \end{pmatrix}\lambda^2 + \begin{pmatrix} \frac{1}{25} & 0 & \frac{24}{25} \\ 0 & 0 & 0 \\ \frac{1}{25} & 0 & -\frac{1}{25} \end{pmatrix}\lambda + \begin{pmatrix} -\frac{59}{25} & 1 & -\frac{41}{25} \\ -1 & 0 & 1 \\ \frac{31}{25} & 0 & -\frac{6}{25} \end{pmatrix}.$$

Let $P = Q(B)_R$. Then $P = \begin{pmatrix} -\frac{2}{5} & \frac{1}{25} & \frac{7}{25} \\ -1 & 0 & 1 \\ \frac{4}{5} & \frac{1}{25} & -\frac{18}{25} \end{pmatrix}$, and hence P is invertible and $B = P^{-1}AP$ by Corollary 6.4.6. In fact, $|P| = \frac{1}{125}$,

$$P^{-1} = \begin{pmatrix} -5 & 5 & 5 \\ 10 & 8 & 15 \\ -5 & 6 & 5 \end{pmatrix}, \text{ and}$$

$$\begin{pmatrix} -5 & 5 & 5 \\ 10 & 8 & 15 \\ -5 & 6 & 5 \end{pmatrix} \begin{pmatrix} -1 & -4 & -6 \\ -5 & -13 & -20 \\ 3 & 10 & 15 \end{pmatrix} \begin{pmatrix} -\frac{2}{5} & \frac{1}{25} & \frac{7}{25} \\ -1 & 0 & 1 \\ \frac{4}{5} & \frac{1}{25} & -\frac{18}{25} \end{pmatrix} = \begin{pmatrix} 1 & 0 & 0 \\ 0 & 0 & 1 \\ 0 & -1 & 0 \end{pmatrix}. \quad \square$$

Remark 6.4.8. Let $\alpha_1, \alpha_2, \ldots, \alpha_n$ be a basis for an n-dimensional linear space V and $\sigma \in \text{End}(V)$. Assume that A is the matrix of σ with respect to this basis.

(1) For convenience, we call the determinantal divisors, invariant factors, and elementary divisors of the characteristic matrix $\lambda I_n - A$ of A the **determinantal divisors, invariant factors, and elementary divisors** of the matrix A (linear transformation σ), respectively.

(2) Since $|\lambda I_n - A| = f_A(\lambda) \neq 0$, we have $\text{rank}(\lambda I_n - A) = n$. Hence the matrix A (linear transformation σ) has n determinantal divisors and n invariant factors, $f_A(\lambda) = D_n(\lambda) = \prod_{i=1}^{n} d_i(\lambda)$ is also the product of all elementary divisors of A.

From Theorems 6.3.5, 6.3.11, and 6.4.5, we have the following.

Theorem 6.4.9. *Let $A, B \in \text{M}_n(F)$. Then the following statements are equivalent:*

(1) *A is similar to B.*
(2) *A and B have the same determinantal divisors.*
(3) *A and B have the same invariant factors.*
(4) *A and B have the same elementary divisors.*

Definition 6.4.10. Let

$$R = \text{diag}\,(R_1, R_2, \ldots, R_s) \in \text{M}_n(F), \tag{6.12}$$

where $R_i \in \mathrm{M}_{n_i}(F)$ is the companion matrix of the monic polynomial $d_i(\lambda) \in F[\lambda]$, $1 \leqslant i \leqslant s$, and $d_i(\lambda) | d_{i+1}(\lambda)$, $i = 1, 2, \ldots, s-1$. Then R is called a **rational canonical form** or **Frobenius normal form** over F.

Lemma 6.4.11.

(1) *Let $d(\lambda) = \lambda^n + a_1 \lambda^{n-1} + \cdots + a_n \in F[\lambda]$ and let C be the companion matrix of $d(\lambda)$. Then the invariant factors of C are $d_1(\lambda) = \cdots = d_{n-1}(\lambda) = 1, d_n(\lambda) = d(\lambda)$.*

(2) *In Definition 6.4.10, the rational canonical form R of order n has invariant factors $1, \ldots, 1, d_1(\lambda), \ldots, d_s(\lambda)$, where the number of 1 is $n - s$.*

Proof. (1) Since $C = \begin{pmatrix} 0 & & & -a_n \\ 1 & 0 & & -a_{n-1} \\ & \ddots & \ddots & \vdots \\ & & \ddots & 0 & -a_2 \\ & & & 1 & -a_1 \end{pmatrix}$, it's obvious that

the characteristic matrix $\lambda I_n - C$ has a minor of order $n - 1$:

$$\begin{vmatrix} -1 & \lambda & & \\ & \ddots & \ddots & \\ & & \ddots & \lambda \\ & & & -1 \end{vmatrix} = (-1)^{n-1}.$$ Hence $D_{n-1}(\lambda) = 1$, and so the deter-

minantal divisors of C are $D_1(\lambda) = \cdots = D_{n-1}(\lambda) = 1, D_n(\lambda) = f_C(\lambda) = d(\lambda)$. Therefore, the invariant factors of C are $d_1(\lambda) = \cdots = d_{n-1}(\lambda) = 1, d_n(\lambda) = d(\lambda)$.

(2) For $i = 1, 2, \ldots, s$, by (1), the invariant factors of $\lambda I_{n_i} - R_i$ are $1, \ldots, 1, d_i(\lambda)$, and so $\lambda I_{n_i} - R_i$ is equivalent to $\mathrm{diag}\,(1, \ldots, 1, d_i(\lambda))$. It follows that the λ-matrix

$$\lambda I_n - R = \mathrm{diag}\,(\lambda I_{n_1} - R_1, \lambda I_{n_2} - R_2, \ldots, \lambda I_{n_s} - R_s)$$

is equivalent to

$$\mathrm{diag}\,(1, \ldots, 1, d_1(\lambda), 1, \ldots, 1, d_2(\lambda), \ldots, 1, \ldots, 1, d_s(\lambda)).$$

Furthermore, it's also equivalent to

$$\Lambda(\lambda) = \mathrm{diag}\,(1, \ldots, 1, d_1(\lambda), d_2(\lambda), \ldots, d_s(\lambda)).$$

Since $d_i(\lambda)|d_{i+1}(\lambda)$, $i = 1, 2, \ldots, s - 1$, $\Lambda(\lambda)$ is the Smith normal form of $\lambda I_n - R$. Therefore, $1, \ldots, 1, d_1(\lambda), d_2(\lambda), \ldots, d_s(\lambda)$ are the invariant factors of R (or $\lambda I_n - R$). $\qquad\square$

Theorem 6.4.12. *Let $A \in \mathrm{M}_n(F)$. Then A is similar to a unique rational canonical form R over F. The unique matrix R is called the* **rational canonical form** *of A.*

Proof. Assume that the invariant factors of the matrix A are

$$d_1(\lambda) = \cdots = d_{n-s}(\lambda) = 1, d_{n-s+1}(\lambda), d_{n-s+2}(\lambda), \ldots, d_n(\lambda),$$

where $\deg(d_{n-s+j}(\lambda)) \geqslant 1$, $j = 1, 2, \ldots, s$. Then

$$n = \sum_{j=1}^{s} \deg(d_{n-s+j}(\lambda)).$$

Let R_j be the companion matrix of $d_{n-s+j}(\lambda)$, $j = 1, 2, \ldots, s$. Put $R = \mathrm{diag}\,(R_1, R_2, \ldots, R_s)$. Then R is a rational canonical form over F. From Lemma 6.4.11, R and A have the same invariant factors, and so A is similar to R.

The rational canonical form R is determined uniquely by the invariant factors of A, and so R is determined uniquely by A. $\qquad\square$

Corollary 6.4.13. *Let $A \in \mathrm{M}_n(F)$. Then the minimal polynomial $m_A(\lambda)$ of A is the nth invariant factor $d_n(\lambda)$, i.e., $m_A(\lambda) = d_n(\lambda)$.*

Proof. From Theorem 6.4.12, we may assume that the rational canonical form of A is the matrix R in (6.12). Then from Theorem 5.6.5 (4) (5) and Example 5.6.8, we have

$$m_A(\lambda) = m_R(\lambda) = [m_{R_1}(\lambda), \ldots, m_{R_s}(\lambda)]$$
$$= [d_{n-s+1}(\lambda), \ldots, d_n(\lambda)] = d_n(\lambda). \qquad\square$$

Theorem 6.4.12 can be described as follows using the language of linear transformations.

Theorem 6.4.14. *Let V be an n-dimensional linear space and $\sigma \in \mathrm{End}(V)$. Then the following statements are true:*

(1) *There exists a basis for V such that the matrix R of σ with respect to this basis is a rational canonical form, and R is uniquely determined by σ. The matrix R is called the* **rational canonical form** *of σ.*

(2) *Let the invariant factors of σ be $d_1(\lambda) = \cdots = d_r(\lambda) = 1$, $d_{r+1}(\lambda) \neq 1$, $\ldots$, $d_n(\lambda)$, $0 \leqslant r < n$. Assume that $s = n - r$, $n_i = \deg(d_{r+i}(\lambda))$, and R_i is the companion matrix of $d_{r+i}(\lambda)$, $i = 1, 2, \ldots, s$. Then there exist $\alpha_1, \ldots, \alpha_s \in V$ such that $V = V_1 \oplus V_2 \oplus \cdots \oplus V_s$, where $V_i = W_\sigma(\alpha_i)$ is an n_i-dimensional σ cyclic subspace generated by α_i, and R_i is the matrix of $\sigma|_{V_i}$ with respect to the basis $\alpha_i, \sigma(\alpha_i), \ldots, \sigma^{n_i-1}(\alpha_i)$, $i = 1, 2, \ldots, s$. Furthermore, the matrix of σ with respect to the basis $\alpha_1, \sigma(\alpha_1), \ldots, \sigma^{n_1-1}(\alpha_1), \ldots, \alpha_s, \sigma(\alpha_s), \ldots, \sigma^{n_s-1}(\alpha_s)$ is the rational canonical form R in (6.12).*

6.5   Jordan Canonical Forms of Complex Matrices Under Similarity

A monic irreducible polynomial $p(\lambda)$ over the complex number field $\mathbb{C}$ must be of degree one, i.e., $p(\lambda) = \lambda - c$, $c \in \mathbb{C}$. Hence the elementary divisors of a square complex matrix A must be the powers of monic polynomials of degree one:

$$(\lambda - \lambda_1)^{e_1}, (\lambda - \lambda_2)^{e_2}, \ldots, (\lambda - \lambda_t)^{e_t},$$

where $\lambda_1, \lambda_2, \ldots, \lambda_t \in \mathbb{C}$ (not necessarily different), $e_1, e_2, \ldots, e_t \in \mathbb{N}^*$ and

$$f_A(\lambda) = (\lambda - \lambda_1)^{e_1}(\lambda - \lambda_2)^{e_2} \cdots (\lambda - \lambda_t)^{e_t}.$$

In this section, we use the theory of elementary divisors to prove that every complex square matrix is similar to a Jordan matrix.

From Examples 6.3.7 and 6.3.13, we have the following.

Lemma 6.5.1. *Let $J_n(\lambda_0) = \begin{pmatrix} \lambda_0 & & & \\ 1 & \ddots & & \\ & \ddots & \ddots & \\ & & 1 & \lambda_0 \end{pmatrix}$ be a Jordan block of order n with eigenvalue $\lambda_0 \in \mathbb{C}$. Then*

(1) *the determinantal divisors and invariant factors of $J_n(\lambda_0)$ are both $1, \ldots, 1, (\lambda - \lambda_0)^n$,*
(2) *$(\lambda - \lambda_0)^n$ is the unique elementary divisor of $J_n(\lambda_0)$.*

Then from Theorem 6.3.15, we have the following.

Lemma 6.5.2. *Let $J = \mathrm{diag}\,(J_1, J_2, \ldots, J_s)$ be a Jordan matrix, where J_i is a Jordan block of order n_i with eigenvalue λ_i, $i = 1, 2, \ldots, s$. Then the elementary divisors of J are*

$$(\lambda - \lambda_1)^{n_1}, \ (\lambda - \lambda_2)^{n_2}, \ \ldots, (\lambda - \lambda_s)^{n_s}.$$

Theorem 6.5.3 (Jordan canonical form). *Let $A \in \mathrm{M}_n(\mathbb{C})$. Then A is similar to a Jordan matrix which is uniquely determined by A regardless of the order of Jordan blocks.*

Proof. Let $(\lambda - \lambda_1)^{n_1}, (\lambda - \lambda_2)^{n_2}, \ldots, (\lambda - \lambda_s)^{n_s}$ be the elementary divisors of A. Put

$$J_i = \begin{pmatrix} \lambda_i & & & \\ 1 & \ddots & & \\ & \ddots & \ddots & \\ & & 1 & \lambda_i \end{pmatrix} \in \mathrm{M}_{n_i}(\mathbb{C}), \quad i = 1, \ldots, s,$$

$$J = \mathrm{diag}\,(J_1, J_2, \ldots, J_s).$$

Then, the matrices J and A have the same elementary divisors by Lemma 6.5.2, so they are similar.

If A is similar to another Jordan matrix J', then J' and A have the same elementary divisors. Therefore, J' and J are uniquely determined by the elementary divisors of A regardless of the order of Jordan blocks. $\square$

Theorem 6.5.3 can be stated as follows using the language of linear transformations.

Theorem 6.5.4. *Let V be an n-dimensional complex linear space and $\sigma \in \mathrm{End}(V)$.*

(1) *The matrix of σ with respect to some basis for V is the Jordan matrix $J = \mathrm{diag}\,(J_1, J_2, \ldots, J_s)$, where $J_i = J_{n_i}(\lambda_i)$ is the Jordan*

block of order n_i with eigenvalue λ_i, $i = 1, 2, \ldots, s$, and J is determined uniquely by σ regardless of the order of Jordan blocks.

(2) *$V = \bigoplus\limits_{i=1}^{s} V_i$, where V_i is an n_i-dimensional σ-subspace, and there exists a basis $\alpha_{i1}, \ldots, \alpha_{in_i}$ for V_i such that the matrix of $\sigma|_{V_i}$ with respect to this basis is the Jordan block $J_i = J_{n_i}(\lambda_i)$, $i = 1, 2, \ldots, s$, thus the matrix of σ with respect to the basis $\alpha_{11}, \ldots, \alpha_{1n_1}, \ldots, \alpha_{s1}, \ldots, \alpha_{sn_s}$ for V is the Jordan matrix $J = \mathrm{diag}\,(J_1, J_2, \ldots, J_s)$.*

Note that a diagonal matrix is a special Jordan matrix whose Jordan blocks are all of order one, i.e., the elementary divisors are all of degree one. From Lemma 6.4.13, we know that the minimal polynomial of a matrix is its nth invariant factor, therefore we have the following result.

Theorem 6.5.5. *Let $A \in \mathrm{M}_n(\mathbb{C})$. Then the following statements are equivalent:*

(1) *A is diagonalizable.*
(2) *The elementary divisors of A are of degree one.*
(3) *The invariant factors of A have no multiple roots.*
(4) *The nth invariant factor $d_n(\lambda)$ of A has no multiple roots.*
(5) *The minimal polynomial $m_A(\lambda)$ of A has no multiple roots.*

Given a complex square matrix A of order n, how do we find the Jordan canonical form J of A and the invertible matrix P of order n such that $P^{-1}AP = J$? We give two methods to solve this problem.

The first one is suitable for matrices of small order n. We can determine the Jordan canonical form J of A by the eigenvalues of A and then calculate the column vectors of P by the equality $AP = PJ$ (see Method 1 in Example 6.5.6).

The second method is to apply the theory of λ-matrices to determine the Jordan canonical form by elementary divisors and then to calculate the invertible matrix P by Corollary 6.4.6. The process is as follows:

(1) Find the invariant factors of A. First of all, we apply elementary operations to $\lambda I_n - A$ to get the Smith normal form

$$\Lambda(\lambda) = \mathrm{diag}(d_1(\lambda), d_2(\lambda), \ldots, d_n(\lambda)).$$

Then we have the invariant factors $d_1(\lambda)$, $d_2(\lambda)$, $\ldots$, $d_n(\lambda)$, and find $P_1(\lambda), Q_1(\lambda) \in \mathrm{GL}_n(\mathbb{C}[\lambda])$ such that

$$P_1(\lambda)(\lambda I_n - A)Q_1(\lambda) = \Lambda(\lambda).$$

(2) Find the elementary divisors of A. Factorizing invariant factors $d_i(\lambda) \neq 1$ into powers of monic polynomials of degree one, we get all elementary divisors: $(\lambda - \lambda_1)^{n_1}, (\lambda - \lambda_2)^{n_2}, \ldots, (\lambda - \lambda_s)^{n_s}$.

(3) Find the Jordan canonical form of A. Write out the Jordan blocks $J_i = J_{n_i}(\lambda_i)$ corresponding to the elementary divisor $(\lambda - \lambda_i)^{n_i}$, $i = 1, 2, \ldots, s$. Then the Jordan matrix $J = \mathrm{diag}\,(J_1, J_2, \ldots, J_s)$ is the Jordan canonical form of A.

(4) Find the invertible matrix P such that $P^{-1}AP = J$. Transform $\lambda I_n - J$ into the Smith normal form $\Lambda(\lambda)$ by elementary operations and find $P_2(\lambda), Q_2(\lambda) \in \mathrm{GL}_n(\mathbb{C}[\lambda])$ such that $P_2(\lambda)(\lambda I_n - J)Q_2(\lambda) = \Lambda(\lambda)$. Let $Q(\lambda) = Q_1(\lambda)Q_2(\lambda)^{-1}$. Then $P = Q(J)_R \in \mathrm{GL}_n(\mathbb{C})$ is what we want.

Example 6.5.6. Find the Jordan canonical form J of matrix
$$A = \begin{pmatrix} 4 & 5 & -2 \\ -2 & -2 & 1 \\ -1 & -1 & 1 \end{pmatrix} \text{ and } P \in \mathrm{GL}_3(\mathbb{C}) \text{ such that } P^{-1}AP = J.$$

Solution. **Method 1.** It's easy to see that the characteristic polynomial of A is $f_A(\lambda) = |\lambda I_3 - A| = (\lambda - 1)^3$. Hence A has only one eigenvalue 1 with multiplicity 3. Let V_1 be the eigenspace of A associated with the eigenvalue 1. Then $\dim V_1 = 3 - \mathrm{rank}(I_3 - A) = 3 - 2 = 1$. Therefore, the Jordan canonical form of A is $J = \begin{pmatrix} 1 & 0 & 0 \\ 1 & 1 & 0 \\ 0 & 1 & 1 \end{pmatrix}$.

Let $P = (\alpha_1, \alpha_2, \alpha_3) \in \mathrm{GL}_3(\mathbb{C})$ such that $P^{-1}AP = J$. From

$$A(\alpha_1, \alpha_2, \alpha_3) = AP = PJ = (\alpha_1 + \alpha_2, \alpha_2 + \alpha_3, \alpha_3),$$

we get that $A\alpha_3 = \alpha_3$, $A\alpha_2 = \alpha_2 + \alpha_3$, $A\alpha_1 = \alpha_1 + \alpha_2$, i.e., $(A - I_3)\alpha_3 = 0$, $(A - I_3)\alpha_2 = \alpha_3$, $(A - I_3)\alpha_1 = \alpha_2$.

First, solving the homogeneous linear system $(A - I_3)X = 0$ gives the general solution $X = k(-1, 1, 1)'$, $k \in \mathbb{C}$. Take $\alpha_3 = (-1, 1, 1)'$.

Second, solve the nonhomogeneous linear system $(A - I_3)X = \alpha_3$ to get the general solution $X = (-2, 1, 0)' + k\alpha_3$, $k \in \mathbb{C}$. Take $\alpha_2 = (-2, 1, 0)'$.

Finally, solving the nonhomogeneous linear system $(A - I_3)X = \alpha_2$ yields the general solution $X = (1, -1, 0)' + k\alpha_3, k \in \mathbb{C}$. Take $\alpha_1 = (1, -1, 0)'$.

Let $P = (\alpha_1, \alpha_2, \alpha_3) = \begin{pmatrix} 1 & -2 & -1 \\ -1 & 1 & 1 \\ 0 & 0 & 1 \end{pmatrix}$. Then P is invertible and $P^{-1}AP = J$.

Method 2. First, applying elementary operations to the characteristic matrix $\lambda I_3 - A$ of A yields

$$\lambda I_3 - A = \begin{pmatrix} \lambda - 4 & -5 & 2 \\ 2 & \lambda + 2 & -1 \\ 1 & 1 & \lambda - 1 \end{pmatrix}$$

$$\xrightarrow[\substack{(-1)\times c_1 + c_2 \\ (1-\lambda)\times c_1 + c_3}]{(r_1, r_3)} \begin{pmatrix} 1 & 0 & 0 \\ 2 & \lambda & 1 - 2\lambda \\ \lambda - 4 & -1 - \lambda & -\lambda^2 + 5\lambda - 2 \end{pmatrix}$$

$$\xrightarrow[\substack{(4-\lambda)\times r_1 + r_3}]{(-2)\times r_1 + r_2} \begin{pmatrix} 1 & 0 & 0 \\ 0 & \lambda & 1 - 2\lambda \\ 0 & -\lambda - 1 & -\lambda^2 + 5\lambda - 2 \end{pmatrix}$$

$$\xrightarrow{1 \times r_3 + r_2} \begin{pmatrix} 1 & 0 & 0 \\ 0 & -1 & -\lambda^2 + 3\lambda - 1 \\ 0 & -\lambda - 1 & -\lambda^2 + 5\lambda - 2 \end{pmatrix}$$

$$\xrightarrow{(-1)\times r_2} \begin{pmatrix} 1 & 0 & 0 \\ 0 & 1 & \lambda^2 - 3\lambda + 1 \\ 0 & -\lambda - 1 & -\lambda^2 + 5\lambda - 2 \end{pmatrix}$$

$$\xrightarrow[\substack{(\lambda+1)\times r_2 + r_3}]{(-\lambda^2 + 3\lambda - 1)\times c_2 + c_3} \begin{pmatrix} 1 & 0 & 0 \\ 0 & 1 & 0 \\ 0 & 0 & (\lambda - 1)^3 \end{pmatrix} = \Lambda(\lambda). \tag{6.13}$$

Therefore, the invariant factors of A are $1, 1, (\lambda - 1)^3$, and so $(\lambda - 1)^3$ is the unique elementary divisor of A. Hence the Jordan canonical form of A is $J = \begin{pmatrix} 1 & 0 & 0 \\ 1 & 1 & 0 \\ 0 & 1 & 1 \end{pmatrix}$. Writing the process (6.13) as a matrix equation, we have $P_1(\lambda)(\lambda I_3 - A)Q_1(\lambda) = \Lambda(\lambda)$, where

$$Q_1(\lambda) = \begin{pmatrix} 1 & -1 & 0 \\ 0 & 1 & 0 \\ 0 & 0 & 1 \end{pmatrix} \begin{pmatrix} 1 & 0 & 1-\lambda \\ 0 & 1 & 0 \\ 0 & 0 & 1 \end{pmatrix} \begin{pmatrix} 1 & 0 & 0 \\ 0 & 1 & -\lambda^2 + 3\lambda - 1 \\ 0 & 0 & 1 \end{pmatrix}$$

$$= \begin{pmatrix} 1 & -1 & \lambda^2 - 4\lambda + 2 \\ 0 & 1 & -\lambda^2 + 3\lambda - 1 \\ 0 & 0 & 1 \end{pmatrix}.$$

Then, applying elementary operations to $\lambda I_3 - J$, we get

$$\lambda I_3 - J = \begin{pmatrix} \lambda - 1 & 0 & 0 \\ -1 & \lambda - 1 & 0 \\ 0 & -1 & \lambda - 1 \end{pmatrix}$$

$$\xrightarrow[\ (\lambda-1)\times c_1 + c_2\ ]{(r_1, r_2)} \begin{pmatrix} -1 & 0 & 0 \\ \lambda - 1 & (\lambda - 1)^2 & 0 \\ 0 & -1 & \lambda - 1 \end{pmatrix}$$

$$\xrightarrow[\ (-1)\times r_3\ ]{(\lambda-1)\times r_1 + r_2} \begin{pmatrix} -1 & 0 & 0 \\ 0 & (\lambda - 1)^2 & 0 \\ 0 & 1 & 1 - \lambda \end{pmatrix}$$

$$\xrightarrow[\ (-1)\times r_1\ ]{(r_2, r_3)} \begin{pmatrix} 1 & 0 & 0 \\ 0 & 1 & 1 - \lambda \\ 0 & (\lambda - 1)^2 & 0 \end{pmatrix}$$

$$\xrightarrow[\ (\lambda-1)\times c_2 + c_3\ ]{(-(\lambda-1)^2)\times r_2 + r_3} \begin{pmatrix} 1 & 0 & 0 \\ 0 & 1 & 0 \\ 0 & 0 & (\lambda - 1)^3 \end{pmatrix} = \Lambda(\lambda).$$

Writing the process above as a matrix equation, we obtain

$$P_2(\lambda)(\lambda I_3 - J)Q_2(\lambda) = \Lambda(\lambda),$$

where

$$Q_2(\lambda) = \begin{pmatrix} 1 & \lambda - 1 & 0 \\ 0 & 1 & 0 \\ 0 & 0 & 1 \end{pmatrix} \begin{pmatrix} 1 & 0 & 0 \\ 0 & 1 & \lambda - 1 \\ 0 & 0 & 1 \end{pmatrix} = \begin{pmatrix} 1 & \lambda - 1 & (\lambda - 1)^2 \\ 0 & 1 & \lambda - 1 \\ 0 & 0 & 1 \end{pmatrix}.$$

Let $P(\lambda) = P_2(\lambda)^{-1}P_1(\lambda)$, $Q(\lambda) = Q_1(\lambda)Q_2(\lambda)^{-1}$. Then

$$P(\lambda)(\lambda I_3 - A)Q(\lambda) = \lambda I_3 - J,$$

where $Q(\lambda) = \begin{pmatrix} 0 & 0 & 1 \\ 0 & 0 & -1 \\ 0 & 0 & 0 \end{pmatrix} \lambda^2 + \begin{pmatrix} 0 & -1 & -3 \\ 0 & 0 & 2 \\ 0 & 0 & 0 \end{pmatrix} \lambda + \begin{pmatrix} 1 & 0 & 1 \\ 0 & 1 & 0 \\ 0 & 0 & 1 \end{pmatrix}.$

From Corollary 6.4.6,

$$P = Q(J)_R = \begin{pmatrix} 1 & -2 & -1 \\ -1 & 1 & 1 \\ 0 & 0 & 1 \end{pmatrix} \in \mathrm{GL}_3(\mathbb{C}), \quad P^{-1}AP = J.$$

In fact, $|P| = -1$, $P^{-1} = \begin{pmatrix} -1 & -2 & 1 \\ -1 & -1 & 0 \\ 0 & 0 & 1 \end{pmatrix}$, and

$$\begin{pmatrix} -1 & -2 & 1 \\ -1 & -1 & 0 \\ 0 & 0 & 1 \end{pmatrix} \begin{pmatrix} 4 & 5 & -2 \\ -2 & -2 & 1 \\ -1 & -1 & 1 \end{pmatrix} \begin{pmatrix} 1 & -2 & -1 \\ -1 & 1 & 1 \\ 0 & 0 & 1 \end{pmatrix} = \begin{pmatrix} 1 & 0 & 0 \\ 1 & 1 & 0 \\ 0 & 1 & 1 \end{pmatrix}. \qquad \square$$

6.6* A Brief Introduction to Integer Matrices

In this section, we briefly introduce some definitions and results about integer matrices, and readers are encouraged to give the corresponding proofs following the theory of λ-matrices. From now on we denote by $\mathrm{M}_{m \times n}(\mathbb{Z})$ the set of all $m \times n$ **integer matrices**, and write $\mathrm{M}_n(\mathbb{Z})$ for $\mathrm{M}_{n \times n}(\mathbb{Z})$.

Definition 6.6.1. Let $A \in \mathrm{M}_n(\mathbb{Z})$. If there exists $B \in \mathrm{M}_n(\mathbb{Z})$ such that $AB = BA = I_n$, then we say that A is **invertible** over $\mathbb{Z}$ and B is called the **inverse** of A.

Theorem 6.6.2. *Let $A \in \mathrm{M}_n(\mathbb{Z})$.*

(1) *If A is invertible over $\mathbb{Z}$ then the inverse is unique, denoted by A^{-1}.*

(2) *A is invertible over $\mathbb{Z}$ if and only if $|A| = \pm 1$, and in this case, $A^{-1} = \frac{1}{|A|} A^*$, where A^* is the adjoint matrix of A.*

The set of all $n \times n$ invertible matrices over $\mathbb{Z}$ is often denoted by $\mathrm{GL}_n(\mathbb{Z})$. It's clear that $\mathrm{GL}_1(\mathbb{Z}) = \{1, -1\}$, i.e., the invertible integers in $\mathbb{Z}$ are only ± 1. Therefore, the **elementary row (column) operations** of integer matrices are as follows:

(1) Multiply a row (column) by an integer $k \in \mathrm{GL}_1(\mathbb{Z}) = \{1, -1\}$.
(2) Add b times a row (column) to another row (column), where $b \in \mathbb{Z}$.
(3) Interchange two rows (columns).

Similarly, we can define the **elementary matrices** of integer matrices. These elementary matrices are all invertible. Applying an elementary row (column) operation to an integer matrix A is equivalent to multiplying A on the left (right) by the corresponding elementary matrix.

Definition 6.6.3. Let $A, B \in \mathrm{M}_{m \times n}(\mathbb{Z})$. A is said to be **equivalent** to B if B can be obtained by applying a finite sequence of elementary operations to A.

Theorem 6.6.4. *The equivalence of integer matrices has the following properties:*

(1) *The equivalence of integer matrices is an equivalence relation.*
(2) *Two equivalent integer matrices have the same rank.*
(3) *Let $A, B \in \mathrm{M}_{m \times n}(\mathbb{Z})$ be equivalent. Then there exist elementary matrices $P_1, \ldots, P_s \in \mathrm{GL}_m(\mathbb{Z})$ and $Q_1, \ldots, Q_t \in \mathrm{GL}_n(\mathbb{Z})$ such that $B = P_s \cdots P_1 A Q_1 \cdots Q_t$.*

Definition 6.6.5. Let $A \in \mathrm{M}_{m \times n}(\mathbb{Z})$ and $r = \mathrm{rank} A \geqslant 1$. For any $1 \leqslant k \leqslant r$, the positive greatest common divisor of all minors of order k, denoted by D_k, is called the kth **determinantal divisor** of A.

Theorem 6.6.6. *Let $A \in \mathrm{M}_{m \times n}(\mathbb{Z})$ and $r = \mathrm{rank} A \geqslant 1$.*

(1) *The integer matrices equivalent to each other have the same determinantal divisors.*

(2) *A is equivalent to $\Lambda = \begin{pmatrix} d_1 & & & & \\ & d_2 & & & \\ & & \ddots & & \\ & & & d_r & \\ & & & & 0_{(m-r) \times (n-r)} \end{pmatrix}$, which is*

*called the **Smith normal form** of A, where $d_1, d_2, \ldots, d_r$ are positive integers and $d_i | d_{i+1}$, $i = 1, 2, \ldots, r - 1$.*

(3) *The Smith normal form of the integer matrix A is unique, the positive integers $d_1, d_2, \ldots, d_r$ are called the **invariant factors** of A, and $d_1 = D_1$, $d_k = \frac{D_k}{D_{k-1}}$, $k = 2, 3, \ldots, r$.*

Definition 6.6.7. Let $d_1, d_2, \ldots, d_r$ be the invariant factors of an integer matrix A. Factorizing the invariant factors larger than 1 into a product of powers of different prime numbers, all these powers of prime numbers (counted by the number of occurrences) are called the **elementary divisors** of A.

Theorem 6.6.8. *Let $A, B \in \mathrm{M}_{m \times n}(\mathbb{Z})$ and $\mathrm{rank}A = \mathrm{rank}B \geqslant 1$. Then the following statements are equivalent:*

(1) *A is equivalent to B as integer matrices.*
(2) *A and B have the same determinantal divisors.*
(3) *A and B have the same invariant factors.*
(4) *A and B have the same elementary divisors.*

Example 6.6.9. Let $m, n \in \mathbb{N}^*$ and

$$\mathrm{M}_n(m) = \{A \in \mathrm{M}_n(\mathbb{Z}) \mid |A| = m\}.$$

If $A \in \mathrm{M}_2(4)$, then the Smith normal form of A has two possibilities:

$$D(1, 4) = \begin{pmatrix} 1 & 0 \\ 0 & 4 \end{pmatrix} \text{ or } D(2, 2) = \begin{pmatrix} 2 & 0 \\ 0 & 2 \end{pmatrix}.$$

Put

$$S(1, 4) = \left\{ A \in \mathrm{M}_2(4) \,\middle|\, \text{the Smith normal form of } A \text{ is } \begin{pmatrix} 1 & 0 \\ 0 & 4 \end{pmatrix} \right\},$$

$$S(2, 2) = \left\{ A \in \mathrm{M}_2(4) \,\middle|\, \text{the Smith normal form of } A \text{ is } \begin{pmatrix} 2 & 0 \\ 0 & 2 \end{pmatrix} \right\}.$$

Obviously, $S(1, 4) \cap S(2, 2) = \emptyset$, $\mathrm{M}_2(4) = S(1, 4) \cup S(2, 2)$.

    A natural question is as follows: How do we compare the "size" of the sets $S(1, 4)$ and $S(2, 2)$? Or how much is the "proportion" of $S(1, 4)$ and $S(2, 2)$ respectively in $\mathrm{M}_2(4)$? To this end, for $A = \begin{pmatrix} a & b \\ c & d \end{pmatrix} \in \mathrm{M}_2(\mathbb{Z})$, we define the **Euclidean norm** of A as

$$\|A\| = \sqrt{a^2 + b^2 + c^2 + d^2}.$$

Theorem 6.6.10. *Let r be a positive real number and*

$$T_2(1, 4, r)$$

$$= \left| \left\{ A \in \mathrm{M}_2(4) \,\middle|\, \text{the Smith normal form of } A \text{ is } \begin{pmatrix} 1 & 0 \\ 0 & 4 \end{pmatrix}, \ \|A\| \leqslant r \right\} \right|,$$

$$T_2(2, 2, r)$$

$$= \left| \left\{ A \in \mathrm{M}_2(4) \,\middle|\, \text{the Smith normal form of } A \text{ is } \begin{pmatrix} 2 & 0 \\ 0 & 2 \end{pmatrix}, \ \|A\| \leqslant r \right\} \right|,$$

$$T_2(4, r) = |\{ A \in \mathrm{M}_2(4) \mid \|A\| \leqslant r \}|, \quad \rho(2, 2, r) = \frac{T_2(2, 2, r)}{T_2(4, r)}.$$

Then $\displaystyle\lim_{r \to +\infty} \rho(2, 2, r) = \frac{1}{7}$, *and so* $\displaystyle\lim_{r \to +\infty} \frac{T_2(2, 2, r)}{T_2(1, 4, r)} = \frac{1}{6}$.

Since the proof of the abovementioned theorem is beyond this book's scope, interested readers may refer to Example 1.6 in [13].

Question. From Theorem 6.6.6, the number of Smith normal forms of integer matrices of order n with determinant being a positive integer m is finite, denoted by $N_n(m)$. Partition the set $\mathrm{M}_n(m)$ by the Smith normal forms into $N_n(m)$ disjoint subsets. Similar to Theorem 6.6.10, consider the proportion of each such subset in $\mathrm{M}_n(m)$.

Exercises

1. Find the Smith normal forms of the following λ-matrices:

$$(1) \ \begin{pmatrix} \lambda^3 - \lambda & 2\lambda^2 \\ \lambda^2 + 5\lambda & 3\lambda \end{pmatrix}, \quad (2) \ \begin{pmatrix} \lambda - 2 & -1 & 0 \\ 0 & \lambda - 2 & -1 \\ 0 & 0 & \lambda - 2 \end{pmatrix}.$$

2. Let $A = \begin{pmatrix} \lambda & 0 & 0 \\ 1 & \lambda & 0 \\ 0 & 1 & \lambda \end{pmatrix}$. Find A^k, where k is a positive integer.

3. Find the Smith normal forms and invariant factors of the characteristic matrices of the following matrices:

$$(1)\ \begin{pmatrix} 3 & 2 & -5 \\ 2 & 6 & -10 \\ 1 & 2 & -3 \end{pmatrix},\quad (2)\ \begin{pmatrix} 4 & 6 & -15 \\ 1 & 3 & -5 \\ 1 & 2 & -4 \end{pmatrix}.$$

4. Find the rational canonical forms of the following matrices:

$$(1)\ \begin{pmatrix} 37 & -20 & -4 \\ 34 & -17 & -4 \\ 119 & -70 & -11 \end{pmatrix},\quad (2)\ \begin{pmatrix} 1 & -3 & 3 \\ -2 & -6 & 13 \\ -1 & -4 & 8 \end{pmatrix}.$$

5. Let $A \in M_n(\mathbb{C})$. Prove that the following statements are equivalent:

   (1) There exists $\alpha \in \mathbb{C}^n$ such that $\alpha, A\alpha, A^2\alpha, \dots, A^{n-1}\alpha$ are linearly independent.
   (2) All eigenspaces of A have dimension 1.
   (3) For every eigenvalue of A, the Jordan canonical form of A has only one Jordan block corresponding to it.
   (4) The minimal polynomial and characteristic polynomial of A are equal, i.e., $m_A(\lambda) = f_A(\lambda)$.
   (5) The rational canonical form of A is the companion matrix of the characteristic polynomial of A.

6. Let $d_n(\lambda)$ be the nth invariant factor of a matrix $A \in M_n(F)$ and $g(\lambda) \in F[\lambda]$. Prove the following:

   (1) $\operatorname{rank}(g(A)) = \operatorname{rank}(d(A))$, where $d(\lambda) = (g(\lambda), d_n(\lambda))$,

   (2) $g(A)$ is invertible if and only if $(g(\lambda), d_n(\lambda)) = 1$.

7. Let $A \in M_n(\mathbb{C})$. Assume $\operatorname{rank}A^k = \operatorname{rank}A^{k+1}$ for some natural number k. Prove that any elementary divisor of A corresponding to the eigenvalue zero has a degree less than or equal to k.

8. Let $A, B \in M_n(\mathbb{C})$. Assume that the elementary divisors of A and B both have degree one and $AB = BA$. Prove that there exists an invertible matrix P of order n such that $P^{-1}AP$ and $P^{-1}BP$ are both diagonal matrices.

9. Let $A \in M_n(\mathbb{C})$. Prove that A is a scalar matrix if and only if all invariant factors of A have a degree greater than 0.

10. Let $A, B \in M_n(\mathbb{C})$.

    (1) Let $n = 2$. Prove that A is similar to B if and only if A and B have the same minimal polynomial.

(2) Let $n = 3$. Prove that A is similar to B if and only if A and B have the same minimal polynomial and characteristic polynomial.

(3) Find $A, B \in \mathrm{M}_4(\mathbb{C})$ such that A and B have the same characteristic polynomial and minimal polynomial, but they are not similar.

11. Let $A \in \mathrm{M}_n(F)$. Prove that A and A' have the same rational canonical form.

12. Let L/F be an extension of number fields (i.e., F, L are both number fields and $F \subseteq L$) and let $A, B \in \mathrm{M}_n(F)$. Prove that if A and B are similar over L, then they are similar over F.

13. Let $A(\lambda) \in \mathrm{M}_n(F[\lambda])$ and $\mathrm{rank}(A(\lambda)) = n$. Prove that there exists the unique decomposition $A(\lambda) = P(\lambda)Q(\lambda)$, where $P(\lambda) \in \mathrm{GL}_n(F[\lambda])$, $Q(\lambda) = (q_{ij}(\lambda))$ is an upper triangular λ-matrix with $q_{ii}(\lambda)$ a monic polynomial for any $i = 1, 2, \ldots, n$, and $q_{ij}(\lambda) = 0$ or $\deg(q_{ij}(\lambda)) < \deg(q_{jj}(\lambda))$ for any $1 \leqslant i < j \leqslant n$.

14. Let A be a complex square matrix with $f_A(\lambda) = (\lambda - 1)^n$. Prove that A^k is similar to A for any positive integer k.

15. Let $A \in \mathrm{M}_5(\mathbb{C})$. Assume that the characteristic polynomial and minimal polynomial of A are respectively

$$f_A(\lambda) = (\lambda - 2)^3(\lambda + 7)^2 \text{ and } m_A(\lambda) = (\lambda - 2)^2(\lambda + 7).$$

Find the Jordan canonical form of A.

16. Find the Jordan canonical form of the matrix

$$A = \begin{pmatrix} 2 & 0 & 0 & 0 & 0 & 0 \\ 1 & 2 & 0 & 0 & 0 & 0 \\ -1 & 0 & 2 & 0 & 0 & 0 \\ 0 & 1 & 0 & 2 & 0 & 0 \\ 1 & 1 & 1 & 1 & 2 & 0 \\ 0 & 0 & 0 & 0 & 1 & -1 \end{pmatrix}.$$

17. Let $J_n(\lambda_1)$ be the Jordan block of order n with eigenvalue λ_1 and $J_m(\lambda_2)$ the Jordan block of order m with eigenvalue λ_2 (where λ_1, λ_2 may be the same or not). Put

$$J = \begin{pmatrix} J_n(\lambda_1) & 0 \\ 0 & J_m(\lambda_2) \end{pmatrix}, \quad C(J) = \{A \in \mathrm{M}_{n+m}(\mathbb{C}) \,|\, AJ = JA\}.$$

Prove that $C(J)$ is a subspace of $\mathrm{M}_{n+m}(\mathbb{C})$ and find its dimension.

18. Prove that every Jordan matrix J of order n has a decomposition $J = S_1 S_2 = S_3 S_4$, where S_1, S_4 are both real symmetric invertible matrices of order n, and S_2, S_3 are both complex symmetric matrices of order n.

19.* Prove that every complex matrix A of order n has decompositions $A = C_1 C_2 = C_3 C_4$, where C_1, C_2, C_3, C_4 are all complex symmetric matrices of order n, and C_1, C_4 are both invertible matrices.

20.* Let $A \in M_n(\mathbb{C})$. Prove that there exists a symmetric invertible matrix P of order n such that $P^{-1}AP = A'$.

21.* Let $A \in GL_n(\mathbb{C})$. Prove that there exists $B \in GL_n(\mathbb{C})$ such that $A = B^2$.

22.* (Weyl[2]) Let $A, B \in M_n(\mathbb{C})$. Prove that A is similar to B if and only if $\mathrm{rank}(aI_n - A)^k = \mathrm{rank}(aI_n - B)^k$ for any $a \in \mathbb{C}$ and positive integer k.

---

[2]Hermann Weyl, 1885–1955, German mathematician and physicist.

Chapter 7

Quadratic Forms

7.1　Introduction

The study of quadratic forms has a rich and illustrious history. Initially, the quadratic forms had integer coefficients, and their indeterminates took integer values. Such forms are known as integral quadratic forms. Fermat,[1] Leonhard Euler, Joseph-Louis Lagrange, Legendre,[2] Carl Friedrich Gauss, and many other great mathematicians have made significant contributions to the study of integral quadratic forms. When Gauss studied the composition law of binary quadratic forms, he found that the equivalence class of binary quadratic forms with the same discriminant has a group structure. In the 19th century, Hermann Minkowski and others realized that replacing integers with rational numbers would simplify the theory of integral quadratic forms and provide helpful information for studying integral quadratic forms. Thus, they established the general theory of rational quadratic forms.

For the quadratic forms over the real number field, James Joseph Sylvester proposed the inertia theorem in 1852 without proof. In 1857, Jacobi[3] independently discovered and strictly proved the inertia theorem. This chapter first introduces the theory of quadratic

---

[1]Pierre de Fermat, 1607–1665, French mathematician.
[2]Adrien Marie Legendre, 1752–1833, French mathematician.
[3]Carl Gustav Jacob Jacobi, 1804–1851, German mathematician.

forms over a general number field and then focuses on the quadratic forms over the real and complex number fields.

7.2 The Matrices of Quadratic Forms and Congruences of Matrices

Definition 7.2.1. Let F be a number field. A quadratic homogenous polynomial $f(x_1, x_2, \ldots, x_n)$ is called a **quadratic form** over F in n indeterminates. The zero element of F is regarded as the trivial quadratic form.

Henceforth, all quadratic forms discussed in this section are over a number field F.

Example 7.2.2. The polynomial $f(x_1, x_2, x_3) = 5x_1^2 + 3x_2^2 - 12x_3^2$ is a quadratic form in three indeterminates, $g(x_1, x_2, x_3, x_4) = x_1^2 + 3x_1x_2 - 4x_2^2 + x_2x_3 + x_1x_4$ is a quadratic form in four indeterminates, while $x_1^2 + 1$, $x_1^2 + x_2$ are not quadratic forms.

Remark 7.2.3. The general form of a quadratic form in n indeterminates is

$$f(x_1, x_2, \ldots, x_n) = \sum_{1 \leqslant i \leqslant j \leqslant n} c_{ij} x_i x_j. \tag{7.1}$$

To write (7.1) in a symmetric form, let $a_{ij} = a_{ji} = \frac{c_{ij}}{2}$ if $i < j$; $a_{ii} = c_{ii}$ if $i = j$. Hence, (7.1) can be written out as

$$f(x_1, x_2, \ldots, x_n) = X'AX, \tag{7.2}$$

where $X = (x_1, x_2, \ldots, x_n)'$, $A = (a_{ij})_{n \times n}$.

Definition 7.2.4. We call the $n \times n$ symmetric matrix A in (7.2) the **matrix of the quadratic form** $f(x_1, x_2, \ldots, x_n)$.

By Definition 7.2.4, the matrix of the quadratic form $f(x_1, x_2, \ldots, x_n)$ is uniquely determined by $f(x_1, x_2, \ldots, x_n)$. Conversely, for any given $n \times n$ symmetric matrix A, there is only one quadratic form whose matrix is A. Hence, there is a one-to-one correspondence between the set of quadratic forms in n indeterminates and the set of $n \times n$ symmetric matrices. If A is an $n \times n$ diagonal matrix, then $f(x_1, x_2, \ldots, x_n) = X'AX$ is called a **diagonal quadratic form**.

Example 7.2.5. Find the matrix of the quadratic form

$$f(x, y) = (x, y) \begin{pmatrix} 1 & 2 \\ 3 & 4 \end{pmatrix} \begin{pmatrix} x \\ y \end{pmatrix}.$$

Solution. Since $f(x, y) = x^2 + 5xy + 4y^2 = (x, y) \begin{pmatrix} 1 & \frac{5}{2} \\ \frac{5}{2} & 4 \end{pmatrix} \begin{pmatrix} x \\ y \end{pmatrix}$, the

matrix of the quadratic form $f(x, y)$ is $\begin{pmatrix} 1 & \frac{5}{2} \\ \frac{5}{2} & 4 \end{pmatrix}$. $\qquad\square$

Remark 7.2.6. For any $B \in M_n(F)$, the polynomial $f(x_1, x_2, \ldots, x_n) = X'BX$ is a quadratic form, however, the matrix of $f(x_1, x_2, \ldots, x_n)$ is $\frac{B+B'}{2}$. When we write $f(x_1, x_2, \ldots, x_n)$ as $f(x_1, x_2, \ldots, x_n) = X'AX$, the matrix A is always symmetric.

Definition 7.2.7. Let $f(x_1, x_2, \ldots, x_n) = X'AX$. The rank of A is called the **rank** of $f(x_1, x_2, \ldots, x_n)$, denoted by $\mathrm{rank}(f)$.

Example 7.2.8. Find the rank of the quadratic form

$$f(x_1, x_2, x_3) = x_1^2 + 5x_1 x_2 - 2x_2^2 + 4x_2 x_3.$$

Solution. The matrix of the quadratic form $f(x_1, x_2, x_3)$ is $A = \begin{pmatrix} 1 & \frac{5}{2} & 0 \\ \frac{5}{2} & -2 & 2 \\ 0 & 2 & 0 \end{pmatrix}$. Since the rank of A is 3, it follows that the rank of $f(x_1, x_2, x_3)$ is 3. $\qquad\square$

Definition 7.2.9. Let $f(x_1, x_2, \ldots, x_n)$ be a quadratic form in n indeterminates and

$$\begin{cases} x_1 = c_{11} y_1 + c_{12} y_2 + \cdots + c_{1n} y_n, \\ x_2 = c_{21} y_1 + c_{22} y_2 + \cdots + c_{2n} y_n, \\ \quad \vdots \qquad\qquad \vdots \\ x_n = c_{n1} y_1 + c_{n2} y_2 + \cdots + c_{nn} y_n, \end{cases}$$

i.e., $X = CY$, where $C = (c_{ij})_{n \times n}$, $X = (x_1, x_2, \ldots, x_n)'$, $Y = (y_1, y_2, \ldots, y_n)'$. Then, we call $X = CY$ a **linear substitution**.

If the matrix C is invertible, we call $X = CY$ an **invertible linear substitution**, or a **nondegenerate linear substitution**.

Two quadratic forms $f(x_1, x_2, \ldots, x_n)$ and $g(y_1, y_2, \ldots, y_n)$ are said to be **equivalent** if $f(x_1, x_2, \ldots, x_n)$ can be transformed to $g(y_1, y_2, \ldots, y_n)$ by an invertible linear substitution $X = CY$.

It is easy to see that the equivalence of quadratic forms is an equivalence relation.

Example 7.2.10. The quadratic form $f(x_1, x_2) = x_1 x_2$ can be transformed to $g(y_1, y_2) = y_1^2 - y_2^2$ by the invertible linear substitution $\begin{pmatrix} x_1 \\ x_2 \end{pmatrix} = \begin{pmatrix} 1 & 1 \\ 1 & -1 \end{pmatrix} \begin{pmatrix} y_1 \\ y_2 \end{pmatrix}$. So $f(x_1, x_2)$ is equivalent to $g(y_1, y_2)$.

By the uniqueness of the matrix of a quadratic form, we have the following proposition.

Proposition 7.2.11. *If a quadratic form $f(x_1, x_2, \ldots, x_n) = X'AX$ is transformed to $g(y_1, y_2, \ldots, y_n) = Y'C'ACY$ by an invertible linear substitution $X = CY$, then the matrix of the quadratic form $g(y_1, y_2, \ldots, y_n)$ is $C'AC$.*

By Proposition 7.2.11, we give the following definition.

Definition 7.2.12. Assume $A, B \in M_n(F)$. If there is an $n \times n$ invertible matrix C such that $A = C'BC$, then we say that A and B are **congruent**, or A is congruent to B.

It is easy to see that the congruence of matrices is an equivalence relation. By Definition 7.2.12, the matrices congruent to a symmetric matrix are also symmetric. If A is congruent to B, then they are equivalent. The following example shows that the converse is not true, i.e., if A is equivalent to B, they may not be congruent.

Example 7.2.13. Let $A = \begin{pmatrix} 1 & 0 \\ 0 & 0 \end{pmatrix}$ and $B = \begin{pmatrix} 0 & 1 \\ 0 & 0 \end{pmatrix}$. Then $\operatorname{rank} A = \operatorname{rank} B = 1$, and hence A is equivalent to B. Since A is a symmetric matrix, so is $C'AC$ for any invertible matrix C. However, B is not a symmetric matrix; it follows that B is not congruent to A.

Remark 7.2.14. We have learned in Chapter 2 that any $n \times n$ matrix A is equivalent to its transpose A'. By Exercise 11 of Chapter 6, any $n \times n$ matrix A is similar to its transpose A'. In fact, any $n \times n$ matrix A is also congruent to A' (see [10]).

All $n \times n$ symmetric matrices can be partitioned into equivalence classes according to the congruence relation. In each equivalence class, one can always choose a simple representative. In Theorem 7.2.15, we prove that every symmetric matrix is congruent to a diagonal matrix, i.e., every quadratic form is equivalent to a diagonal quadratic form.

Theorem 7.2.15. *Any nonzero quadratic form $f(x_1, x_2, \ldots, x_n) = X'AX$ can be transformed to the quadratic form $d_1 y_1^2 + d_2 y_2^2 + \cdots + d_r y_r^2$ by an invertible linear substitution $X = CY$, where $d_1 d_2 \cdots d_r \neq 0$, $r = rank(f)$, so $C'AC = \mathrm{diag}(d_1, d_2, \ldots, d_r, 0, \ldots, 0)$ The quadratic form $d_1 y_1^2 + d_2 y_2^2 + \cdots + d_r y_r^2$ is called the* **standard form** *of $f(x_1, x_2, \ldots, x_n)$.*

Proof. We proceed by induction on the number n of indeterminates. If $n = 1$, then the theorem holds obviously. If $n > 1$, we assume the theorem holds for any quadratic form in $n - 1$ indeterminates. Next, we consider a nonzero quadratic form $f(x_1, x_2, \ldots, x_n)$ in n indeterminates.

If $f(x_1, x_2, \ldots, x_n)$ doesn't have a square term, then we assume that $f(x_1, x_2, \ldots, x_n)$ has a cross term $x_i x_j$. Let

$$x_i = y_i + y_j,$$

$$x_j = y_i - y_j,$$

$$x_k = y_k, k \neq i, j.$$

After this invertible linear substitution, $f(x_1, x_2, \ldots, x_n)$ is transformed to a quadratic form with square terms.

Without loss of generality, we may assume that the quadratic form $f(x_1, x_2, \ldots, x_n)$ has a square term $d_1 x_1^2$, where $d_1 \neq 0$. Now

$$f(x_1, x_2, \ldots, x_n) = d_1 x_1^2 + 2g(x_2, \ldots, x_n)x_1 + h(x_2, \ldots, x_n),$$

where $g(x_2, \ldots, x_n)$ is a linear combination of $x_2, \ldots, x_n$, and $h(x_2, \ldots, x_n)$ is a quadratic form in the indeterminates $x_2, \ldots, x_n$.

Hence

$$f(x_1, x_2, \ldots, x_n) = d_1 \left(x_1 + \frac{g(x_2, \ldots, x_n)}{d_1} \right)^2$$

$$+ h(x_2, \ldots, x_n) - \frac{g(x_2, \ldots, x_n)^2}{d_1}.$$

Let $\begin{cases} y_1 = x_1 + \dfrac{g(x_2, \ldots, x_n)}{d_1}, \\ y_i = x_i, i \geqslant 2, \end{cases}$ or $\begin{cases} x_1 = y_1 - \dfrac{g(y_2, \ldots, y_n)}{d_1}, \\ x_i = y_i, i \geqslant 2, \end{cases}$

i.e., $X = C_1 Y$, where C_1 is an upper triangular matrix with diagonal elements 1. Thus C_1 is an invertible matrix, and so $f(x_1, x_2, \ldots, x_n)$ is transformed by the invertible linear substitution $X = C_1 Y$ to

$$\widetilde{f}(y_1, y_2, \ldots, y_n) = d_1 y_1^2 + f_1(y_2, \ldots, y_n),$$

where $f_1(y_2, \ldots, y_n) = h(y_2, \ldots, y_n) - \frac{g(y_2, \ldots, y_n)^2}{d_1}$ is a quadratic form in the $n - 1$ indeterminates $y_2, \ldots, y_n$. By the inductive hypothe-

sis, there is an invertible linear substitution $\begin{pmatrix} y_2 \\ \vdots \\ y_n \end{pmatrix} = C_2 \begin{pmatrix} z_2 \\ \vdots \\ z_n \end{pmatrix}$ trans-

forming $f_1(y_2, \ldots, y_n)$ to $d_2 z_2^2 + \cdots + d_r z_r^2$, where $r = \operatorname{rank}(f)$. Let $Z = (z_1, z_2, \ldots, z_n)'$, $X = C_1 \begin{pmatrix} 1 & 0 \\ 0 & C_2 \end{pmatrix} Z$. Then the quadratic form $f(x_1, x_2, \ldots, x_n)$ is transformed to the standard form $d_1 z_1^2 + d_2 z_2^2 + \cdots + d_r z_r^2$. $\qquad\qquad\square$

Example 7.2.16. Let $f(x_1, x_2, x_3) = 2x_1^2 - 4x_1 x_2 + x_2^2 - 4x_2 x_3$. Use an invertible linear substitution to transform the quadratic form $f(x_1, x_2, x_3)$ to the standard form.

Solution. We use the method of completing the squares to perform linear substitution,

$$f(x_1, x_2, x_3) = 2x_1^2 - 4x_1 x_2 + x_2^2 - 4x_2 x_3$$

$$= 2(x_1 - x_2)^2 - x_2^2 - 4x_2 x_3$$

$$= 2(x_1 - x_2)^2 - (x_2 + 2x_3)^2 + (2x_3)^2$$

$$= 2y_1^2 - y_2^2 + y_3^2,$$

$$\text{where } \begin{cases} y_1 = x_1 - x_2, \\ y_2 = x_2 + 2x_3, \\ y_3 = 2x_3, \end{cases} \quad \text{i.e.,} \quad \begin{cases} x_1 = y_1 + y_2 - y_3, \\ x_2 = y_2 - y_3, \\ x_3 = \dfrac{y_3}{2}. \end{cases} \qquad \square$$

By Theorem 7.2.15 and Example 7.2.16, we know how to use invertible linear substitutions to transform quadratic forms to standard forms.

In fact, for a given symmetric matrix A, we can use elementary operations to find an invertible matrix C such that $C'AC = D$ is a diagonal matrix.

Definition 7.2.17. Assume that $A \in \mathrm{M}_{m \times n}(F)$, $1 \leqslant i \neq j \leqslant \min(m, n)$. The following three operations are called **elementary congruent operations**:

(1) Multiply row i of A by a nonzero number k to obtain a matrix A_1, and then multiply column i of A_1 by the same number k, where $k \in F$.
(2) Add b times row i of A to row j of A to obtain a matrix A_2, and then add b times column i of A_2 to column j of A_2, where $b \in F$.
(3) Interchange rows i and j of A to obtain a matrix A_3, and then interchange columns i and j of A_3.

In the following, we outline a specific operational method. This approach is akin to the method used to obtain the inverse of an invertible matrix through elementary operations, as discussed in Section 2.9. This method employs a series of elementary congruent operations to convert a symmetric matrix into a diagonal matrix.

Let A be an $n \times n$ symmetric matrix. We apply elementary congruent operations to the $2n \times n$ matrix $\begin{pmatrix} A \\ \hline I_n \end{pmatrix}$. When the A at the top of $\begin{pmatrix} A \\ \hline I_n \end{pmatrix}$ is transformed to a diagonal matrix D, the I_n at the bottom of $\begin{pmatrix} A \\ \hline I_n \end{pmatrix}$ is transformed to an invertible matrix C we need, i.e.,

$$\begin{pmatrix} A \\ \hline I_n \end{pmatrix} \longrightarrow \begin{pmatrix} C' & 0 \\ 0 & I_n \end{pmatrix} \begin{pmatrix} A \\ \hline I_n \end{pmatrix} C = \begin{pmatrix} C'AC \\ \hline C \end{pmatrix} = \begin{pmatrix} D \\ \hline C \end{pmatrix}.$$

Example 7.2.18. Let $A = \begin{pmatrix} 2 & -2 & 0 \\ -2 & 1 & -2 \\ 0 & -2 & 0 \end{pmatrix}$. Find an invertible matrix C such that $C'AC$ is a diagonal matrix.

Solution. Applying elementary congruent operations to $\left(\dfrac{A}{I_3}\right)$, we have

$$\begin{pmatrix} 2 & -2 & 0 \\ -2 & 1 & -2 \\ 0 & -2 & 0 \\ \hline 1 & 0 & 0 \\ 0 & 1 & 0 \\ 0 & 0 & 1 \end{pmatrix} \xrightarrow[c_1+c_2]{r_1+r_2} \begin{pmatrix} 2 & 0 & 0 \\ 0 & -1 & -2 \\ 0 & -2 & 0 \\ \hline 1 & 1 & 0 \\ 0 & 1 & 0 \\ 0 & 0 & 1 \end{pmatrix} \xrightarrow[(-2)\times c_2+c_3]{(-2)\times r_2+r_3} \begin{pmatrix} 2 & 0 & 0 \\ 0 & -1 & 0 \\ 0 & 0 & 4 \\ \hline 1 & 1 & -2 \\ 0 & 1 & -2 \\ 0 & 0 & 1 \end{pmatrix}.$$

Hence, $C = \begin{pmatrix} 1 & 1 & -2 \\ 0 & 1 & -2 \\ 0 & 0 & 1 \end{pmatrix}$, and $C'AC = \begin{pmatrix} 2 & 0 & 0 \\ 0 & -1 & 0 \\ 0 & 0 & 4 \end{pmatrix}$. $\qquad\square$

Remark 7.2.19. In Example 7.2.16, the matrix of the quadratic form $f(x_1, x_2, x_3)$ is just the matrix A in Example 7.2.18, and the matrix of the linear substitution to transform $f(x_1, x_2, x_3)$ to the standard form is $C = \begin{pmatrix} 1 & 1 & -1 \\ 0 & 1 & -1 \\ 0 & 0 & \frac{1}{2} \end{pmatrix}$ which is different from the C in 7.2.18. The diagonal matrix $\begin{pmatrix} 2 & 0 & 0 \\ 0 & -1 & 0 \\ 0 & 0 & 1 \end{pmatrix}$ of the standard form in Example 7.2.16 is different from the diagonal matrix in Example 7.2.18, which implies that both $2y_1^2 - y_2^2 + y_3^2$ and $2y_1^2 - y_2^2 + 4y_3^2$ are the standard forms of the quadratic form $f(x_1, x_2, x_3) = 2x_1^2 - 4x_1x_2 + x_2^2 - 4x_2x_3$. Hence, the standard form of a quadratic form is not unique. It depends on the choice of the invertible linear substitution.

7.3 The Canonical Forms of Quadratic Forms

Section 7.2 presents the standard form theory of quadratic forms over general number fields. In this section, we introduce canonical forms that offer more detail than the standard forms when applied to real and complex fields.

Theorem 7.3.1. *Let $f(x_1, x_2, \ldots, x_n)$ be a complex quadratic form. Then there exists an invertible linear substitution $X = CY$ such that $f(x_1, x_2, \ldots, x_n)$ is transformed to $g(y_1, y_2, \ldots, y_n) = y_1^2 + y_2^2 + \cdots + y_r^2$, where $r = \mathrm{rank}(f)$, $g(y_1, y_2, \ldots, y_n)$ is called the* **canonical form** *of the complex quadratic form $f(x_1, x_2, \ldots, x_n)$. The canonical form of a complex quadratic form is completely determined by its rank.*

Proof. By Theorem 7.2.15, $f(x_1, x_2, \ldots, x_n)$ is equivalent to a standard form. Without loss of generality, we can assume $f(x_1, x_2, \ldots, x_n) = d_1 x_1^2 + d_2 x_2^2 + \cdots + d_r x_r^2$. Let $y_1 = \sqrt{d_1} x_1, y_2 = \sqrt{d_2} x_2, \ldots, y_r = \sqrt{d_r} x_r, y_{r+1} = x_{r+1}, y_{r+2} = x_{r+2}, \ldots, y_n = x_n$. Then $f(x_1, x_2, \ldots, x_n)$ is transformed to $g(y_1, y_2, \ldots, y_n) = y_1^2 + \cdots + y_r^2$. It is easy to see that the standard form of $f(x_1, x_2, \ldots, x_n)$ is completely determined by its rank r. $\qquad\square$

By Theorem 7.3.1, for any $n \times n$ complex symmetric matrix A, one can always find an $n \times n$ invertible complex matrix C such that $C'AC = \begin{pmatrix} I_r & 0 \\ 0 & 0 \end{pmatrix}$, where $r = \mathrm{rank}(A)$. We call $\begin{pmatrix} I_r & 0 \\ 0 & 0 \end{pmatrix}$ the **canonical form** of A.

The key point in the proof of Theorem 7.3.1 is that each coefficient d_i has a square root in the field of complex numbers, but negative numbers have no square roots in the field of real numbers, so we have the following.

Theorem 7.3.2 (Law of inertia). *Let $f(x_1, x_2, \ldots, x_n)$ be a real quadratic form of rank r. Then there is an invertible linear substitution $X = CY$ such that $f(x_1, x_2, \ldots, x_n)$ is transformed to*

$$g(y_1, y_2, \ldots, y_n) = y_1^2 + y_2^2 + \cdots + y_p^2 - y_{p+1}^2 - y_{p+2}^2 - \cdots - y_r^2. \tag{7.3}$$

The standard form (7.3) is called the **canonical form** *of the real quadratic form $f(x_1, x_2, \ldots, x_n)$. The canonical form of $f(x_1, x_2, \ldots, x_n)$ is uniquely determined by $f(x_1, x_2, \ldots, x_n)$. The number p in (7.3) is called the* **positive index of inertia** *of $f(x_1, x_2, \ldots, x_n)$, the number $q = r - p$ is called the* **negative index**

of inertia *of* $f(x_1, x_2, \ldots, x_n)$, *and the difference* $p - q$ *is called the* ***signature*** *of* $f(x_1, x_2, \ldots, x_n)$.

Proof. Any quadratic form can be transformed into the canonical form specified in the theorem using an invertible linear substitution. To see this, one can follow the proof for the case of the complex number field. The only distinction in the real number field is that only nonnegative numbers possess square roots.

We demonstrate that the positive index of inertia is uniquely determined by the original quadratic form. Consequently, the original quadratic form also uniquely determines the negative index of inertia.

Assume that the quadratic form $f(x_1, x_2, \ldots, x_n)$ can be transformed not only by an invertible linear substitution $X = C_1 Y$ to

$$g(y_1, y_2, \ldots, y_n) = y_1^2 + \cdots + y_p^2 - y_{p+1}^2 - \cdots - y_r^2 \qquad (7.4)$$

but also by an invertible linear substitution $X = C_2 Z$ to

$$h(z_1, z_2, \ldots, z_n) = z_1^2 + \cdots + z_s^2 - z_{s+1}^2 - \cdots - z_r^2. \qquad (7.5)$$

Next, we show that $p = s$. Note that all y_i, z_i are linear combinations of $x_1, x_2, \ldots, x_n$. If $p < s$, then the number of equations in the homogeneous linear system

$$y_1 = 0, \ \ldots, \ y_p = 0, \ z_{s+1} = 0, \ \ldots, \ z_n = 0 \qquad (7.6)$$

is $n - s + p < n$, which implies that (7.6) has a nonzero solution $(x_1, x_2, \ldots, x_n)' = (a_1, a_2, \ldots, a_n)'$. Assume that $\begin{pmatrix} b_1 \\ b_2 \\ \vdots \\ b_n \end{pmatrix} =$

$C_1^{-1} \begin{pmatrix} a_1 \\ a_2 \\ \vdots \\ a_n \end{pmatrix}, \begin{pmatrix} c_1 \\ c_2 \\ \vdots \\ c_n \end{pmatrix} = C_2^{-1} \begin{pmatrix} a_1 \\ a_2 \\ \vdots \\ a_n \end{pmatrix}$. Then $b_1 = b_2 = \cdots = b_p = 0, c_{s+1} =$

$c_{s+2} = \cdots = c_n = 0$, so $c_1, c_2, \ldots, c_s$ are not all zero. By (7.4) and (7.5), $g(b_1, b_2, \ldots, b_n) = -b_{p+1}^2 - \cdots - b_r^2 \leqslant 0$, $h(c_1, c_2, \ldots, c_n) = c_1^2 + c_2^2 + \cdots + c_s^2 > 0$, contradicting $g(b_1, b_2, \ldots, b_n) = f(a_1, a_2, \ldots, a_n) = h(c_1, c_2, \ldots, c_n)$. Hence $p \geqslant s$.

By the same argument, $p \leqslant s$, so $p = s$. Hence, the positive index of inertia of the quadratic form $f(x_1, x_2, \ldots, x_n)$ is uniquely determined. $\qquad\square$

Remark 7.3.3. The "inertia" in Theorem 7.3.2 refers to something that remains unchanged under certain operations.

By the law of inertia above, for any $n \times n$ real symmetric matrix A, we can always find an $n \times n$ real invertible matrix C such that

$$C'AC = \begin{pmatrix} I_p & 0 & 0 \\ 0 & -I_q & 0 \\ 0 & 0 & 0 \end{pmatrix}, \text{ which is called the } \textbf{canonical form} \text{ of } A.$$

7.4 Positive Definite Quadratic Forms

In this section, we discuss the real quadratic forms. The real quadratic forms in n indeterminates can be seen as functions on $\mathbb{R}^n$.

Definition 7.4.1. Let $f(x_1, x_2, \ldots, x_n)$ be a real quadratic form. $f(x_1, x_2, \ldots, x_n)$ is called **positive definite** if $f(a_1, a_2, \ldots, a_n) > 0$ for any real numbers $a_1, a_2, \ldots, a_n$ that are not all zero. $f(x_1, x_2, \ldots, x_n)$ is called **positive semi-definite** if $f(a_1, a_2, \ldots, a_n) \geqslant 0$ for any real numbers $a_1, a_2, \ldots, a_n$. Similarly, we can also define **negative definite** and **negative semi-definite** quadratic forms. If a quadratic form can take positive and negative values, we call it an **indefinite** quadratic form.

An $n \times n$ symmetric matrix A is called **positive definite** if $f(x_1, x_2, \ldots, x_n) = X'AX$ is a positive definite quadratic form. Similarly, we can also define a **positive semi-definite matrix, negative definite matrix, negative semi-definite matrix**, and an **indefinite matrix**.

It is easy to see that if $f(x_1, x_2, \ldots, x_n)$ is equivalent to $g(x_1, x_2, \ldots, x_n)$, then $f(x_1, x_2, \ldots, x_n)$ is positive definite (negative definite, positive semi-definite, negative semi-definite, indefinite) if and only if $g(x_1, x_2, \ldots, x_n)$ is positive definite (negative definite, positive semi-definite, negative semi-definite, indefinite).

Let A be an $n \times n$ matrix and $1 \leqslant k \leqslant n$. The minor consisting of the first k rows and k columns is called the **leading principal minor of order k** of A.

Theorem 7.4.2. *Let* $f(x_1, x_2, \ldots, x_n) = X'AX$ *be a real quadratic form. Then the following statements are equivalent:*

(1) $f(x_1, x_2, \ldots, x_n)$ *is a positive definite quadratic form.*
(2) *The positive index of inertia of* $f(x_1, x_2, \ldots, x_n)$ *is* n.
(3) A *is congruent to* I_n.
(4) *There is a* $D \in GL_n(\mathbb{R})$ *such that* $A = D'D$.
(5) *All leading principal minors of* A *are positive.*
(6) *All principal minors of* A *are positive.*

Proof. (1) $\implies$ (2). If the positive index of inertia of $f(x_1, x_2, \ldots, x_n)$ is $p < n$, then there is an invertible linear substitution transforming $f(x_1, x_2, \ldots, x_n)$ to the canonical form $y_1^2 + \cdots + y_p^2 + d_{p+1}y_{p+1}^2 + \cdots + d_n y_n^2$, where $d_{p+1}, \ldots, d_n$ are each -1 or 0.

We consider the following system of linear equations

$$y_1 = 0, \ \ldots, \ y_p = 0, \ y_{p+1} = 1, \ \ldots, \ y_n = 1$$

in the n unknowns $x_1, x_2, \ldots, x_n$. Since the linear substitution from $x_1, x_2, \ldots, x_n$ to $y_1, y_2, \ldots, y_n$ is invertible, the matrix of the above system of linear equations is also invertible. Thus the linear system above has a unique nonzero solution $(a_1, a_2, \ldots, a_n)'$, and hence $f(a_1, a_2, \ldots, a_n) = d_{p+1} + \cdots + d_n \leqslant 0$, which contradicts the assumption that $f(x_1, x_2, \ldots, x_n)$ is a positive definite quadratic form.

(2) $\implies$ (3) $\implies$ (4) $\implies$ (1) are obvious. Hence the first four statements are equivalent.

(1) $\implies$ (6). Suppose that $1 \leqslant i_1 < i_2 < \cdots < i_t \leqslant n$ and that B is the $t \times t$ square matrix obtained by keeping the entries in the intersections of the i_1th, i_2th, $\ldots$, i_tth rows and the i_1th, i_2th, $\ldots$, i_tth columns of A.

Let $x_{i_1} = y_1, \ldots, x_{i_t} = y_t$, and all the other indeterminates $x_j = 0$. Then, we get a positive definite quadratic form

$$g(y_1, \ldots, y_t) = (y_1, \ldots, y_t)B \begin{pmatrix} y_1 \\ \vdots \\ y_t \end{pmatrix}$$

in t indeterminates. By (4), there is a $C \in GL_t(\mathbb{R})$ such that $B = C'C$. So $|B| = |C'C| = |C|^2 > 0$.

$(6) \implies (5)$ is obvious.

$(5) \implies (1)$. We prove by induction on n.

When $n = 1$, the theorem obviously holds. We assume that $n \geqslant 2$, and the theorem holds for $(n-1) \times (n-1)$ symmetric matrices. Now consider an $n \times n$ real symmetric matrix A. We can write the matrix $A = (a_{ij})$ in the form $A = \begin{pmatrix} B & Y \\ Y' & a_{nn} \end{pmatrix}$, where B is an $(n-1) \times (n-1)$ symmetric matrix. By (5), all the leading principal minors of A are positive, so $|A| > 0, |B| > 0$. Hence B is invertible. Since

$$\begin{pmatrix} I_{n-1} & -B^{-1}Y \\ 0 & 1 \end{pmatrix}' \begin{pmatrix} B & Y \\ Y' & a_{nn} \end{pmatrix} \begin{pmatrix} I_{n-1} & -B^{-1}Y \\ 0 & 1 \end{pmatrix} = \begin{pmatrix} B & 0 \\ 0 & a_{nn} - Y'B^{-1}Y \end{pmatrix},$$

$b = a_{nn} - Y'B^{-1}Y = \frac{|A|}{|B|} > 0$. Hence $f(x_1, x_2, \ldots, x_n)$ is equivalent to the quadratic form

$$g(y_1, \ldots, y_{n-1}, y_n) = (y_1, \ldots, y_{n-1}, y_n) \begin{pmatrix} B & 0 \\ 0 & b \end{pmatrix} \begin{pmatrix} y_1 \\ \vdots \\ y_{n-1} \\ y_n \end{pmatrix}$$

$$= (y_1, \ldots, y_{n-1}) B \begin{pmatrix} y_1 \\ \vdots \\ y_{n-1} \end{pmatrix} + b y_n^2.$$

Since the $(n-1) \times (n-1)$ matrix B is positive definite, we have

$(y_1, \ldots, y_{n-1}) B \begin{pmatrix} y_1 \\ \vdots \\ y_{n-1} \end{pmatrix} \geqslant 0$, and the equality holds if and only if

$y_1 = \cdots = y_{n-1} = 0$. Since $b > 0$, we have $g(y_1, y_2, \ldots, y_{n-1}, y_n) \geqslant 0$, and the equality holds if and only if $y_1 = \cdots = y_{n-1} = y_n = 0$. So $g(y_1, y_2, \ldots, y_{n-1}, y_n)$ is a positive definite quadratic form, which implies that $f(x_1, x_2, \ldots, x_n) = X'AX$ is a positive definite quadratic form. $\qquad\square$

Remark 7.4.3. A symmetric matrix A is negative definite if and only if $-A$ is positive definite, if and only if the leading principal minors of odd order of A are negative and those of even order of A are positive.

Theorem 7.4.4. *Let $f(x_1, x_2, \ldots, x_n) = X'AX$ be a quadratic form. Then the following statements are equivalent:*

(1) $f(x_1, x_2, \ldots, x_n)$ is a positive semi-definite quadratic form.

(2) The negative index of inertia of $f(x_1, x_2, \ldots, x_n)$ is 0.

(3) A is congruent to $\begin{pmatrix} I_r & 0 \\ 0 & 0 \end{pmatrix}$.

(4) There is a matrix $D \in \mathrm{M}_n(\mathbb{R})$ such that $A = D'D$.

(5) All the principal minors of A are nonnegative.

Proof. The equivalences of the first four statements are obvious. We only need to prove that they are equivalent to (5).

We proceed by induction. The theorem holds for $n = 1$. For $n > 1$, we assume that the theorem holds for matrices of order less than n.

(1) $\implies$ (5). Suppose that $1 \leqslant i_1 < i_2 < \cdots < i_t \leqslant n$ and that B is the $t \times t$ square matrix obtained by keeping the entries in the intersections of the i_1th, i_2th, $\ldots$, i_tth rows and the i_1th, i_2th, $\ldots$, i_tth columns of A.

Let $x_{i_1} = y_1, \ldots, x_{i_t} = y_t$, and all the other indeterminates $x_j = 0$. Then, we get a positive semi-definite quadratic form

$$g(y_1, \ldots, y_t) = (y_1, \ldots, y_t)B \begin{pmatrix} y_1 \\ \vdots \\ y_t \end{pmatrix}$$

in t indeterminates.

By (4), there is a matrix $C \in \mathrm{M}_t(\mathbb{R})$ such that $B = C'C$. So $|B| = |C'C| = |C|^2 \geqslant 0$.

(5) $\implies$ (1). We prove that if all the principal minors of A are nonnegative, then A is positive semi-definite.

Let us consider the polynomial $|\lambda I_n + A| = \lambda^n + s_1 \lambda^{n-1} + \cdots + s_n$, where the coefficient s_k is the sum of all principal minors of order k of A by Lemma 5.5.7. By (5), every s_k is nonnegative, hence $|\lambda I_n + A|$ is positive for any positive real number λ. Similarly, all leading principal minors of $\lambda I_n + A$ are positive, and so $\lambda I_n + A$ is positive definite for any positive real number λ by Theorem 7.4.2.

Note that $\lambda I_n + A$ is positive definite for an arbitrarily small positive real number λ. If there is a nonzero vector Y such that $Y'AY < 0$, then $Y'(\lambda I_n + A)Y < 0$ holds for $0 < \lambda < -\frac{Y'AY}{Y'Y}$, contradicting the fact that $\lambda I_n + A$ is positive definite. Thus there is no nonzero vectors Y such that $Y'AY < 0$, and hence A is positive semi-definite. $\qquad \square$

By Theorem 7.4.2, a real symmetric matrix is positive definite if and only if all leading principal minors are positive, if and only if all principal minors are positive. By Theorem 7.4.4, a real symmetric matrix is positive semi-definite if and only if all principal minors are nonnegative. Only the leading principal minors being nonnegative is not enough to guarantee that the matrix is positive semi-definite. For example, the leading principal minors of $A = \begin{pmatrix} 0 & 0 & 0 \\ 0 & 1 & 0 \\ 0 & 0 & -1 \end{pmatrix}$ are 0. However, A is indefinite.

Exercises

1. Show that a symmetric matrix of rank r can be written as the sum of r symmetric matrices of rank 1.
2. Show that a necessary and sufficient condition for a real quadratic form $f(x_1, x_2, \ldots, x_n)$ to be factored into the product of two homogeneous polynomials of degree 1 with real coefficients is $\mathrm{rank}(f) = 2$ with the signature 0, or $\mathrm{rank}(f) = 1$.
3. Find the standard form of each of the following quadratic forms:

   (1) $\displaystyle \sum_{i=1}^{n} x_i^2 + \sum_{1 \leqslant i < j \leqslant n} x_i x_j$;

   (2) $\displaystyle \sum_{i=1}^{n} x_i^2 + \sum_{i=1}^{n-1} x_i x_{i+1}$;

   (3) $x_1 x_2 + x_2 x_3 + \cdots + x_{n-1} x_n$;

   (4) $x_1 x_{2n} + x_2 x_{2n-1} + \cdots + x_n x_{n+1}$;

   (5) $(x_1 - \bar{x})^2 + (x_2 - \bar{x})^2 + \cdots + (x_n - \bar{x})^2$, where $\bar{x} = \dfrac{x_1 + x_2 + \cdots + x_n}{n}$.

4. Assume that the real quadratic form

$$f(x_1, x_2, \ldots, x_n) = \sum_{i=1}^{s} (a_{i1} x_1 + a_{i2} x_2 + \cdots + a_{in} x_n)^2.$$

Show that the rank of f is equal to the rank of the matrix $A = (a_{ij})_{s \times n}$.

5. If the $n \times n$ real symmetric matrices are classified according to the congruence relation, how many types are there?

6. Show that the $n \times n$ real matrix $\mathrm{diag}(\lambda_1, \lambda_2, \ldots, \lambda_n)$ is congruent to $\mathrm{diag}(\lambda_{i_1}, \lambda_{i_2}, \ldots, \lambda_{i_n})$, where $i_1 i_2 \ldots i_n$ is a permutation of $1, 2, \ldots, n$.

7. Assume that A is an $n \times n$ real matrix such that there are α, $\beta \in \mathbb{R}^n$ satisfying $\alpha' A \alpha > 0$, $\beta' A \beta < 0$. Show that there is a nonzero vector $\gamma \in \mathbb{R}^n$ such that $\gamma' A \gamma = 0$.

8. Find the canonical forms of the following real symmetric matrices:

$$(1)\ A = \begin{pmatrix} 1 & 1 & 4 \\ 1 & 0 & 6 \\ 4 & 6 & 4 \end{pmatrix}; \quad (2)\ A = \begin{pmatrix} & & & & 1 \\ & & & 1 & \\ & & \cdot^{\,\cdot^{\,\cdot}} & & \\ & 1 & & & \\ 1 & & & & \end{pmatrix}; \quad (3)\ A = \begin{pmatrix} 0 & I_n \\ I_n & 0 \end{pmatrix}.$$

9. Let $X = \begin{pmatrix} x_{11} & x_{12} & \cdots & x_{1n} \\ x_{21} & x_{22} & \cdots & x_{2n} \\ \vdots & \vdots & & \vdots \\ x_{n1} & x_{n2} & \cdots & x_{nn} \end{pmatrix}$. Find the positive index of inertia and the negative index of inertia of the quadratic form $f(x_{11}, x_{12}, \ldots, x_{nn}) = \mathrm{Tr}(X'X)$ in n^2 indeterminates.

10. Let A be an $n \times n$ real symmetric matrix with $|A| < 0$. Show that there is an n-dimensional real vector $X \neq 0$ such that $X'AX < 0$.

11. Under what condition for the real number t, the following quadratic forms are positive definite:

(1) $x_1^2 + x_2^2 + 5x_3^2 + 2tx_1x_2 - 2x_1x_3 + 4x_2x_3$;
(2) $x_1^2 + x_2^2 + x_3^2 + tx_1x_2 + 10x_1x_3 + 3x_2x_3$;
(3) $t(x_1^2 + x_2^2 + x_3^2) + 2x_1x_2 - 2x_1x_3 + 2x_2x_3$.

12. Under what condition for the real numbers a and b, the matrix

$$\begin{pmatrix} a & 1 & b \\ 1 & -1 & 0 \\ b & 0 & -1 \end{pmatrix}$$

is positive definite, negative definite, positive semi-definite, negative semi-definite, indefinite.

13. Find the matrices and canonical forms of the following real quadratic forms and judge whether they are positive definite, negative definite, positive semi-definite, semi-negative definite, indefinite:

    (1) $-x_1^2 - 2x_2^2 - 2x_3^2 + 2x_1 x_2$;

    (2) $-4x_1^2 - x_2^2 + 6x_1 x_3 - 2x_3^2 + 2x_2 x_3$;

    (3) $x_1^2 - 2x_1 x_2 + x_1 x_3 + 2x_2 x_3 + 2x_3^2 + 3x_3 x_1$;

    (4) $-x_1^2 - x_2^2 + 2x_1 x_3 + 4x_2 x_3 + 2x_3^2$;

    (5) $-x_1^2 + 2x_1 x_2 - 2x_2^2 + 2x_1 x_3 - 5x_3^2 + 4x_2 x_3$;

    (6) $2x_1^2 + 2x_1 x_2 + 2x_2^2 + 4x_3^2$;

    (7) $x_1^2 + 2x_2^2 + 3x_3^2 + 4x_1 x_2 + 2x_1 x_3 + 2x_2 x_3$.

14. Show that the quadratic form $f(x_1, x_2, \ldots, x_n) = n \sum\limits_{i=1}^{n} x_i^2 - \left(\sum\limits_{i=1}^{n} x_i \right)^2$ is positive semi-definite.

15. Assume that the real quadratic form

$$f(x_1, x_2, \ldots, x_n) = l_1^2 + \cdots + l_p^2 - l_{p+1}^2 - \cdots - l_{p+q}^2,$$

    where l_i is a homogeneous polynomial of degree 1 in $x_1, x_2, \ldots, x_n$. Show that the positive index of inertia of $f(x_1, x_2, \ldots, x_n)$ is less than or equal to p, and the negative index of inertia of $f(x_1, x_2, \ldots, x_n)$ is less than or equal to q.

16. Show that the inverse A^{-1} of a positive definite matrix A is also a positive definite matrix.

17. Show that any real symmetric matrix is the difference between two positive definite matrices.

18. Let A be an $n \times n$ real matrix. Show that

    (1) A is a skew-symmetric matrix if and only if $X'AX = 0$ for any n-dimensional column vector X;

    (2) if A is a symmetric matrix and $X'AX = 0$ for any n-dimensional column vector X, then $A = 0$.

19. Let $A = (a_{ij})_{n \times n}$ be a positive definite matrix. Show that

    (1) $f(y_1, y_2, \ldots, y_n) = \begin{vmatrix} A & Y \\ Y' & 0 \end{vmatrix}$ is negative definite, where $Y = (y_1, y_2, \ldots, y_n)'$;

(2) $|A| \leqslant a_{nn}P_{n-1}$, where P_{n-1} is the leading principal minor of order $n-1$ of A;

(3) $|A| \leqslant \prod\limits_{i=1}^{n} a_{ii}$;

(4) if $T = (t_{ij})_{n \times n}$ is an invertible matrix, then
$$|T|^2 \leqslant \prod_{i=1}^{n} (t_{1i}^2 + \cdots + t_{ni}^2).$$

20. Let $A \in \mathrm{GL}_n(F)$. Show that there is a lower triangular matrix L and an upper triangular matrix U whose diagonal elements are all 1 such that $A = LU$ if and only if all the leading principal minors of A are not 0, and L and U are unique in this decomposition. This decomposition is called the LU **decomposition** of the matrix A.

21. Let $A \in \mathrm{GL}_n(F)$. Show that A can be decomposed as $A = LDU$, where L is a lower triangular matrix whose diagonal elements are all 1, D is a diagonal matrix, U is an upper triangular matrix whose diagonal elements are all 1, if and only if all the leading principal minors of A are not 0, and L, D, and U are unique in this decomposition. This decomposition is called the LDU **decomposition** of the matrix A.

22. Let A be an $n \times n$ positive semi-definite matrix of rank 1. Show that there is an n-dimensional nonzero real vector α such that $A = \alpha\alpha'$.

23. Let A be an $n \times n$ skew-symmetric matrix over F. Show that there is an invertible matrix C over F such that
$$C'AC = \mathrm{diag}\left(\begin{pmatrix} 0 & 1 \\ -1 & 0 \end{pmatrix}, \ldots, \begin{pmatrix} 0 & 1 \\ -1 & 0 \end{pmatrix}, 0, \ldots, 0 \right).$$

24. Let A be an $n \times n$ skew-symmetric matrix over $\mathbb{Z}$. Show that there is an $m \in \mathbb{Z}$ such that $|A| = m^2$.

25. Let $A = (a_{ij})_{n \times n}$ be a positive semi-definite matrix and $a_{ii} = 0$ for some i with $1 \leqslant i \leqslant n$. Show that $a_{ij} = a_{ji} = 0$ for any $j \neq i$.

26. Let $A = (a_{ij})_{n \times n}$, where $a_{ij} = \min\{i, j\}, 1 \leqslant i, j \leqslant n$. Show that A is a positive definite matrix.

27. Let $A = (a_{ij})_{n \times n}$ be a positive definite matrix. Show that $B = \left(\dfrac{a_{ij}}{\sqrt{a_{ii}a_{jj}}} \right)_{n \times n}$ is also a positive definite matrix.

28.* Let $A = (x_{ij})_{n \times n}$ be a skew-symmetric matrix of even order. Show that there is a polynomial

$$f(x_{12}, x_{13}, \ldots, x_{1n}, x_{23}, \ldots, x_{2n}, \ldots, x_{n-1,n})$$

in the indeterminates x_{ij} $(i < j)$ such that

$$|A| = (f(x_{12}, x_{13}, \ldots, x_{1n}, x_{23}, \ldots, x_{2n}, \ldots, x_{n-1,n}))^2.$$

29.* Let $A = (a_{ij})_{n \times n}$, where $a_{ij} = (i, j)$ is the positive greatest common divisor of i and j, $1 \leqslant i, j \leqslant n$. Show that A is a positive definite matrix.

30.* Let $A = (a_{ij})_{n \times n}$, where $a_{ij} = \frac{1}{i+j}$, $1 \leqslant i, j \leqslant n$. Show that A is a positive definite matrix.

Chapter 8

Inner Product Spaces

8.1 Introduction

By definition, a linear space only has algebraic operations and lacks geometric concepts, such as the length of a vector and the angle between vectors. However, an inner product space, a linear space with an inner product, can provide geometric concepts, such as length and angle. All the linear spaces we encountered earlier are inner product spaces. The 3-dimensional space is a special inner product space that the ancient Greeks studied in-depth more than two thousand years ago, and they made significant mathematical achievements with it, such as the Pythagorean theorem. In 1637, René Descartes introduced analytic geometry, which transformed geometric problems of inner product spaces into algebraic problems, making it possible to study them using algebraic methods. The modern concept of inner product spaces was first introduced by Giuseppe Peano in 1898. In the 20th century, Hilbert and Banach spaces in functional analysis are special inner product spaces.

A finite-dimensional real linear space with an inner product is called a Euclidean space, and a finite-dimensional complex linear space with an inner product is called a unitary space. This chapter introduces these inner product spaces, with a focus on studying orthogonal transformations — linear transformations that preserve the lengths of vectors in Euclidean spaces. Rotations are studied using quaternions, and least squares solutions are studied through the Moore[1]–Penrose[2] inverse of matrices.

8.2 Euclidean Spaces

In this section, we give the definition and properties of Euclidean spaces.

Definition 8.2.1. Let V be a real linear space. A map

$$f : V \times V \longrightarrow \mathbb{R}$$

is called an **inner product** on V if it satisfies the following conditions:

(1) $f(\alpha, \beta) = f(\beta, \alpha)$,
(2) $f(\alpha + \beta, \gamma) = f(\alpha, \gamma) + f(\beta, \gamma)$,
(3) $f(k\alpha, \beta) = kf(\alpha, \beta)$,
(4) $f(\alpha, \alpha) \geqslant 0$, and the equality holds if and only if $\alpha = 0$,

where α, β, γ are arbitrary vectors of V, $k \in \mathbb{R}$.

For each ordered pair of vectors α, β of V, $f(\alpha, \beta)$ is called the **inner product** of α, β, abbreviated as (α, β).

A finite-dimensional real linear space with an inner product is called a **Euclidean space**.

Remark 8.2.2.

(1) Euclid was a great mathematician in ancient Greece in the 3rd and 4th centuries BCE and the author of the book "Elements". This book has been used as a mathematics textbook by many countries for a long time, and it can be called a miracle in the history of mathematics. During the Ming Dynasty, Chinese mathematician Xu[3] and Italian mathematician Ricci[4] collaborated to translate the first six volumes of this book into Chinese.

---

[1] Eliakim Hastings Moore, 1862–1932, American physicist.
[2] Roger Penrose, 1931 to present, English mathematician and 2020 Nobel Laureate in Physics.
[3] Guangqi Xu, 1562–1633, Chinese scientist, thinker, statesman, and military strategist in Ming Dynasty.
[4] Matteo Ricci, 1552–1610, Italian missionary and scholar.

When published, it was named "The Elements of Geometry", from which the name of Geometry in Chinese comes.

(2) In some literature, an infinite-dimensional real linear space with an inner product is also called a Euclidean space.

By Definition 5.2.1, a linear map σ from a real linear space V to another real linear space W satisfies $\sigma(k\alpha + l\beta) = k\sigma(\alpha) + l\sigma(\beta)$, where $\alpha, \beta \in V$, $k, l \in \mathbb{R}$. For a linear map between Euclidean spaces, we hope that the linear map σ can still maintain the inner products unchanged.

Definition 8.2.3. Let V and W be Euclidean spaces. A linear map $\sigma : V \longrightarrow W$ is called an **isometry** from V to W if $(\sigma(\alpha), \sigma(\beta)) = (\alpha, \beta)$ for any $\alpha, \beta \in V$.

By Definition 8.2.1 (4), an isometry $\sigma : V \longrightarrow W$ is an injection.

Definition 8.2.4. Let V and W be Euclidean spaces. An isometry $\sigma : V \longrightarrow W$ is called an **isomorphism** from V to W if σ is a surjection. Two Euclidean spaces V and W are said to be **isomorphic** if there is an isomorphism from V to W.

Let V and W be Euclidean spaces. By definition, $\sigma : V \longrightarrow W$ is an isomorphism of Euclidean spaces if and only if $\sigma : V \longrightarrow W$ is an isomorphism of linear spaces and $(\sigma(\alpha), \sigma(\beta)) = (\alpha, \beta)$ for any $\alpha, \beta \in V$.

Example 8.2.5. Let $\alpha = (x_1, x_2, \ldots, x_n)'$, $\beta = (y_1, y_2, \ldots, y_n)' \in \mathbb{R}^n$. We define the inner product on $\mathbb{R}^n$ as

$$(\alpha, \beta) = \alpha'\beta = (x_1, x_2, \ldots, x_n) \begin{pmatrix} y_1 \\ y_2 \\ \vdots \\ y_n \end{pmatrix} = x_1 y_1 + x_2 y_2 + \cdots + x_n y_n.$$

$$(8.1)$$

In the following, unless specified otherwise, the inner product on the Euclidean space $\mathbb{R}^n$ will refer to the inner product defined by Equation (8.1).

The Cauchy–Bunyakovsky–Schwarz inequality is well known (see Example 2.11.6). This inequality can be generalized to Euclidean spaces.

Theorem 8.2.6. *Let α, β be two vectors of a Euclidean space V. Then*

$$(\alpha, \beta)^2 \leqslant (\alpha, \alpha)(\beta, \beta),$$

and the equality holds if and only if α, β are linearly dependent.

Proof. If one of α, β is zero, then the theorem obviously holds.

Assume neither of α, β is zero. Let $f(t) = (\alpha, \alpha)t^2 + 2(\alpha, \beta)t + (\beta, \beta)$. Then $f(t)$ is a quadratic polynomial with a positive leading coefficient. For any $t \in \mathbb{R}$, by Definition 8.2.1, $f(t) = (t\alpha+\beta, t\alpha+\beta) \geqslant 0$. Hence the discriminant of $f(t)$ is $\Delta = 4(\alpha, \beta)^2 - 4(\alpha, \alpha)(\beta, \beta) \leqslant 0$. So $(\alpha, \beta)^2 \leqslant (\alpha, \alpha)(\beta, \beta)$, and the equality holds if and only if there is t such that $t\alpha + \beta = 0$, i.e., α, β are linearly dependent. $\square$

Definition 8.2.7. Let V be a Euclidean space and $\alpha \in V$. We define $|\alpha| = \sqrt{(\alpha, \alpha)}$ as the **length**, or **norm** of α. A vector of length 1 is called a **unit vector**.

By Theorem 8.2.6, $|(\alpha, \beta)| \leqslant |\alpha||\beta|$ for any $\alpha, \beta \in V$. So

$$\begin{aligned}
(\alpha + \beta, \alpha + \beta) &= (\alpha, \alpha) + (\beta, \beta) + 2(\alpha, \beta) \\
&\leqslant (\alpha, \alpha) + (\beta, \beta) + 2|\alpha||\beta| \\
&= |\alpha|^2 + |\beta|^2 + 2|\alpha||\beta| \\
&= (|\alpha| + |\beta|)^2.
\end{aligned}$$

After taking the square root, the above formula becomes $|\alpha + \beta| \leqslant |\alpha| + |\beta|$, which is the triangle inequality we are familiar with.

Definition 8.2.8. Let α and β be nonzero vectors of V. The **angle** between α and β is defines as

$$\theta = \arccos \frac{(\alpha, \beta)}{|\alpha||\beta|}.$$

Definition 8.2.9. Let $\alpha_1, \alpha_2, \ldots, \alpha_n$ be a basis for a Euclidean space V. We define the **metric matrix of V with respect to the basis**

$\alpha_1, \alpha_2, \ldots, \alpha_n$ as

$$\begin{pmatrix} (\alpha_1, \alpha_1) & (\alpha_1, \alpha_2) & \cdots & (\alpha_1, \alpha_n) \\ (\alpha_2, \alpha_1) & (\alpha_2, \alpha_2) & \cdots & (\alpha_2, \alpha_n) \\ \vdots & \vdots & & \vdots \\ (\alpha_n, \alpha_1) & (\alpha_n, \alpha_2) & \cdots & (\alpha_n, \alpha_n) \end{pmatrix}.$$

The metric matrix is also called the **Gram**[5] **matrix**

Proposition 8.2.10. *Let $\alpha_1, \alpha_2, \ldots, \alpha_n$ be a basis for a Euclidean space V. Then the metric matrix A of V with respect to the basis $\alpha_1, \alpha_2, \ldots, \alpha_n$ is a positive definite matrix. Suppose that the metric matrix of V with respect to another basis $\beta_1, \beta_2, \ldots, \beta_n$ for V is B. If the transition matrix from $\alpha_1, \alpha_2, \ldots, \alpha_n$ to $\beta_1, \beta_2, \ldots, \beta_n$ is C, then $B = C'AC$.*

Proof. Let $0 \neq X = (c_1, c_2, \ldots, c_n)' \in \mathbb{R}^n$ and $\gamma = c_1\alpha_1 + c_2\alpha_2 + \cdots + c_n\alpha_n \in V$. Then $\gamma \neq 0$. Hence $X'AX = (\gamma, \gamma) > 0$. So A is a positive definite matrix.

From the given condition, we have $(\beta_1, \ldots, \beta_n) = (\alpha_1, \ldots, \alpha_n)C$. Let $C = (c_{ij})$. Then $\beta_i = \sum_{k=1}^{n} c_{ki}\alpha_k$, $1 \leqslant i \leqslant n$. Hence the (i, j) entry of B is

$$(\beta_i, \beta_j) = \left(\sum_{k=1}^{n} c_{ki}\alpha_k, \sum_{l=1}^{n} c_{lj}\alpha_l \right) = (c_{1i}, c_{2i}, \ldots, c_{ni})A \begin{pmatrix} c_{1j} \\ c_{2j} \\ \vdots \\ c_{nj} \end{pmatrix}.$$

So, $B = C'AC$. $\qquad\square$

Definition 8.2.11. Let V be a Euclidean space and $\alpha, \beta \in V$. If $(\alpha, \beta) = 0$, then we say that α and β are **orthogonal**, or α is orthogonal to β, denoted by $\alpha \perp \beta$. A sequence of nonzero vectors of V is said to be a **sequence of orthogonal vectors** if all vectors in the sequence are mutually orthogonal.

Let $\alpha \in V$ and W be a subspace of V. If $(\alpha, \beta) = 0$ for any $\beta \in W$, then we say α is orthogonal to W, denoted by $\alpha \perp W$. Let V_1 and

---

[5] Jørgen Pedersen Gram, 1850–1916, Danish mathematician.

V_2 be subspaces of V. If $(\alpha_1, \alpha_2) = 0$ for any $\alpha_1 \in V_1$ and $\alpha_2 \in V_2$, then we say that V_1 is orthogonal to V_2, denoted by $V_1 \perp V_2$.

Proposition 8.2.12. *Let V_1 and V_2 be subspaces of V. If $V_1 \perp V_2$, then $V_1 + V_2$ is a direct sum.*

Proof. Let $\alpha \in V_1 \cap V_2$. Since $V_1 \perp V_2$, we have $(\alpha, \alpha) = 0$ which implies that $\alpha = 0$. So $V_1 + V_2$ is a direct sum. $\square$

Proposition 8.2.13. *Let $\alpha_1, \alpha_2, \ldots, \alpha_k$ be a sequence of orthogonal vectors. Then $\alpha_1, \alpha_2, \ldots, \alpha_k$ are linearly independent.*

Proof. Let c_1, c_2, $\ldots$, c_k be real numbers such that $\sum\limits_{i=1}^{k} c_i \alpha_i = 0$. Then

$$0 = \left(\sum_{i=1}^{k} c_i \alpha_i, \sum_{i=1}^{k} c_i \alpha_i \right) = \sum_{i=1}^{k} c_i^2 (\alpha_i, \alpha_i),$$

which implies that $c_1 = c_2 = \cdots = c_k = 0$. Hence $\alpha_1, \alpha_2, \ldots, \alpha_k$ are linearly independent. $\square$

Definition 8.2.14. Let V be an n-dimensional Euclidean space.
If $\alpha_1, \alpha_2, \ldots, \alpha_k$ is a sequence of vectors of V such that $(\alpha_i, \alpha_j) = \delta_{ij}$ for $1 \leqslant i, j \leqslant k$, then we call the sequence $\alpha_1, \alpha_2, \ldots, \alpha_k$ a **sequence of orthonormal vectors** of V.
If $e_1, e_2, \ldots, e_n$ is a sequence of orthonormal (respectively, orthogonal) vectors of V, then we call the sequence $e_1, e_2, \ldots, e_n$ an **orthonormal (respectively, orthogonal) basis** for V.

Example 8.2.15. Let $n = 2$ in Example 8.2.5. Then the sequence $\begin{pmatrix} \cos \theta \\ \sin \theta \end{pmatrix}$, $\begin{pmatrix} \sin \theta \\ -\cos \theta \end{pmatrix}$ is an orthonormal basis for V, where $\theta \in \mathbb{R}$.

By Proposition 8.2.10, the metric matrix A of a Euclidean space V with respect to a basis $\alpha_1, \alpha_2, \ldots, \alpha_n$ for V is a positive definite matrix, so there is an $n \times n$ invertible real matrix C such that $C'AC = I_n$. Let $(\beta_1, \beta_2, \ldots, \beta_n) = (\alpha_1, \alpha_2, \ldots, \alpha_n)C$. Then the sequence $\beta_1, \beta_2, \ldots, \beta_n$ is a basis for V. By Proposition 8.2.10, the metric matrix of V with respect to the basis $\beta_1, \beta_2, \ldots, \beta_n$ is the identity matrix. Hence the sequence $\beta_1, \beta_2, \ldots, \beta_n$ is an orthonormal basis for V. The following theorem tells us that the transition matrix C can be chosen as an upper triangular matrix.

Theorem 8.2.16. *Let $\alpha_1, \alpha_2, \ldots, \alpha_n$ be linearly independent vectors of a Euclidean space V and $W = L(\alpha_1, \alpha_2, \ldots, \alpha_n)$. Then there is an orthonormal basis $e_1, e_2, \ldots, e_n$ for W such that $L(e_1, e_2, \ldots, e_m) = L(\alpha_1, \alpha_2, \ldots, \alpha_m)$ for $1 \leqslant m \leqslant n$.*

Proof. We first construct an orthogonal basis $\beta_1, \beta_2, \ldots, \beta_n$ for W such that

$$L(\beta_1, \beta_2, \ldots, \beta_m) = L(\alpha_1, \alpha_2, \ldots, \alpha_m), \ 1 \leqslant m \leqslant n. \qquad (8.2)$$

To achieve this aim, let

$$\beta_1 = \alpha_1,$$

$$\beta_2 = \alpha_2 - \frac{(\alpha_2, \beta_1)}{(\beta_1, \beta_1)}\beta_1,$$

$$\beta_3 = \alpha_3 - \frac{(\alpha_3, \beta_1)}{(\beta_1, \beta_1)}\beta_1 - \frac{(\alpha_3, \beta_2)}{(\beta_2, \beta_2)}\beta_2,$$

$$\vdots \qquad \vdots$$

$$\beta_m = \alpha_m - \frac{(\alpha_m, \beta_1)}{(\beta_1, \beta_1)}\beta_1 - \cdots - \frac{(\alpha_m, \beta_{m-1})}{(\beta_{m-1}, \beta_{m-1})}\beta_{m-1},$$

$$\vdots \qquad \vdots$$

$$\beta_n = \alpha_n - \frac{(\alpha_n, \beta_1)}{(\beta_1, \beta_1)}\beta_1 - \cdots - \frac{(\alpha_n, \beta_{n-1})}{(\beta_{n-1}, \beta_{n-1})}\beta_{n-1}.$$

It is easy to verify that the sequence $\beta_1, \beta_2, \ldots, \beta_n$ is an orthogonal basis for W and satisfies equation (8.2).

Let $e_m = \frac{\beta_m}{|\beta_m|}$, $1 \leqslant m \leqslant n$. Then the sequence $e_1, e_2, \ldots, e_n$ is an orthonormal basis for W such that $L(e_1, e_2, \ldots, e_m) = L(\alpha_1, \alpha_2, \ldots, \alpha_m)$ for $1 \leqslant m \leqslant n$. $\qquad \square$

In the proof of the above theorem, the process of finding the orthonormal vectors is called the **Gram–Schmidt**[6] **orthonormalization process**, or abbreviated as **Schmidt orthonormalization**.

---

[6]Erhard Schmidt, 1876–1959, German mathematician.

Remark 8.2.17.

(1) In the Gram–Schmidt orthonormalization process, the method for transforming a sequence of linearly independent vectors into a sequence of orthonormal vectors was published by Schmidt in 1907. This method is essentially the same as the one Gram used in a paper published in 1883.

(2) The Schmidt orthonormalization can also be executed by performing orthogonalization followed by normalization at each step:

$$
\begin{cases}
\beta_1 = \alpha_1, & e_1 = \dfrac{\beta_1}{|\beta_1|}; \\[2ex]
\beta_2 = \alpha_2 - (\alpha_2, e_1)e_1, & e_2 = \dfrac{\beta_2}{|\beta_2|}; \\[2ex]
\quad\vdots \qquad\qquad \vdots & \qquad \vdots \\[2ex]
\beta_n = \alpha_n - (\alpha_n, e_1)e_1 - \cdots - (\alpha_n, e_{n-1})e_{n-1}, & e_n = \dfrac{\beta_n}{|\beta_n|}.
\end{cases}
$$

Example 8.2.18. Let $\alpha_1 = \begin{pmatrix} 1 \\ 1 \\ 1 \end{pmatrix}$, $\alpha_2 = \begin{pmatrix} 1 \\ 0 \\ 2 \end{pmatrix}$, $\alpha_3 = \begin{pmatrix} 0 \\ 0 \\ 1 \end{pmatrix} \in \mathbb{R}^3$.

Show that the sequence $\alpha_1, \alpha_2, \alpha_3$ is a basis for $\mathbb{R}^3$, and use Schmidt orthonormalization to make it an orthonormal basis for $\mathbb{R}^3$.

Proof. By $\begin{vmatrix} 1 & 1 & 0 \\ 1 & 0 & 0 \\ 1 & 2 & 1 \end{vmatrix} = -1$, one can see that the sequence $\alpha_1, \alpha_2, \alpha_3$ is a basis for $\mathbb{R}^3$.

We first perform the orthogonalization:

$$
\beta_1 = \alpha_1 = \begin{pmatrix} 1 \\ 1 \\ 1 \end{pmatrix}, \quad \beta_2 = \alpha_2 - \frac{(\alpha_2, \beta_1)}{(\beta_1, \beta_1)}\beta_1 = \begin{pmatrix} 0 \\ -1 \\ 1 \end{pmatrix},
$$

$$
\beta_3 = \alpha_3 - \frac{(\alpha_3, \beta_1)}{(\beta_1, \beta_1)}\beta_1 - \frac{(\alpha_3, \beta_2)}{(\beta_2, \beta_2)}\beta_2 = \begin{pmatrix} -\dfrac{1}{3} \\[1ex] \dfrac{1}{6} \\[1ex] \dfrac{1}{6} \end{pmatrix}.
$$

Then we perform the normalization:

$$e_1 = \frac{\beta_1}{|\beta_1|} = \begin{pmatrix} \frac{1}{\sqrt{3}} \\ \frac{1}{\sqrt{3}} \\ \frac{1}{\sqrt{3}} \end{pmatrix}, \quad e_2 = \frac{\beta_2}{|\beta_2|} = \begin{pmatrix} 0 \\ -\frac{1}{\sqrt{2}} \\ \frac{1}{\sqrt{2}} \end{pmatrix}, \quad e_3 = \frac{\beta_3}{|\beta_3|} = \begin{pmatrix} -\frac{2}{\sqrt{6}} \\ \frac{1}{\sqrt{6}} \\ \frac{1}{\sqrt{6}} \end{pmatrix}.$$

Hence, the sequence e_1, e_2, e_3 is an orthonormal basis for $\mathbb{R}^3$. $\qquad\square$

Theorem 8.2.19. *Let V and W be Euclidean spaces. Then V is isomorphic to W if and only if* $\dim V = \dim W$.

Proof. Necessity. If V is isomorphic to W, then $\dim V = \dim W$ by Theorem 5.2.11.

Sufficiency. Assume that $\dim V = \dim W = n$, and let $\alpha_1, \alpha_2, \ldots, \alpha_n$ and $\beta_1, \beta_2, \ldots, \beta_n$ be orthonormal bases for V and W, respectively. We define the linear map $\sigma : V \longrightarrow W$ such that $\sigma(\alpha_i) = \beta_i$, $i = 1, 2, \ldots, n$. It is obvious that σ is an isomorphism from the Euclidean space V to the Euclidean space W. Hence, V is isomorphic to W. $\qquad\square$

The above theorem shows that Euclidean spaces with the same dimension are all isomorphic. In particular, all n-dimensional Euclidean spaces are isomorphic to $\mathbb{R}^n$.

Definition 8.2.20. Let W be a subspace of a Euclidean space V. The **orthogonal complement** of W is defined as

$$W^\perp = \{x \in V \mid (x, \alpha) = 0 \text{ for all } \alpha \in W\}.$$

It is obvious that the orthogonal complement $W^\perp$ of a subspace W of V is also a subspace of V, and W is orthogonal to $W^\perp$.

Theorem 8.2.21. *Let U and W be subspaces of a Euclidean space V. Then the following statements hold:*

(1) $V^\perp = 0$, $0^\perp = V$.
(2) *If* $U \subseteq W$, *then* $W^\perp \subseteq U^\perp$.
(3) $V = W \oplus W^\perp$.
(4) $W = (W^\perp)^\perp$.

(5) $(U + W)^\perp = U^\perp \cap W^\perp$.
(6) $(U \cap W)^\perp = U^\perp + W^\perp$.

Proof.   It is easy to see that (1) and (2) hold by Definition 8.2.20.
(3) If $W = 0$ or $W = V$, then (3) holds by (1).

Next, we assume that W is a nontrivial subspace of dimension k.
To prove that $V = W \oplus W^\perp$, we only need to prove that $V = W + W^\perp$
by Proposition 8.2.12. For any $\alpha \in V$ and an orthonormal basis
$\alpha_1, \alpha_2, \ldots, \alpha_k$ for W, let $x_1 = (\alpha, \alpha_1)$, $x_2 = (\alpha, \alpha_2)$, $\ldots$, $x_k = (\alpha, \alpha_k)$.
Then

$$
\begin{aligned}
(\alpha - x_1\alpha_1 - x_2\alpha_2 - \cdots - x_k\alpha_k, \alpha_i) & \\
&= (\alpha, \alpha_i) - x_1(\alpha_1, \alpha_i) - x_2(\alpha_2, \alpha_i) - \cdots - x_k(\alpha_k, \alpha_i) \\
&= (\alpha, \alpha_i) - x_i(\alpha_i, \alpha_i) \\
&= x_i - x_i \\
&= 0
\end{aligned}
$$

for $1 \leqslant i \leqslant k$. Let $\beta = \alpha - x_1\alpha_1 - x_2\alpha_2 - \cdots - x_k\alpha_k$. Then β is
orthogonal to all $\alpha_1, \alpha_2, \ldots, \alpha_k$, and hence $\beta \in W^\perp$. Since $\alpha - \beta \in W$,
$\alpha = (\alpha - \beta) + \beta \in W + W^\perp$. Thus $V \subseteq W + W^\perp \subseteq V$, and so
$V = W + W^\perp$.

(4) By Definition 8.2.20, $W \subseteq (W^\perp)^\perp$. By (3), $\dim W = \dim(W^\perp)^\perp$. So $W = (W^\perp)^\perp$.

(5) It is easy to see that the statement holds by Definition 8.2.20.
(6) comes from (4) and (5). $\square$

Remark 8.2.22. For an infinite-dimensional inner product space,
Theorem 8.2.21 does not hold in general. For example, let $V = \mathbb{R}[x]$
be the infinite-dimensional real linear space consisting of polynomials
with real coefficients. Then it is obvious that

$$
\left(\sum_{i=0}^{n} a_i x^i, \sum_{j=0}^{m} b_j x^j \right) = \sum_{k=0}^{\min(m,n)} a_k b_k
$$

is an inner product on V and $W = \{ f(x) \in V \mid f(1) = 0 \}$ is a nontrivial subspace of V. Assume that $0 \neq g(x) = \sum_{j=0}^{m} b_j x^j \in W^\perp$,
where $b_m \neq 0$. Let $f(x) = x^{m+1} - x^m$. Then $f(x) \in W$. Since

$(g(x), f(x)) = 0$, $b_m = 0$, a contradiction! Hence $W^\perp = 0$, which implies that $V \neq W \oplus W^\perp$.

Definition 8.2.23. Let W be a subspace of a Euclidean space V. If a linear map $\sigma : V \longrightarrow W$ satisfies $\sigma|_W = 1_W$ and $\sigma|_{W^\perp} = 0$, then we call σ an **orthogonal projection** from V to W.

By Theorem 8.2.21, the orthogonal projection from V to W is uniquely determined by W. For example, let $W = \{(t, t)| \ t \in \mathbb{R}\}$. Then W is a subspace of the Euclidean space $\mathbb{R}^2$, and the orthogonal projection from V to W is

$$\sigma : V \longrightarrow W, \ (x, y) \mapsto \left(\frac{x + y}{2}, \frac{x + y}{2} \right).$$

8.3 Orthogonal Matrices and Orthogonal Transformations

The previous section discussed the method of finding orthonormal bases, and now we are discussing the transition matrix from an orthonormal basis to another orthonormal basis.

Let $e_1, e_2, \ldots, e_n$ and $f_1, f_2, \ldots, f_n$ be two orthonormal bases for V. The transition matrix from $e_1, e_2, \ldots, e_n$ to $f_1, f_2, \ldots, f_n$ is $A = (a_{ij})_{n \times n}$, i.e., $(f_1, f_2, \ldots, f_n) = (e_1, e_2, \ldots, e_n)A$. So $(f_i, f_j) = \sum_{k=1}^{n} a_{ki} a_{kj}, 1 \leqslant i, j \leqslant n$. Since $f_1, f_2, \ldots, f_n$ is an orthonormal basis, $(f_i, f_j) = \delta_{ij}$. It follows that $\sum_{k=1}^{n} a_{ki} a_{kj} = \delta_{ij}, 1 \leqslant i, j \leqslant n$, i.e., $A'A = I_n$. Therefore, we introduce the following definition.

Definition 8.3.1. Let $A \in \mathrm{M}_n(\mathbb{R})$. If $A'A = I_n$, then we call A an **orthogonal matrix**. The set of all $n \times n$ orthogonal matrices is denoted by $\mathrm{O}(n)$, called the **orthogonal group** of order n.

It is easy to see that the determinant of an orthogonal matrix is equal to ± 1, the transition matrix from an orthonormal basis to another orthonormal basis is orthogonal, and the column (row) vectors of an $n \times n$ orthogonal matrix form an orthonormal basis for $\mathbb{R}^n$.

By Definition 5.2.1, a linear transformation σ of a real linear space V satisfies $\sigma(k\alpha + l\beta) = k\sigma(\alpha) + l\sigma(\beta)$, where $\alpha, \beta \in V, \ k, l \in \mathbb{R}$.

For a Euclidean space V, we hope that the linear transformation σ also preserves the inner product, so we have the following.

Definition 8.3.2. Let V be an n-dimensional Euclidean space, $\sigma \in \text{End}(V)$. If $(\alpha, \beta) = (\sigma(\alpha), \sigma(\beta))$ for any $\alpha, \beta \in V$, then we call σ an **orthogonal transformation** of V.

In the following, we give several equivalent characterizations of an orthogonal transformation.

Theorem 8.3.3. *Let V be an n-dimensional Euclidean space and $\sigma \in End(V)$. Then the following statements are equivalent:*

(1) *σ is an orthogonal transformation.*
(2) *$|\sigma(\alpha)| = |\alpha|$ for any $\alpha \in V$.*
(3) *The matrix of σ with respect to an orthonormal basis for V is an orthogonal matrix.*
(4) *σ transforms an orthonormal basis for V to another orthonormal basis for V.*

Proof. (1) $\Longrightarrow$ (2). Since σ is an orthogonal transformation, $(\sigma(\alpha), \sigma(\alpha)) = (\alpha, \alpha)$ for any $\alpha \in V$ by Definition 8.3.2. Hence $|\sigma(\alpha)| = |\alpha|$.

(2) $\Longrightarrow$ (1). Assume that $|\sigma(\alpha)| = |\alpha|$ for any $\alpha \in V$. By Definition 8.2.1, for any $\alpha, \beta \in V$,

$$(\alpha, \beta) = \frac{1}{2}((\alpha + \beta, \alpha + \beta) - (\alpha, \alpha) - (\beta, \beta))$$

$$= \frac{1}{2}(|\alpha + \beta|^2 - |\alpha|^2 - |\beta|^2).$$

Similarly, $(\sigma(\alpha), \sigma(\beta)) = \frac{1}{2}(|\sigma(\alpha + \beta)|^2 - |\sigma(\alpha)|^2 - |\sigma(\beta)|^2)$. Since

$$|\sigma(\alpha + \beta)| = |\alpha + \beta|, |\sigma(\alpha)| = |\alpha|, |\sigma(\beta)| = |\beta|,$$

we have $(\alpha, \beta) = (\sigma(\alpha), \sigma(\beta))$ for any $\alpha, \beta \in V$. Hence, σ is an orthogonal transformation.

(1) $\Longleftrightarrow$ (3). Let $A = (a_{ij})_{n \times n}$ be the matrix of σ with respect to the orthonormal basis $e_1, e_2, \ldots, e_n$ for V. Then $\sigma(e_1, e_2, \ldots, e_n) = (e_1, e_2, \ldots, e_n)A$. Thus, for any $1 \leqslant i \leqslant n$, $\sigma(e_i) = \sum_{k=1}^{n} a_{ki} e_k$, and

hence

$$(\sigma(e_i), \sigma(e_j)) = \left(\sum_{k=1}^{n} a_{ki} e_k, \sum_{l=1}^{n} a_{lj} e_l \right) = \sum_{k,l=1}^{n} a_{ki} a_{lj} (e_k, e_l)$$

$$= \sum_{k,l=1}^{n} a_{ki} a_{lj} \delta_{kl} = \sum_{k=1}^{n} a_{ki} a_{kj}, 1 \leqslant i, j \leqslant n.$$

Let $A'A = (c_{ij})$. Then $c_{ij} = \sum_{k=1}^{n} a_{ki} a_{kj}$. Hence

$$(\sigma(e_i), \sigma(e_j)) = c_{ij}, 1 \leqslant i, j \leqslant n. \tag{8.3}$$

If σ is an orthogonal transformation, then $(\sigma(e_i), \sigma(e_j)) = (e_i, e_j) = \delta_{ij}, 1 \leqslant i, j \leqslant n$. By (8.3), $c_{ij} = \delta_{ij}$. Hence $A'A = I_n$, i.e., A is an orthogonal matrix.

If A is an orthogonal matrix, then $c_{ij} = \sum_{k=1}^{n} a_{ki} a_{kj} = \delta_{ij}$ since $A'A = I_n$. By (8.3), we have $(\sigma(e_i), \sigma(e_j)) = (e_i, e_j), 1 \leqslant i, j \leqslant n$. It follows that $(\sigma(\alpha), \sigma(\beta)) = (\alpha, \beta)$ for any $\alpha, \beta \in V$, i.e., σ is an orthogonal transformation of V.

(1) $\Longrightarrow$ (4). Let $e_1, e_2, \ldots, e_n$ be an orthonormal basis for V. Since σ is an orthogonal transformation of V by (1),

$$(\sigma(e_i), \sigma(e_j)) = (e_i, e_j) = \delta_{ij}, 1 \leqslant i, j \leqslant n.$$

Hence, $\sigma(e_1), \sigma(e_2), \ldots, \sigma(e_n)$ is an orthonormal basis for V.

(4) $\Longrightarrow$ (3). Let A be the matrix of σ with respect to an orthonormal basis $e_1, e_2, \ldots, e_n$ for V. Then

$$(\sigma(e_1), \sigma(e_2), \ldots, \sigma(e_n)) = (e_1, e_2, \ldots, e_n)A.$$

By (4), the sequence $\sigma(e_1), \sigma(e_2), \ldots, \sigma(e_n)$ is an orthonormal basis for V, so A is the transition matrix from the orthonormal basis $e_1, e_2, \ldots, e_n$ to the orthonormal basis $\sigma(e_1), \sigma(e_2), \ldots, \sigma(e_n)$. Hence, A is an orthogonal matrix. $\qquad \square$

Remark 8.3.4. Let V be an n-dimensional Euclidean space and $\sigma \in \text{End}(V)$. By definition and Theorem 8.3.3, σ is an orthogonal transformation of V if and only if σ is an automorphism of V (i.e., σ is an isomorphism from V to itself).

Definition 8.3.5. An orthogonal matrix with determinant 1 is called a **special orthogonal matrix**. The set of all $n \times n$ special orthogonal matrices is denoted by $\mathrm{SO}(n)$, called **special orthogonal group** of order n. An orthogonal transformation with determinant 1 is called a **rotation**.

Obviously, an orthogonal transformation of a Euclidean space is a rotation if and only if the matrix of the orthogonal transformation with respect to an orthonormal basis is a special orthogonal matrix.

Theorem 8.3.6. *Let σ be an orthogonal transformation of a Euclidean space V. If W is a σ-subspace of V, then the orthogonal complement $W^{\perp}$ of W is also a σ-subspace of V.*

Proof. For any $\alpha \in W^{\perp}$, we prove $(\sigma(\alpha), \beta) = 0$ for any $\beta \in W$. Since σ is an isomorphism of V, it is also an isomorphism of W when it is restricted to the σ-subspace W. Thus there is a $\gamma \in W$ such that $\sigma(\gamma) = \beta$, and hence $(\sigma(\alpha), \beta) = (\sigma(\alpha), \sigma(\gamma)) = (\alpha, \gamma)$. Since $\alpha \in W^{\perp}$ and $\gamma \in W$, we have $(\alpha, \gamma) = 0$. Therefore, $(\sigma(\alpha), \beta) = 0$, and so $\sigma(\alpha) \in W^{\perp}$. Hence $W^{\perp}$ is also a σ-subspace of V. $\qquad\square$

Note that if σ is not an orthogonal transformation, then the orthogonal complement of a σ-subspace is not necessarily a σ-subspace (see Exercise 7).

Proposition 8.3.7. *Let $A \in \mathrm{SO}(2)$. Then there is a θ with $0 \leqslant \theta < 2\pi$ such that $A = \begin{pmatrix} \cos\theta & -\sin\theta \\ \sin\theta & \cos\theta \end{pmatrix}$. If $A \in O(2)$ with $|A| = -1$, then there is a θ with $0 \leqslant \theta < 2\pi$ such that $A = \begin{pmatrix} \cos\theta & \sin\theta \\ \sin\theta & -\cos\theta \end{pmatrix}$.*

Proof. Assume that $A = \begin{pmatrix} a & b \\ c & d \end{pmatrix}$. From $A'A = I_2$, we have $a^2 + c^2 = 1$, $b^2 + d^2 = 1$ and $ab + cd = 0$.

Since $a^2 + c^2 = 1$, there is a θ such that $a = \cos\theta$, $c = \sin\theta$, $0 \leqslant \theta < 2\pi$. Similarly, there is a φ such that $b = \cos\varphi$, $d = \sin\varphi$, $0 \leqslant \varphi < 2\pi$. Since $ab + cd = 0$, we have $\cos\theta\cos\varphi + \sin\theta\sin\varphi = \cos(\theta - \varphi) = 0$. So $\theta - \varphi = \pm\frac{\pi}{2}$ or $\pm\frac{3\pi}{2}$.

If $\theta - \varphi = -\frac{\pi}{2}$ or $\frac{3\pi}{2}$, then $A = \begin{pmatrix} \cos\theta & -\sin\theta \\ \sin\theta & \cos\theta \end{pmatrix}$, $|A| = 1$.

If $\theta - \varphi = \frac{\pi}{2}$ or $-\frac{3\pi}{2}$, then $A = \begin{pmatrix} \cos\theta & \sin\theta \\ \sin\theta & -\cos\theta \end{pmatrix}$, $|A| = -1$. $\qquad\square$

Definition 8.3.8. Let $A, B \in \mathrm{M}_n(\mathbb{R})$. Then A and B are said to be **orthogonally similar** if there is a $C \in \mathrm{O}(n)$ such that $A = C^{-1}BC$.

If A and B are orthogonally similar, then A and B are both equivalent and congruent.

Example 8.3.9. Let $A \in \mathrm{O}(2)$ with $|A| = -1$. Then A is orthogonally similar to $\begin{pmatrix} 1 & 0 \\ 0 & -1 \end{pmatrix}$.

Proof. By Proposition 8.3.7, there is a θ with $0 \leqslant \theta < 2\pi$ such that $A = A(\theta) = \begin{pmatrix} \cos\theta & \sin\theta \\ \sin\theta & -\cos\theta \end{pmatrix}$. Note that the characteristic polynomial of $A(\theta)$ is $\lambda^2 - 1$, and so the eigenvalues of $A(\theta)$ are 1 and -1. It is obvious that $A(0) = \begin{pmatrix} 1 & 0 \\ 0 & -1 \end{pmatrix}$ and $A(\pi) = \begin{pmatrix} -1 & 0 \\ 0 & 1 \end{pmatrix}$ are all diagonal.

If $\theta \neq 0, \pi$, then by solving the system of linear equations
$$\begin{pmatrix} 1 - \cos\theta & -\sin\theta \\ -\sin\theta & 1 + \cos\theta \end{pmatrix} \begin{pmatrix} x_1 \\ x_2 \end{pmatrix} = \begin{pmatrix} 0 \\ 0 \end{pmatrix},$$
we see that $\alpha = \begin{pmatrix} 1 + \cos\theta \\ \sin\theta \end{pmatrix}$ is an eigenvector associated with the eigenvalue 1. Similarly, we can see that $\beta = \begin{pmatrix} 1 - \cos\theta \\ -\sin\theta \end{pmatrix}$ is an eigenvector associated with the eigenvalue -1. Since $(1+\cos\theta)(1-\cos\theta) - \sin^2\theta = 0$, α, β are orthogonal vectors. Let
$$C(\theta) = \left(\frac{\alpha}{|\alpha|}, \frac{\beta}{|\beta|} \right) = \begin{pmatrix} \dfrac{1 + \cos\theta}{\sqrt{2 + 2\cos\theta}} & \dfrac{1 - \cos\theta}{\sqrt{2 - 2\cos\theta}} \\ \dfrac{\sin\theta}{\sqrt{2 + 2\cos\theta}} & \dfrac{-\sin\theta}{\sqrt{2 - 2\cos\theta}} \end{pmatrix}, \quad \theta \neq 0, \pi.$$
Then $C(\theta)$ is an orthogonal matrix, and $C(\theta)^{-1}A(\theta)C(\theta) = \begin{pmatrix} 1 & 0 \\ 0 & -1 \end{pmatrix}$, $\theta \neq 0, \pi$, so $A(\theta)$ is orthogonally similar to $\begin{pmatrix} 1 & 0 \\ 0 & -1 \end{pmatrix}$. $\qquad\square$

Proposition 8.3.10. *Let* $A = (a_{ij})_{n\times n}, B = (b_{ij})_{n\times n} \in \mathrm{M}_n(\mathbb{R})$, *and let* A *be orthogonally similar to* B. *Then*
$$\sum_{i,j=1}^{n} a_{ij}^2 = \sum_{i,j=1}^{n} b_{ij}^2.$$

Proof. Note that $\mathrm{Tr}(AA') = \sum\limits_{i,j=1}^{n} a_{ij}^2$ and $\mathrm{Tr}(BB') = \sum\limits_{i,j=1}^{n} b_{ij}^2$. If there is an orthogonal matrix T such that $T^{-1}AT = B$, then $\mathrm{Tr}(BB') = \mathrm{Tr}(T^{-1}AA'T) = \mathrm{Tr}(AA')$. Hence the proposition holds.

$\square$

Example 8.3.11. Let $A = \begin{pmatrix} 3 & 1 \\ -2 & 0 \end{pmatrix}$ and $B = \begin{pmatrix} 1 & 1 \\ 0 & 2 \end{pmatrix}$. Then the characteristic polynomials of both A and B are $(\lambda-1)(\lambda-2)$. Hence both A and B are similar to $\begin{pmatrix} 1 & 0 \\ 0 & 2 \end{pmatrix}$. However, $\mathrm{Tr}(AA') = 14$, $\mathrm{Tr}(BB') = 6$. By Proposition 8.3.10, A and B are not orthogonally similar.

Example 8.3.12. Let $A = \begin{pmatrix} 0 & 1 \\ 0 & 0 \end{pmatrix}$ and $B = \begin{pmatrix} 0 & 2 \\ 0 & 0 \end{pmatrix}$.

Since $\begin{pmatrix} 2 & 0 \\ 0 & 1 \end{pmatrix}\begin{pmatrix} 0 & 1 \\ 0 & 0 \end{pmatrix}\begin{pmatrix} 2 & 0 \\ 0 & 1 \end{pmatrix}^{-1} = \begin{pmatrix} 0 & 2 \\ 0 & 0 \end{pmatrix}$, A is similar to B. Since $\begin{pmatrix} 1 & 0 \\ 0 & 2 \end{pmatrix}\begin{pmatrix} 0 & 1 \\ 0 & 0 \end{pmatrix}\begin{pmatrix} 1 & 0 \\ 0 & 2 \end{pmatrix} = \begin{pmatrix} 0 & 2 \\ 0 & 0 \end{pmatrix}$, they are also congruent. By Proposition 8.3.10, they are not orthogonally similar.

Remark 8.3.13. Let $A, B \in \mathrm{M}_n(\mathbb{R})$. If there is an $n \times n$ invertible real matrix P such that $P^{-1}AP = B$, $P^{-1}A'P = B'$, then A is orthogonally similar to B (see Exercise 39).

Theorem 8.3.14. *Any $A \in O(n)$ is orthogonally similar to a block diagonal matrix $\mathrm{diag}(A_1, A_2, \ldots, A_k)$, where each $A_i = \pm 1$ or $A_i = \begin{pmatrix} \cos\theta_i & -\sin\theta_i \\ \sin\theta_i & \cos\theta_i \end{pmatrix}$, $0 < \theta_i < 2\pi$, $\theta_i \neq \pi$.*

Proof. We prove the theorem by induction. It is easy to see that the theorem holds for the 1×1 orthogonal matrices. Assume that $n > 1$ and the theorem holds for any orthogonal matrix of order less than n. Now consider the theorem for an orthogonal matrix of order n. The theorem is equivalent to the statement that for an orthogonal transformation σ of the Euclidean space $\mathbb{R}^n$, there is an orthonormal basis for $\mathbb{R}^n$ such that the matrix of σ with respect to this basis is of the form given in the theorem.

Let $e_1, e_2, \ldots, e_n$ be an orthonormal basis for $\mathbb{R}^n$. An orthogonal matrix A can determine an orthogonal transformation σ with respect to this basis such that $\sigma(e_1, e_2, \ldots, e_n) = (e_1, e_2, \ldots, e_n)A$.

If σ has real eigenvalues, it has a 1-dimensional σ-subspace W. If σ doesn't have real eigenvalues, then, by Exercise 33 of Chapter 5, σ has a 2-dimensional σ-subspace W.

By Theorem 8.3.6, the orthogonal complement $W^\perp$ is also a σ-subspace of dimension less than n. By the inductive hypothesis, the theorem holds for $W^\perp$, i.e., the matrix of σ with respect to some orthonormal basis for $W^\perp$ is the block diagonal matrix of the form in the theorem. Since $\dim W \leqslant 2$, the matrix of an orthogonal transformation of W with respect to an orthonormal basis for W is ± 1

or $\begin{pmatrix} \cos\theta & -\sin\theta \\ \sin\theta & \cos\theta \end{pmatrix}, 0 \leqslant \theta < 2\pi$. By merging the orthonormal bases

for W and $W^\perp$ together, we get an orthonormal basis for V, and the matrix of σ with respect to this orthonormal basis is the block diagonal matrix in the theorem. $\qquad\square$

Definition 8.3.15. Let σ be an orthogonal transformation of an n-dimensional Euclidean space V. If the matrix of σ with respect

to some orthonormal basis for V is $\begin{pmatrix} I_{n-1} & 0 \\ 0 & -1 \end{pmatrix}$, then we call σ a

mirror reflection. The matrix of a mirror reflection of a Euclidean space with respect to some orthonormal basis is called a **matrix of a mirror reflection**.

Proposition 8.3.16. *Let σ be an orthogonal transformation of an n-dimensional Euclidean space V. If σ is the identity transformation, then σ is the product of two mirror reflections. If σ is not the identity transformation, then σ is the product of $n - k$ mirror reflections, where k is the multiplicity of the eigenvalue 1 of σ (if 1 is not an eigenvalue of σ, then $k = 0$).*

Proof. If σ is then identity transformation, then σ is the product of two mirror reflections since $I_n = \begin{pmatrix} I_{n-1} & 0 \\ 0 & -1 \end{pmatrix} \begin{pmatrix} I_{n-1} & 0 \\ 0 & -1 \end{pmatrix}$.

Next, we assume that σ is not the identity transformation. By Theorem 8.3.14, there is a basis for V such that the matrix of σ with respect to this basis is a block diagonal matrix $\mathrm{diag}(A_1, A_2, \ldots, A_t)$, where each A_i $(1 \leqslant i \leqslant t)$ is ± 1 or $\begin{pmatrix} \cos\theta & -\sin\theta \\ \sin\theta & \cos\theta \end{pmatrix}, 0 < \theta < 2\pi, \theta \neq \pi$.

Note that $\begin{pmatrix} \cos\theta & -\sin\theta \\ \sin\theta & \cos\theta \end{pmatrix} = \begin{pmatrix} \cos\theta & \sin\theta \\ \sin\theta & -\cos\theta \end{pmatrix}\begin{pmatrix} 1 & 0 \\ 0 & -1 \end{pmatrix}$. By Example 8.3.9, $\begin{pmatrix} 1 & 0 \\ 0 & -1 \end{pmatrix}$ and $\begin{pmatrix} \cos\theta & \sin\theta \\ \sin\theta & -\cos\theta \end{pmatrix}$ are all mirror reflections. So $\begin{pmatrix} \cos\theta & -\sin\theta \\ \sin\theta & \cos\theta \end{pmatrix}$ is the product of two mirror reflections. Furthermore, since each eigenvalue -1 corresponds to a mirror reflection, σ can be written in the form of the product of $n - k$ mirror reflections, where k is the multiplicity of the eigenvalue 1 of σ (if 1 is not an eigenvalue of σ, then $k = 0$). $\qquad\square$

8.4   Rotations in 3-Dimensional Spaces and Quaternions

Proposition 8.3.7 describes the matrices of rotations with respect to orthonormal bases for $\mathbb{R}^2$. Next, we prove that the rotation in Definition 8.3.5 with $n = 3$ is just the rotation $\sigma = R_{\alpha,\varphi}$ in Examples 5.2.5 and 5.3.7.

Theorem 8.4.1. *Assume that the linear transformation* $\sigma : \mathbb{R}^3 \longrightarrow \mathbb{R}^3$ *is a rotation. Then there is an orthonormal basis* α, β, γ *for* $\mathbb{R}^3$ *such that the matrix of* σ *with respect to this basis is* $\begin{pmatrix} 1 & 0 & 0 \\ 0 & \cos\varphi & -\sin\varphi \\ 0 & \sin\varphi & \cos\varphi \end{pmatrix}$, *i.e.,* $\sigma = R_{\alpha,\varphi}$.

Proof. Since σ is a rotation, σ has eigenvalue 1 by Theorem 8.3.14, it follows that there is a vector $\alpha \in \mathbb{R}^3$ with length 1 such that $\sigma(\alpha) = \alpha$. Extend α to an orthonormal basis α, β, γ for $\mathbb{R}^3$. Let $W = L(\beta, \gamma)$. Then W is the orthogonal complement of $L(\alpha)$. By Theorem 8.3.6, W is also a σ-subspace. So, σ is also a rotation of W. By Proposition 8.3.7, there is a φ such that $\sigma(\beta, \gamma) = (\beta, \gamma)\begin{pmatrix} \cos\varphi & -\sin\varphi \\ \sin\varphi & \cos\varphi \end{pmatrix}$. Furthermore, since $\sigma(\alpha) = \alpha$, we have $\sigma(\alpha, \beta, \gamma) = (\alpha, \beta, \gamma)\begin{pmatrix} 1 & 0 & 0 \\ 0 & \cos\varphi & -\sin\varphi \\ 0 & \sin\varphi & \cos\varphi \end{pmatrix}$. So $\sigma = R_{\alpha,\varphi}$. $\qquad\square$

Remark 8.4.2. Since the product of two orthogonal matrices of order 3 with determinant 1 is still an orthogonal matrix with determinant 1, the composition of rotations is still a rotation. In particular, if σ_1 is a rotation around the vector α_1 through the angle θ_1 by the right-hand rule (right fingers curled in the direction of rotation and the right thumb pointing in the positive direction of the axis), and σ_2 is a rotation around the vector α_2 through the angle θ_2 by the right-hand rule, then $\sigma_2\sigma_1$ is still a rotation, i.e., there must be an $\alpha_3 \in V, \theta_3 \in \mathbb{R}$ such that $\sigma_2\sigma_1$ is a rotation around the vector α_3 through the angle θ_3 by the right-hand rule.

In physics, we often use the **Euler angles** to describe rotations. Let the coordinate system of $\mathbb{R}^3$ be xyz and the vectors in a basis for $\mathbb{R}^3$ be

$$\varepsilon_1 = \begin{pmatrix} 1 \\ 0 \\ 0 \end{pmatrix}, \quad \varepsilon_2 = \begin{pmatrix} 0 \\ 1 \\ 0 \end{pmatrix}, \quad \varepsilon_3 = \begin{pmatrix} 0 \\ 0 \\ 1 \end{pmatrix}. \tag{8.4}$$

Assume that the coordinate system xyz is changed to uvw after the rotation σ and the basis vectors in (8.4) are changed to $\alpha = \sigma(\varepsilon_1)$, $\beta = \sigma(\varepsilon_2)$, $\gamma = \sigma(\varepsilon_3)$.

We decompose σ into three steps. Each step is to rotate around a certain coordinate axis through an angle, and these three angles together are called the Euler angles (θ, φ, ψ) of the rotation σ. When the xy plane coincides with the uv plane, let $\alpha_1 = \alpha$. Otherwise, let the intersection line of the xy plane and the uv plane be N, i.e., $L(\varepsilon_1, \varepsilon_2) \cap L(\alpha, \beta) = N$. Let $\alpha_1 = \frac{\varepsilon_3 \times \gamma}{|\varepsilon_3 \times \gamma|}$, where $\varepsilon_3 \times \gamma$ is the exterior product of ε_3 and γ. Then $|\alpha_1| = 1$.

(1) Keeping the z-axis fixed and rotating the xy plane so that ε_1 rotates to the position of α_1. We assume that the rotation angle is θ, β_1 is the position where ε_2 rotates to, and $\gamma_1 = \varepsilon_3$. Denote this rotation by σ_1. Then we have

$$(\alpha_1, \beta_1, \gamma_1) = \sigma_1(\varepsilon_1, \varepsilon_2, \varepsilon_3) = (\varepsilon_1, \varepsilon_2, \varepsilon_3) \begin{pmatrix} \cos\theta & -\sin\theta & 0 \\ \sin\theta & \cos\theta & 0 \\ 0 & 0 & 1 \end{pmatrix}. \tag{8.5}$$

(2) Keeping N still and rotating the zw plane so that z rotates to the position of w. Let the rotation angle be φ. The original xy plane has been rotated to coincide with the uv plane. Let $\alpha_2 = \alpha_1$,

and assume that β_1 and γ_1 rotate to β_2 and γ_2, respectively. Denote this rotation by σ_2. Then we have

$$(\alpha_2, \beta_2, \gamma_2) = \sigma_2(\alpha_1, \beta_1, \gamma_1) = (\alpha_1, \beta_1, \gamma_1) \begin{pmatrix} 1 & 0 & 0 \\ 0 & \cos\varphi & -\sin\varphi \\ 0 & \sin\varphi & \cos\varphi \end{pmatrix}. \quad (8.6)$$

(3) Keeping w still and rotating the xy plane so that xy rotates to the position of uv. Let the rotation angle be ψ, and denote this rotation by σ_3. Then we have

$$(\alpha, \beta, \gamma) = \sigma_3(\alpha_2, \beta_2, \gamma_2) = (\alpha_2, \beta_2, \gamma_2) \begin{pmatrix} \cos\psi & -\sin\psi & 0 \\ \sin\psi & \cos\psi & 0 \\ 0 & 0 & 1 \end{pmatrix}. \quad (8.7)$$

By (8.5), (8.6), and (8.7), σ can be decomposed as $\sigma = \sigma_3\sigma_2\sigma_1$, i.e., an special orthogonal matrix A of order 3 can be written as

$$A = \begin{pmatrix} \cos\psi & -\sin\psi & 0 \\ \sin\psi & \cos\psi & 0 \\ 0 & 0 & 1 \end{pmatrix} \begin{pmatrix} 1 & 0 & 0 \\ 0 & \cos\varphi & -\sin\varphi \\ 0 & \sin\varphi & \cos\varphi \end{pmatrix} \begin{pmatrix} \cos\theta & -\sin\theta & 0 \\ \sin\theta & \cos\theta & 0 \\ 0 & 0 & 1 \end{pmatrix}.$$

The equality above decomposes a rotation into the composite of three rotations around the coordinate axis, as shown in the following figure:

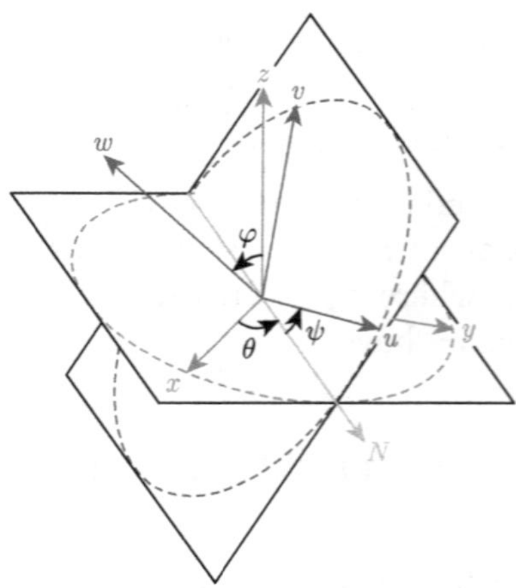

Figure of the Euler angles (θ, φ, ψ).

We can also use quaternions to describe rotations in the 3-dimensional space.

The field $\mathbb{C}$ of complex numbers (see Section 1.4) can be regarded as a 2-dimensional linear space spanned by 1, i over the field $\mathbb{R}$ of real numbers, i.e., $\mathbb{C} = \{x + yi \mid x, y \in \mathbb{R}\}$. Quaternions are higher-dimensional generalizations of complex numbers.

Definition 8.4.3. Let $\mathbb{H} = \{w + x\mathbf{i} + y\mathbf{j} + z\mathbf{k} \mid w, x, y, z \in \mathbb{R}\}$ be a 4-dimensional real linear space with basis 1, $\mathbf{i}$, $\mathbf{j}$, $\mathbf{k}$. For any $q_1 = w_1 + x_1\mathbf{i} + y_1\mathbf{j} + z_1\mathbf{k}$, $q_2 = w_2 + x_2\mathbf{i} + y_2\mathbf{j} + z_2\mathbf{k} \in \mathbb{H}$, the product q_1q_2 of q_1 and q_2 is defined by

$$q_1q_2$$

$$= (w_1 + x_1\mathbf{i} + y_1\mathbf{j} + z_1\mathbf{k})(w_2 + x_2\mathbf{i} + y_2\mathbf{j} + z_2\mathbf{k})$$

$$= (w_1w_2 - x_1x_2 - y_1y_2 - z_1z_2) + (w_1x_2 + x_1w_2 + y_1z_2 - z_1y_2)\mathbf{i}$$

$$+ (w_1y_2 + y_1w_2 + z_1x_2 - x_1z_2)\mathbf{j} + (w_1z_2 + z_1w_2 + x_1y_2 - y_1x_2)\mathbf{k}.$$
$$(8.8)$$

With the above operations, $\mathbb{H}$ is called a **quaternion algebra**, and the elements of $\mathbb{H}$ are called **quaternions**.

Remark 8.4.4. It is easy to verify the following facts:

(1) **Associative law of multiplication:** $q_1(q_2q_3) = (q_1q_2)q_3$.
(2) **Left distributive law of multiplication with respect to addition:** $q_1(q_2 + q_3) = q_1q_2 + q_1q_3$.
(3) **Right distributive law of multiplication with respect to addition:** $(q_1 + q_2)q_3 = q_1q_3 + q_2q_3$.
(4) **Every nonzero element has a multiplication inverse:** Let $\alpha \neq 0$. Then there is a unique β such that $\alpha\beta = \beta\alpha = 1$, β is called the **inverse** of α, denoted by α^{-1}. In fact, if $\alpha = w + x\mathbf{i} + y\mathbf{j} + z\mathbf{k} \neq 0$, then $\alpha^{-1} = \frac{w - x\mathbf{i} - y\mathbf{j} - z\mathbf{k}}{w^2 + x^2 + y^2 + z^2}$.
(5) **Multiplication doesn't satisfy commutative law:**

$$\mathbf{ij} = \mathbf{k} \neq \mathbf{ji} = -\mathbf{k}.$$

(6) $\mathbf{i}^2 = \mathbf{j}^2 = \mathbf{k}^2 = \mathbf{ijk} = -1$.

Remark 8.4.5. In the early 19th century, the theory of complex numbers was well-developed. Since complex numbers correspond bijectively to points in the 2-dimensional real plane, mathematicians hoped to find "triple numbers" corresponding to points in the 3-dimensional real space, but it has been unsuccessful. On October 16, 1843, when Hamilton and his wife were walking along a river, he suddenly realized that "triple numbers" do not exist. Still, the multiplication of quaternions can be defined according to the formula (8.8). Hamilton took out a knife in ecstasy and carved the multiplication formula of quaternions on a boulder under Brougham Bridge. What he carved is "$\mathbf{i}^2 = \mathbf{j}^2 = \mathbf{k}^2 = \mathbf{ijk} = -1$".

The discovery of quaternions significantly promoted the development of mathematics and physics. Arthur Cayley discovered octonions based on quaternions and developed matrix theory. Physicist Maxwell[7] formulated the electromagnetic equations in terms of quaternions.

Before explaining quaternions with matrices, let's see how to describe complex numbers using matrices.

Consider the set $V = \left\{ \begin{pmatrix} x & y \\ -y & x \end{pmatrix} \,\middle|\, x,\, y \in \mathbb{R} \right\}$.

By Exercise 5 of Chapter 5, the map

$$\sigma : \mathbb{C} \longrightarrow V,\, x + y\mathrm{i} \longmapsto x \begin{pmatrix} 1 & 0 \\ 0 & 1 \end{pmatrix} + y \begin{pmatrix} 0 & 1 \\ -1 & 0 \end{pmatrix}$$

is an isomorphism of real linear spaces and satisfies $\sigma(\alpha\beta) = \sigma(\alpha)\sigma(\beta)$ for $\alpha,\, \beta \in \mathbb{C}$.

Quaternions can also be defined similarly. Consider the set

$$H = \left\{ \begin{pmatrix} u & v \\ -\overline{v} & \overline{u} \end{pmatrix} \,\middle|\, u,\, v \in \mathbb{C} \right\}.$$

Let $u = w + x\mathrm{i}$, $v = y + z\mathrm{i}$, where $w, x, y, z \in \mathbb{R}$. Then

$$\begin{pmatrix} u & v \\ -\overline{v} & \overline{u} \end{pmatrix} = w \begin{pmatrix} 1 & 0 \\ 0 & 1 \end{pmatrix} + x \begin{pmatrix} \mathrm{i} & 0 \\ 0 & -\mathrm{i} \end{pmatrix} + y \begin{pmatrix} 0 & 1 \\ -1 & 0 \end{pmatrix} + z \begin{pmatrix} 0 & \mathrm{i} \\ \mathrm{i} & 0 \end{pmatrix}.$$

---

[7]James Clerk Maxwell, 1831–1879, British physicist.

Hence, H is a 4-dimensional linear space over the field of real numbers. Define the linear transformation $\sigma : \mathbb{H} \longrightarrow H$ such that

$$\sigma(1) = \begin{pmatrix} 1 & 0 \\ 0 & 1 \end{pmatrix}, \quad \sigma(\mathbf{i}) = \begin{pmatrix} i & 0 \\ 0 & -i \end{pmatrix},$$

$$\sigma(\mathbf{j}) = \begin{pmatrix} 0 & 1 \\ -1 & 0 \end{pmatrix}, \sigma(\mathbf{k}) = \begin{pmatrix} 0 & i \\ i & 0 \end{pmatrix}.$$

Then, σ is an isomorphism of real linear spaces.

Let $\mathbb{H}^0 = \{x\mathbf{i} + y\mathbf{j} + z\mathbf{k} \mid x, \, y, \, z \in \mathbb{R}\}$. Then $\mathbb{H}^0$ is a 3-dimensional real space. We call every element of $\mathbb{H}^0$ a **pure quaternion**. For any $\alpha, \beta \in \mathbb{H}^0$, define

$$(\alpha, \beta) = \frac{-\alpha\beta - \beta\alpha}{2}, \quad \alpha \times \beta = \frac{\alpha\beta - \beta\alpha}{2}. \tag{8.9}$$

Let $\alpha = x_1\mathbf{i} + y_1\mathbf{j} + z_1\mathbf{k}$, $\beta = x_2\mathbf{i} + y_2\mathbf{j} + z_2\mathbf{k}$. Then

$$(\alpha, \beta) = x_1x_2 + y_1y_2 + z_1z_2,$$

$$\alpha \times \beta = (y_1z_2 - y_2z_1)\mathbf{i} + (x_2z_1 - x_1z_2)\mathbf{j} + (x_1y_2 - x_2y_1)\mathbf{k}.$$

It is obvious that (8.9) defines the **inner product** and **exterior product** of $\mathbb{H}^0$, and they are the same as the inner product $\alpha \cdot \beta$ and exterior product $\alpha \times \beta$ of the 3-dimensional space $\mathbb{R}^3$, respectively. Hence $\mathbb{H}^0$ is isomorphic to $\mathbb{R}^3$ as Euclidean spaces.

Definition 8.4.6. Let $q = w + x\mathbf{i} + y\mathbf{j} + z\mathbf{k}$. Then we call $\bar{q} = w - x\mathbf{i} - y\mathbf{j} - z\mathbf{k}$ the **conjugate** of q, and call $|q| = \sqrt{q\bar{q}} = \sqrt{w^2 + x^2 + y^2 + z^2}$ the **length** of q.

For $q_1, q_2 \in \mathbb{H}$, it is easy to see $|q_1q_2| = |q_1||q_2|$.

Theorem 8.4.7. *Any quaternion q can be written in the form of*

$$|q|(\cos\theta + \alpha\sin\theta),$$

where θ is a real number, $\alpha \in \mathbb{H}^0$ and $|\alpha| = 1$.

Proof. Let $q = w + x\mathbf{i} + y\mathbf{j} + z\mathbf{k} \in \mathbb{H}$. If $q = 0$, then the theorem holds trivially. Next, we assume $q \neq 0$. Choose $\theta \in \mathbb{R}$ such that

$$\cos\theta = \frac{w}{|q|}, \quad \sin\theta = \frac{\sqrt{x^2 + y^2 + z^2}}{|q|}.$$

Let $\alpha = \dfrac{x\mathbf{i} + y\mathbf{j} + z\mathbf{k}}{\sqrt{x^2 + y^2 + z^2}}$. Then the theorem follows. $\qquad\square$

Theorem 8.4.8. *Assume $0 \neq \delta \in \mathbb{H}$. If $\delta q = q\delta$ for any $q \in \mathbb{H}^0$, then $\delta \in \mathbb{R}$.*

Proof. Let $\delta = w + x\mathbf{i} + y\mathbf{j} + z\mathbf{k}$. If $q = \mathbf{i}$, then $\delta\mathbf{i} = -x + w\mathbf{i} + z\mathbf{j} - y\mathbf{k}$, $\mathbf{i}\delta = -x + w\mathbf{i} - z\mathbf{j} + y\mathbf{k}$. So $\delta\mathbf{i} = \mathbf{i}\delta$ if and only if $y = z = 0$. Similarly, letting $q = \mathbf{j}$, we have $\delta\mathbf{j} = \mathbf{j}\delta$ if and only if $x = z = 0$. Hence $\delta = w \in \mathbb{R}$. $\qquad\square$

Next, we use quaternions to describe rotations.

Theorem 8.4.9. *Let $\alpha \in \mathbb{H}^0$ and $|\alpha| = 1$, $\sigma = R_{\alpha,\varphi} \in End(\mathbb{H}^0)$. Then there is a $\delta \in \mathbb{H}$ of length 1 such that $\sigma(\xi) = \delta\xi\delta^{-1}$ for $\xi \in \mathbb{H}^0$.*

Proof. Let $\beta \in \mathbb{H}^0$ be a vector of length 1 such that $(\alpha, \beta) = 0$. Then $\alpha\beta = -\beta\alpha$. If $\gamma = \alpha\beta$, then $|\gamma| = |\alpha||\beta| = 1$ and $\alpha\gamma = \alpha^2\beta = -\alpha\beta\alpha = -\gamma\alpha$. So $(\alpha, \gamma) = 0$. Similarly, $(\beta, \gamma) = 0$. Thus α, β, γ are three pairwise orthogonal vectors of length 1, and they form an orthonormal basis for $\mathbb{H}^0$. Assume $0 \leqslant \theta < 2\pi$ and $\delta = \cos\theta + \alpha\sin\theta$. Then $|\delta| = 1$ and $\delta^{-1} = \cos\theta - \alpha\sin\theta$. Consider the map $\tau : \mathbb{H}^0 \longrightarrow \mathbb{H}^0$, $\xi \longmapsto \delta\xi\delta^{-1}$. It is easy to see that τ is a linear transformation, and

$$\tau(\alpha) = (\cos\theta + \alpha\sin\theta)\alpha(\cos\theta - \alpha\sin\theta) = \alpha,$$

$$\tau(\beta) = (\cos\theta + \alpha\sin\theta)\beta(\cos\theta - \alpha\sin\theta) = \beta\cos(2\theta) + \gamma\sin(2\theta),$$

$$\tau(\gamma) = (\cos\theta + \alpha\sin\theta)\gamma(\cos\theta - \alpha\sin\theta) = \gamma\cos(2\theta) - \beta\sin(2\theta).$$

Thus, $\tau(\alpha, \beta, \gamma) = (\alpha, \beta, \gamma) \begin{pmatrix} 1 & 0 & 0 \\ 0 & \cos(2\theta) & -\sin(2\theta) \\ 0 & \sin(2\theta) & \cos(2\theta) \end{pmatrix}$, and so τ is an orthogonal transformation rotating around α through the angle 2θ by the right-hand rule. Let $\theta = \frac{\varphi}{2}$, then $\sigma = \tau$. $\qquad\square$

By the above proof, we see that every quaternion $\delta = \cos\frac{\varphi}{2} + \alpha\sin\frac{\varphi}{2}$ of length 1 corresponds to the rotation $R_{\alpha,\varphi}$ in $\mathbb{H}^0$. Representing rotations in quaternions has great advantages. First, after knowing the rotation axis α and the rotation angle φ, it is easy to obtain $\delta = \cos\frac{\varphi}{2} + \alpha\sin\frac{\varphi}{2}$. Second, it is straightforward to use quaternions to calculate the combination of two rotations; one only needs to calculate the product of two quaternions. However, it isn't straightforward to calculate the combination of rotations with the Euler angles.

Finally, you can also use Formula (8.9) and Theorem 8.4.9 to prove that Rodrigue's rotation formula holds (see Exercise 16).

8.5 Real Symmetric Matrices and the Polar Decompositions of Real Matrices

Real symmetric matrices have many special properties. We prove that the eigenvalues of real symmetric matrices are all real numbers and that real symmetric matrices are all diagonalizable. This section also covers the polar and singular value decompositions of matrices.

Theorem 8.5.1. *The complex eigenvalues of a real symmetric matrix are all real numbers, and the eigenvectors associated with different eigenvalues must be orthogonal.*

Proof. Let A be a real symmetric matrix. If $\lambda \in \mathbb{C}$ is a complex eigenvalue of A and $\alpha = (a_1, a_2, \ldots, a_n)' \in \mathbb{C}^n$ is an eigenvector of A associated with λ, then $A\alpha = \lambda\alpha$, and so $\overline{\alpha}' A\alpha = \lambda \, \overline{\alpha}'\alpha$. Take the conjugate transpose of both sides to get $\overline{\alpha}' A\alpha = \overline{\lambda} \, \overline{\alpha}'\alpha$, which implies that $\lambda\overline{\alpha}'\alpha = \overline{\lambda} \, \overline{\alpha}'\alpha$. Since $\alpha \neq 0$, $\overline{\alpha}'\alpha = |a_1|^2 + |a_2|^2 + \cdots + |a_n|^2 > 0$. Hence $\lambda = \overline{\lambda}$, i.e., λ is a real number.

Let $A\alpha = \lambda\alpha$ and $A\beta = \mu\beta$, where α and β are nonzero vectors, $\lambda \neq \mu$. Then $\beta' A\alpha = \beta'\lambda\alpha = \lambda\beta'\alpha$. Note that $\beta' A\alpha = (A\beta)'\alpha = (\mu\beta)'\alpha = \mu\beta'\alpha$, so $(\lambda - \mu)\beta'\alpha = 0$. Since $\lambda \neq \mu$, $\beta'\alpha = 0$, i.e., α and β are orthogonal. $\qquad\square$

Theorem 8.5.2. *Any real symmetric matrix A of order n is orthogonally similar to a diagonal matrix.*

Proof. The theorem holds trivially fort $n = 1$. We assume that $n > 1$, and the theorem holds for any real symmetric matrix of order $n - 1$. Now, we consider a real symmetric matrix A of order n.

By Theorem 8.5.1, A has a real eigenvalue λ. Assume that $A\alpha = \lambda\alpha$, where α is real eigenvector with length 1. Extend α to an orthonormal basis $\alpha_1 = \alpha, \alpha_2, \ldots, \alpha_n$ for $\mathbb{R}^n$. Let $C = (\alpha_1, \alpha_2, \ldots, \alpha_n)$. Then C is an orthogonal matrix, and $AC = C\begin{pmatrix} \lambda & \star \\ 0 & \star \end{pmatrix}$, i.e., $C'AC = \begin{pmatrix} \lambda & \star \\ 0 & \star \end{pmatrix}$. Since A is a symmetric matrix, $C'AC$

is also a symmetric matrix, which implies that $C'AC = \begin{pmatrix} \lambda & 0 \\ 0 & B \end{pmatrix}$, where B is a real symmetric matrix of order $n - 1$. By the inductive hypothesis, B is orthogonally similar to a diagonal matrix, so A is orthogonally similar to a diagonal matrix. $\qquad \square$

Definition 8.5.3. In Definition 7.2.9, if $F = \mathbb{R}$ and $C \in O(n)$, then we call a linear substitution $X = CY$ an **orthogonal linear substitution**.

Theorem 8.5.2 can be expressed as the following theorem by using the language of quadratic forms.

Theorem 8.5.4 (Sylvester's law of inertia). *If $f(x_1, x_2, \ldots, x_n)$ $= X'AX$ is a real quadratic form in n indeterminates with rank r, then there is an orthogonal linear substitution $X = CY$ such that $f(x_1, x_2, \ldots, x_n)$ is transformed to*

$$g(y_1, y_2, \ldots, y_n) = d_1 y_1^2 + \cdots + d_p y_p^2 - d_{p+1} y_{p+1}^2 - \cdots - d_r y_r^2,$$

where d_1, ..., d_p are the positive eigenvalues of A, $-d_{p+1}$, ..., $-d_r$ are the negative eigenvalues of A.

The process of transforming a quadratic form $f(x_1, x_2, \ldots, x_n)$ to a standard form by orthogonal linear substitution can be divided into the following four steps:

(1) Write out the matrix A of the quadratic form, and find all the different eigenvalues $\lambda_1, \lambda_2, \ldots, \lambda_k$ of A.
(2) For each λ_i, solve the linear system $(\lambda_i I_n - A)X = 0$, and find a basis for the solution set of this system, that is, a basis $\alpha_{i1}, \ldots, \alpha_{ia_i}$ for the eigenspace V_{λ_i} associated with λ_i. Note that real symmetric matrices are diagonalizable by Theorem 8.5.2, so $\mathbb{R}^n = V_{\lambda_1} \oplus \cdots \oplus V_{\lambda_k}$.
(3) Applying Schmidt orthonormalization to the sequence $\alpha_{i1}, \ldots, \alpha_{ia_i}$, we get an orthonormal basis $\beta_{i1}, \ldots, \beta_{ia_i}$ for V_{λ_i}. By Theorem 8.5.1, the sequence $\beta_{11}, \ldots, \beta_{1a_1}, \ldots, \beta_{k1}, \ldots, \beta_{ka_k}$ is an orthonormal basis for $\mathbb{R}^n$.
(4) Let $C = (\beta_{11}, \ldots, \beta_{1a_1}, \ldots, \beta_{k1}, \ldots, \beta_{ka_k})$. Then $C'AC$ is a diagonal matrix, and $X = CY$ is an orthogonal linear substitution.

Example 8.5.5. Transform the quadratic form

$$f(x_1, x_2, x_3) = 7x_1^2 + x_2^2 + 7x_3^2 - 8x_1x_2 - 4x_1x_3 - 8x_2x_3$$

to a standard form by an orthogonal linear substitution.

Solution. The matrix of the quadratic form $f(x_1, x_2, x_3)$ is $A = \begin{pmatrix} 7 & -4 & -2 \\ -4 & 1 & -4 \\ -2 & -4 & 7 \end{pmatrix}$, whose characteristic polynomial is

$$|\lambda I_3 - A| = \lambda^3 - 15\lambda^2 + 27\lambda + 243 = (\lambda + 3)(\lambda - 9)^2,$$

and the eigenvalues of A are -3, 9.

By solving the linear system $(-3I_3 - A)X = 0$, one gets an eigenvector $\alpha_1 = \begin{pmatrix} 1 \\ 2 \\ 1 \end{pmatrix}$ associated with the eigenvalue -3. By solving the linear system $(9I_3 - A)X = 0$, one gets two linearly independent eigenvectors $\alpha_2 = \begin{pmatrix} -2 \\ 1 \\ 0 \end{pmatrix}$ and $\alpha_3 = \begin{pmatrix} -1 \\ 0 \\ 1 \end{pmatrix}$ associated with the eigenvalue 9. Applying Schmidt orthonormalization to the sequence $\alpha_1, \alpha_2, \alpha_3$, we get

$$e_1 = \begin{pmatrix} \dfrac{\sqrt{6}}{6} \\ \dfrac{\sqrt{6}}{3} \\ \dfrac{\sqrt{6}}{6} \end{pmatrix}, \quad e_2 = \begin{pmatrix} -\dfrac{2\sqrt{5}}{5} \\ \dfrac{\sqrt{5}}{5} \\ 0 \end{pmatrix}, \quad e_3 = \begin{pmatrix} -\dfrac{\sqrt{30}}{30} \\ -\dfrac{\sqrt{30}}{15} \\ \dfrac{\sqrt{30}}{6} \end{pmatrix}.$$

Let $C = (e_1, e_2, e_3)$. Then $X = CY$ is an orthogonal linear substitution; this orthogonal linear substitution transforms the original quadratic form to the standard form $-3y_1^2 + 9y_2^2 + 9y_3^2$. $\square$

Example 8.5.6. Complex symmetric matrices are not necessarily diagonalizable. For example, the Jordan canonical form of $A = \begin{pmatrix} 1 & i \\ i & -1 \end{pmatrix}$ is $\begin{pmatrix} 0 & 0 \\ 1 & 0 \end{pmatrix}$. So A cannot be diagonalized.

Theorem 8.5.7. *Let A be a symmetric matrix. Then*

(1) *A is positive definite if and only all eigenvalues of A are positive;*
(2) *A is positive semi-definite if and only all eigenvalues of A are nonnegative.*

Proof. Theorem 8.5.2 shows that a real symmetric matrix A is orthogonally similar to a diagonal matrix. Note that a diagonal matrix is positive definite (positive semi-definite) if and only if all the diagonal elements of the diagonal matrix are positive (nonnegative). So, a real symmetric matrix is definite (positive semi-definite) if and only if its eigenvalues are positive (nonnegative). $\square$

Definition 8.5.8. Let σ be a linear transformation of a Euclidean space V. If $(\sigma(\alpha), \beta) = (\alpha, \sigma(\beta))$ for any $\alpha, \beta \in V$, then we call σ a **symmetric transformation**.

Theorem 8.5.9. *If σ is a linear transformation of an n-dimensional Euclidean space V, then σ is a symmetric transformation if and only if the matrix of σ with respect to an orthonormal basis for V is a real symmetric matrix.*

Proof. Let $\alpha_1, \alpha_2, \ldots, \alpha_n$ be an orthonormal basis for V, and let $A = (a_{ij})_{n \times n}$ be the matrix of σ with respect to this orthonormal basis. Then $\sigma(\alpha_k) = a_{1k}\alpha_1 + a_{2k}\alpha_2 + \cdots + a_{nk}\alpha_n, 1 \leqslant k \leqslant n$. It is obvious that σ is a symmetric transformation if and only if $(\sigma(\alpha_i), \alpha_j) = (\alpha_i, \sigma(\alpha_j))$, i.e., $a_{ji} = a_{ij}$ for all $1 \leqslant i, j \leqslant n$. Hence, σ is a symmetric transformation if and only if A is a real symmetric matrix. $\square$

Remark 8.5.10. Since the matrices of a linear transformation of a Euclidean space V with respect to different orthonormal bases are orthogonally similar to each other, to show that the matrix of a linear transformation σ of V with respect to any orthonormal basis for V is a symmetric matrix, we only need to show that the matrix of σ with respect to one orthonormal basis for V is symmetric. Note that the matrix of a symmetric transformation with respect to a nonorthonormal basis is not necessarily symmetric (leave it as an exercise).

Proposition 8.5.11. *Let A and B be symmetric matrices with B positive definite. Then A is positive definite if and only if all the eigenvalues of AB are positive.*

Proof. Since B is positive definite, there is a real invertible matrix T such that $B = T'T$ by Theorem 7.4.2. Hence $AB = AT'T = T^{-1}(TAT')T$. So all the eigenvalues of AB are positive if and only if all the eigenvalues of TAT' are positive. By Theorem 8.5.7, all the eigenvalues of TAT' are positive if and only if TAT' is positive definite, i.e., A is positive definite. $\square$

Theorem 8.5.12. *If A is a positive semi-definite matrix and k is a positive integer, then there is a unique positive semi-definite matrix B such that $B^k = A$. If A is positive definite, then B is also positive definite.*

Proof. First, we prove the existence. Since A is a positive semi-definite matrix, there is an orthogonal matrix T such that $T'AT = \operatorname{diag}(\lambda_1, \lambda_2, \ldots, \lambda_n)$, where $\lambda_1, \lambda_2, \ldots, \lambda_n$ are nonnegative real numbers. Let

$$B = T\operatorname{diag}(\lambda_1^{\frac{1}{k}}, \lambda_2^{\frac{1}{k}}, \ldots, \lambda_n^{\frac{1}{k}})T'.$$

Then B is positive semi-definite and $B^k = A$.

Next, we prove the uniqueness. Let $A = B^k = C^k$, where B and C are both positive semi-definite matrices. Since B and C are positive semi-definite matrices with the same eigenvalues, they are orthogonally similar to the same diagonal matrix. Assume $B = U'\operatorname{diag}(\lambda_1^{\frac{1}{k}}, \lambda_2^{\frac{1}{k}}, \ldots, \lambda_n^{\frac{1}{k}})U$, $C = V'\operatorname{diag}(\lambda_1^{\frac{1}{k}}, \lambda_2^{\frac{1}{k}}, \ldots, \lambda_n^{\frac{1}{k}})V$, where $\lambda_1, \lambda_2, \ldots, \lambda_n$ are n eigenvalues of A and U, V are orthogonal matrices. Let $\lambda_1, \ldots, \lambda_t$ be the different eigenvalues of A. Then $\operatorname{diag}(\lambda_1^{\frac{1}{k}}, \ldots, \lambda_n^{\frac{1}{k}})$ is orthogonally similar to $\Lambda = \operatorname{diag}(\lambda_1^{\frac{1}{k}} I_{n_1}, \ldots, \lambda_t^{\frac{1}{k}} I_{n_t})$. Hence we can assume $B = U'\Lambda U, C = V'\Lambda V$.

From $B^k = C^k$, we know that $U'\Lambda^k U = V'\Lambda^k V$, and so $\Lambda^k UV' = UV'\Lambda^k$. By Exercise 43 of Chapter 2, $UV' = \operatorname{diag}(N_1, N_2, \ldots, N_t)$, where $N_i \in \mathrm{M}_{n_i}(\mathbb{R})$, $i = 1, \ldots, t$. So $\Lambda UV' = \operatorname{diag}(\lambda_1^{\frac{1}{k}} N_1, \lambda_2^{\frac{1}{k}} N_2, \ldots, \lambda_t^{\frac{1}{k}} N_t) = UV'\Lambda$, which implies that $U'\Lambda U = V'\Lambda V$, i.e., $B = C$. $\square$

Definition 8.5.13. Let A be an $m \times n$ real matrix. Then, the arithmetic square roots of the positive eigenvalues of the positive semi-definite matrix $A'A$ are called the **singular value** of A.

Theorem 8.5.14. *Let $A \in \mathrm{M}_{m \times n}(\mathbb{R})$. Then there are $U \in O(m)$, $V \in O(n)$ and the block matrix $S = \begin{pmatrix} \mathrm{diag}(\lambda_1, \lambda_2, \ldots, \lambda_r) & 0 \\ 0 & 0 \end{pmatrix}$ such that $A = USV$, where $\lambda_1, \lambda_2, \ldots, \lambda_r$ are the singular values of A. This decomposition of A is called the* **singular value decomposition** *of A.*

Proof. By Theorem 8.5.2, there is a $C \in O(n)$ such that

$$A'A = C\mathrm{diag}(\lambda_1^2, \lambda_2^2, \ldots, \lambda_r^2, 0, 0, \ldots, 0)C'.$$

Let $D = \mathrm{diag}(\lambda_1^2, \lambda_2^2, \ldots, \lambda_r^2)$, $C = (C_1, C_2)$, where $C_1 \in \mathrm{M}_{n \times r}(\mathbb{R})$, $C_2 \in \mathrm{M}_{n \times (n-r)}(\mathbb{R})$. Then $A'A = C_1 D C_1'$. Since C is an orthogonal matrix, $C_1'C_1 = I_r$ and $C_1'C_2 = 0$. It follows that

$$(AC_2)'AC_2 = C_2'A'AC_2 = C_2'C_1 D C_1'C_2 = 0.$$

Hence, by Exercise 23 of Chapter 3, $AC_2 = 0$. Let $D_1 = \mathrm{diag}(\lambda_1, \lambda_2, \ldots, \lambda_r)$, $S = \begin{pmatrix} D_1 & 0 \\ 0 & 0 \end{pmatrix}$ and $U_1 = AC_1 D_1^{-1}$. Then $U_1'U_1 = (AC_1 D_1^{-1})'AC_1 D_1^{-1} = I_r$. So the column vectors $\beta_1, \beta_2, \ldots, \beta_r$ of U_1 form a sequence of orthonormal vectors. Extend $\beta_1, \beta_2, \ldots, \beta_r$ to an orthonormal basis $\beta_1, \beta_2, \ldots, \beta_r, \beta_{r+1}, \ldots, \beta_m$ for $\mathbb{R}^m$. Let $U_2 = (\beta_{r+1}, \ldots, \beta_m)$, $U = (U_1, U_2)$ and $V = C'$. Then $U \in O(m)$ and $V \in O(n)$. Since

$$I_n = CC' = (C_1, C_2)\begin{pmatrix} C_1' \\ C_2' \end{pmatrix} = C_1 C_1' + C_2 C_2',$$

we have

$$U\begin{pmatrix} D_1 & 0 \\ 0 & 0 \end{pmatrix}V = (U_1, U_2)\begin{pmatrix} D_1 & 0 \\ 0 & 0 \end{pmatrix}\begin{pmatrix} C_1' \\ C_2' \end{pmatrix} = U_1 D_1 C_1'$$

$$= AC_1 C_1' = A(I_n - C_2 C_2') = A.$$

So $A = USV$. $\qquad\qquad\qquad\qquad\qquad\qquad\qquad\qquad\qquad\qquad\square$

Theorem 8.5.15. *Let $A \in \mathrm{M}_n(\mathbb{R})$. Then there is an $n \times n$ orthogonal matrix R and a positive semi-definite matrix P such that $A = RP$ (or $A = PR$), where P is uniquely determined by A. If A is invertible, R is also uniquely determined by A and P is positive definite. This decomposition is called the* **polar decomposition** *of a matrix.*

Proof. In writing the matrix A as a product of an orthogonal matrix and a positive semi-definite matrix, the proofs for $A = RP$ and $A = PR$ are symmetric, so we only need to prove one. Here, we show that $A = RP$.

If there is a matrix R and a positive semi-definite P such that $A = RP$, then $A'A = P'R'RP = P^2$. By Theorem 8.5.12, P is unique.

By Theorem 8.5.14, there are $U \in O(m)$, $V \in O(n)$ and the block matrix

$$S = \begin{pmatrix} \mathrm{diag}(\lambda_1, \lambda_2, \ldots, \lambda_r) & 0 \\ 0 & 0 \end{pmatrix}$$

such that $A = USV$. So $A = (UV)(V'SV)$, where UV is an orthogonal matrix, and $V'SV$ is a positive semi-definite matrix. $\square$

The polar decomposition $A = PR$ of a matrix A can be compared to the exponential form of a complex number. Any complex number α can be written as $\alpha = re^{i\theta}$. The positive semi-definite matrix P of the polar decomposition can be compared to the modulus r of α, and the orthogonal matrix R can be compared to $e^{i\theta}$.

Definition 8.5.16. Let $A \in M_{m \times n}(\mathbb{R})$. If there is a matrix $B \in M_{n \times m}(\mathbb{R})$ such that

$$ABA = A, \quad BAB = B, \quad (AB)' = AB, \quad (BA)' = BA, \quad (8.10)$$

then B is called the **Moore–Penrose inverse** of A, denoted by A^+.

The Moore–Penrose inverse of a matrix was independently discovered by Moore in 1920 and Penrose in 1951 and has important applications in mathematical statistics and computational mathematics.

Theorem 8.5.17. *If $A \in M_{m \times n}(\mathbb{R})$, then A^+ exists and is unique.*

Proof. We begin with the existence portion of the theorem. Assume that the singular decomposition of A is $A = USV$, where $U \in O(m)$, $V \in O(n)$, and $S = \begin{pmatrix} D & 0 \\ 0 & 0 \end{pmatrix}$, D is a diagonal matrix whose diagonal elements are the singular values of A. Let $B = V'S^+U'$, where $S^+ = \begin{pmatrix} D^{-1} & 0 \\ 0 & 0 \end{pmatrix}$. Then

$$ABA = USVV'S^+U'USV = USS^+SV = USV = A.$$

Similarly, we can prove that B also satisfy the remaining three equalities of (8.10).

Next, we establish the uniqueness. If both B and C are the Moore–Penrose inverses of A, then

$$B = (BA)B = (A'B')B = A'(B'B) = (A'C'A')(B'B)$$
$$= (A'C')(A'B')B = (CA)(BA)B = C(ABA)B = CAB.$$

Similarly, we have

$$C = C(AC) = C(C'A') = (CC')A' = (CC')(A'B'A')$$
$$= C(C'A')(B'A') = C(AC)(AB) = C(ACA)B = CAB,$$

which implies that $B = C$. Hence, A^+ exists and is unique. $\qquad\square$

Remark 8.5.18. It is easy to verify that the Moore–Penrose inverse satisfies the following properties:

(1) If $A \in \mathrm{GL}_n(\mathbb{R})$, then $A^+ = A^{-1}$. So, the Moore–Penrose inverse is the generalization of the inverse of a square matrix.
(2) If $A \in \mathrm{M}_{m \times n}(\mathbb{R})$, then $(A^+)^+ = A$.
(3) If $A \in \mathrm{M}_{m \times n}(\mathbb{R})$, then $(A^+)' = (A')^+$.
(4) If $A \in \mathrm{M}_{m \times n}(\mathbb{R})$, $B \in \mathrm{M}_{n \times p}(\mathbb{R})$, then

$$(AB)^+ = (A^+AB)^+(ABB^+)^+.$$

8.6   Method of Least Squares

We know that the linear system $AX = \beta$ has a solution if and only if $\mathrm{rank}A = \mathrm{rank}(A, \beta)$, where $A \in \mathrm{M}_{m \times n}(F)$, $X = (x_1, x_2, \ldots, x_n)'$ and $\beta \in F^m$. We set $F = \mathbb{R}$ in this section. When the linear system $AX = \beta$ has no solutions, we hope to find an approximate solution $X_0 \in \mathbb{R}^n$ such that $|AX_0 - \beta|$ is minimal.

Theorem 8.6.1. *Let W be a nonzero subspace of $\mathbb{R}^m$ and $\beta \in \mathbb{R}^m$. Then there is a $w \in W$ such that $|w - \beta| = min\{|\tilde{w} - \beta| \mid \tilde{w} \in W\}$.*

Proof. Let $w_1, w_2, \ldots, w_k$ be an orthonormal basis for W, $a_i = (\beta, w_i)$, where $i = 1, 2, \ldots, k$, and let $w = \sum_{i=1}^{k} a_i w_i$. Then $w - \beta \in W^{\perp}$. For any $\widetilde{w} \in W$, we have

$$
\begin{aligned}
|\widetilde{w} - \beta|^2 &= |\widetilde{w} - w + w - \beta|^2 \\
&= |\widetilde{w} - w|^2 + |w - \beta|^2 + 2(\widetilde{w} - w, w - \beta) \\
&= |\widetilde{w} - w|^2 + |w - \beta|^2 \geqslant |w - \beta|^2.
\end{aligned}
$$

Hence $|w - \beta| = \min\{|\widetilde{w} - \beta| \mid \widetilde{w} \in W\}$. $\qquad\square$

Definition 8.6.2. Let W be a nonzero subspace of $\mathbb{R}^m$ and $\beta \in \mathbb{R}^m$. We call $\min\{|\beta - \gamma| \mid \gamma \in W\}$ the **distance** between β and W.

Let $A \in \mathrm{M}_{m \times n}(F)$ and $W = \{A\alpha \mid \alpha \in \mathbb{R}^n\}$. Then W is a subspace of $\mathbb{R}^m$, and its elements are the linear combination of the column vectors of A. Let $\beta \notin W$. Next, we show how to find the distance between β and W. By the proof of Theorem 8.6.1, we need to find $X_0 \in \mathbb{R}^n$ such that $AX_0 - \beta \in W^{\perp}$, i.e., $A'(AX_0 - \beta) = 0$. In this way, the minimum distance problem is transformed into solving the nonhomogeneous linear system $A'AX = A'\beta$. By Exercise 25 of Chapter 3, this nonhomogeneous linear system always has a solution. Note that if $AX = \beta$ has a solution, then $AX = \beta$ and $A'AX = A'\beta$ have the same solution set.

Definition 8.6.3. Let $A \in \mathrm{M}_{m \times n}(\mathbb{R})$ and $\beta \in \mathbb{R}^m$. When $AX = \beta$ has no solutions, we call the solution of $A'AX = A'\beta$ the **least squares solution** of $AX = \beta$. The method of using the least squares solution as the approximate solution of $AX = \beta$ is called the **method of least squares**.

We can use the Moore–Penrose inverse to give the least squares solution of $AX = \beta$ directly.

Theorem 8.6.4. *Let $A \in \mathrm{M}_{m \times n}(\mathbb{R})$ and $\beta \in \mathbb{R}^m$. Then the set of least squares solutions of $AX = \beta$ is $\{A^+\beta + (I_n - A^+A)w \mid w \in \mathbb{R}^n\}$.*

Proof. To prove this theorem, it is equivalent to prove that the set

$$
\{A^+\beta + (I_n - A^+A)w \mid w \in \mathbb{R}^n\}
$$

is the set of solutions of $A'AX = A'\beta$.

First, $A^+\beta$ satisfy $A'AX = A'\beta$. In fact,

$$A'A(A^+\beta) = A'(AA^+)'\beta = (AA^+A)'\beta = A'\beta.$$

Second, let $V_1 = \{(I_n - A^+A)w \mid w \in \mathbb{R}^n\}$ and V_2 be the space of solutions of $A'AX = 0$. According to the theorem on the structure of the solutions of a nonhomogeneous system, it is enough to prove that $V_1 = V_2$. On the one hand, V_1 is clearly a subspace of $\mathbb{R}^n$, and for any $w \in \mathbb{R}^n$, $A'A(I_n - A^+A)w = (A'A - A'AA^+A)w = (A'A - A'A)w = 0$, which implies that $V_1 \subseteq V_2$. On the other hand, since $A = AA^+A$, $\operatorname{rank}(A^+A) = \operatorname{rank}A$ and $(A^+A)^2 = A^+A$. So, by Exercise 57 of Chapter 2,

$$\dim V_1 = \operatorname{rank}(I_n - A^+A) = n - \operatorname{rank}(A^+A)$$

$$= n - \operatorname{rank}A = n - \operatorname{rank}(A'A) = \dim V_2.$$

Hence, $V_1 = V_2$. $\square$

By Theorem 8.6.4, the least squares solution is unique if and only if $A^+A = I_n$.

Theorem 8.6.5. *Let $A \in \mathrm{M}_{m\times n}(\mathbb{R})$ and $\beta \in \mathbb{R}^m$. If X_0 is a least squares solution of $AX = \beta$, then $|X_0| \geqslant |A^+\beta|$, and the equality holds if and only if $X_0 = A^+\beta$.*

Proof. By Theorem 8.6.4, we can assume

$$X_0 = A^+\beta + (I_n - A^+A)w,$$

where $w \in \mathbb{R}^n$. Since

$$(A^+\beta)'(I_n - A^+A)w = \beta'(A^+)'(I_n - A^+A)w$$

$$= \beta'(A^+)'(I_n - (A^+A)')w = 0,$$

$A^+\beta$ is orthogonal to $(I_n - A^+A)w$. So

$$|X_0|^2 = |A^+\beta|^2 + |(I_n - A^+A)w|^2 \geqslant |A^+\beta|^2,$$

and the equality holds if and only if $X_0 = A^+\beta$. $\square$

By the theorem above, $A^+\beta$ is the least squares solution with minimal norm, hence $A^+\beta$ is called the **minimal norm least squares solution** of $AX = \beta$, or the **optimal solution**.

Remark 8.6.6. The method of least squares was discovered by Carl Friedrich Gauss in 1795, but it was not officially published. Adrien

Marie Legendre first published his work on the method of least squares in 1805. The method of least squares has important practical applications. On January 1, 1801, Piazzi[8] discovered Ceres. He announced his discovery in a letter to a friend on January 24, 1801. In April 1801, Piazzi gave a complete observation report and officially published it in September 1801. However, the position of Ceres changed at this time because it was too close to the sun to observe, which made it difficult for other astronomers to confirm Piazzi's discovery. Ceres was supposed to be observed at the end of the year, but it is difficult to predict its exact position because of the long time that has passed. Astronomers needed to know in which area of the sky to look for Ceres. At this time, 24-year-old Gauss used the least square method, based on the observation data, to accurately predict where Ceres would appear, and he sent the predicted route of Ceres to the astronomer Zach.[9] On December 31, 1801, Zach really saw Ceres at the position predicted by Gauss.

Example 8.6.7. It is known that a physical quantity $f(t)$ is a linear function of the time t, and the actual measurement found that $f(1) = 1$, $f(2) = 2$, $f(3) = 2$. Find the approximate expression of $f(t)$ by the method of least squares.

Solution. We assume that $f(t) = at + b$, where a and b are undetermined. It is easy to see that the system of linear equations

$$\begin{cases} a + b - 1 = 0, \\ 2a + b - 2 = 0, \\ 3a + b - 2 = 0 \end{cases}$$

has no solutions. We need to find a and b such that the length of the vector $\begin{pmatrix} a + b - 1 \\ 2a + b - 2 \\ 3a + b - 2 \end{pmatrix}$ reaches the minimal value, i.e., find the least squares solution of $\begin{pmatrix} 1 & 1 \\ 2 & 1 \\ 3 & 1 \end{pmatrix} \begin{pmatrix} a \\ b \end{pmatrix} = \begin{pmatrix} 1 \\ 2 \\ 2 \end{pmatrix}$. By Definition 8.6.3, we need

---

[8] Giuseppe Piazzi, 1746–1826, Italian astronomer.
[9] Franz Xaver von Zach, 1754–1832, Hungarian astronomer.

to find the solution of the matrix equation $\begin{pmatrix} 1 & 2 & 3 \\ 1 & 1 & 1 \end{pmatrix} \begin{pmatrix} 1 & 1 \\ 2 & 1 \\ 3 & 1 \end{pmatrix} \begin{pmatrix} a \\ b \end{pmatrix} =$

$\begin{pmatrix} 1 & 2 & 3 \\ 1 & 1 & 1 \end{pmatrix} \begin{pmatrix} 1 \\ 2 \\ 2 \end{pmatrix}$, i.e., $\begin{pmatrix} 14 & 6 \\ 6 & 3 \end{pmatrix} \begin{pmatrix} a \\ b \end{pmatrix} = \begin{pmatrix} 11 \\ 5 \end{pmatrix}$. Obviously, $\begin{cases} a = \dfrac{1}{2}, \\ b = \dfrac{2}{3}. \end{cases}$

Hence, the line closest to the three points $(1,1), (2,2), (3,2)$ is $y = \frac{1}{2}t + \frac{2}{3}$. Hence the approximate expression of $f(t)$ is $f(t) = \frac{1}{2}t + \dfrac{2}{3}$. $\quad\square$

8.7* Unitary Spaces

Definition 8.7.1. Let V be a complex linear space. A map

$$f : V \times V \longrightarrow \mathbb{C}$$

is called an **inner product** on V if it satisfies the following conditions:

(1) $f(\alpha, \beta) = \overline{f(\beta, \alpha)}$,
(2) $f(\alpha + \beta, \gamma) = f(\alpha, \gamma) + f(\beta, \gamma)$,
(3) $f(k\alpha, \beta) = kf(\alpha, \beta)$,
(4) $f(\alpha, \alpha) \geqslant 0,$ and the equality holds if and only if $\alpha = 0$,

where $\alpha, \beta, \gamma \in V$, $k \in \mathbb{C}$.

For each ordered pair of vectors α, β of V, $f(\alpha, \beta)$ is called the inner product of α and β, abbreviated as (α, β).

A finite-dimensional complex linear space V with an inner product is called a **unitary space**.

Remark 8.7.2.

(1) Although the first component of the inner product of Definition 8.7.1 is linear, the second component is only **semi-linear**, i.e.,

$$f(\alpha, \beta_1 + \beta_2) = f(\alpha, \beta_1) + f(\alpha, \beta_2), \ \ f(\alpha, k\beta) = \overline{k}f(\alpha, \beta),$$

where $\alpha, \beta, \beta_1, \beta_2 \in V, k \in \mathbb{C}$.
(2) By the same argument as in a Euclidean space, we can define the length of a vector α in a unitary space as $|\alpha| = \sqrt{(\alpha, \alpha)}$.

This length also satisfies the Cauchy–Bunyakovsky–Schwarz inequality. We can similarly define the concepts such as orthogonality and orthogonal complement.

Example 8.7.3. Let $V = \mathbb{C}^n$. For any $\alpha = (a_1, a_2, \ldots, a_n)' \in V$, $\beta = (b_1, b_2 \ldots, b_n)' \in V$, define

$$(\alpha, \beta) = a_1\overline{b_1} + a_2\overline{b_2} + \cdots + a_n\overline{b_n}.$$

It is not difficult to verify that the inner product defined in this way satisfies the conditions in Definition 8.7.1.

Definition 8.7.4. If an $n \times n$ complex matrix A satisfies $\overline{A'}A = I_n$, then A is called a **unitary matrix**. The set of all $n \times n$ unitary matrices is denoted by $\mathrm{U}(n)$. Let V be an n-dimensional unitary space and $\sigma \in \mathrm{End}(V)$. If $(\alpha, \beta) = (\sigma(\alpha), \sigma(\beta))$ for any $\alpha, \beta \in V$, then σ is called a **unitary transformation** of V.

Example 8.7.5. We can prove by definition that

$$\mathrm{U}(1) = \{e^{i\theta} \mid \theta \in \mathbb{R}\},$$

$$\mathrm{U}(2) = \left\{ \begin{pmatrix} \alpha & -\overline{\beta} \\ e^{i\theta}\beta & e^{i\theta}\overline{\alpha} \end{pmatrix} \, \middle| \, \alpha, \beta \in \mathbb{C}, \theta \in \mathbb{R}, |\alpha|^2 + |\beta|^2 = 1 \right\}.$$

The set of 2×2 unitary matrices with determinant 1 is denoted by $\mathrm{SU}(2)$. One can prove that

$$\mathrm{SU}(2) = \left\{ \begin{pmatrix} \alpha & -\overline{\beta} \\ \beta & \overline{\alpha} \end{pmatrix} \, \middle| \, \alpha, \beta \in \mathbb{C}, |\alpha|^2 + |\beta|^2 = 1 \right\}.$$

Unitary transformations and orthogonal transformations have similar properties, for example, corresponding to Theorem 8.3.3, unitary transformations have the following equivalent characterizations.

Theorem 8.7.6. *Let V be an n-dimensional unitary space and $\sigma \in End(V)$. Then the following statements are equivalent:*

(1) *σ is a unitary transformation.*
(2) *$|\sigma(\alpha)| = |\alpha|$ for any $\alpha \in V$.*
(3) *The matrix of σ with respect to an orthonormal basis for V is a unitary matrix.*
(4) *σ transforms an orthonormal basis for V to another orthonormal basis for V.*

Definition 8.7.7. Let $A \in M_n(\mathbb{C})$. If $A = \overline{A'}$, then A is called a **Hermitian matrix**. If $A = -\overline{A'}$, then A is called a **skew-Hermitian matrix**.

If A is a real matrix of order n, then A is a Hermitian matrix if and only if A is a symmetric matrix. As we saw in Chapter 7, the properties of symmetric and skew-symmetric matrices over the real number field are very different. However, a complex matrix A is a skew-Hermitian matrix if and only if iA is a Hermitian matrix; that is, the distinction between a skew-Hermitian matrix and a Hermitian matrix is merely the coefficient i.

Definition 8.7.8. Let V be an n-dimensional unitary space and $\sigma \in \text{End}(V)$. If $(\sigma(\alpha), \beta) = (\alpha, \sigma(\beta))$ for any $\alpha, \beta \in V$, then σ is called a **Hermitian transformation** of V.

One can see from the definition above that a Hermitian transformation of a unitary space is similar to a symmetric transformation of a Euclidean space, for example, similar to Theorem 8.5.9, a linear transformation σ of an n-dimensional unitary space V is a Hermitian transformation if and only if the matrix of σ with respect to an orthonormal basis for V is a Hermitian matrix.

Definition 8.7.9. Let $A, B \in M_n(\mathbb{C})$. If there is a unitary matrix U such that $U^{-1}AU = B$, A and B are called **unitarily similar**.

It is easy to see that unitary similarity is an equivalence relation. Note that two similar matrices are not necessarily unitarily similar; see Example 8.3.11.

Proposition 8.7.10. *Assume that* $A = (a_{ij})_{n \times n}$, $B = (b_{ij})_{n \times n} \in M_n(\mathbb{C})$, *and let* A *be unitarily similar to* B. *Then*

$$\sum_{i,j=1}^{n} |a_{ij}|^2 = \sum_{i,j=1}^{n} |b_{ij}|^2.$$

Proof. By the given condition, there is a unitary matrix U such that $U^{-1}AU = B$. Hence,

$$\sum_{i,j=1}^{n} |b_{ij}|^2 = \text{Tr}(B\overline{B'}) = \text{Tr}(U^{-1}A\overline{A'}U) = \text{Tr}(A\overline{A'}) = \sum_{i,j=1}^{n} |a_{ij}|^2. \quad \square$$

Theorem 8.7.11. *The eigenvalues of a Hermitian matrix are all real numbers, and the eigenvectors associated with different eigenvalues must be orthogonal.*

Proof. One can prove it by the same argument as in Theorem 8.5.1.

$\square$

Theorem 8.7.12. *Every $n \times n$ Hermitian matrix is unitarily similar to a diagonal matrix.*

Proof. We proceed by induction on n. The theorem obviously holds when $n = 1$. Assume that $n \geqslant 2$ and the theorem holds for any Hermitian matrix of order $n - 1$, that is, any Hermitian matrix of order $n - 1$ is unitarily similar to a diagonal matrix.

Let A be an $n \times n$ Hermitian matrix, and let λ be an eigenvalue of A and α a unit eigenvector associated with the eigenvalue λ. Then $A\alpha = \lambda\alpha$. Extend α to an orthonormal basis $\alpha_1 = \alpha, \alpha_2, \ldots, \alpha_n$ for the unitary space $\mathbb{C}^n$. Let $U = (\alpha_1, \alpha_2, \ldots, \alpha_n)$. Then U is a unitary matrix, and $AU = U \begin{pmatrix} \lambda & \star \\ 0 & B \end{pmatrix}$, so $\overline{U}'AU = \begin{pmatrix} \lambda & \star \\ 0 & B \end{pmatrix}$. Since A is a Hermitian matrix, $\begin{pmatrix} \lambda & \star \\ 0 & B \end{pmatrix}$ is also a Hermitian matrix, which implies that the "$\star$"-position must be a zero matrix and B is an $(n-1) \times (n-1)$ Hermitian matrix. Hence A is unitarily similar to $\begin{pmatrix} \lambda & 0 \\ 0 & B \end{pmatrix}$. By the inductive hypothesis, B is unitarily similar to an $(n-1) \times (n-1)$ diagonal matrix. So A is unitarily similar to an $n \times n$ diagonal matrix. $\square$

Let A be an $n \times n$ Hermitian matrix. By Theorem 8.7.12, there is a sequence of eigenvectors of A such that it is an orthonormal basis for $\mathbb{C}^n$.

The following theorem shows that the relative phases between the components of any eigenvectors of an $n \times n$ Hermitian matrix A can be computed from the eigenvalues of A and its n submatrices.

Theorem 8.7.13 (Eigenvector-eigenvalue identity). *Let A be an $n \times n$ Hermitian matrix. Assume that the n eigenvalues of A are $\lambda_1(A), \lambda_2(A), \ldots, \lambda_n(A)$, $v_i = (v_{i1}, \ldots, v_{ij}, \ldots, v_{in})'$ is a unit eigenvector associated with the eigenvalue $\lambda_i(A)$, where $i = 1, 2, \ldots, n$,* *and $v_1, v_2, \ldots, v_n$ are pairwise orthogonal. For $1 \leqslant j \leqslant n$, let M_j be the $(n-1) \times (n-1)$ submatrix of A after deleting the jth row and jth column of A and the $n-1$ eigenvalues of M_j be $\lambda_1(M_j), \lambda_2(M_j), \ldots, \lambda_{n-1}(M_j)$. Then*

$$|v_{ij}|^2 \prod_{\substack{k=1 \\ k \neq i}}^{n} (\lambda_i(A) - \lambda_k(A)) = \prod_{k=1}^{n-1} (\lambda_i(A) - \lambda_k(M_j)), \quad 1 \leqslant i, j \leqslant n.$$

$$(8.11)$$

Proof. Let $C = (v_1, v_2, \ldots, v_n)$. Then, C is a unitary matrix. By Corollary 5.5.18 and Exercise 28 of Chapter 5, the column vectors $v_1, v_2, \ldots, v_n$ of C are the eigenvectors of A^* associated with the eigenvalues $\lambda_1^*(A), \lambda_2^*(A), \ldots, \lambda_n^*(A)$ respectively, where

$$\lambda_k^*(A) = \prod_{\substack{t=1 \\ t \neq k}}^{n} \lambda_t(A), \quad k = 1, 2, \ldots, n.$$

Thus,

$$A^*C = C\mathrm{diag}(\lambda_1^*(A), \lambda_2^*(A), \ldots, \lambda_n^*(A)),$$

and hence

$$A^* = \sum_{k=1}^{n} \lambda_k^*(A) v_k \overline{v_k'}. \tag{8.12}$$

Let λ be a complex number. Note that $v_1, v_2, \ldots, v_n$ are the eigenvectors of $\lambda I_n - A$ associated with the eigenvalues $\lambda - \lambda_1(A), \lambda - \lambda_2(A), \ldots, \lambda - \lambda_n(A)$, respectively. Replacing A by $\lambda I_n - A$ in (8.12), we get

$$(\lambda I_n - A)^* = \sum_{k=1}^{n} \lambda_k^*(\lambda I_n - A) v_k \overline{v_k'}. \tag{8.13}$$

Letting $\lambda = \lambda_i(A)$ in (8.13), $i = 1, 2, \ldots, n$, we have

$$(\lambda_i(A) I_n - A)^* = \sum_{k=1}^{n} \lambda_k^*(\lambda_i(A) I_n - A) v_k \overline{v_k'} = \lambda_i^*(\lambda_i(A) I_n - A) v_i \overline{v_i'}.$$

$$(8.14)$$

Since the (j, j) entries of both sides of (8.14) are equal, we obtain

$$|\lambda_i(A)I_{n-1} - M_j| = \left(\prod_{\substack{k=1 \\ k \neq i}}^{n} (\lambda_i(A) - \lambda_k(A)) \right) |v_{ij}|^2.$$

Note that $|\lambda_i(A)I_{n-1} - M_j| = \sum_{k=1}^{n-1} (\lambda_i(A) - \lambda_k(M_j))$. Thus the theorem holds. $\square$

Remark 8.7.14. In August 2019, three physicists who studied neutrino scattering wrote an email to 2006 Fields Medalist Tao[10] about their discovery of the identity (8.11) in Theorem 8.7.13, which is referred to as the eigenvector-eigenvalue identity in [8]. This identity allows them to use the eigenvalues of the matrix to calculate the eigenvectors very simply, and calculating the eigenvector is very important for studying the behavior of neutrinos. They knew that the identity (8.11) is correct through a lot of calculations, and it is very efficient to calculate the eigenvectors, but it cannot be proved. So, the three physicists turned to the world-recognized mathematical genius Tao for help.

Tao said his first reaction when he saw this formula was "it looked too good to be true". If such a simple, beautiful, powerful identity existed, it should have appeared in the linear algebra textbook. However, as far as he knows, no linear algebra textbook has this identity. But this identity cannot be wrong because physicists have verified it is correct.

Only two hours later, Tao replied emails to three physicists. He admitted that he had never seen the identity, but the identity was correct, and he had strictly proved the identity to be correct and gave three completely different proofs. This wonderful story of mathematics and physics supporting each other spread rapidly worldwide, and then everyone discovered that this identity already existed in the literature. After carefully searching the literature, Tao and others found that Theorem 8.7.13 has a surprisingly rich history. It was the great mathematician Carl Gustav Jacob Jacobi who first discovered this theorem. In a paper on quadratic forms in 1834, Jacobi

---

[10]Terence Tao, Born in 1975, Australian-American mathematician.

proved an identity that is essentially the identity (8.11), when the term "matrix" has not yet appeared. In the next two hundred years, identity (8.11) was continuously discovered independently by mathematicians in various fields and then disappeared, waiting to be discovered next time. It was not until 2019, because of the cooperation of Tao and three physicists, that this theorem became widely known.

A detailed historical overview and various proofs of Theorem 8.7.13 can be found in the literature [8].

Definition 8.7.15. Let $A \in M_n(\mathbb{C})$. If $A\overline{A'} = \overline{A'}A$, then A is called a **normal matrix**.

Example 8.7.16. Note that Hermitian and unitary matrices are normal. For any real numbers a and b, the 2×2 real matrix $\begin{pmatrix} a & b \\ -b & a \end{pmatrix}$ is a normal matrix.

Theorem 8.7.17. *Let $A \in M_n(\mathbb{C})$. Then A is a normal matrix if and only if A is unitarily similar to a diagonal matrix.*

Proof. We only need to prove that if A is a normal matrix, then A is unitarily similar to a diagonal matrix. The converse holds by Definition 8.7.15.

Let $B = \frac{1}{2}(A + \overline{A'})$ and $C = -\frac{i}{2}(A - \overline{A'})$. Then $A = B + iC$. It is obvious that B and C are both Hermitian matrices. Since $A\overline{A'} = \overline{A'}A$, $(B + iC)(\overline{B'} - i\overline{C'}) = (\overline{B'} - i\overline{C'})(B + iC)$. By $\overline{B'} = B, \overline{C'} = C$, we have $2i(BC - CB) = 0$ which implies that $BC = CB$. Thus B and C can be simultaneously unitarily similar to diagonal matrices (see Exercise 45). So, A is unitarily similar to a diagonal matrix. $\square$

Theorem 8.7.18 (Schur[11] inequality). *Assume that $A = (a_{ij})_{n \times n} \in M_n(\mathbb{C})$ and $\lambda_1, \ldots, \lambda_n$ are the eigenvalues of A. Then*

$$\sum_{i,j=1}^{n} |a_{ij}|^2 \geqslant \sum_{i=1}^{n} |\lambda_i|^2, \tag{8.15}$$

and the equality holds if and only if A is a normal matrix.

---

[11]Issai Schur, 1875–1941, Russian mathematician who worked in Germany for most of his life.

Proof. In the equation (8.15), $\sum\limits_{i,j=1}^{n} |a_{ij}|^2 = \text{Tr}\,(\overline{A'}A)$. Since any complex matrix is unitarily similar to an upper triangular matrix (see Exercise 43), by Proposition 8.7.10, we may assume that A is an upper triangular matrix and $a_{11} = \lambda_1, a_{22} = \lambda_2, \ldots, a_{nn} = \lambda_n$. Then

$$
\begin{aligned}
\sum_{i,j=1}^{n} |a_{ij}|^2 &= \sum_{i=1}^{n} |a_{ii}|^2 + \sum_{1 \leqslant i < j \leqslant n} |a_{ij}|^2 \\
&= \sum_{i=1}^{n} |\lambda_i|^2 + \sum_{1 \leqslant i < j \leqslant n} |a_{ij}|^2 \\
&\geqslant \sum_{i=1}^{n} |\lambda_i|^2,
\end{aligned}
$$

and the equality holds if and only if A is a diagonal matrix. By Theorem 8.7.17, A is unitarily similar to a diagonal matrix if and only if A is a normal matrix. Hence the equality holds in Schur inequality if and only if A is a normal matrix. $\qquad\square$

The previous chapter defined positive definite matrices over the real number field. Similarly, over the field of complex numbers, we have the following definition.

Definition 8.7.19. Let H be an $n \times n$ Hermitian matrix. If $\overline{\alpha'}H\alpha > 0$ $(\overline{\alpha'}H\alpha \geqslant 0)$ holds for any nonzero vector $\alpha \in \mathbb{C}^n$, then H is called **positive definite (positive semi-definite)**; if $\overline{\alpha'}H\alpha < 0$ $(\overline{\alpha'}H\alpha \leqslant 0)$ holds for any nonzero vector $\alpha \in \mathbb{C}^n$, then H is called **negative definite (negative semi-definite)**.

The proofs of the following three theorems are similar to those of real matrices and are left as exercises.

Theorem 8.7.20. *Let H be an $n \times n$ Hermitian matrix. Then the following statements are equivalent:*

(1) *H is positive definite.*
(2) *There is an invertible matrix C such that $H = \overline{C'}C$.*
(3) *All leading principal minors of H are positive.*
(4) *All principal minors of H are positive.*
(5) *All eigenvalues of H are positive.*

Theorem 8.7.21. *Let A be an $n \times n$ complex matrix. Then there is a unitary matrix U and a positive semi-definite (Hermitian) matrix H such that $A = UH$, where H is uniquely determined by A. If A is invertible, then U is also uniquely determined by A and H is positive definite. This decomposition is called the* **polar decomposition** *of a complex matrix.*

Theorem 8.7.22. *Let $A \in M_{m \times n}(\mathbb{C})$. Then there are $U \in U(m), V \in U(n)$ and the block matrix $S = \begin{pmatrix} \mathrm{diag}(\lambda_1, \lambda_2, \ldots, \lambda_r) & 0 \\ 0 & 0 \end{pmatrix}$ such that $A = USV$, where $\lambda_1, \lambda_2, \ldots, \lambda_r$ are positive numbers, called the* **singular value** *of A. This decomposition is called the* **singular value decomposition** *of A.*

The Moore–Penrose inverse A^+ of a complex matrix A can also be defined as follows.

Definition 8.7.23. Let $A \in M_{m \times n}(\mathbb{C})$. If $A^+ \in M_{n \times m}(\mathbb{C})$ satisfies

$$AA^+A = A, \quad A^+AA^+ = A^+, \quad (\overline{AA^+})' = AA^+, \quad (\overline{A^+A})' = A^+A,$$

then A^+ is called the **Moore–Penrose inverse** of A.

Similar to 8.5.17, one can prove that the Moore–Penrose inverse of A exists and is unique. Assume that $A \in M_{m \times n}(\mathbb{C})$ has the singular value decomposition $A = USV$, where $U \in U(m)$, $V \in U(n)$ and $S = \begin{pmatrix} D & 0 \\ 0 & 0 \end{pmatrix}$, where D is a diagonal matrix whose diagonal elements are the singular values of A. Then $A^+ = \overline{V}'S^+\overline{U}'$, where $S^+ = \begin{pmatrix} D^{-1} & 0 \\ 0 & 0 \end{pmatrix}$ is the Moore–Penrose inverse of S.

Similar to Remark 8.5.18, it is easy to prove that the Moore–Penrose inverse of a complex matrix has the following properties:

(1) If $A \in GL_n(\mathbb{C})$, then $A^+ = A^{-1}$.
(2) If $A \in M_{m \times n}(\mathbb{C})$, then $(A^+)^+ = A$.
(3) If $A \in M_{m \times n}(\mathbb{C})$, $(\overline{A^+})' = (\overline{A}')^+$.
(4) If $A \in M_{m \times n}(\mathbb{C})$, $B \in M_{n \times p}(\mathbb{C})$, then

$$(AB)^+ = (A^+AB)^+(ABB^+)^+.$$

Exercises

1. Assume that the inner product on $\mathbb{R}[x]_4$ is defined as $(f(x), g(x)) = \int_0^1 f(x)g(x)\mathrm{d}x$.

   (1) Show that $\mathbb{R}[x]_4$ is a Euclidean space.
   (2) Find the polynomials of $\mathbb{R}[x]_4$ that are orthogonal to all of $1, x, x^2$.

2. Show that any real square matrix A can be decomposed as $A = QR$, where Q is an orthogonal matrix, R is an upper triangular matrix. This decomposition is called the QR **decomposition** of a real matrix.

3. Show that any positive definite matrix A can be decomposed as $A = T'T$, where T is an upper triangular matrix with positive diagonal elements. This decomposition is called the **Cholesky**[12] **decomposition**.

4. Use Schmidt orthonormalization to transform the following linearly independent vectors to orthonormal vectors:

   (1) $(1, 2, 2)$, $(-1, 0, 2) \in \mathbb{R}^3$;
   (2) $(1, -1, 1, -1)$, $(1, 1, 3, -1)$, $(3, 7, 1, -3) \in \mathbb{R}^4$.

5. If the row vectors of a real matrix $A = (a_{ij})_{n \times n}$ are pairwise orthogonal, i.e., for any $1 \leqslant i \neq j \leqslant n$, $a_{i1}a_{j1} + a_{i2}a_{j2} + \cdots + a_{in}a_{jn} = 0$, then are the column vectors of A also pairwise orthogonal? Equivalently, does $a_{1i}a_{1j} + a_{2i}a_{2j} + \cdots + a_{ni}a_{nj} = 0$ hold for any $1 \leqslant i \neq j \leqslant n$? Please prove your assertion or give a counterexample.

6. Let $W = \{A \in \mathrm{M}_n(\mathbb{R}) \mid A = A'\}$. We define $f : W \times W \longrightarrow \mathbb{R}$ such that $f(A, B) = \mathrm{Tr}(AB)$.

   (1) Show that f is an inner product on W.
   (2) Let $S = \{A \in W \mid \mathrm{Tr}(A) = 0\}$. Find the orthogonal complement $S^\perp$ of S in W.

7. Find an example to illustrate that if the linear transformation σ of a Euclidean space is not an orthogonal transformation, then the orthogonal complement of a σ-subspace is not necessarily a σ-subspace.

---

[12] André-Louis Cholesky, 1875–1918, French mathematician.

8. Show that the modulus of the complex eigenvalue of an orthogonal transformation is equal to 1.

9. Let $A \in SO(3)$. Show that there are three real numbers θ, ψ, ϕ such that

$$A = \begin{pmatrix} \cos\theta\cos\psi & -\cos\phi\sin\psi + \sin\phi\sin\theta\cos\psi & \sin\phi\sin\psi + \cos\phi\sin\theta\cos\psi \\ \cos\theta\sin\psi & \cos\phi\cos\psi + \sin\phi\sin\theta\sin\psi & -\sin\phi\cos\psi + \cos\phi\sin\theta\sin\psi \\ -\sin\theta & \sin\phi\cos\theta & \cos\phi\cos\theta \end{pmatrix}.$$

10. Let $\alpha_1, \alpha_2, \ldots, \alpha_m$ and $\beta_1, \beta_2, \ldots, \beta_m$ be two sequences of vectors of an n-dimensional Euclidean space V. Prove that there is an orthogonal transformation σ such that $\sigma(\alpha_i) = \beta_i, i = 1, 2, \ldots, m$ if and only if $(\alpha_i, \alpha_j) = (\beta_i, \beta_j), i, j = 1, 2, \ldots, m$.

11. Let A be an $n \times n$ orthogonal matrix and k a positive odd integer. Prove that there is an $n \times n$ orthogonal matrix B such that $A = B^k$.

12. Let $e_1 = (1, \ 0, \ 0)', \ e_2 = (0, \ 1, \ 0)', \ e_3 = (0, \ 0, \ 1)'$ be an orthonormal basis for $\mathbb{R}^3$. Rotate all the vectors in $\mathbb{R}^3$ around the fixed vector $(1, \ 1, \ 1)'$ through $120°$ by the right-hand rule. Find the matrix of the rotation with respect to the basis $e_1, \ e_2, \ e_3$.

13. Calculate $(1 + 2\mathbf{i} + 3\mathbf{j} + 4\mathbf{k})(5 + 6\mathbf{i} + 7\mathbf{j} + 8\mathbf{k})$.

14. Find the inverse of the quaternion $1 + \mathbf{i} + \mathbf{j} + \mathbf{k}$.

15. If a natural number $n = a^2 + b^2 + c^2 + d^2$, where $a, b, c, d \in \mathbb{Z}$, then we say that n is a sum of four squares. Show that if m, n are both sums of squares, then mn is a sum of four squares.

16. Use quaternions to prove the Rodrigues rotation formula in Example 5.2.5.

17. Let A be an $n \times n$ real symmetric matrix and B an $n \times n$ positive definite matrix. Prove that there is an $n \times n$ invertible matrix C such that both $C'AC$ and $C'BC$ are diagonal matrices.

18.* Assume that A, B, C are all positive definite matrices. Prove that if ABC is a symmetric matrix, i.e., $ABC = CBA$, then ABC is a positive definite matrix.

19.* Assume that A and B are $n \times n$ positive semi-definite matrices. Prove that there is an $n \times n$ invertible matrix C such that both $C'AC$ and $C'BC$ are diagonal matrices.

20. Prove that a linear transformation σ of a Euclidean space V is a mirror reflection if and only if there is a unit vector of V such that $\sigma(\alpha) = \alpha - 2(\eta, \alpha)\eta$ for any vector $\alpha \in V$.

21. Let α, β be two unit vectors of a Euclidean space V. Prove that there is a mirror reflection σ such that $\sigma(\alpha) = \beta$.

22. Let σ be a transformation of a Euclidean space V. Prove that if for any $\alpha, \beta \in V$, we have $(\sigma(\alpha), \sigma(\beta)) = (\alpha, \beta)$, then σ is a linear transformation which implies that σ is an orthogonal transformation.

23. Find an orthogonal matrix T such that $T'AT$, where A is

$$(1)\ \begin{pmatrix} -3 & 4 \\ 4 & 3 \end{pmatrix}; \quad (2)\ \begin{pmatrix} -1 & 1 & 1 \\ 1 & -1 & 1 \\ 1 & 1 & -1 \end{pmatrix}; \quad (3)\ \begin{pmatrix} 1 & 0 & 1 \\ 0 & 1 & 0 \\ 1 & 0 & 1 \end{pmatrix};$$

$$(4)\ \begin{pmatrix} 0 & 1 & 0 \\ 1 & 0 & 1 \\ 0 & 1 & 0 \end{pmatrix}; \quad (5)\ \begin{pmatrix} 2 & 2 & -2 \\ 2 & 5 & -4 \\ -2 & -4 & 5 \end{pmatrix}; \quad (6)\ \begin{pmatrix} 1 & 1 & 1 & 1 \\ 1 & 1 & 1 & 1 \\ 1 & 1 & 1 & 1 \\ 1 & 1 & 1 & 1 \end{pmatrix}.$$

24. Use orthogonal linear substitutions to transform the following quadratic forms to standard forms:

(1) $2x_1^2 + 2x_2^2 - 2x_1x_2$;

(2) $x_1x_2 + x_2x_3$;

(3) $8x_1x_3 + 2x_1x_4 + 2x_2x_3 + 8x_2x_4$;

(4) $\displaystyle\sum_{1 \leqslant i < j \leqslant n} x_ix_j$.

25. Assume that the sum of the first-order principal minors and the sum of the second-order principal minors of a real symmetric matrix A are both 0. Prove that $A = 0$.

26. Give an example to illustrate that the matrix of a symmetric transformation with respect to a nonorthonormal basis may not be symmetric.

27. Let A be an $n \times n$ real symmetric matrix. Prove that there is a real number $c > 0$ such that $|X'AX| \leqslant cX'X$ for any n-dimensional real vector X.

28. (Rayleigh[13] Theorem) Let $S^{n-1} = \{X \in \mathbb{R}^n \mid |X| = 1\}$ be the $(n-1)$-dimensional sphere and A an $n \times n$ real symmetric matrix. Prove that $\lambda \leqslant X'AX \leqslant \mu$ for any $X \in S^{n-1}$, where λ is the least eigenvalue of A and μ is the greatest eigenvalue of A.

---

[13] John William Strutt, 3rd Baron Rayleigh, 1842–1919, British physicist and 1904 Nobel Prize winner.

29. Let V_1 and V_2 be two subspaces of an n-dimensional Euclidean space V. Prove that if $\dim V_1 < \dim V_2$, then there is a nonzero $\alpha \in V_2$ such that $\alpha \perp V_1$.

30. Let A and B be two $n \times n$ positive semi-definite matrices such that A^2 is orthogonally similar to B^2. Prove that A is orthogonally similar to B.

31.* Let $A \in M_{m \times n}(\mathbb{R})$. Prove that $A^+ = \lim\limits_{\varepsilon \to 0} (A'A + \varepsilon I_n)^{-1} A'$.

32.* Let $A = GH \in M_{m \times n}(\mathbb{R})$ and $\mathrm{rank} A = r \geqslant 1$, where $G \in M_{m \times r}(\mathbb{R})$, $H \in M_{r \times n}(\mathbb{R})$. Prove that $A^+ = H'(HH')^{-1}(G'G)^{-1}G'$.

33. Find the least squares solution of the following system of linear equations (calculate with three significant figures):
$$\begin{cases} a + b + c = 10, \\ 4a + 2b + c = 5.49, \\ 9a + 3b + c = 0.89, \\ 16a + 4b + c = -0.14, \\ 25a + 5b + c = -1.07, \\ 36a + 6b + c = 0.84. \end{cases}$$

34. There are five points $(-2, 0)$, $(-1, 0)$, $(0, 1)$, $(1, 0)$, $(2, 0)$ on the xy plane. Find the closest parabola to these 5 points using the method of least squares.

35. Prove that a 2×2 real normal matrix must be symmetric or of the form $\begin{pmatrix} a & b \\ -b & a \end{pmatrix}$.

36. Prove that $A = \begin{pmatrix} 5 & -3 \\ 4 & -2 \end{pmatrix}$ and $B = \begin{pmatrix} 1 & -1 \\ 0 & 2 \end{pmatrix}$ are similar but not orthogonally similar.

37. (1) Assume that the real matrix $A = \begin{pmatrix} a_{11} & a_{12} \\ a_{21} & a_{22} \end{pmatrix}$ is similar to the real matrix $B = \begin{pmatrix} b_{11} & b_{12} \\ b_{21} & b_{22} \end{pmatrix}$, and $\sum\limits_{i,j=1}^{2} a_{ij}^2 = \sum\limits_{i,j=1}^{2} b_{ij}^2$. Prove that A is orthogonally similar to B.

    (2) Give an example to illustrate that the conclusion in (1) is not valid for 3×3 matrices.

38. Let $A \in M_n(\mathbb{R})$ and B be an $n \times n$ positive definite matrix. If k is a positive integer and $AB^k = B^k A$, prove that $AB = BA$.

39.* (1) Let $A, B \in M_n(\mathbb{R})$. Prove that A is orthogonally similar to B if and only if there is a $P \in GL_n(\mathbb{R})$ such that $P^{-1}AP = B$, $P^{-1}A'P = B'$.

(2) Let $A, B \in M_n(\mathbb{C})$. Prove that A is unitarily similar to B if and only if there is a $P \in GL_n(\mathbb{C})$ such that $P^{-1}AP = B$, $P^{-1}\overline{A'}P = \overline{B'}$.

40. Let $A \in M_n(\mathbb{R})$ be a real normal matrix. Prove that A is symmetric if and only if all the eigenvalues of A are real numbers.

41. Let S be an $n \times n$ positive semi-definite matrix and O an $n \times n$ orthogonal matrix. If the characteristic polynomial of SO is equal to the characteristic polynomial of S, prove that $SO = S$.

42.* Let $A, B \in M_n(\mathbb{R})$. If A is unitarily similar to B, prove that A is orthogonally similar to B.

43. Prove that any $n \times n$ complex matrix is unitarily similar to an upper triangular matrix.

44. Assume that an $n \times n$ matrix $A = B + iC$, where B and C are $n \times n$ real matrices. Prove that A is a unitary matrix if and only if $B'C$ is symmetric and $B'B + C'C = I_n$.

45. Assume that A, B are two $n \times n$ complex matrices satisfying $AB = BA$, and both of them are unitarily similar to diagonal matrices. Prove that A and B can be simultaneously unitarily similar to diagonal matrices.

46. Prove Theorem 8.7.20.

47. Prove Theorem 8.7.21.

48. Prove Theorem 8.7.22.

Chapter 9

Bilinear Forms

9.1 Introduction

At the beginning of the 17th century, René Descartes and Pierre de Fermat conducted in-depth research on bilinear forms on 2-dimensional linear spaces. Later in the 18th century, Leonhard Euler extended the study of bilinear forms to 3-dimensional spaces. At the beginning of 19th century, Monge[1] studied the bilinear forms in the modern sense. Camille Jordan *et al.* systematically studied the profound connection between bilinear forms and classical group theory in the late 19th century. Skew-symmetric bilinear forms are widely used in geometry and physics. In 1930, Hermann Weyl named linear spaces with skew-symmetric bilinear forms symplectic spaces, and the group consisting of the linear transformations that preserve the metric of a symplectic space is called the symplectic group.

9.2 Bilinear Forms

Unless otherwise specified, a linear space is a finite-dimensional linear space over a number field F.

---

[1]Gaspard Monge, 1746–1818, French mathematician.

Definition 9.2.1. Let V be a linear space. A map $f : V \times V \longrightarrow F$ is called a **bilinear form** on V if it satisfies the following conditions:

(1) $f(k\alpha + l\beta, \gamma) = kf(\alpha, \gamma) + lf(\beta, \gamma)$,
(2) $f(\alpha, k\beta + l\gamma) = kf(\alpha, \beta) + lf(\alpha, \gamma)$,

where $\alpha, \beta, \gamma \in V, \ k, l \in F$.

Let f be a bilinear form on V. If $f(\alpha, \beta) = f(\beta, \alpha)$ for $\alpha, \beta \in V$, then we call f a **symmetric bilinear form**. If $f(\alpha, \beta) = -f(\beta, \alpha)$ for $\alpha, \beta \in V$, then we call f a **skew-symmetric bilinear form**. If $F = \mathbb{C}$, and $f(\alpha, \beta) = \overline{f(\beta, \alpha)}$ for $\alpha, \beta \in V$, then we call f a **Hermitian form**.

Example 9.2.2. The function $f\left((x_1, y_1), (x_2, y_2)\right) = x_1 y_2 - x_2 y_1$ on $\mathbb{R}^2$ is a skew-symmetric bilinear form, and the function $g\left((x_1, y_1), (x_2, y_2)\right) = x_1 x_2 - y_1 y_2$ is a symmetric bilinear form.

Definition 9.2.3. Let f be a bilinear form on V, and let $\alpha_1, \alpha_2, \ldots, \alpha_n$ be a basis for V. Then the matrix

$$\begin{pmatrix} f(\alpha_1, \alpha_1) & f(\alpha_1, \alpha_2) & \cdots & f(\alpha_1, \alpha_n) \\ f(\alpha_2, \alpha_1) & f(\alpha_2, \alpha_2) & \cdots & f(\alpha_2, \alpha_n) \\ \vdots & \vdots & & \vdots \\ f(\alpha_n, \alpha_1) & f(\alpha_n, \alpha_2) & \cdots & f(\alpha_n, \alpha_n) \end{pmatrix}$$

is called the **metric matrix** of f with respect to the basis $\alpha_1, \alpha_2, \ldots, \alpha_n$.

Example 9.2.4. In Example 9.2.2, the matrices of the bilinear forms f and g with respect to the basis $\alpha_1 = (1, 0), \alpha_2 = (0, 1)$ for $\mathbb{R}^2$ are the skew-symmetric matrix $\begin{pmatrix} 0 & 1 \\ -1 & 0 \end{pmatrix}$ and the symmetric matrix $\begin{pmatrix} 1 & 0 \\ 0 & -1 \end{pmatrix}$, respectively.

Theorem 9.2.5. *Assume that the metric matrices of a bilinear form f with respect to bases $\alpha_1, \alpha_2, \ldots, \alpha_n$ and $\beta_1, \beta_2, \ldots, \beta_n$ for a linear space V are A and B respectively. If the transition matrix from $\alpha_1, \alpha_2, \ldots, \alpha_n$ to $\beta_1, \beta_2, \ldots, \beta_n$ is C, then $B = C'AC$.*

Proof. By the given condition, $(\beta_1, \beta_2, \ldots, \beta_n) = (\alpha_1, \alpha_2, \ldots, \alpha_n)C$, where $C \in \mathrm{GL}_n(F)$. Let $C = (c_{ij})_{n \times n}$. Then

$$(\beta_i, \beta_j) = \sum_{k,l=1}^{n} c_{ki}(\alpha_k, \alpha_l)c_{lj}$$

for any $1 \leqslant i, j \leqslant n$. By the multiplication of matrices, $B = C'AC$.
$\square$

By the theorem above, the metric matrices of a bilinear form f with respect to different bases for a linear space V are congruent. Since congruent matrices have the same rank, we call the rank of the metric matrix of the bilinear form f with respect to a basis for V the **rank** of f, denoted by $\mathrm{rank}(f)$.

Definition 9.2.6. Let f be a bilinear form on a linear space V. We define

$$\mathrm{rad}_L(f) = \{v \in V \mid f(v, w) = 0 \ \text{for any } w \in V\},$$

$$\mathrm{rad}_R(f) = \{w \in V \mid f(v, w) = 0 \ \text{for any } \ v \in V\}.$$

$\mathrm{rad}_L(f)$ and $\mathrm{rad}_R(f)$ are called the **left radical** and **right radical** of f, respectively.

Theorem 9.2.7. *Let f be a bilinear form on an n-dimensional linear space V. Then the following statements are equivalent:*

(1) The metric matrix of f with respect to a basis for V is invertible.
(2) $\mathrm{rad}_L(f) = 0$.
(3) $\mathrm{rad}_R(f) = 0$.

*If f satisfies the above equivalence conditions, then f is called a **non-degenerate bilinear form**; otherwise, f is called a **degenerate bilinear form**.*

Proof. Let $\alpha_1, \alpha_2, \ldots, \alpha_n$ be a basis for V.

$(2) \implies (1)$. If the metric matrix $(f(\alpha_i, \alpha_j))_{n \times n}$ of f with respect to the basis $\alpha_1, \alpha_2, \ldots, \alpha_n$ is not invertible, then there are $c_1, c_2, \ldots, c_n \in F$, not all zero, such that

$$\sum_{i=1}^{n} c_i(f(\alpha_i, \alpha_1), \ldots, f(\alpha_i, \alpha_n)) = 0,$$

so $\left(f\left(\sum_{i=1}^{n} c_i\alpha_i, \alpha_1 \right), \ldots, f\left(\sum_{i=1}^{n} c_i\alpha_i, \alpha_n \right) \right) = 0$. Since $c_1, c_2, \ldots, c_n$ are not all zero, $\alpha = \sum_{i=1}^{n} c_i\alpha_i \neq 0$, and $f(\alpha, \alpha_i) = 0$, $i = 1, 2, \ldots, n$.

Let $\beta \in V$. Then there are $x_1, x_2, \ldots, x_n \in F$ such that $\beta = \sum_{i=1}^{n} x_i \alpha_i$. So,

$$f(\alpha, \beta) = f\left(\alpha, \sum_{i=1}^{n} x_i \alpha_i\right) = \sum_{i=1}^{n} x_i f(\alpha, \alpha_i) = 0.$$

Hence, $\alpha \in \mathrm{rad}_L(f)$, contradicting (2).

(1) $\Longrightarrow$ (2). If there is a nonzero vector $\alpha = \sum_{i=1}^{n} c_i \alpha_i \in \mathrm{rad}_L(f)$, then

$$\left(f\left(\sum_{i=1}^{n} c_i \alpha_i, \alpha_1\right), \ldots, f\left(\sum_{i=1}^{n} c_i \alpha_i, \alpha_n\right)\right) = 0,$$

i.e., $\sum_{i=1}^{n} c_i(f(\alpha_i, \alpha_1), \ldots, f(\alpha_i, \alpha_n)) = 0$. Thus the row vectors of the metric matrix $(f(\alpha_i, \alpha_j))_{n \times n}$ of f with respect to the basis $\alpha_1, \alpha_2, \ldots, \alpha_n$ are linearly dependent, and so $(f(\alpha_i, \alpha_j))_{n \times n}$ is not invertible, a contradiction.

Similarly, we can prove that (1) $\Longleftrightarrow$ (3). $\qquad\qquad\square$

It is easy to see that if f is a symmetric (skew-symmetric, Hermitian) bilinear form on a linear space V, then the metric matrix of f with respect to any basis for V is also a symmetric (skew-symmetric, Hermitian) matrix.

9.3*   Quadratic Spaces

Definition 9.3.1. Let f be a symmetric bilinear form on a linear space V. Define $Q : V \longrightarrow F$ such that $Q(\alpha) = f(\alpha, \alpha)$ for any $\alpha \in V$. Then Q is called a **quadratic form** on V, and V is called a **quadratic space** with the quadratic form Q, denoted by (V, Q). If there is no confusion, (V, Q) is abbreviated as V. If f is a nondegenerate symmetric bilinear form, then V is called a **nondegenerate quadratic space**; otherwise, V is called a **degenerate quadratic space**.

Remark 9.3.2. Let (V, Q) be a nonzero quadratic space, and let $\alpha_1, \alpha_2, \ldots, \alpha_n$ be a basis for V.

(1) Let $g(x_1, x_2, \ldots, x_n) = Q\left(\sum\limits_{i=1}^{n} x_i \alpha_i\right)$. Then

$$g(x_1, x_2, \ldots, x_n) = f\left(\sum_{i=1}^{n} x_i \alpha_i, \sum_{i=1}^{n} x_i \alpha_i\right) = \sum_{i,j=1}^{n} f(\alpha_i, \alpha_j) x_i x_j$$

is the quadratic form in n indeterminates we introduced in Chapter 7, called the quadratic form on V with respect to the basis $\alpha_1, \alpha_2, \ldots, \alpha_n$.

(2) By Definition 9.3.1, the quadratic form Q is uniquely determined by the symmetric bilinear form f. On the contrary, the quadratic form Q also uniquely determines the bilinear form f. In fact,

$$f(\alpha, \beta) = \frac{1}{2}(f(\alpha + \beta, \alpha + \beta) - f(\alpha, \alpha) - f(\beta, \beta))$$

$$= \frac{1}{2}(Q(\alpha + \beta) - Q(\alpha) - Q(\beta)),$$

where $\alpha, \beta \in V$. It follows that the quadratic form Q on V is in one-to-one correspondence to the symmetric bilinear form f on V.

Definition 9.3.3. Let V be a quadratic space with the corresponding symmetric bilinear form f. If $\alpha, \beta \in V$ satisfy $f(\alpha, \beta) = 0$, then we say that α is **orthogonal** to β, or α and β are orthogonal. If W is a subspace of V, the orthogonal complement of W is defined by

$$W^{\perp} = \{\alpha \in V \mid f(\alpha, \beta) = 0 \text{ for any } \beta \in W\}.$$

$V^{\perp}$ is called the **radical space** of V, denoted by $\mathrm{rad}(V)$. Let V_1 and V_2 be two subspaces of V. If all vectors of V_1 are orthogonal to all vectors of V_2, then we say that V_1 is **orthogonal** to V_2.

By Theorem 9.2.7, $\mathrm{rad}(V) = 0$ if and only if V is nondegenerate.

Definition 9.3.4. Let V be a quadratic space. If a nonzero vector $\alpha \in V$ satisfies $Q(\alpha) = 0$, then α is called an **isotropic vector**. A subspace containing an isotropic vector is called an **isotropic subspace**, and a subspace consisting of zero and isotropic vectors is called a **totally isotropic subspace**.

Definition 9.3.5. Let $\alpha_1, \alpha_2, \ldots, \alpha_n$ be a basis for a quadratic space V. If $\alpha_1, \alpha_2, \ldots, \alpha_n$ are pairwise orthogonal, then we say that the basis $\alpha_1, \alpha_2, \ldots, \alpha_n$ is an **orthogonal basis**. If every vector α_i in an orthogonal basis $\alpha_1, \alpha_2, \ldots, \alpha_n$ satisfies $Q(\alpha_i) = 1, 1 \leqslant i \leqslant n$, then the basis $\alpha_1, \alpha_2, \ldots, \alpha_n$ is called an **orthonormal basis**.

Proposition 9.3.6. *Let W be a nondegenerate nonzero subspace of a quadratic space V. Then $V = W \oplus W^\perp$.*

Proof. First, if there is a vector α such that $0 \neq \alpha \in W \cap W^\perp$, then $\alpha \in W^\perp$, i.e., α is orthogonal to W, which implies that W is degenerate, a contradiction. Hence, $W + W^\perp$ is a direct sum.

Second, let $\alpha \in V$, and let $\alpha_1, \alpha_2, \ldots, \alpha_k$ be a basis for W. Next, we will prove that there are $c_1, c_2, \ldots, c_k \in F$ such that

$$\alpha - c_1 \alpha_1 - c_2 \alpha_2 - \cdots - c_k \alpha_k \in W^\perp.$$

Let f be the symmetric bilinear form on V, and let A be the metric matrix of the restriction of f to W with respect to the basis $\alpha_1, \alpha_2, \ldots, \alpha_k$. By the given condition, A is invertible. It is easy to see that

$$f(\alpha - c_1 \alpha_1 - c_2 \alpha_2 - \cdots - c_k \alpha_k, \alpha_i) = 0 \quad \text{for } i = 1, \ldots, k$$

if and only if $(c_1, c_2, \ldots, c_k)'$ is a solution of the linear system $AX = \beta$, where $\beta = (f(\alpha, \alpha_1), f(\alpha, \alpha_2), \ldots, f(\alpha, \alpha_n))'$. Thus $(c_1, c_2, \ldots, c_k)' = A^{-1}\beta$ is what we need.

Let $\gamma = c_1 \alpha_1 + c_2 \alpha_2 + \cdots + c_k \alpha_k$. Then $\alpha = \gamma + (\alpha - \gamma) \in W + W^\perp$. So $V = W \oplus W^\perp$. $\qquad\qquad\square$

Proposition 9.3.7. *Any nonzero quadratic space has an orthogonal basis.*

Proof. Let V be a nonzero quadratic space. We proceed by induction on the dimension n of V. The case $n = 1$ is trivial. For any $n \geqslant 2$, we assume that proposition holds for any quadratic space of dimension less than n.

Let f be the symmetric bilinear form on V. If the space V is a totally isotropic space, then

$$f(\alpha, \beta) = \frac{1}{2}(f(\alpha + \beta, \alpha + \beta) - f(\alpha, \alpha) - f(\beta, \beta)) = 0$$

for any $\alpha, \beta \in V$. Hence, any basis is orthogonal.

If the quadratic space V is not a totally isotropic space, then there is a non-isotropic nonzero vector $\alpha \in V$. By Proposition 9.3.6, $V = L(\alpha) \oplus L(\alpha)^{\perp}$. Note that $L(\alpha)^{\perp}$ has an orthogonal basis $\alpha_1, \ldots, \alpha_{n-1}$ by induction. So the sequence $\alpha, \alpha_1, \ldots, \alpha_{n-1}$ is an orthogonal basis for V. $\qquad\square$

Remark 9.3.8.

(1) Proposition 9.3.7 is equivalent to Theorem 7.2.15 by Remark 9.3.2.

(2) Although nonzero quadratic space V has an orthogonal basis, it may not have an orthogonal basis consisting of nonisotropic vectors. In fact, a quadratic space V has an orthogonal basis consisting of nonisotropic vectors if and only if V is nondegenerate. The proof is left to the reader.

(3) Any sequence of orthogonal vectors of a Euclidean space can be extended to an orthogonal basis for the Euclidean space. However, this conclusion is not valid for a general quadratic space. For example, if f is the symmetric bilinear form on the 2-dimensional quadratic space V and the metric matrix of f with respect to a basis α_1, α_2 for V is $\begin{pmatrix} 1 & 0 \\ 0 & -1 \end{pmatrix}$, then $f(\alpha_1 + \alpha_2, \alpha_1 + \alpha_2) = 0$. Next, we prove that $\alpha_1 + \alpha_2$ cannot be extended to an orthogonal basis for V. In fact, if $f(x\alpha_1 + y\alpha_2, \alpha_1 + \alpha_2) = 0$, where $x, y \in F$, then $x = y$. Hence, the vector orthogonal to $\alpha_1 + \alpha_2$ must be in $L(\alpha_1 + \alpha_2)$. So $\alpha_1 + \alpha_2$ can not be extended to an orthogonal basis for V.

Theorem 9.3.9. *Let $n \geqslant 1$ be an integer, and let U and W be two subspaces of an n-dimensional non-degenerate quadratic space V. Then the following statements hold:*

(1) $V^{\perp} = 0$, $0^{\perp} = V$.

(2) *If $U \subseteq W$, then $W^{\perp} \subseteq U^{\perp}$.*

(3) $n = \dim U + \dim U^{\perp}$.

(4) $U = (U^{\perp})^{\perp}$.

(5) $(U + W)^{\perp} = U^{\perp} \cap W^{\perp}$.

(6) $(U \cap W)^{\perp} = U^{\perp} + W^{\perp}$.

(7) $\operatorname{rad}(U) = \operatorname{rad}(U^{\perp}) = U \cap U^{\perp}$.

Proof. First, we prove (3). By Proposition 9.3.7, V has an orthogonal basis $\alpha_1, \alpha_2, \ldots, \alpha_n$. If $U = 0$, the conclusion holds. For $U \neq 0$, let $\beta_j = \sum_{i=1}^{n} b_{ij}\alpha_i, j = 1, 2, \ldots, k$, be a basis for U. Assume that f is the symmetric bilinear form on V. Then

$$\gamma = \sum_{l=1}^{n} c_l \alpha_l \in U^{\perp} \iff f(\beta_j, \gamma) = 0, \quad j = 1, 2, \ldots, k$$

$$\iff \sum_{l,i=1}^{n} b_{ij} f(\alpha_i, \alpha_l) c_l = 0, \quad j = 1, 2, \ldots, k$$

$$\iff B'A(c_1, c_2, \ldots, c_n)' = 0,$$

where $A = (f(\alpha_i, \alpha_l))_{n \times n} \in \mathrm{GL}_n(F)$ is the metric matrix of f with respect to the basis $\alpha_1, \alpha_2, \ldots, \alpha_n$, $B = (b_{ij})_{n \times k} \in \mathrm{M}_{n \times k}(F)$ is a column full rank matrix.

By Theorem 3.4.2, the dimension of the solution space of $B'AX = 0$ is $n - \mathrm{rank}(B'A) = n - k$, so $\dim U^{\perp} = n - k$, which implies that $n = \dim U + \dim U^{\perp}$.

(7) follows from the definition of a radical space. The proofs of the other statements are similar to those of Theorem 8.2.21. $\qquad\square$

Definition 9.3.10. Let (V, Q) and (V', Q') be two quadratic spaces, and let φ be a linear map from (V, Q) to (V', Q'). If $Q'(\varphi(\alpha)) = Q(\alpha)$ for any $\alpha \in V$, then φ is called an **isometric map** from (V, Q) to (V', Q'), abbreviated as **isometry**. If $(V, Q) = (V', Q')$, then we call φ an **isometric transformation** of V. If an isometry φ is bijective, then we call φ an **isometric isomorphism**. Two quadratic spaces (V, Q) and (V', Q') are said to be **isomorphic** if there is an isometric isomorphism from (V, Q) to (V', Q').

Remark 9.3.11. Let (V, Q) be a nondegenerate quadratic space and φ an isometry from (V, Q) to (V', Q'). Then φ is injective. In fact, if φ is not injective, then there is a nonzero vector α such that $\varphi(\alpha) = 0$. Let Q and Q' correspond to the symmetric bilinear forms f and f', respectively. Then for any $\beta \in V$, $f(\alpha, \beta) = f'(\varphi(\alpha), \varphi(\beta)) = 0$, contradicting the fact that V is a nondegenerate quadratic space. Hence, an isometric transformation of a (finite-dimensional) nondegenerate quadratic space must be an isomorphism.

Theorem 9.3.12 (Witt's[2] extension theorem). *Let (V, Q) and (V', Q') be two isomorphic nondegenerate quadratic spaces. If U is a subspace of V and $\sigma : U \longrightarrow V'$ is an isometry, then there is an isometric isomorphism $\tau : V \longrightarrow V'$ such that $\tau|_U = \sigma$, i.e., σ can be extended to an isometric isomorphism from V to V'.*

Proof. Without loss of generality, we can assume that $(V, Q) = (V', Q')$, for if $\varphi : V \longrightarrow V'$ is an isomorphism, then $\varphi^{-1}\sigma : U \longrightarrow V$ is an isometry. If we can prove that $\varphi^{-1}\sigma$ can be extended to an isometric isomorphism $\tau : V \longrightarrow V$, then $\varphi\tau$ is an extension of σ and an isometric isomorphism from (V, Q) to (V', Q').

If U is degenerate, then there is a nonzero element $\alpha \in \mathrm{rad}(U)$. Let W be a subspace of U such that $U = L(\alpha) \oplus W$. Then $L(\alpha)$ is orthogonal to W (since $\alpha \in \mathrm{rad}(U)$, α is orthogonal to any element of U). By Theorem 9.3.9, $W^\perp \supsetneq U^\perp$. So there is a $\gamma \in W^\perp$ such that $\gamma \notin U^\perp$. Let f be the symmetric bilinear form on V. If $f(\alpha, \gamma) = 0$, then $\gamma \in W^\perp \cap L(\alpha)^\perp = U^\perp$, a contradiction. So $f(\alpha, \gamma) \neq 0$. Multiplying γ by an appropriate number, we may assume $f(\alpha, \gamma) = 1$. Note that any vector of U is orthogonal to α, so $\gamma \notin U$. Let $\beta = \gamma - \frac{f(\gamma, \gamma)}{2}\alpha \in W^\perp$. By direct calculation, $f(\alpha, \beta) = 1$, $f(\beta, \beta) = 0$. So β is an isotropic vector and $\beta \notin U$. Similarly, we can find $\gamma_1 \in \sigma(W)^\perp$ such that $f(\sigma(\alpha), \gamma_1) = 1$. Let $\beta_1 = \gamma_1 - \frac{f(\gamma_1, \gamma_1)}{2}\sigma(\alpha) \in \sigma(W)^\perp$. Then $f(\sigma(\alpha), \beta_1) = 1$, $f(\beta_1, \beta_1) = 0$. Hence β_1 is an isotropic vector and $\beta_1 \notin \sigma(U)$.

Define $\widetilde{\sigma} : U \oplus L(\beta) \longrightarrow V$ such that $\widetilde{\sigma}(\alpha_1 + c\beta) = \sigma(\alpha_1) + c\beta_1$, where $\alpha_1 \in U$, $c \in F$. Then $\widetilde{\sigma}|_U = \sigma$, $f(\widetilde{\sigma}(\beta), \widetilde{\sigma}(\beta)) = f(\beta_1, \beta_1) = 0 = f(\beta, \beta)$. For any $a\alpha + w \in U$, where $a \in F, w \in W$, we have $f(a\alpha + w, \beta) = af(\alpha, \beta) + f(w, \beta) = a$, $f(\sigma(a\alpha + w), \beta_1) = af(\sigma(\alpha), \beta_1) + f(\sigma(w), \beta_1) = a$, which implies that for any $\alpha_1 \in U$, we have $f(\widetilde{\sigma}(\alpha_1), \widetilde{\sigma}(\beta)) = f(\sigma(\alpha_1), \beta_1) = f(\alpha_1, \beta)$. Hence, $\widetilde{\sigma}$ is an isometry. In this way, we extend σ to the space $U \oplus L(\beta)$ whose dimension is equal to $\dim U + 1$. So, we only need to prove the theorem for a nondegenerate U.

Next, we assume U is nondegenerate. We proceed by induction on $\dim U$.

---

[2]Ernst Witt, 1911–1991, German mathematician.

If $\dim U = 1$, then U is spanned by a nonisotropic vector α. Let $\beta = \sigma(\alpha)$. Then $f(\alpha, \alpha) = f(\beta, \beta)$. If

$$f(\alpha + \beta, \alpha + \beta) = f(\alpha - \beta, \alpha - \beta) = 0,$$

then $f(\alpha, \alpha) = f(\alpha, \beta) = 0$, contradicting the fact that α is non-isotropic. Assume that $f(\alpha + \beta, \alpha + \beta)$ and $f(\alpha - \beta, \alpha - \beta)$ are not both 0. Let $f(\alpha + \varepsilon\beta, \alpha + \varepsilon\beta) \neq 0$, where $\varepsilon = 1$ or -1, and $\gamma = \alpha + \varepsilon\beta$. By Proposition 9.3.6, $V = L(\gamma)^\perp \oplus L(\gamma)$. Define $\tau : V \longrightarrow V$ such that $\tau(\omega_1 + \omega_2) = \omega_1 - \omega_2$, where $\omega_1 \in L(\gamma)^\perp$, $\omega_2 \in L(\gamma)$. Obviously, τ is an isometric transformation, $\alpha - \varepsilon\beta \in L(\gamma)^\perp$, so $\tau(\alpha - \varepsilon\beta) = \alpha - \varepsilon\beta$, $\tau(\alpha + \varepsilon\beta) = -\alpha - \varepsilon\beta$. Hence $\tau(\alpha) = \frac{1}{2}(\tau(\alpha - \varepsilon\beta) + \tau(\alpha + \varepsilon\beta)) = -\varepsilon\beta$, which implies that $-\varepsilon\tau(\alpha) = \beta$. So $-\varepsilon\tau$ is an extension of σ.

Assume that $\dim(U) > 1$, and the theorem holds for any space of dimension less than $\dim(U)$. By Remark 9.3.8 (2), there are two orthogonal nondegenerate subspaces U_1, U_2 of U such that $U = U_1 \oplus U_2$. By the inductive hypothesis, the restriction of σ to U_1 can be extended to an automorphism ρ of V. Replacing σ by $\rho^{-1}\sigma$, we can assume that the restriction of σ to U_1 is an identity. Since σ is an isometry, σ maps the subspace U_2 which is orthogonal to U_1 to a subspace that is still orthogonal to U_1, i.e., $\sigma(U_2) \subseteq U_1^\perp$.

By the inductive hypothesis, the restriction $\sigma|_{U_2}$ of σ to U_2 can be extended to an isometric transformation ϕ of $U_1^\perp$. By Proposition 9.3.6, $V = U_1^\perp \oplus U_1$. Define the isometric transformation τ of V such that $\tau(\alpha + \beta) = \phi(\alpha) + \beta$, where $\alpha \in U_1^\perp$, $\beta \in U_1$. It is easy to see that τ is an extension of σ.

In summary, σ can be extended to an isometric isomorphism from (V, Q) to (V', Q'). $\qquad\square$

Theorem 9.3.13 (Witt's cancellation theorem). *Let (V, Q) and (V', Q') be two isomorphic nondegenerate quadratic spaces. Assume that U is a subspace of V and U' is a subspace of V'. If U is isomorphic to U', then $U^\perp$ is isomorphic to $U'^\perp$.*

Proof. If $\sigma : U \longrightarrow U'$ is an isometric isomorphism, then $\sigma : U \longrightarrow V'$ is an isometry. By Theorem 9.3.12, σ can be extended to an isometric isomorphism $\tau : V \longrightarrow V'$. Since τ is an isometry, $\tau(U^\perp) \subseteq \tau(U)^\perp = U'^\perp$. Furthermore, since τ is injective, $\tau|_{U^\perp}$ is also injective. It follows that $\dim\tau(U^\perp) = \dim U^\perp$. From $\dim U = \dim U'$, we have $\dim U^{\perp} = \dim U'^{\perp}$. Thus $\dim \tau(U^{\perp}) = \dim U'^{\perp}$, and hence $\tau(U^{\perp}) = U'^{\perp}$. So $U^{\perp}$ is isomorphic to $U'^{\perp}$. $\qquad\square$

9.4* Symplectic Spaces

In Chapter 8, we discussed Euclidean and unitary spaces; now, we discuss another type of metric spaces — symplectic spaces. Symplectic spaces have critical applications in many fields, such as mathematics and physics.

Definition 9.4.1. If V is a linear space and f is a nondegenerate skew-symmetric bilinear form on V, then (V, f) is called a **symplectic space**, abbreviated as V. Let (V_1, f_1), (V_2, f_2) be two symplectic spaces and $\sigma : V_1 \longrightarrow V_2$ a linear map. If $f_1(\alpha, \beta) = f_2(\sigma(\alpha), \sigma(\beta))$ for any $\alpha, \beta \in V_1$, then we call σ a **symplectic map**. If σ is bijective, then σ is called a **symplectic isomorphism**. If $\alpha, \beta \in V$ satisfy $f(\alpha, \beta) = 0$, then we say that α and β are **symplectically orthogonal**. Let $\alpha \in V$ and W be a subspace of V. If $f(\alpha, \beta) = 0$ for any $\beta \in W$, then we say that α is symplectically orthogonal to W. If W_1, W_2 are two subspaces of V and $f(\alpha, \beta) = 0$ for any $\alpha \in W_1, \beta \in W_2$, then we say that W_1 is symplectically orthogonal to W_2.

Theorem 9.4.2. *Let (V, f) be a symplectic space. Then the dimension of V must be an even number, and there is a basis $e_1, e_2, \ldots, e_{2n}$ for V such that the metric matrix of f with respect to this basis is $J_n = \begin{pmatrix} 0 & I_n \\ -I_n & 0 \end{pmatrix}$. We call $e_1, e_2, \ldots, e_{2n}$ a **symplectically orthogonal basis** for V.*

Proof. Since (V, f) is a symplectic space, f is a skew-symmetric nondegenerate bilinear form. So, the metric matrix A of f with respect to a basis for V is an invertible skew-symmetric matrix. By Example 2.3.2, A is of even order. From Exercise 23 of Chapter 7, we know that A is congruent to $\widetilde{J_n} = \operatorname{diag}\left(\begin{pmatrix} 0 & 1 \\ -1 & 0 \end{pmatrix}, \begin{pmatrix} 0 & 1 \\ -1 & 0 \end{pmatrix}, \ldots, \begin{pmatrix} 0 & 1 \\ -1 & 0 \end{pmatrix}\right)$, so there is a basis $\alpha_1, \alpha_2, \ldots, \alpha_n, \alpha_{n+1}, \alpha_{n+2}, \ldots, \alpha_{2n}$ for V such that the metric matrix of f with respect to this basis is $\widetilde{J_n}$. Let

$$e_1 = \alpha_1, e_2 = \alpha_3, \ldots, e_n = \alpha_{2n-1}, e_{n+1} = \alpha_2, e_{n+2} = \alpha_4, \ldots, e_{2n} = \alpha_{2n}.$$

Then the metric matrix of f with respect to the basis $e_1, e_2, \ldots, e_{2n}$ is $J_n = \begin{pmatrix} 0 & I_n \\ -I_n & 0 \end{pmatrix}$.

$\square$

Definition 9.4.3. Let $S \in M_{2n}(F)$. If $S'J_nS = J_n$, then S is called a **symplectic matrix**. If a linear transformation σ of a symplectic space (V, f) satisfies that $f(\alpha, \beta) = f(\sigma(\alpha), \sigma(\beta))$ for any $\alpha, \beta \in V$, then σ is called a **symplectic transformation**.

It is easy to prove that the transpose of a symplectic matrix is a symplectic matrix; the product of two symplectic matrices of the same order is also a symplectic matrix (see Exercise 16).

Proposition 9.4.4. *The determinant of a symplectic matrix is equal to 1.*

Proof. Let S be a $2n \times 2n$ symplectic matrix and $S = \begin{pmatrix} A & B \\ C & D \end{pmatrix}$, where A, B, C, D are all $n \times n$ matrices. By Definition 9.4.3,

$$\begin{pmatrix} A' & C' \\ B' & D' \end{pmatrix} \begin{pmatrix} 0 & I_n \\ -I_n & 0 \end{pmatrix} \begin{pmatrix} A & B \\ C & D \end{pmatrix} = \begin{pmatrix} 0 & I_n \\ -I_n & 0 \end{pmatrix},$$

so $C'A = A'C$, $D'B = B'D$, $A'D - C'B = I_n$, $D'A - B'C = I_n$.

If A is invertible, then

$$\begin{vmatrix} A & B \\ C & D \end{vmatrix} = \begin{vmatrix} A & B \\ 0 & D - CA^{-1}B \end{vmatrix} = |A||D - CA^{-1}B|$$

$$= |A'||D - CA^{-1}B| = |A'D - A'CA^{-1}B|$$

$$= |A'D - C'AA^{-1}B| = |A'D - C'B| = 1.$$

Now assume A is not invertible. By Exercise 48 of Chapter 2, there is an invertible matrix R such that $R'C$ is a symmetric matrix, i.e., $R'C = C'R$. Note that there are only finitely many $\lambda \in F$ such that $|A + \lambda R| = |R||\lambda I_n + R^{-1}A| = 0$. Hence we can find infinitely many $\lambda \in F$ such that $|A + \lambda R| \neq 0$, so

$$\begin{vmatrix} A + \lambda R & B \\ C & D \end{vmatrix} = \begin{vmatrix} A + \lambda R & B \\ 0 & D - C(A + \lambda R)^{-1}B \end{vmatrix}$$

$$= |A + \lambda R||D - C(A + \lambda R)^{-1}B|$$

$$= |(A + \lambda R)'||D - C(A + \lambda R)^{-1}B|$$

$$= |A'D + \lambda R'D - (A + \lambda R)'C(A + \lambda R)^{-1}B|$$

$$= |A'D + \lambda R'D - C'(A + \lambda R)(A + \lambda R)^{-1}B|$$

$$= |A'D - C'B + \lambda R'D| = |I_n + \lambda R'D|.$$

Thus, both sides of the above equality as polynomials in λ are equal, whence they are equal for $\lambda = 0$. So $|S| = 1$. $\qquad\square$

Proposition 9.4.5. *Let $e_1, e_2, \ldots, e_{2n}$ be a symplectically orthogonal basis for a symplectic space V and σ a symplectic transformation of V. Then the matrix S of σ with respect to the basis $e_1, e_2, \ldots, e_{2n}$ is a symplectic matrix.*

Proof. By the definition of a symplectic transformation, the sequence $\sigma(e_1), \sigma(e_2), \ldots, \sigma(e_{2n})$ is a symplectically orthogonal basis for V. By the given condition, $\sigma(e_1, e_2, \ldots, e_{2n}) = (e_1, e_2, \ldots, e_{2n})S$. Hence S is the transition matrix from the symplectically orthogonal basis $e_1, e_2, \ldots, e_{2n}$ to the symplectically orthogonal basis $\sigma(e_1), \sigma(e_2), \ldots, \sigma(e_{2n})$. Thus $S'J_nS = J_n$ by Theorem 9.2.5, and so S is a the symplectic matrix. $\qquad\square$

Definition 9.4.6. Let W be a subspace of a symplectic space (V, f). We call $W^{\perp} = \{\alpha \in V \mid f(\alpha, \beta) = 0$ for any $\beta \in W\}$ the **symplectically orthogonal complement** of W.

Theorem 9.4.7. *Let W be a subspace of a symplectic space (V, f). Then $W^{\perp}$ is also a subspace of V and $\dim W + \dim W^{\perp} = \dim V$.*

Proof. It is easy to prove that $W^{\perp}$ is a subspace of V. Assume that the metric matrix of f with respect to a basis $e_1, e_2, \ldots, e_{2n}$ for V is A, the sequence $\alpha_1, \alpha_2, \ldots, \alpha_k$ is a basis for W, and $(\alpha_1, \alpha_2, \ldots, \alpha_k) = (e_1, e_2, \ldots, e_{2n})C$, where $C = (C_1, C_2, \ldots, C_k) \in \mathrm{M}_{2n \times k}(F)$. Then $\mathrm{rank}(C) = k$.

Let $\beta = \sum_{i=1}^{2n} x_i e_i = (e_1, e_2, \ldots, e_{2n})X_\beta \in V$. Then

$$\beta \in W^{\perp} \iff f(\alpha_i, \beta) = 0, \quad i = 1, 2, \ldots, k,$$

$$\iff C_i'AX_\beta = 0, \quad i = 1, 2, \ldots, k,$$

$$\iff C'AX_\beta = 0,$$

$$\iff X_\beta \text{ is a solution of } C'AX = 0.$$

Since A is invertible, $\mathrm{rank}(C'A) = k$. Let $\eta_1, \eta_2, \ldots, \eta_{2n-k}$ be a basis for the solution set of $C'AX = 0$ and $\gamma_i = (e_1, e_2, \ldots, e_{2n})\eta_i$, $i = 1, 2, \ldots, 2n - k$. Then $\gamma_1, \gamma_2, \ldots, \gamma_{2n-k}$ is a basis for $W^\perp$. So $\dim W + \dim W^\perp = \dim V$. $\qquad\qquad\square$

Remark 9.4.8. Let W be a subspace of a symplectic space (V, f). Note that the sum of W and $W^\perp$ may not be a direct sum.

Definition 9.4.9. Let W be a subspace of a symplectic space (V, f). If $W \subseteq W^\perp$, then we call W an **isotropic subspace** of (V, f); if $W = W^\perp$, then we call W a **maximal isotropic subspace** (not unique), or **Lagrangian subspace** of (V, f); if $W \cap W^\perp = 0$, then we call W a **symplectic subspace** of (V, f).

Example 9.4.10. Let (V, f) be a finite-dimensional symplectic space, and let $e_1, e_2, \ldots, e_{2n}$ be a symplectically orthogonal basis for V. Then, for any positive integer k with $1 \leqslant k \leqslant n$, the subspace $L(e_1, e_2, \ldots, e_k)$ is an isotropic subspace, the subspace $L(e_1, e_2, \ldots, e_n)$ is a Lagrangian subspace, and the subspace $L(e_1, e_2, \ldots, e_k, e_{n+1}, \ldots, e_{n+k})$ is a symplectic subspace.

Theorem 9.4.11. *Let (V, f) be a finite-dimensional symplectic space. If U and W are two subspaces of V, then the following statements hold:*

(1) $(W^\perp)^\perp = W$.
(2) *If $U \subset W$, then $W^\perp \subset U^\perp$.*
(3) *If W is a symplectic subspace of (V, f), then $W \oplus W^\perp = V$.*
(4) *If W is an isotropic subspace of (V, f), then $\dim W \leqslant \frac{1}{2} \dim V$.*
(5) *If W is a Lagrangian subspace (V, f), then $\dim W = \frac{1}{2} \dim V$.*

Proof. This follows from Theorem 9.4.7 and the definition. $\qquad\square$

Theorem 9.4.12. *Let (V, f) be a symplectic space of dimension $2n$ and W a Lagrangian subspace of V. Then any basis $e_1, e_2, \ldots, e_n$ for W can be extended to a symplectically orthogonal basis for V.*

Proof. Let $W_i = L(e_1, \ldots, e_{i-1}, e_{i+1}, \ldots, e_n)$. Then $W_i^\perp$ is a subspace with dimension $n + 1$. By Theorem 9.4.11, $W \subsetneq W_i^\perp$. Take $e_{n+1} \in W_1^\perp$ such that $e_{n+1} \notin W$ and $f(e_1, e_{n+1}) = 1$ (one can multiply an appropriate number so that the right-hand side is exactly 1).

Let $e_{n+2} \in W_2^{\perp}$ such that $e_{n+2} \notin W$ and $f(e_2, e_{n+2}) = 1$. Although e_{n+2} may not be orthogonal to e_{n+1}, we can find a proper x such that $f(e_{n+2} + xe_1, e_{n+1}) = 0$. Replacing e_{n+2} with $e_{n+2} + xe_1$, we have $f(e_2, e_{n+2}) = 1$ and e_{n+2} is orthogonal to $W_2 \oplus L(e_{n+1})$. By the same argument, we can find vectors $e_{n+3}, \ldots, e_{2n}$ such that the sequence $e_1, e_2, \ldots, e_n, e_{n+1}, \ldots, e_{2n}$ is a symplectically orthogonal basis for V.
$\square$

Corollary 9.4.13. *A basis for an isotropic subspace of a symplectic space can be extended to a symplectically orthogonal basis for V.*

Proof. A basis for an isotropic subspace of a symplectic space can be first extended to a basis for a Lagrangian subspace and then extended to a symplectically orthogonal basis for V.
$\square$

Theorem 9.4.14. *Let W be a symplectic subspace of a symplectic space (V, f). Then a symplectically orthogonal basis for W can be extended to a symplectically orthogonal basis for V.*

Proof. This result follows from Theorem 9.4.11 (3).
$\square$

Theorem 9.4.15. *Let S be a $2n \times 2n$ symplectic matrix. Then the characteristic polynomial $f(\lambda)$ of S satisfies*

$$f(\lambda) = \lambda^{2n} f\left(\frac{1}{\lambda}\right).$$

Proof. We assume that the characteristic polynomial $f(\lambda)$ of S can be factored as $f(\lambda) = \prod_{i=1}^{2n} (\lambda - \lambda_i)$, where $\lambda_i \in \mathbb{C}, i = 1, 2, \ldots, 2n$. Then

$$\lambda^{2n} f\left(\frac{1}{\lambda}\right) = \prod_{i=1}^{2n}(1 - \lambda\lambda_i) = \left(\prod_{i=1}^{2n} \lambda_i\right)\left(\prod_{i=1}^{2n}(\lambda - \lambda_i^{-1})\right).$$

By Proposition 9.4.4, the determinant of the symplectic matrix S is 1, which implies that the product of all eigenvalues is 1, i.e., $\prod_{i=1}^{2n} \lambda_i = 1$.

So $\lambda^{2n} f\left(\frac{1}{\lambda}\right) = \prod_{i=1}^{2n}(\lambda - \lambda_i^{-1})$ is the characteristic polynomial of S^{-1}. Thus, we only need to prove that S and S^{-1} have the same characteristic polynomial.

By Definition 9.4.3, $S = J_n^{-1}(S^{-1})'J_n$. So S and $(S^{-1})'$ have the same characteristic polynomial, which implies that S and S^{-1} have the same characteristic polynomial. $\qquad\square$

Theorem 9.4.16. *Let σ be a symplectic transformation of a symplectic space (V, f), and let λ, μ be two different eigenvalues of σ such that $\lambda\mu \neq 1$. If V_λ and V_μ are the eigenspaces of σ associated with λ and μ respectively, then V_λ is symplectically orthogonal to V_μ.*

Proof. Let $\alpha \in V_\lambda$ and $\beta \in V_\mu$. Then $\sigma(\alpha) = \lambda\alpha$, $\sigma(\beta) = \mu\beta$. Since σ is a symplectic transformation of the symplectic space (V, f),

$$f(\alpha, \beta) = f(\sigma(\alpha), \sigma(\beta)) = f(\lambda\alpha, \mu\beta) = \lambda\mu f(\alpha, \beta).$$

and so $(\lambda\mu - 1)f(\alpha, \beta) = 0$. Note that $\lambda\mu \neq 1$, and hence $f(\alpha, \beta) = 0$. Thus V_λ is symplectically orthogonal to V_μ. $\qquad\square$

Exercises

1. Determine whether the following functions defined on $\mathbb{R}^2$ are bilinear forms:

   (1) $f((x_1, y_1), (x_2, y_2)) = (x_1 - x_2)^2 + y_1 y_2;$

   (2) $f((x_1, y_1), (x_2, y_2)) = (x_1 + x_2)^2 - (y_1 - y_2)^2.$

2. Determine whether the following functions defined on $\mathbb{R}^3$ are bilinear forms. If the answer is yes, write out the metric matrices of the bilinear forms with respect to the standard basis $\alpha_1 = (1, 0, 0), \alpha_2 = (0, 1, 0), \alpha_3 = (0, 0, 1)$.

   (1) $f((x_1, y_1, z_1), (x_2, y_2, z_2)) = x_1 y_2 - x_2 y_1 + z_1 z_2;$

   (2) $f((x_1, y_1, z_1), (x_2, y_2, z_2)) = x_1 x_2 + y_1 y_2 + z_1^2;$

   (3) $f((x_1, y_1, z_1), (x_2, y_2, z_2)) = 1;$

   (4) $f((x_1, y_1, z_1), (x_2, y_2, z_2)) = 0.$

3. Let $A = \begin{pmatrix} 0 & 2 & -1 & 3 \\ -2 & 0 & 4 & -2 \\ 1 & -4 & 0 & 6 \\ -3 & 2 & -6 & 0 \end{pmatrix}$. Find a rational number matrix C such that

$$C'AC = \begin{pmatrix} 0 & 1 & 0 & 0 \\ -1 & 0 & 0 & 0 \\ 0 & 0 & 0 & 1 \\ 0 & 0 & -1 & 0 \end{pmatrix}.$$

4. Let $A \in M_m(F)$ and $V = M_{m \times n}(F)$. Define the function f on $V \times V$ such that $f(X, Y) = \mathrm{Tr}(X'AY)$, where $X, Y \in V$. Prove that f is a bilinear form on V, and find the metric matrix of f with respect to the basis $E_{11}, E_{12}, \ldots, E_{mn}$ for V, where E_{ij} is the matrix unit defined in Exercise 25 of Chapter 2.

5. Let V be a finite-dimensional linear space, and let $\alpha_1, \alpha_2, \ldots, \alpha_n$ be n nonzero vectors of V. Prove that there is an $f \in V^*$ such that $f(\alpha_i) \neq 0$, $i = 1, 2, \ldots, n$.

6. Let V be a finite-dimensional linear space, and let $f_1, f_2, \ldots, f_n$ be n nonzero vectors of V^*. Prove that there is an $\alpha \in V$ such that $f_i(\alpha) \neq 0$, $i = 1, 2, \ldots, n$.

7. Define the inner product on $V = \mathbb{R}[x]_4$ by $(f(x), g(x)) = \int_0^1 f(x)g(x)\mathrm{d}x$. Find an orthonormal basis for V.

8.* Let f be a bilinear form on an n-dimensional linear space V. Suppose that $f(\alpha, \beta) = 0$ if and only if $f(\beta, \alpha) = 0$ for any $\alpha, \beta \in V$, then we say that the orthogonality of vectors of V about f is symmetric. Prove that the orthogonality of vectors of V about f is symmetric if and only if f is symmetric or skew-symmetric.

9. Let f be a nondegenerate symmetric bilinear form on an n-dimensional linear space V. Define the map $\varepsilon : V \longrightarrow V^*$ such that $\varepsilon(\alpha)(\beta) = f(\alpha, \beta)$ for any $\alpha, \beta \in V$. Prove that ε is an isomorphism.

10. Let V be an n-dimensional linear space and $f(\alpha, \beta)$ a skew-symmetric bilinear form of V. Prove that $\mathrm{rank}(f) = 2$ if and only if there are linearly independent $g_1, g_2 \in V^*$ such that $f(\alpha, \beta) = g_1(\alpha)g_2(\beta) - g_1(\beta)g_2(\alpha)$.

11.* Assume that two symmetric matrices $A = \begin{pmatrix} A_1 & 0 \\ 0 & A_2 \end{pmatrix}$ and $B = \begin{pmatrix} B_1 & 0 \\ 0 & B_2 \end{pmatrix}$ over F are congruent, and A_1 is congruent to B_1. Prove that A_2 is congruent to B_2.

12. Let V be an n-dimensional linear space and V^* the dual space of V. Suppose that the sequence $e_1, e_2, \ldots, e_n$ is a basis for V and the sequence $e_1^*, e_2^*, \ldots, e_n^*$ is the dual basis of $e_1, e_2, \ldots, e_n$ (see Exercise 17 of Chapter 5).

   (1) Prove that for any linear transformation σ of V and any $f \in V^*$, we have $f\sigma \in V^*$.

   (2) For any linear transformation σ of V, define $\sigma^* : V^* \longrightarrow V^*$ such that $\sigma^*(f) = f\sigma$, where $f \in V^*$. Prove that σ^* is a linear transformation of V^*.

   (3) Let σ be a linear transformation of V. If the matrix of σ with respect to the basis $e_1, e_2, \ldots, e_n$ for V is A, prove that the matrix of σ^* with respect to the basis $e_1^*, e_2^*, \ldots, e_n^*$ for V^* is A'.

13. Let V be an n-dimensional linear space and f a bilinear form on V. Prove that there are bases $\alpha_1, \alpha_2, \ldots, \alpha_n$ and $\beta_1, \beta_2, \ldots, \beta_n$ for V such that $(f(\alpha_i, \beta_j))_{n \times n} = \mathrm{diag}(1, \ldots, 1, 0, \ldots, 0)$.

14. Let V be an n-dimensional real linear space and f a bilinear form on V such that $f(\alpha, \alpha) > 0$ for any nonzero α. Prove that there is a basis $\alpha_1, \alpha_2, \ldots, \alpha_n$ for V such that the metric matrix of f with respect to this basis is $\mathrm{diag}\left(\begin{pmatrix} 1 & t_1 \\ -t_1 & 1 \end{pmatrix}, \begin{pmatrix} 1 & t_2 \\ -t_2 & 1 \end{pmatrix}, \ldots, \begin{pmatrix} 1 & t_k \\ -t_k & 1 \end{pmatrix}, I_{n-2k} \right)$, where $t_1, t_2, \ldots, t_k$ are all positive real numbers.

15.* Let $V = \mathbb{Q}[x]$ be the ring of polynomials over the rational number field. Prove that V is a countable set, while V^* is an uncountable set. Thus, no bijection exists between V and V^*.

16. Let $S,\ T \in \mathrm{M}_{2n}(F)$ be symplectic matrices. Prove that S^{-1}, S' and ST are all symplectic matrices.

17.* Let U and W be subspaces of a symplectic space V. If $\sigma : U \longrightarrow W$ is a symplectic isomorphism, prove that σ can be extended to a symplectic transformation of V.

Bibliography

[1] Algebra Group in the Department of Mathematics at Beijing University (2019), *Advanced Algebra* (in Chinese). Revised by Wang, E.F. and Shi, S.M., 5th edn. Higher Education Press, Beijing.

[2] Artin, M. (2004). *Algebra*. Pearson Education Asia Limited and China Machine Press, Beijing.

[3] Chen, J.L., Zhou, J.H. and Zhang, X.X., *et al.* (2016). *Linear Algebra* (in Chinese), 2nd edn. Science Press, Beijing.

[4] Chen, Z.J. (2008). *Advanced Algebra and Analytic Geometry* (in Chinese), 2nd edn. Higher Education Press, Beijing.

[5] Childs, L.N. (2009). *A Concrete Introduction to Higher Algebra*. Undergraduate Texts in Mathematics. Springer-Verlag, New York.

[6] Chong, C.T. (1985). Some Remarks on the History of Linear Algebra, *Mathematical Medley*, 13(11), 59–73.

[7] Cox, D.A. (2011). Why Eisenstein proved the Eisenstein criterion and why Schönemann discovered it first, *American Mathematical Monthly*, 118, 3–21.

[8] Denton, P.B., Parke, S.J. Tao, T., *et al.* (2022). Eigenvectors from eigenvalues: A survey of a basic identity in linear algebra, *Bulletin of the American Mathematical Society*, 59(1), 31–58.

[9] Derbyshire, J. (2010). *Unknown quantity: a real and imaginary history of algebra* (in Chinese). Translated by Feng, S. Posts & Telecom Press, Beijing.

[10] Doković, D. and Ikramov, K. (2002). A square matrix is congruent to its transpose, *Journal of Algebra*, 257, 97–105.

[11] Dorwart, H.L. (1935). Irreducibility of polynomials, *American Mathematical Monthly*, 42, 369–381.

[12]   Du, X.K., Xu, X.W. Ma, J., *et al.* (2017). *Advanced Algebra* (in Chinese). Science Press, Beijing.

[13]   Duke, W., Rudnick, Z. and Sarnak, P. (1993). Density of integer points on affine homogeneous varieties, *Duke Mathematical Journal*, 71, 143–179.

[14]   Guo, R.Q., Cen, J.P. and Wang, Z.P. (2014). *A Course in Advanced Algebra* (in Chinese). Science Press, Beijing.

[15]   Halmos, P. R. (1980/81). Does mathematics have elements?, *The Mathematical Intelligencer*, 3(4), 147–153.

[16]   Hoffman, K. and Kunze, R. (2012). *Linear Algebra*, 2nd edn. World Publishing Corporation, Beijing.

[17]   Hua, L.G. and Su, B.Q. (1988). *Encyclopedia of China — Mathematics* (in Chinese). Encyclopedia of China Publishing House, Beijing.

[18]   Huang, Z.D., Li, F. Wen, D.W., *et al.* (2013). *Advanced Algebra (1)* (in Chinese). Zhejiang University Press, Hangzhou.

[19]   Hungerford, T.W. (1985). *Algebra* (in Chinese). Translated by Feng, K.Q., proofread by Nie, L.Z. Hunan Education Press, Changsha.

[20]   Jacobson, N. (1974). *Basic Algebra I.* W. H. Freeman and Company, San Francisco.

[21]   Kline, M. (1990). *Mathematical Thought from Ancient to Modern Times.* 2nd edn. The Clarendon Press, Oxford University Press, New York.

[22]   Kostrikin, A.I. (2006). *Introduction to Algebra, Part I: Fundamentals of Algebra* (in Chinese). Translated by Zhang, Y.B., 2nd edn. Higher Education Press, Beijing.

[23]   Lam, T.Y. (2003). *Exercises in Classical Ring Theory.* 2nd edn. Problem Books in Mathematics. Springer-Verlag, New York.

[24]   Lam, T.Y. and Nielsen, P.P. (2019). Jacobson pairs and Bott-Duffin decompositions in rings, Contemporary Mathematics, *American Mathematical Society*, Providence, RI, 727, 249–267.

[25]   Lam, T.Y. and Nielsen, P.P. (2014). Jacobson's lemma for Drazin inverses. ring theory and its applications, Contemporary Mathematics, *American Mathematical Society*, Providence, RI, 609, 185–195.

[26]   Lan, Y.Z. (2007). *A Coincise Course of Advanced Algebra* (in Chinese), 2nd edn. Beijing University Press, Beijing.

[27]   Li, F., Huang, Z.D. Wen, D.W., *et al.* (2013). *Advanced Algebra (2)* (in Chinese). Zhejiang University Press, Hangzhou.

[28]   Li, J.S., Zha, J.G. and Wang, X.M. (2010). *Linear Algebra* (in Chinese), 2nd edn. University of Science and Technology of China Press, Hefei.

[29]   Li, S.Z. (2011). *Linear Algebra* (in Chinese). Higher Education Press, Beijing.

[30] Lin, Y.N. (2013). *Advanced Algebra* (in Chinese). Higher Education Press, Beijing.

[31] Meng, D.J. (2014). *Advanced Algebra and Analytic Geometry* (in Chinese). 3rd edn. Science Press, Beijing.

[32] Moore, G.H. (1995). The axiomatization of linear algebra: 1875–1940, *Historia Mathematica*, 22(3), 262–303.

[33] Nowicki, A. and Światek, A. (1998). *Irreducible Polynomials and Prime Numbers*. 22 November, https://www-users.mat.umk.pl//~anow/ps-dvi/pol1adaa.pdf.

[34] Qiu, W.S. (1996). *Advanced Algebra* (in Chinese). Higher Education Press, Beijing.

[35] Ram Murty, M. (2002). Prime numbers and irreducible polynomials, *American Mathematical Monthly*, 109, 452–458.

[36] Rodrigues' rotation formula, https://infogalactic.com/info/Rodrigues%27_rotation_formula.

[37] Rotman, J.J. (2004). *A First Course in Abstract Algebra*. 2nd edn. Pearson Education Asia Limited and China Machine Press, Beijing.

[38] Shi, W.J. and Dai, G.S. (2005). *Advanced Algebra* (in Chinese). Higher Education Press, Beijing.

[39] Wang, Q.W. (2012). *Main Idea and Applications of Linear Algebra* (in Chinese). Science Press, Beijing.

[40] Xi, N.H. (2016). *Basic Algebra, Part I* (in Chinese). Science Press, Beijing.

[41] Xi, N.H. (2018). *Basic Algebra, Part II* (in Chinese). Science Press, Beijing.

[42] Xu, Y.C. (2008). *Linear Algebra and Matrix Theory* (in Chinese). 2nd edn. Higher Education Press, Beijing.

[43] Yao, M.S., Wu, Q.S. and Xie, Q.H. (2014). *Advanced Algebra* (in Chinese), 3rd edn. Fudan University Press, Shanghai.

[44] Zhang, H.R. (1958). *An Introduction to Abstract Algebra* (in Chinese), 5th edn. (revision). Commercial Press, Beijing.

[45] Zhang, X.K. and Xu, P.H. (2017). *Advanced Algebra* (in Chinese), 2nd edn. Qinghua University Press, Beijing.

[46] Zhou, B.X. (1988). *An Introduction to Advanced Algebra* (in Chinese). Higher Education Press, Beijing.

[47] Zhou, S.F., Wang, X.H. Gu, M.Y., *et al.* (1985). *Problem Solving Analysis in Advanced Algebra* (in Chinese). Jiangsu Science and Technology Press, Nanjing.

[48] Zhu, F.H. and Chen, Z.Q. (2017). *Advanced Algebra and Analytic Geometry* (in Chinese). Science Press, Beijing.

Index of Symbols

A^*, 148
$\mathrm{Aut}(V)$, 265
$\mathrm{Aut}_F(V)$, 265

$W^\perp$, 417, 463, 471
$\mathbb{C}$, 19
$\equiv$, 11, 40

$\deg(f(x))$, 22
$f'(x)$, 46
$(f(x))'$, 46
$f''(x)$, 46
$f^{(k)}(x)$, 46
$\det(\sigma)$, 294
$|\sigma|$, 294
$|a_{ij}|_n$, 91
$|A|$, 147
$\det(A)$, 147
$\dim V$, 245
$\dim_F V$, 245
$n\,|\,m$, 5
$n \nmid m$, 5
$g(x)\,|\,f(x)$, 29
$g(x) \nmid f(x)$, 29
V^*, 265

$E(i(k))$, 160, 350
$E(i, j(b))$, 160, 161
$E(i, j(f(\lambda)))$, 350
$E(i, j)$, 161, 351
$\mathrm{End}(V)$, 265
$\mathrm{End}_F(V)$, 265
$A \sim B$, 158

(m, n), 6
$(m_1, m_2, \ldots, m_k)$, 6
$(f(x), g(x))$, 31
$(f_1(x), f_2(x), \ldots, f_n(x))$, 40
$\mathrm{GL}_n(F)$, 147
$\mathrm{GL}_n(F[\lambda])$, 348
$\mathrm{GL}_n(\mathbb{Z})$, 382

$\mathrm{Hom}(V, W)$, 265
$\mathrm{Hom}_F(V, W)$, 265

I_n, 144
$\mathrm{Im}(\sigma)$, 270
A^{-1}, 147
A^+, 439

$\mathrm{Ker}(\sigma)$, 270

Index of Terms

N

O

Index of Mathematicians